# An Integrated Introduction to

# *Computer Graphics and Geometric Modeling*

# An Integrated Introduction to
# *Computer Graphics and Geometric Modeling*

## Ronald Goldman

CRC Press
Taylor & Francis Group
Boca Raton London New York

CRC Press is an imprint of the
Taylor & Francis Group, an **informa** business

A CHAPMAN & HALL BOOK

CRC Press
Taylor & Francis Group
6000 Broken Sound Parkway NW, Suite 300
Boca Raton, FL 33487-2742

First issued in paperback 2018

© 2009 by Taylor and Francis Group, LLC
CRC Press is an imprint of Taylor & Francis Group, an Informa business

No claim to original U.S. Government works

ISBN-13: 978-1-4398-0334-9 (hbk)
ISBN-13: 978-1-138-38147-6 (pbk)

---

**Library of Congress Cataloging-in-Publication Data**

---

Goldman, Ron, 1947-
   An integrated introduction to computer graphics and geometric modeling / Ronald Goldman.
      p. cm.
   Includes bibliographical references and index.
   ISBN 978-1-4398-0334-9 (alk. paper)
   1. Computer graphics. 2. Three-dimensional display systems. 3. Curves on surfaces--Mathematical models. I. Title.

T385.G6397 2009
006.6--dc22
                                       2008054783

---

**Visit the Taylor & Francis Web site at**
**http://www.taylorandfrancis.com**

**and the CRC Press Web site at**
**http://www.crcpress.com**

# Contents

Foreword    **xv**

Dedication    **xvii**

Preface    **xix**

Author    **xxix**

## I   Two-Dimensional Computer Graphics: From Common Curves to Intricate Fractals

**1   Turtle Graphics** ............................................................ **3**
   1.1   Turtle Graphics ......................................................... 3
   1.2   Turtle Commands ..................................................... 4
   1.3   Turtle Programs ....................................................... 7
   1.4   Summary ............................................................... 9
   Exercises ..................................................................... 9

**2   Fractals from Recursive Turtle Programs** ..................... **13**
   2.1   Fractals .................................................................. 13
   2.2   Looping Lemmas ..................................................... 13
   2.3   Fractal Curves and Recursive Turtle Programs ............. 17
      2.3.1   Fractal Gaskets ............................................ 17
      2.3.2   Bump Fractals .............................................. 19
   2.4   Summary: Fractals—Recursion Made Visible .............. 20
   Exercises ..................................................................... 21
   Programming Projects .................................................... 23

**3   Some Strange Properties of Fractal Curves** ................. **29**
   3.1   Fractal Strangeness ................................................. 29
   3.2   Dimension .............................................................. 29
      3.2.1   Fractal Dimension ......................................... 31
      3.2.2   Computing Fractal Dimension from Recursive Turtle Programs ....... 32
   3.3   Differentiability ...................................................... 32
   3.4   Attraction .............................................................. 34
      3.4.1   Base Cases for the Sierpinski Gasket ............... 34
      3.4.2   Base Cases for the Koch Curve ...................... 35
      3.4.3   Attractors ................................................... 36
   3.5   Summary ............................................................... 36
   Exercises ..................................................................... 37

**4   Affine Transformations** ........................................................................... **39**
    4.1   Transformations ...................................................................................... 39
    4.2   Conformal Transformations ..................................................................... 40
        4.2.1   Translation..................................................................................... 40
        4.2.2   Rotation ......................................................................................... 41
        4.2.3   Uniform Scaling ............................................................................ 42
    4.3   The Algebra of Affine Transformations .................................................... 43
    4.4   The Geometry of Affine Transformations ................................................. 44
    4.5   Affine Coordinates and Affine Matrices.................................................... 45
    4.6   Conformal Transformations: Revisited ..................................................... 46
    4.7   General Affine Transformations ................................................................ 46
        4.7.1   Image of One Point and Two Lineary Independent Vectors.............. 47
        4.7.2   Nonuniform Scaling........................................................................ 48
        4.7.3   Image of Three Noncollinear Points ................................................ 50
    4.8   Summary .................................................................................................. 51
        4.8.1   Affine Transformations and Affine Coordinates............................... 51
        4.8.2   Matrices for Affine Transformations in the Plane ........................... 53
    Exercises ........................................................................................................... 54

**5   Affine Geometry: A Connect-the-Dots Approach to Two-Dimensional**
**Computer Graphics** .......................................................................................... **61**
    5.1   Two Shortcomings of Turtle Graphics ...................................................... 61
    5.2   Affine Graphics......................................................................................... 62
        5.2.1   The CODO Language ..................................................................... 62
        5.2.2   Sample CODO Programs ................................................................ 63
    5.3   Summary .................................................................................................. 65
    Exercises ........................................................................................................... 67

**6   Fractals from Iterated Function Systems** ....................................................... **71**
    6.1   Generating Fractals by Iterating Transformations ..................................... 71
    6.2   Fractals as Fixed Points of Iterated Function Systems............................... 73
    6.3   Fractals as Attractors................................................................................ 74
    6.4   Fractals with Condensation Sets ............................................................... 74
    6.5   Summary .................................................................................................. 76
    Exercises ........................................................................................................... 77
    Programming Project ......................................................................................... 79

**7   The Fixed-Point Theorem and Its Consequences** ........................................... **81**
    7.1   Fixed Points and Iteration......................................................................... 81
    7.2   The Trivial Fixed Point Theorem .............................................................. 82
    7.3   Consequences of the Trivial Fixed-Point Theorem..................................... 84
        7.3.1   Root Finding Methods .................................................................... 84
        7.3.2   Relaxation Methods ....................................................................... 87
        7.3.3   Fractals ......................................................................................... 89
            7.3.3.1   Compact Sets and the Haussdorf Metric ........................... 90
            7.3.3.2   Contractive Transformations and Iterated Function Systems ............. 91
            7.3.3.3   Fractal Theorem, Fractal Algorithm, and Fractal Strategy ............... 92
    7.4   Summary................................................................................................... 93
    Exercises ........................................................................................................... 94
    Programming Projects.......................................................................................... 98

**8 Recursive Turtle Programs and Conformal Iterated Function Systems**......................................................................... **101**

8.1  Motivating Questions............................................................. 101

8.2  The Effect of Changing the Turtle's Initial State ....................... 101

8.3  Equivalence Theorems........................................................... 103

8.4  Conversion Algorithms........................................................... 106

  8.4.1  Ron's Algorithm.......................................................... 106

  8.4.2  Tao's Algorithm.......................................................... 107

8.5  Bump Fractals ...................................................................... 109

8.6  Summary ............................................................................. 110

Exercises ..................................................................................... 111

Programming Projects.................................................................... 113

## II  Mathematical Methods for Three-Dimensional Computer Graphics

**9  Vector Geometry: A Coordinate-Free Approach** ...................... **117**

9.1  Coordinate-Free Methods ..................................................... 117

9.2  Vectors and Vector Spaces .................................................... 118

9.3  Points and Affine Spaces ...................................................... 119

9.4  Vector Products.................................................................... 120

  9.4.1  Dot Product ............................................................... 121

  9.4.2  Cross Product ............................................................ 122

  9.4.3  Determinant ............................................................... 123

9.5  Summary ............................................................................. 123

Appendix A: The Nonassociativity of the Cross Product ................... 124

Appendix B: The Algebra of Points and Vectors .............................. 126

Exercises ..................................................................................... 128

**10  Coordinate Algebra** ................................................................ **131**

10.1  Rectangular Coordinates ...................................................... 131

10.2  Addition, Subtraction, and Scalar Multiplication ...................... 131

10.3  Vector Products.................................................................. 132

  10.3.1  Dot Product ............................................................. 133

  10.3.2  Cross Product .......................................................... 133

  10.3.3  Determinant ............................................................. 134

10.4  Summary ........................................................................... 134

Exercises ..................................................................................... 135

**11  Some Applications of Vector Geometry**.................................. **139**

11.1  Introduction....................................................................... 139

11.2  Trigonometric Laws............................................................ 139

  11.2.1  Law of Cosines ......................................................... 139

  11.2.2  Law of Sines ............................................................ 140

11.3  Representations for Lines and Planes ..................................... 141

  11.3.1  Lines....................................................................... 141

  11.3.2  Planes ..................................................................... 141

11.4   Metric Formulas ............................................................................ 142
       11.4.1   Distance ............................................................................ 142
                11.4.1.1   Distance between Two Points ........................ 142
                11.4.1.2   Distance between a Point and a Line ............ 143
                11.4.1.3   Distance between a Point and a Plane .......... 143
                11.4.1.4   Distance between Two Lines .......................... 144
       11.4.2   Area .................................................................................... 145
                11.4.2.1   Triangles and Parallelograms ........................ 145
                11.4.2.2   Polygons: Newell's Formula .......................... 146
       11.4.3   Volume ................................................................................ 147
11.5   Intersection Formulas for Lines and Planes ................................ 148
       11.5.1   Intersecting Two Lines .................................................... 148
       11.5.2   Intersecting Three Planes ................................................ 149
       11.5.3   Intersecting Two Planes .................................................. 150
11.6   Spherical Linear Interpolation .................................................... 151
11.7   Inside–Outside Tests .................................................................. 153
       11.7.1   Ray Casting ...................................................................... 153
       11.7.2   Winding Number .............................................................. 154
11.8   Summary ...................................................................................... 156
       11.8.1   Trigonometric Laws ........................................................ 156
       11.8.2   Metric Formulas .............................................................. 157
                11.8.2.1   Distance .......................................................... 157
                11.8.2.2   Area ................................................................ 157
                11.8.2.3   Volume ............................................................ 158
       11.8.3   Intersections .................................................................... 158
       11.8.4   Interpolation .................................................................... 158
       11.8.5   Winding Number .............................................................. 158
Exercises .................................................................................................. 159


12   **Coordinate-Free Formulas for Affine and Projective**
     **Transformations** ............................................................................ **163**
12.1   Transformations for Three-Dimensional Computer Graphics ...... 163
12.2   Affine and Projective Transformations ........................................ 163
12.3   Rigid Motions .............................................................................. 164
       12.3.1   Translation ........................................................................ 165
       12.3.2   Rotation ............................................................................ 165
       12.3.3   Mirror Image .................................................................... 167
12.4   Scaling .......................................................................................... 168
       12.4.1   Uniform Scaling .............................................................. 169
       12.4.2   Nonuniform Scaling ........................................................ 169
12.5   Projections .................................................................................... 170
       12.5.1   Orthogonal Projection ...................................................... 171
       12.5.2   Perspective ...................................................................... 171
12.6   Summary ...................................................................................... 173
       12.6.1   Affine and Projective Transformations without Matrices ...... 173
       12.6.2   Formulas for Affine and Projective Transformations ...... 174
Exercises .................................................................................................. 174

**13 Matrix Representations for Affine and Projective Transformations** ........................................................................................ **179**

13.1 Matrix Representations for Affine Transformations .................... 179

13.2 Linear Transformation Matrices and Translation Vectors ........... 181

    13.2.1 Linear Transformation Matrices ...................................... 182

    13.2.2 Translation Vectors .......................................................... 183

13.3 Rigid Motions ............................................................................ 183

    13.3.1 Translation ....................................................................... 183

    13.3.2 Rotation ........................................................................... 184

    13.3.3 Mirror Image .................................................................... 185

13.4 Scaling ....................................................................................... 187

    13.4.1 Uniform Scaling ............................................................... 187

    13.4.2 Nonuniform Scaling ......................................................... 187

13.5 Projections .................................................................................. 188

    13.5.1 Orthogonal Projection ...................................................... 189

13.6 Perspective ................................................................................. 189

    13.6.1 Projective Transformations and Homogeneous Coordinates .... 190

    13.6.2 Matrices for Perspective Projections ................................ 191

13.7 Summary ..................................................................................... 193

    13.7.1 Matrix Representations for Affine and Projective Transformations .................................................. 193

    13.7.2 Matrices for Affine and Projective Transformations ......... 194

Exercises ............................................................................................ 195

Programming Projects ........................................................................ 199

**14 Projective Space versus the Universal Space of Mass-Points** .......... **205**

14.1 Algebra and Geometry ............................................................... 205

14.2 Projective Space: The Standard Model ...................................... 206

14.3 Mass-Points: The Universal Model ............................................ 210

14.4 Perspective and Pseudoperspective ............................................ 213

    14.4.1 Perspective and the Law of the Lever ............................... 213

    14.4.2 Pseudoperspective and Pseudodepth ................................ 214

14.5 Summary ..................................................................................... 218

Exercises ............................................................................................ 219

**15 Quaternions: Multiplication in the Space of Mass-Points** ............... **223**

15.1 Vector Spaces and Division Algebras ......................................... 223

15.2 Complex Numbers ...................................................................... 224

15.3 Quaternions ................................................................................ 227

    15.3.1 Quaternion Multiplication ................................................ 227

    15.3.2 Quaternion Representations for Conformal Transformations .... 230

    15.3.3 Quaternions versus Matrices ............................................ 232

    15.3.4 Avoiding Distortion .......................................................... 233

    15.3.5 Key Frame Animation ....................................................... 234

    15.3.6 Conversion Formulas ........................................................ 235

15.4 Summary ..................................................................................... 238

Exercises ............................................................................................ 239

Programming Projects ........................................................................ 245

## III   Three-Dimensional Computer Graphics: Realistic Rendering

**16  Color and Intensity** ............................................................................ **249**
  16.1  Motivation ................................................................................... 249
  16.2  The RGB Color Model ................................................................. 249
  16.3  Ambient Light .............................................................................. 250
  16.4  Diffuse Reflection ........................................................................ 251
  16.5  Specular Reflection ...................................................................... 252
  16.6  Total Intensity .............................................................................. 254
  16.7  Summary ...................................................................................... 255
  Exercises ................................................................................................ 256

**17  Recursive Ray Tracing** ...................................................................... **257**
  17.1  Raster Graphics ............................................................................ 257
  17.2  Recursive Ray Tracing .................................................................. 257
  17.3  Shadows ....................................................................................... 259
  17.4  Reflection ..................................................................................... 260
  17.5  Refraction ..................................................................................... 261
  17.6  Summary ...................................................................................... 264
  Exercises ................................................................................................ 265

**18  Surfaces I: The General Theory** ...................................................... **267**
  18.1  Surface Representations ................................................................ 267
      18.1.1  Implicit Surfaces .................................................................. 267
      18.1.2  Parametric Surfaces ............................................................. 267
      18.1.3  Deformed Surfaces .............................................................. 268
      18.1.4  Procedural Surfaces ............................................................. 268
  18.2  Surface Normals ........................................................................... 269
      18.2.1  Implicit Surfaces .................................................................. 269
      18.2.2  Parametric Surfaces ............................................................. 269
      18.2.3  Deformed Surfaces .............................................................. 270
  18.3  Ray–Surface Intersections ........................................................... 272
      18.3.1  Implicit Surfaces .................................................................. 272
      18.3.2  Parametric Surfaces ............................................................. 272
      18.3.3  Deformed Surfaces .............................................................. 273
  18.4  Mean and Gaussian Curvature ..................................................... 274
      18.4.1  Implicit Surfaces .................................................................. 274
      18.4.2  Parametric Surfaces ............................................................. 275
      18.4.3  Deformed Surfaces .............................................................. 275
  18.5  Summary ...................................................................................... 275
      18.5.1  Implicit Surfaces .................................................................. 276
      18.5.2  Parametric Surfaces ............................................................. 276
      18.5.3  Deformed Surfaces .............................................................. 278
  Exercises ................................................................................................ 278

**19  Surfaces II: Simple Surfaces** ............................................................ **281**
  19.1  Simple Surfaces ........................................................................... 281
  19.2  Intersection Strategies ................................................................. 281
  19.3  Planes and Polygons .................................................................... 282

19.4    Natural Quadrics .................................................................................. 284

    19.4.1    Spheres .................................................................................. 284

          19.4.1.1    Intersecting a Line and a Circle ................................ 285

          19.4.1.2    Inversion Formulas for the Line .............................. 285

    19.4.2    Cylinders ................................................................................ 287

          19.4.2.1    Intersecting a Line and an Infinite Cylinder .......... 287

          19.4.2.2    Intersecting a Line and a Bounded Cylinder .......... 289

    19.4.3    Cones .................................................................................... 290

    19.4.4    Ellipsoids, Elliptical Cylinders, and Elliptical Cones ............. 292

19.5    General Quadric Surfaces .................................................................. 292

19.6    Tori ..................................................................................................... 295

    19.6.1    Bounding the Torus ................................................................ 298

19.7    Surfaces of Revolution ....................................................................... 299

19.8    Summary ............................................................................................. 303

Exercises ....................................................................................................... 304

Programming Projects .................................................................................... 306

**20    Solid Modeling** ................................................................................... **309**

20.1    Solids .................................................................................................. 309

20.2    Constructive Solid Geometry (CSG) ................................................. 309

20.3    Boundary Representations (B-Rep) .................................................... 313

20.4    Octrees ................................................................................................ 317

20.5    Summary ............................................................................................. 319

Exercises ....................................................................................................... 319

Programming Projects .................................................................................... 322

**21    Shading** ............................................................................................... **325**

21.1    Polygonal Models .............................................................................. 325

    21.1.1    Newell's Formula for the Normal to a Polygon ..................... 326

21.2    Uniform Shading ................................................................................ 326

21.3    Gouraud Shading ............................................................................... 327

21.4    Phong Shading ................................................................................... 331

    21.4.1    Naive Phong Shading ............................................................. 331

    21.4.2    Fast Phong Shading and Diffuse Reflection ........................... 332

    21.4.3    Fast Phong Shading and Specular Reflection ......................... 334

    21.4.4    Phong Shading and Spherical Linear Interpolation ................ 335

21.5    Summary ............................................................................................. 337

Exercises ....................................................................................................... 339

Programming Project ...................................................................................... 339

**22    Hidden Surface Algorithms** ............................................................. **341**

22.1    Hidden Surface Algorithms ............................................................... 341

22.2    The Heedless Painter .......................................................................... 342

22.3    *z*-Buffer (Depth Buffer) .................................................................... 342

22.4    Scan Line ............................................................................................ 343

22.5    Ray Casting ........................................................................................ 346

22.6    Depth Sort .......................................................................................... 347

    22.6.1    Polygon Splitting ................................................................... 350

22.7    BSP-Tree ............................................................................................ 351

22.8   Summary ................................................................................................ 352
Exercises ............................................................................................................ 352
Programming Projects ....................................................................................... 353

**23   Radiosity** ................................................................................................. **355**
23.1   Radiosity ................................................................................................. 355
23.2   The Radiosity Equations ......................................................................... 355
        23.2.1   The Rendering Equation ............................................................. 356
        23.2.2   The Radiosity Equation: Continuous Form ................................ 356
        23.2.3   The Radiosity Equation: Discrete Form ..................................... 359
23.3   Form Factors ........................................................................................... 361
        23.3.1   Hemi-Cubes ................................................................................ 363
23.4   The Radiosity Rendering Algorithm ....................................................... 366
23.5   Solving the Radiosity Equations ............................................................. 368
        23.5.1   Gathering .................................................................................... 368
        23.5.2   Shooting: Progressive Refinement ............................................. 370
23.6   Summary ................................................................................................ 372
Exercises ............................................................................................................ 373
Programming Project ......................................................................................... 375

**IV   Geometric Modeling: Freeform Curves and Surfaces**

**24   Bezier Curves and Surfaces** .................................................................... **379**
24.1   Interpolation and Approximation ............................................................ 379
24.2   The de Casteljau Evaluation Algorithm .................................................. 380
24.3   The Bernstein Representation .................................................................. 383
24.4   Geometric Properties of Bezier Curves .................................................. 384
        24.4.1   Affine Invariance ....................................................................... 385
        24.4.2   Convex Hull Property ................................................................ 386
        24.4.3   Variation Diminishing Property ................................................. 386
        24.4.4   Interpolation of the First and Last Control Points .................... 387
24.5   Differentiating the de Casteljau Algorithm ............................................ 388
        24.5.1   Smoothly Joining Two Bezier Curves ...................................... 389
        24.5.2   Uniqueness of the Bezier Control Points .................................. 390
24.6   Tensor Product Bezier Patches ............................................................... 391
24.7   Summary ................................................................................................ 395
Exercises ............................................................................................................ 397

**25   Bezier Subdivision** ................................................................................... **401**
25.1   Divide and Conquer ................................................................................ 401
25.2   The de Casteljau Subdivision Algorithm ................................................ 401
25.3   Rendering and Intersection Algorithms .................................................. 405
        25.3.1   Rendering and Intersecting Bezier Curves ............................... 405
        25.3.2   Rendering and Intersecting Bezier Surfaces ............................ 407
25.4   The Variation Diminishing Property of Bezier Curves ........................... 409
25.5   Joining Bezier Curves Smoothly ............................................................ 410
25.6   Summary ................................................................................................ 411
Exercises ............................................................................................................ 412
Programming Projects ....................................................................................... 414

**26  Blossoming** .................................................................................................... **417**
26.1   Motivation ...................................................................................... 417
26.2   The Blossom .................................................................................. 418
26.3   Blossoming and the de Casteljau Algorithm ................................. 419
    26.3.1   Bezier Subdivision from Blossoming ................................. 422
26.4   Differentiation and the Homogeneous Blossom ............................ 423
    26.4.1   Homogenization and the Homogeneous Blossom ................ 423
    26.4.2   Differentiating the de Casteljau Algorithm ...................... 427
    26.4.3   Conversion Algorithms between Monomial and Bezier Form ... 430
26.5   Summary ........................................................................................ 431
Exercises .................................................................................................. 434

**27  B-Spline Curves and Surfaces** ...................................................................... **437**
27.1   Motivation ...................................................................................... 437
27.2   Blossoming and the Local de Boor Algorithm .............................. 438
27.3   B-Spline Curves and the Global de Boor Algorithm ..................... 441
27.4   Smoothness .................................................................................... 443
27.5   Labeling and Locality in the Global de Boor Algorithm ............... 445
27.6   Every Spline is a B-Spline ............................................................ 446
27.7   Geometric Properties of B-Spline Curves ..................................... 448
27.8   Tensor Product B-Spline Surfaces ................................................. 449
27.9   Non-Uniform Rational B-Splines (NURBS) ................................. 451
27.10  Summary ........................................................................................ 452
Exercises .................................................................................................. 453

**28  Knot Insertion Algorithms for B-Spline Curves and Surfaces** ................. **457**
28.1   Motivation ...................................................................................... 457
28.2   Knot Insertion ................................................................................ 457
28.3   Local Knot Insertion Algorithms .................................................. 458
    28.3.1   Boehm's Knot Insertion Algorithm .................................. 458
    28.3.2   The Oslo Algorithm .......................................................... 460
    28.3.3   Conversion from B-Spline to Piecewise Bezier Form ....... 461
    28.3.4   The Variation Diminishing Property for B-Spline Curves ... 461
    28.3.5   Algorithms for Rendering and Intersecting B-Spline Curves and Surfaces ... 463
28.4   Global Knot Insertion Algorithms ................................................. 464
    28.4.1   The Lane–Riesenfeld Algorithm ....................................... 464
    28.4.2   Schaefer's Knot Insertion Algorithm ................................ 467
    28.4.3   Convergence of Knot Insertion Algorithms ...................... 468
    28.4.4   Algorithms for Rendering and Intersecting B-Spline Curves and Surfaces Revisited ... 470
28.5   Summary ........................................................................................ 471
Exercises .................................................................................................. 474
Programming Project ............................................................................... 475

**29  Subdivision Matrices and Iterated Function Systems** ............................... **477**
29.1   Subdivision Algorithms and Fractal Procedures ........................... 477
29.2   Subdivision Matrices ..................................................................... 478
    29.2.1   Subdivision Matrices for Bezier Curves ........................... 479
    29.2.2   Subdivision Matrices for Uniform B-Spline Curves ......... 481

29.3    Iterated Function Systems Built from Subdivision Matrices ............ 485
        29.3.1  Lifting the Control Points to Higher Dimensions ............... 485
        29.3.2  Normal Curves ..................................................... 489
29.4    Fractals with Control Points ............................................. 491
29.5    Summary ................................................................. 493
        29.5.1  Bezier Curves ..................................................... 494
        29.5.2  Uniform B-Splines ................................................ 495
Exercises ........................................................................... 496
Programming Projects .............................................................. 497

**30  Subdivision Surfaces** ...................................................... **499**
30.1    Motivation .............................................................. 499
30.2    Box Splines ............................................................. 500
        30.2.1  Split and Average ................................................. 500
        30.2.2  A Subdivision Procedure for Box Spline Surfaces ................. 500
30.3    Quadrilateral Meshes .................................................... 503
        30.3.1  Centroid Averaging ................................................ 505
                30.3.1.1  Uniform Bicubic B-Spline Surfaces ..................... 505
                30.3.1.2  Arbitrary Quadrilateral Meshes ........................ 507
        30.3.2  Stencils .......................................................... 509
                30.3.2.1  Stencils for Uniform B-Splines ........................ 509
                30.3.2.2  Stencils for Extraordinary Vertices ................... 511
30.4    Triangular Meshes ....................................................... 512
        30.4.1  Centroid Averaging for Triangular Meshes ........................ 512
                30.4.1.1  Three-Direction Quartic Box Splines ................... 512
                30.4.1.2  Arbitrary Triangular Meshes ........................... 514
        30.4.2  Stencils for Triangular Meshes .................................. 516
                30.4.2.1  Stencils for Three-Direction Quartic Box Splines ...... 516
                30.4.2.2  Stencils for Extraordinary Vertices ................... 517
30.5    Summary ................................................................. 518
        30.5.1  Bicubic Tensor Product B-Splines and Three-Direction
                Quartic Box Splines .............................................. 518
                30.5.1.1  Split and Average ...................................... 519
                30.5.1.2  Centroid Averaging ..................................... 519
                30.5.1.3  Stencils ............................................... 521
        30.5.2  Centroid Averaging for Meshes of Arbitrary Topology ............. 521
        30.5.3  Stencils for Extraordinary Vertices ............................. 523
Exercises ........................................................................... 524
Programming Projects .............................................................. 527

**Further Readings** ............................................................... **529**

**Index** .......................................................................... **533**

# *Foreword*

The field of computer graphics has reached a level of maturity, as evidenced by the graphics capability that is ubiquitous on even entry-level personal computers and by the prevalence of graphics software that is used to create stunning images and animations by artists who need no knowledge of what is going on under the hood. At the same time, the field of computer graphics continues to expand and evolve, as evidenced by the increasing number of research papers written on the topic from one year to the next. This rapid growth poses a challenge for the author of a new book on computer graphics, who must assess what subset of the ever-expanding body of knowledge will be of greatest benefit to the students, and will have the longest shelf life.

Several good books have been written on computer graphics over the years, and many of them are currently available in advanced editions. They span a spectrum from encyclopedic to nuts-and-bolts programming. This book offers a fresh approach. It is discretized into "lectures," organized to fill a semester-long introductory course with one chapter for each of the 30 class periods. The topics chosen cover most of the key concepts of computer graphics. The lectures on mathematical foundations and on geometric modeling are particularly strong. The book is void of programming examples, since the transitory nature of graphics languages would soon render such material outdated.

The author has a distinguished career as a developer, researcher, and educator in computer graphics. After earning a PhD in mathematics, he worked for many years in the young computer graphics and computer-aided design (CAD) industries, where he contributed to early graphics software development. While thus employed, he took an interest in mentoring several PhD students at various universities across the country, even though it was not formally part of his job. He did this partly because he loves research, but even more because he loves helping students succeed. As one of those fortunate students, I can attest to his infectious enthusiasm for his subject matter, his lucid explanations, his noise-free writing style, and his mathematical rigor. Dr. Goldman has dedicated the past 20 years of his career to teaching and research as a professor of computer science at the University of Waterloo and Rice University. The pedagogical style of this book has been refined during his many years of teaching this material. He is an excellent mentor of students and I am pleased that his reach will be extended through the publication of this book.

**Thomas W. Sederberg**
*Brigham Young University*

*To the Logos,*
*−Wordsmith and Mathematician Incarnate−*
*Co-Eternal Consort of the Creator,*
*Who on the First Day spoke the Word and made Light,*
*And saw that it was Good.*

**יהי אור**

*Fiat Lux.*
*Let there be Light.*

*Genesis 1:3*

# Preface

*Behold, the days come, saith the LORD, that I will make a new covenant...*

– Jeremiah 31:31

The good news is that Computer Graphics is fun: fun to look at, fun to use, and when done properly even fun to program and debug. There are also many fun applications of Computer Graphics, ranging from video games, to animated cartoons, to full-length feature movies. If you learn Computer Graphics and Geometric Modeling, you might even get a job in a field where you can have lots of fun. Art and architecture, biomedical imaging, computational photography: whatever you can see, or whatever you imagine you can see, you can design with Geometric Modeling and you can display with Computer Graphics.

Yet for a long time now and for many hapless college students, university courses on Computer Graphics and Geometric Modeling seem to oscillate from tedious to abstruse. Ponderous books and pedantic professors appear to focus endlessly on low-level techniques—line drawing, polygon filling, antialiasing, and clipping—and they elicit little or no connection between Computer Graphics and the rest of Computer Science. Good pedagogy is slighted: levels are mixed; mathematics abused; intuition neglected; elegance avoided; and mysteries ignored. Fun and excitement are drained from the subject. Sadly, I admit, I have taught such courses myself. Collectively, the field must do better. This book is intended as a new testament, a new inspiration for teachers and students alike.

This canon is intentionally short. This book is not an encyclopedia of Computer Graphics, but rather a brief introduction to the subject, essentially what can be taught to advanced undergraduate students and beginning graduate students majoring in Computer Science in a 15 week one semester course. Nevertheless, this book does cover many of the major themes of the discipline.

Broadly, these major themes can be divided into three categories: graphics, modeling, and mathematical foundations. Graphics consists of lighting and shading—reflection and refraction, recursive ray tracing, radiosity, illumination models, polygon shading, and hidden surface procedures. Modeling is the theory of curves, surfaces, and solids—planes and polygons, spheres and quadrics, algebraics and parametrics, constructive solid geometry, boundary files, and octrees, interpolation and approximation, Bezier and B-spline methods, fractal algorithms, and subdivision techniques. The mathematical foundations are mostly linear algebra, but from a somewhat idiosyncratic perspective not typically encountered in standard linear algebra classes—vector geometry and vector algebra, affine spaces and the space of mass-points, affine maps and projective transformations, matrices, and quaternions.

The subject of Computer Graphics is still relatively new. I shall not put new wine in old bottles. Rather I have deliberately sought innovative techniques for presenting this material to Computer Science students. I have borrowed extensively from other authors, but I have rearranged and reordered the topics and I have developed approaches that I believe are elegant to present, appealing to learn, and sensible pedagogically. In contrast to the standard concentration on low-level graphics algorithms, I focus instead on more advanced graphics, modeling, and mathematical methods—ray tracing, polygon shading, radiosity, fractals, freeform curves and surfaces, vector methods, and transformation techniques. Here is how I have organized this material.

*Fractals first.* The course begins with a topic that is visually appealing, intellectually deep, and naturally connected to the rest of Computer Science. Fractal shapes are *recursion made visible.* Computer Science students are used to writing recursive programs; now, perhaps for the first time, these students get to see recursion in visual art as well as in computational science. Fractals are exciting at many levels: visually, intellectually, and computationally. Students can be hooked into learning lots of important technical material in order to generate neat fractals.

I introduce fractals using turtle graphics. LOGO and turtle graphics have been popularized as systems for teaching programming to young children. But LOGO is also a powerful vehicle for studying Computer Graphics and Computational Geometry. This book promotes the turtle as a simple and effective way of introducing many of the fundamental concepts that underlie contemporary Computer Graphics.

In the version of LOGO employed here, the turtle's state is represented by a point $P$ specifying the turtle's location and a direction vector $v$ specifying the turtle's heading in the plane. There are commands such as FORWARD and TURN for altering the turtle's position and orientation. The turtle draws curves by joining with straight lines consecutive points along its path as it moves around in the plane under the control of a turtle programmer.

Much can be learned from this simple turtle. The FORWARD and TURN commands are equivalent to translation and rotation, so students get an early introduction to these fundamental transformations of the graphics pipeline, albeit in two dimensions. (There is also a RESIZE command to change the length of the direction vector, so students get to see scaling along with translation and rotation.) Internally, computations are performed using rectangular coordinates, but turtle programmers have no access to these coordinates. Thus high-level commands are kept distinct from low-level computations, a standard theme of Computer Science that will be stressed repeatedly throughout this book. Students can also write a simple interpreter for LOGO to hone skills learned in other Computer Science courses.

*Points translate, vectors rotate and scale.* Thus, although internally points and vectors in the plane may both be represented by pairs of rectangular coordinates, points and vectors are treated differently in turtle graphics. This distinction between points and vectors persists throughout this book and indeed throughout Computer Graphics; the turtle is a convenient model for introducing this very basic, but often overlooked, distinction. The turtle also anticipates the use later in this book of affine coordinates to distinguish between points and vectors.

Students can generate fractals such as the Sierpinski triangle and the Koch snowflake via simple recursive turtle programs. Thinking about recursive programs to generate concrete visual effects often enhances students' deep understanding of recursion.

Recursive turtle programs to generate fractals are typically easy to write; there is usually an obvious base case, and the body of the recursion consists of recursive calls connected by distinct sequences of the basic turtle commands. Nevertheless, in many such fractal programs a mystery soon appears. Although the choice of the base case seems constrained, changing the base case does not appear to alter the fractal generated in the limit of the recursion. This mystery demands an explanation, and this explanation leads to an important alternative approach to generating fractals— iterated functions systems.

An *iterated function system* is just a finite collection of contractive transformations. Repeatedly applying a fixed set of contractive transformations to a compact set generates a fractal. For particular fractals such as the Sierpinski gasket, one can often easily guess the transformations that generate the fractal. Nevertheless, this general mathematical approach to fractals is much more difficult to motivate to Computer Science students than recursive turtle programs. However, a straightforward analysis of recursive turtle programs reveals that the fractal generated by a recursive turtle program is equivalent to applying the iterated function system generated by the turtle commands in the recursion to the turtle geometry generated by turtle commands in the base case. Thus the analysis of recursive turtle programs motivates the study of iterated function systems.

Moreover, understanding iterated function systems is the key to understanding the mystery students encounter when studying fractals generated by recursive turtle programs—that changing the base case does not appear to alter the fractal generated in the limit of the recursion. The central result is the *trivial fixed point theorem*, which states that iteration of a contractive map on a point in a complete metric space always converges to a unique fixed point. There are some technical mathematical difficulties to overcome in order for students to understand the statement and proof of this fixed point theorem, so strong motivation is essential. The fractal mystery of recursive turtle programs provides this motivation.

Understanding this theorem is well worth the effort because this fixed point theorem will be invoked again later in the course of this book in the chapter on radiosity. Both the Jacobi and the Gauss-Seidel relaxation methods for solving large systems of linear equations are based on this trivial fixed point theorem. Recursive root finding algorithms for certain transcendental equations can also be derived from this theorem. Toward the end of this book, using subdivision, students will discover that Bezier and B-spline curves are also fixed points of iterated function systems. Thus, somewhat surprisingly, smooth polynomial and piecewise polynomial curves are also intimately related to fractals.

The study of fractals from the perspective of turtle graphics and iterated function systems typically takes about four weeks, more that one-quarter of a fifteen-week semester. This time is well spent. In addition to rendering visually exciting graphics, student are introduced to points and vectors, affine coordinates and affine transformations, matrix computations, and an important fixed point theorem. They also learn to distinguish clearly between high-level concepts and low-level computations. With this preparation, students are now ready to move up to three dimensions.

Three dimensions require new mathematical foundations. In two dimensions, students can get by with coordinate geometry, but in three dimensions coordinates can be confusing and often actually get in the way of the analysis. Therefore Part II of this book begins with a thorough review of three-dimensional vector geometry: addition, subtraction, scalar multiplication, dot product, cross product, and determinant. Most Computer Science students have seen vector algebra before either in courses on physics or linear algebra, but their geometric understanding of the vector operations is still tenuous at best. A thorough review of vector methods from a geometric perspective is in order here to prepare the students for modeling and analyzing geometry in three dimensions.

Computer Graphics deals with points as well as with vectors—points, not vectors, are typically displayed on the graphics terminal—so this book discusses affine spaces (spaces of points) and affine transformations along with vector spaces and linear transformations. Affine spaces are new to most students—the restriction to affine combinations may appear artificial at first—but this unfamiliarity is all the more reason to adopt coordinate-free methods. When the levels are not mixed, when theory is kept separate from computation, these concepts are much easier to explain and simpler to understand. Later, students will see that the distinction between points and vectors is computational as well as theoretical: points translate, vector do not.

Vector techniques are applied right away to derive coordinate-free vector formulations for all the affine and projective transformations commonly used in Computer Graphics—translation, rotation, mirror image, scaling, shearing, and orthogonal and perspective projections—before matrix methods are introduced. These vector formulas emphasize the distinction between transformations (high-level concepts) and their matrix representations (low-level computational tools), notions that are too often confused in the minds of the students. When matrices are introduced later to speed up calculations, these vector formulas are invoked to derive matrix representations for each of the corresponding transformations. Because the original vector formulas are coordinate free, these matrix representations are not confined to describing projections into coordinate planes or rotations around coordinate axes, but work for planes and axes in general position.

Vector geometry replaces coordinate computations throughout this book. Coordinates are confined to low-level subroutines for calculating addition, subtraction, scalar multiplication, dot product, cross product, and determinant. Vector methods are applied ubiquitously to derive metric formulas, surface equations, and intersection algorithms; vector algebra is applied to calculate normal vectors for lighting models, and to generate reflection and refraction vectors for recursive ray tracing.

The proper uses of coordinates are to communicate with a computer and to speed up certain common computations. When coordinates are introduced, the theoretical distinction between points and vectors in affine space leads naturally to the insertion of a fourth coordinate, an *affine coordinate*, to distinguish between the rectangular coordinates for points and vectors: the affine coordinate is one for points and zero for vectors. This affine coordinate leads to the adoption of $4 \times 4$ matrices to represent affine transformations.

Affine transformations on affine spaces are similar to linear transformations on vector spaces. Affine transformations map points to points, vectors to vectors, and preserve affine combinations. Thus affine transformations are represented by matrices that preserve the fourth, affine coordinate. Translation, rotation, mirror image, scaling, shearing, and orthogonal projection are all affine transformations.

Regrettably, affine spaces and affine transformations are not sufficient to model all the geometry encountered in Computer Graphics. Perspective projection is not an affine transformation. Perspective is not a well-defined transformation on vectors; moreover, points on the plane through the center of projection parallel to the plane of projection are not mapped to affine points but seem rather to be mapped to infinity by perspective projection. Thus a more general ambient space incorporating a more general collection of transformations is required to accommodate perspective projections.

Most books on Computer Graphics adopt projective spaces and projective transformations. Projective transformations include all the affine transformations as well as perspective projection. Nevertheless, projective spaces have several major drawbacks that make them unsuitable either as an algebraic or as a geometric foundation for Computer Graphics.

Projective space contains two types of points: *affine points and points at infinity*. The points at infinity complete the geometry of affine space—parallel lines in affine space meet at points at infinity in projective space—so perspective projection is defined at all points in projective space except the center of projection. But there is a big price to pay for these points at infinity.

The points at infinity in front of an observer are identical to the points at infinity behind the observer. Therefore gazing along an unobstructed direction in projective space, an observer would see the back of their head! Orientation—up and down, left and right, front and back—plays a fundamental role in the visual world, but there is no notion of orientation in projective space. This peculiar geometry of projective spaces is nonintuitive and difficult for most mathematically unsophisticated students of Computer Science to understand.

Worse yet, the points at infinity in projective space supplant the vectors in affine space. To adopt projective geometry, Computer Graphics would have to abandon the vector geometry that provides the foundation for most of the required mathematical analysis. But realistic lighting models depend on vectors normal to surfaces: diffuse and specular illumination, reflection and refraction computations all make use of surface normal vectors.

Finally, projective space is not a linear space. One cannot add points, or even take affine combinations of points, in projective space. Thus projective space is not a suitable model for most computer computations. Matrix multiplication is allowed, so the standard transformations in the graphics pipeline can be accommodated in projective space. But without an algebra for points it is impossible to construct many of classical parametric curves and surfaces, such as Bezier or B-spline curves and surfaces, inside projective space. Similarly, with no notion of addition or scalar multiplication, projective spaces fail to support shading algorithms based on linear interpolation.

All of these basic problems—theoretical and computational, geometric and algebraic—can be overcome by replacing projective space with the space of mass-points.

Mass-points form a vector space. To multiply a mass-point by a scalar, simply multiply the mass by the scalar (masses are allowed to be negative); to add two mass-points, apply Archimedes' law of the lever—place a mass equal to the sum of the two masses at the center of mass of the original two mass-points. The space of mass-points is a four-dimensional vector space: three dimensions are due to the points, the fourth dimension is induced by the masses. Vectors are incorporated into the space of mass-points as objects with zero mass; affine space is embedded in the space of mass-points by setting the masses of all the points to one.

Replacing projective space by the space of mass-points, replaces a compact, nonorientable, nonlinear three-dimensional manifold by a four-dimensional vector space. Going up in dimension removes the geometric incongruities of projective space and facilitates algebraic computations while modifying only slightly the formal representation.

The space of mass-points has all of the advantages, but none of the disadvantages, of projective space. The space of mass-points incorporates the points and vectors of affine space, so vector algebra and vector geometry make sense in the space of mass-points. The space of mass-points is a linear space, so it is possible to construct Bezier and B-spline curves and surfaces, as well as their rational variants, inside the space of mass-points. Since the space of mass-points is a vector space, the natural transformations on the space of mass-points are linear transformations. The linear transformations that preserve mass are precisely the affine transformations; perspective projections are also linear transformations on the space of mass-points, albeit ones that do not preserve mass.

There are other rewards for working in the space of mass-points. Archimedes' law of the lever can be applied in the space of mass-points to derive the formula for perspective projection from any center of projection into any plane. Also the classical map from the viewing frustum to a rectangular box can be derived simply by adding to the image of an arbitrary point under perspective projection the depth vector from the point to the plane of projection. This addition, however, is the addition of mass-points and vectors in the space of mass-points, not the ordinary addition of points and vectors in affine space. Perspective projection in the space of mass-points introduces additional mass to affine points, and this mass must be taken into account to get the correct formula for the map from the viewing frustum to the rectangular box.

Since the space of mass-points is a four-dimensional vector space, four coordinates are needed to represent mass-points. The fourth coordinate stores the mass; the first three coordinates store the rectangular coordinates of the point multiplied by the mass. Thus the rectangular coordinates of the points can be retrieved from these four coordinates by dividing the first three coordinates by the mass. Notice how coordinates for mass-points extend affine coordinates by permitting the fourth coordinate, the mass, to take on any value.

These coordinates for mass-points resemble homogeneous coordinates for projective points but with this difference: the same affine point with different masses in the space of mass-points represents distinct mass-points; the same affine point with different homogeneous coordinates in projective space represents the same projective point. In fact, projective points are just equivalence classes of mass-points, where two mass-points are equivalent if the masses are located at the same point in affine space.

The space of mass-points is a four-dimensional vector space, so linear transformations on the space of mass-points are represented by $4 \times 4$ matrices. Thus the same matrices are used to represent the same transformations on affine space in both the space of mass-points and projective space. Therefore the familiar computational formalism of homogeneous coordinates and projective transformations remains valid in the space of mass-points.

But there is a big bonus here: we can also multiply vectors in four dimensions. The only real vector spaces that have an associative multiplication with inverses are the real numbers (one dimension), the complex numbers (two dimensions), and the quaternions (four dimensions). Thus quaternion multiplication is multiplication in the space of mass-points. It is not an anomaly that we are working in four dimensions instead of three dimensions; rather we are incredibly lucky, since in four dimensions we can take advantage of this additional multiplicative structure.

Classically, one can use the richness of this quaternion algebra to represent conformal transformations on vectors in three dimensions by sandwiching vectors between quaternions rather than by multiplying vectors by $4 \times 4$ matrices. Thus quaternions provide more compact representations for conformal transformations than $4 \times 4$ matrices. Moreover, composing conformal transformations by multiplying quaternions is faster than composing conformal transformations by matrix multiplication.

Quaternions are used most effectively in Computer Graphics to avoid distortions during conformal transformations (quaternions are easy to renormalize) and to perform key frame animation (by spherical linear interpolation). There is a good deal of folklore about quaternions scattered throughout the literature, but I have yet to find a good textbook treatment of quaternions for Computer Graphics. Fortuitously, the four-dimensional vector space of mass-points incorporates quaternions in a natural way; we do not need to introduce quaternions as an additional *ad hoc* structure unrelated to projective space.

With the proper mathematical foundations—vector geometry and vector algebra, mass points and quaternion multiplication, affine transformations and perspective projections—the technical tools are now in place to study three-dimensional Computer Graphics.

Realistic rendering can be accomplished in three ways: ray tracing, polygon shading, and radiosity. This book covers all three of these topics. Conceptually, the most straightforward method is recursive ray tracing, so the text treats this method first.

Recursive ray tracing is based on the recursive computation of reflection and refraction rays. Two innovations are introduced here by taking advantage of the transformations studied in the previous chapters: reflection rays can be computed from the law of the mirror using the mirror image map for arbitrary mirror planes; refraction rays can be computed from Snell's law using the rotation transformation around arbitrary axes. In fact, these approaches to reflection and refraction are left as easy exercises for the students. The text adopts an even simpler approach, decomposing reflection and refraction vectors into orthogonal components and analyzing each component directly using the appropriate optical laws and the simple mathematics of dot products and cross products. Thus because the proper mathematical foundations are in place, reflection and refraction rays are easy to compute.

Ray casting is a compelling topic only if students have visually interesting surfaces to render. Therefore this subject provides a natural time and place to introduce some classical surfaces. The study of surfaces is confined here to the investigation of those properties and computations necessary for ray tracing: surface equations, surface normals, and ray–surface intersections. Again vector methods and transformation techniques play a key role in an innovative approach to ray tracing surfaces.

The sphere is the simplest nonplanar surface. Vectors normal to the surface of a sphere are parallel to the vectors from the center to points on the surface of the sphere, and the intersection of a line and a sphere can be reduced to the intersection of a line and a coplanar circle. Once the sphere is mastered, several other quadric surfaces can easily be analyzed using affine and projective transformations.

An ellipsoid is the image of a sphere under nonuniform scaling. Thus surface normals on the ellipsoid can be computed by finding surface normals on the sphere and mapping these normals onto the ellipsoid. The intersections of a ray and an ellipsoid can be calculated in a similar fashion by mapping the ray and the ellipsoid to a ray and a sphere, calculating the intersections of the ray and the sphere, and then mapping the resulting intersection points back to the ellipsoid.

Cylinders and cones are next. A sphere is the locus of points equidistant from a point; a cylinder is the locus of points equidistant from a line. From this definition it is easy using vector techniques to compute normals to the surface of the cylinder. Similar definitions and normal vector computations are developed for the cone. The intersections of a ray and a cylinder can be calculated by applying orthogonal projection onto a plane perpendicular to the cylinder axis to map the ray and the cylinder to a ray and a circle, then invoking the algorithm already developed for ray tracing the sphere to calculate the intersections of the ray and the circle, and finally using the parameters of the intersection points on the line to find the intersection points on the cylinder. The intersections of a ray and a cone can be calculated in a similar fashion, just replace orthogonal projection by perspective projection from the vertex of the cone.

Arbitrary algebraic and general parametric surfaces can also be ray traced. General algorithms are presented for computing surfaces normals and ray–surface intersections for these broad surface types. The torus can be represented both as an algebraic and as a parametric surface. Thus these general algorithms can be applied to a geometrically familiar, computationally tractable (fourth-degree) surface.

Solid modeling follows right after surface modeling. Three common approaches to solid modeling are investigated: constructive solid geometry (CSG), boundary file representations (B-REP), and octrees. Constructive solid geometry fits in well with the preceding study of ray tracing and surface modeling. The primitives in CSG trees are typically solids bounded by planes, natural quadrics, and tori. CSG trees can be rendered and their mass properties can be calculated by ray casting algorithms. Boundary file representations introduce the notion of topology—connectivity information—alongside geometry—shape information—to facilitate searching the model for important features. Since conversion algorithms from CSG trees to boundary file representations are often quite complicated, boundary files are frequently restricted to polygonal models. To render these polygonal models to appear as smooth curved surfaces requires polygon-shading techniques.

Polygon shading is the second of the three realistic rendering techniques presented in this book. Both Gouraud and Phong shading are covered here. Although these two shading algorithms are low level, scan line procedures, vector methods are still stressed in order to develop clever, fast, incremental implementations. Spherical linear interpolation is introduced for Phong shading, taking advantage of the students' previous encounter with the use of spherical linear interpolation for quaternions representing rotations in key frame animation.

Hidden surface algorithms are required for realistic rendering of polygonal models. Many hidden surface algorithms are available, and the text surveys four representative procedures: $z$-buffer, scan line, depth sort, and BSP-tree. The $z$-buffer algorithm is the easiest to implement, but the scan line algorithm is the hidden surface procedure that integrates best with Gouraud and Phong shading. Only while describing scan line algorithms does this book intentionally descend into low level, coordinate techniques. It is precisely when speed is at a premium that coordinate techniques are appropriate. Students learn that there are indeed times when it is expedient to descend to coordinate-based methods for fast rendering (*Render unto Caesar* . . . ). Depth sort, in contrast, provides some subtle applications of vector techniques, both for measuring relative depth and for finding obstructing planes. Finally, BSP-trees are a well-known and important data structure in Computer Science; BSP-trees are most useful for finding hidden surfaces when the model is fixed, but the viewpoint or the light sources can change position.

Radiosity is the rendering method that presents the most realistic diffuse images. In contrast to recursive ray tracing, radiosity softens shadows and portrays color bleeding. To introduce radiosity, the text begins with the continuous form of the rendering equation and then successively simplifies this integral into a large discrete system of linear equations. Jacobi and Gauss-Seidel relaxation techniques are recommended for solving these linear systems, techniques already familiar to students from the fixed point methods encountered during the investigation of fractals. Gathering and shooting are both investigated here.

Freeform curves and surfaces constitute the final topic presented in this book. So far, students have rendered only a limited number of rigid surfaces, mostly planes and quadrics. But in order to faithfully represent a rich variety of forms including car bodies, ship hulls, airplane wings, toys, shoes, and even many animated cartoon characters, geometric modeling deals mainly with freeform shapes. Therefore this book closes with techniques for representing, analyzing, and rendering freeform curves and surfaces. Typical textbooks in Computer Graphics provide at best only a cursory *ad hoc* introduction to Bezier and B-spline techniques. In contrast, this book delivers a thorough, unified approach to Bezier and B-spline approximation, as well as to subdivision surfaces. Since these topics generally require some mathematical sophistication, several innovations are introduced here to simplify the presentation.

Linear interpolation is already familiar to students from Gouraud and Phong shading. Dynamic programming procedures based on successive linear interpolation are provided for Bezier and B-spline curves and surfaces: for Bezier curves and surfaces this procedure is called the *de Casteljau algorithm*; for B-splines the method is known as the *de Boor algorithm*.

Figures are often easier to understand than formulas. This book uses simple data flow diagrams—pyramid algorithms—to develop a straightforward, common approach to these dynamic programming procedures. Students are encouraged to bypass formulas and reason directly from these diagrams. A common approach permits common derivations for common properties such as affine invariance and nondegeneracy. Explicit expressions and as well as recursion formulas for the blending functions of Bezier curve and surface schemes can also be generated directly from these diagrams.

Blossoming provides the simplest, most direct approach to deriving many of the unique properties of Bezier and B-spline curves and surfaces. Therefore, following the maxim that students should *use the mathematics most appropriate to the problem at hand*, I have tried to present a clear and unintimidating introduction to blossoming. While blossoming will be new to most students, blossoming is not difficult. In addition to introducing the blossom axiomatically, the text also presents the blossom concretely using pyramid algorithms. The pyramid diagram for the blossom is identical to the pyramid diagram for Bezier curves with one slight variation: a different parameter is introduced on each level of the pyramid algorithm. Thus linear interpolation is also at the heart of blossoming. The standard properties of the blossom—symmetry, multiaffinity, the diagonal property, the dual functional property, as well as existence and uniqueness—are all derived here directly from these pyramid diagrams.

Bezier schemes are intimately related to blossoming because the corresponding pyramid algorithms are so closely connected. It follows from the dual functional property that the blossom of a univariate polynomial evaluated at the end points of a parameter interval yields the corresponding Bezier control points. Therefore blossoming provides a general approach to change of basis procedures for Bezier schemes. Conversion between monomial and Bezier form, degree elevation techniques, and subdivision algorithms are all derived here from blossoming. Moreover, the homogenous variant of the blossom is used to derive a differentiation algorithm for Bezier curves based on the de Casteljau algorithm.

B-splines are also closely related to blossoming. A slight modification of the blossoming interpretation of the de Casteljau algorithm for Bezier curves—starting with the blossom evaluated at consecutive knots instead of at the end points of the parameter domain—generates the de Boor evaluation algorithm for B-spline curves. Many introductory books simply take the de Boor algorithm as the definition of the B-splines. But without appropriate motivation, students are at a loss to understand what is so special about this particular recursion formula. Blossoming provides the proper motivation as well as the natural connection between the de Casteljau algorithm and the de Boor algorithm. As with Bezier curves, the blossom of a spline evaluated at the knots yields the corresponding B-spline control points. Therefore blossoming also provides a general approach to change of basis procedures for B-splines, so knot insertion algorithms for B-splines are readily derived here from blossoming. As with Bezier curves, the homogenous variant of the blossom is used to derive a differentiation algorithm for B-splines curves based on the de Boor algorithm, and this differentiation algorithm is used in turn to prove that the polynomial segments of B-spline curves join smoothly at the knots.

Subdivision is the key to rendering Bezier surfaces. The de Casteljau subdivision algorithm generates a polyhedral approximation that converges to the Bezier surface under recursive subdivision. This polyhedral approximation can then be rendered using either ray tracing, or polygon shading, or radiosity. Knot insertion algorithms play the analogous role for B-spline surfaces.

Subdivision can be implemented using matrices to multiply the control points. For Bezier schemes these subdivision matrices form an iterated function system. Hence, rather remarkably, Bezier curves and surfaces can be rendered as fractals by applying this iterated function system to an arbitrary compact set. Thus the book comes full circle, recapitulating close to the very end the fractal methods introduced almost at the very beginning.

Knot insertion for B-splines is the analogue of subdivision for Bezier curves and surfaces. Knot insertion can also be represented by matrix multiplication and these matrices also form an iterated function system. Thus B-spline curves and surfaces can also be rendered as fractals by applying this iterated function system to an arbitrary compact set. One additional innovation here is that we show how to extend the Lane-Riesenfeld subdivision algorithm for B-splines with uniform knots to a subdivision algorithm due to Scott Schaefer for B-splines with nonuniform knots.

The Lane-Riesenfeld knot insertion algorithm for B-splines with uniformly spaced knots is the principal paradigm for more general subdivision schemas. This book closes with a chapter containing a brief introduction to general subdivision surfaces for rectangular and triangular meshes, including box splines, centroid averaging, Loop subdivision, and Catmull-Clark subdivision, topics still at the frontier of current research. Three simple paradigms are employed to explain each of these methods: split and average (box splines), centroid averaging (meshes of arbitrary topology), and stencils—vertex, edge, and face stencils as well as special stencils for extraordinary vertices (Loop subdivision and Catmull-Clark subdivision).

This concludes the brief survey of the topics covered in this book. Many topics have purposely been omitted from this text, not because they are uninteresting or unimportant, but simply because they are either too advanced—physics-based modeling, scientific visualization, virtual reality—or too specialized—user interfaces, graphics hardware, input devices—for an introductory course.

A good book is a compendium of the abiding past, a snapshot of the transitory present, and a guess at the unknowable future. Writing a good book, much like teaching a compelling course, is about making choices: what to include and what to exclude are dictated by time and taste. I have tried to concentrate on enduring themes and to avoid ephemeral motifs. Graphics hardware and software will certainly soon change, but graphics algorithms based on well-established physical models of light and cogent mathematical methods will last a long time. Therefore I have avoided descriptions of current graphics hardware, and I have omitted from this book special programming languages such as C++ and API's like OpenGL. Low-level graphics algorithms such as line drawing, polygon filling, and clipping are bypassed to give more time and space to high-level graphics techniques such as ray tracing, polygon shading, and radiosity.

I have tried to write a book that is exciting to read without being superficial, rigorous without being pedantic, and innovative without being idiosyncratic. I have kept the manuscript relatively short in the hope that students and lecturers alike will read it in full. Lots of exercises and projects are included to flesh out this book, but there are many topics that I have consciously chosen not to cover. My intention has been to write a *A Guide for the Perplexed*, not a *Summa Theologica*.

Certainly I should thank all those who came before me in the field of Computer Graphics—founders and innovators, scientists and mathematicians, pure academics and practicing engineers, serious students, and conscientious professors. There can be no new testament without an older revelation on which to build. And yet . . . Siggraph is attended by tens of thousands of people, and hundreds of paper are submitted to Siggraph each year. If even only a small percentage of these people and papers are important, I could not hope to list them all here.

Let me close instead by asking forgiveness from my former students, whom I abused with my lackluster teaching in the past. I hope I have done better by them here. I have drawn inspiration from my forebears, encouragement from my colleagues, and constructive criticism from my graduate students. I trust this constellation is sufficient to the task.

The days of any book are numbered. This book, like its predecessors, will inevitably become obsolete due to innovations in theory and technology. Three-dimensional graphics hardware is now technically feasible. If this hardware becomes popular, the current concentration on two-dimensional projections may someday be outdated, and user interfaces will certainly change. Innovations in mathematics such as Clifford algebras or in physics such as quantum computing may eventually make other parts of this book seem as stodgy and old fashioned as the tomes it is written to supplant. Instructors should keep these issues in mind when choosing a textbook for their classes in the future. Authorities argue, barriers breakdown, canons change, consensus crumble, disciplines decline, epistemologies expire, fashions fade, ideologies implode, laws languish, methods mutate, orthodoxies ossify, paradigms perish; *time and chance happen to them all.*

# *Author*

**Ron Goldman** is a Professor of Computer Science at Rice University in Houston, Texas.

Professor Goldman received his BS in Mathematics from the Massachusetts Institute of Technology in 1968 and his MA and PhD in Mathematics from Johns Hopkins University in 1973. He is an Associate Editor of *Computer Aided Geometric Design*. In 2002 he published a book entitled *Pyramid Algorithms: A Dynamic Programming Approach to Curves and Surfaces for Geometric Modeling*.

Dr. Goldman's current research interests lie in the mathematical representation, manipulation, and analysis of shape using computers. His work includes research in computer-aided geometric design, solid modeling, computer graphics, and splines. He is particularly interested in algorithms for polynomial and piecewise polynomial curves and surfaces, and he is currently investigating applications of algebraic and differential geometry to geometric modeling. He has published over a 100 articles in journals, books, and conference proceedings on these and related topics.

Before returning to academia, Dr. Goldman worked for 10 years in industry solving problems in computer graphics, geometric modeling, and computer-aided design. He served as a Mathematician at Manufacturing Data Systems Inc., where he helped to implement one of the first industrial solid modeling systems. Later he worked as a Senior Design Engineer at Ford Motor Company, enhancing the capabilities of their corporate graphics and computer-aided design software. From Ford he moved on to Control Data Corporation, where he was a Principal Consultant for the development group devoted to computer-aided design and manufacture. His responsibilities included database design, algorithms, education, acquisitions, and research.

Dr. Goldman left Control Data Corporation in 1987 to become an Associate Professor of Computer Science at the University of Waterloo in Ontario, Canada. He joined the faculty at Rice University in Houston, Texas as a Full Professor of Computer Science in July 1990.

# Part I

# Two-Dimensional Computer Graphics: From Common Curves to Intricate Fractals

# Chapter 1

## Turtle Graphics

*The turtle and the crane and the swallow observe the time of their coming*

– Jeremiah 8:7

### 1.1 Turtle Graphics

Motion generates geometry. The turtle is a handy paradigm for investigating curves generated by motion. Imagine a turtle that after each step leaves a trail connecting her previous location to her new location. As this turtle crawls around on a flat surface, the turtle may traverse paths that form intriguing geometric patterns (see Figure 1.1).

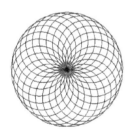 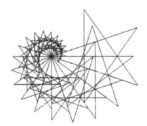 

**FIGURE 1.1:**   Some interesting geometric patterns generated by turtle paths in the plane.

Some of you may have encountered this mythical turtle before in the programming language LOGO, which has been used in primary schools to introduce young children to programming. But the turtle is more than just a toy for young children. In their book *Turtle Geometry*, Abelson and DiSessa use turtles to investigate many advanced topics, ranging from elementary differential geometry and topology to Einstein's General Theory of Relativity. The study of geometry using programming is the branch of Computer Science called *Computational Geometry*.

We shall begin our investigation of Computer Graphics with Turtle Graphics. After developing a simple variant of LOGO, we will see that Turtle Graphics can be used to generate many diverse shapes, ranging from simple polygons to complex fractals.

## 1.2   Turtle Commands

To help understand the ideas behind Turtle Graphics, we introduce a virtual turtle. The virtual turtle is a simple creature: she knows only where she is, in which direction she is facing, and her step size. The virtual turtle obeys solely simple commands to change either her location, or her heading, or her notion of scale.

Consider such a virtual turtle living on a virtual plane. The turtle's location can be represented by a point $P$ (a dot) given by a pair of coordinates $(x,y)$; similarly the turtle's heading can be represented by a vector $w$ (an arrow) given by another pair of coordinates $(u,v)$ (see Figure 1.2). The step size of the turtle is simply the length of the vector $w$. Thus, the step size of the turtle is $|w| = \sqrt{u^2 + v^2}$.

Location is a point; direction is a vector. Points are represented by dots; vectors are represented by arrows. Points have position, but no direction or length; vectors have direction and length, but no fixed position. In many branches of science and engineering, the distinction between points and vectors is often overlooked, but this distinction is very important in Computer Graphics. We shall see shortly that computationally points and vectors are treated quite differently in LOGO.

The pair $(P,w)$ is called the *turtle's state*. Although internally the computer stores the coordinates $(x,y)$ and $(u,v)$, the turtle (and the turtle programmer) has no access to these global coordinates; the turtle knows only the local information $(P,w)$, not the global information $(x,y)$ and $(u,v)$. That is, the turtle knows only that she is here at $P$ facing there in the direction $w$; she does not know how $P$ and $w$ are related to some global origin or coordinate axes, to other heres and theres.

This state model for the turtle is similar to the model of a billiard ball in classical mechanics, where physicists keep track of a ball's position and momentum. Turtle location is analogous to the position of the billiard ball, and turtle direction is analogous to the momentum of the billiard ball. The main difference between these two models is that for billiard balls, the laws of physics (differential equations) govern the position and momentum of the ball; in contrast, we shall write our own programs to change the location and direction of the turtle.

The turtle responds to four basic commands: FORWARD, MOVE, TURN, and RESIZE. These commands affect the turtle in the following ways:

- FORWARD $D$: The turtle moves forward $D$ steps along a straight line from her current position in the direction of her current heading, and draws a straight line from her initial position to her final position.

- MOVE $D$: Same as FORWARD $D$ without drawing a line.

- TURN $A$: The turtle changes her heading by rotating her direction vector in the plane counterclockwise from her current heading by the angle $A$.

- RESIZE $S$: The turtle changes the length of her step size (direction vector) by the factor $S$.

**FIGURE 1.2:**   The virtual turtle is represented by a point $P$ (dot) and a vector $w$ (arrow).

These four turtle commands are implemented internally in the following fashion:

- FORWARD $D$ and MOVE $D$—*Translation* (see Figure 1.3)

$$x_{new} = x + Du$$

$$y_{new} = y + Dv$$

- TURN $A$—*Rotation* (see Figure 1.4)

$$\begin{aligned}u_{new} &= u\cos(A) - v\sin(A)\\ v_{new} &= u\sin(A) + v\cos(A)\end{aligned} \Leftrightarrow (u_{new}\ v_{new}) = (u\ v)\begin{pmatrix}\cos(A) & \sin(A)\\ -\sin(A) & \cos(A)\end{pmatrix}.$$

- RESIZE $S$—*Scaling* (see Figure 1.5)

$$\begin{aligned}u_{new} &= Su\\ v_{new} &= Sv\end{aligned} \Leftrightarrow (u_{new}\ v_{new}) = (u\ v)\begin{pmatrix}S & 0\\ 0 & S\end{pmatrix}.$$

We shall also adopt the following conventions:

- $D < 0 \Rightarrow$ FORWARD $D$ moves the turtle backward.

- $A > 0 \Rightarrow$ TURN $A$ rotates the turtle counterclockwise.

- $A < 0 \Rightarrow$ TURN $A$ rotates the turtle clockwise.

- $S < 0 \Rightarrow$ RESIZE $S$ rotates the turtle by $180°$ and then scales the direction vector by $|S|$.

Notice that the TURN and RESIZE commands can be implemented using matrix multiplication, but that the FORWARD and MOVE commands cannot be implemented in this manner. This distinction arises because rotation and scaling are linear transformations on vectors, but translation is an affine, not a linear, transformation on points. (A point is not a vector, so transformations on points are inherently different from transformations on vectors. We shall discuss linear and affine transformations—their precise meanings as well as their similarities and differences—in detail in Chapter 4.)

The formulas for executing the four turtle commands can be derived easily from simple geometric arguments. We illustrate the effect of each of these turtle commands in Figures 1.3 through 1.5.

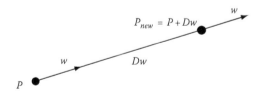

**FIGURE 1.3:** The command FORWARD $D$ changes the turtle's location, but leaves her direction and step size unchanged. The new turtle location is:

$$P_{new} = P + Dw.$$

Thus, in terms of coordinates,

$$x_{new} = x + Du,$$

$$y_{new} = y + Dv.$$

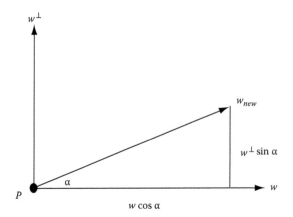

**FIGURE 1.4:**   The command TURN $\alpha$ changes the turtle's heading, but leaves her position and step size unchanged. To derive the turtle's new heading, we work in the turtle's local coordinate system.

Let $(P,w)$ be the turtle's current state, and let $w^{\perp}$ denote the vector perpendicular to $w$ of the same length as $w$. Then,

$$w_{new} = w \cos(\alpha) + w^{\perp} \sin(\alpha).$$

But if $w = (u,v)$, then $w^{\perp} = (-v,u)$ (see Exercise 1.11). Therefore, in terms of coordinates,

$$u_{new} = u \cos(\alpha) - v \sin(\alpha),$$
$$v_{new} = v \cos(\alpha) + u \sin(\alpha),$$

or in matrix notation,

$$(u_{new}\ v_{new}) = (u\ v)\begin{pmatrix} \cos(\alpha) & \sin(\alpha) \\ -\sin(\alpha) & \cos(\alpha) \end{pmatrix}.$$

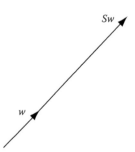

**FIGURE 1.5:**   The command RESIZE $S$ changes the turtle's step size by a factor of $S$, but leaves her position and heading unchanged. The turtle's new direction vector is given by

$$w_{new} = Sw.$$

Thus, in terms of coordinates,

$$u_{new} = Su,$$
$$v_{new} = Sv,$$

or in matrix notation,

$$(u_{new}\ v_{new}) = (u\ v)\begin{pmatrix} S & 0 \\ 0 & S \end{pmatrix}.$$

The turtle state $(P,w)$ is a complete description of what the turtle knows, and the four turtle commands FORWARD, MOVE, TURN, and RESIZE are the only way that a programmer can communicate with the turtle. Yet, with this simple setup, we can use the turtle to draw an amazing variety of interesting patterns in the plane.

---

## 1.3   Turtle Programs

Once the four turtle commands are implemented, we can start to write turtle programs to generate a wide assortment of shapes. The simplest programs just iterate various combinations of the FORWARD, TURN, and RESIZE commands. For example, by iterating the FORWARD and TURN commands, we can create polygons and stars (see Table 1.1). Notice that the angle in the TURN command is the exterior angle, not the interior angle, of the polygon. Circles can be generated by building polygons with lots of sides. Iterating FORWARD and RESIZE, the turtle walks along a straight line, and iterating TURN and RESIZE, the turtle simply spins in place. But by iterating FORWARD, TURN, and RESIZE, the turtle can generate spiral curves (see Figure 1.6).

With a bit more ingenuity (and with some help from the Law of Cosines), we can program the turtle to generate more complicated shapes such as the wheel and the rosette (see Figure 1.7).

The turtle commands FORWARD, TURN, and RESIZE are used to translate, rotate, and scale the turtle. In LOGO, we can also develop turtle programs SHIFT, SPIN, and SCALE to translate, rotate, and scale other turtle programs (see Table 1.2). Examples of SHIFT, SPIN, and SCALE applied to other turtle programs as well as to each other are illustrated in Figures 1.8 through 1.10.

Shapes built by iteration are cute to visualize and fun to build, but by far the most interesting and exciting shapes that can be created using Turtle Graphics are generated by recursion. In Chapter 2, we shall show how to design fractal curves using recursive turtle programs.

**TABLE 1.1:**   Simple turtle programs for generating polygons, stars, and spirals by iterating the basic turtle commands. For the spiral program, $A$ is a fixed angle and $S$ is a fixed scalar.

| POLYGON $N$ | STAR $N$ | SPIRAL $N, A, S$ |
|---|---|---|
| REPEAT $N$ TIMES | REPEAT $N$ TIMES | REPEAT $N$ TIMES |
| FORWARD 1 | FORWARD 1 | FORWARD 1 |
| TURN $2\pi/N$ | TURN $4\pi/N$ | TURN $A$ |
| | | RESIZE $S$ |

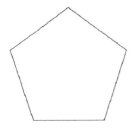
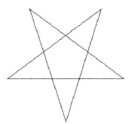

**FIGURE 1.6:**   A pentagon, a five-pointed star, and a spiral generated by the programs in Table 1.1. Here the spiral angle $A = 4\pi/5$ and the scale factor for the spiral is $S = 9/10$.

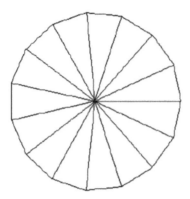

**FIGURE 1.7:**　The wheel and the rosette. The wheel is simply a regular polygon together with the lines connecting its vertices to its center. The rosette is a regular polygon, together with all its diagonals, that is, with the lines joining every vertex to every other vertex. The wheel displayed here has 15 sides and the rosette has 20 sides. We leave it as a challenge to the reader to develop simple turtle programs that generate these shapes, using the Law of Cosines to precalculate the lengths of the radii and the diagonals (see Exercise 1.7).

**TABLE 1.2:**　Turtle programs that translate, rotate, and scale other turtle programs.

| SHIFT *Turtle Program* | SPIN *Turtle Program* | SCALE *Turtle Program* |
|:---:|:---:|:---:|
| REPEAT *N* TIMES | REPEAT *N* TIMES | REPEAT *N* TIMES |
| *Turtle Program* | *Turtle Program* | *Turtle Program* |
| MOVE *D* | TURN *A* | RESIZE *S* |

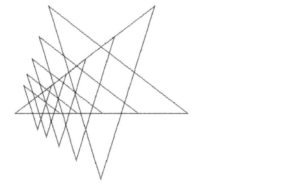

**FIGURE 1.8:**　The effect of applying the SHIFT program to the turtle program for generating a square.

**FIGURE 1.9:**　The effect of SCALE and SPIN. On the left, the SCALE program is applied to the turtle program for generating a star; on the right, the SPIN program is applied to the turtle program for generating a rosette.

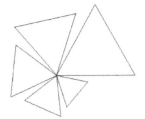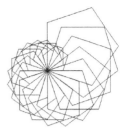

**FIGURE 1.10:** The effect of composing SPIN with SCALE: starting with a triangle (left), a star (center), and a pentagon (right).

## 1.4   Summary

Turtle programming is the main theme of this chapter, but this chapter also includes several important leitmotifs. In addition to gaining some facility with turtle programming, you should also pick up on the following distinctions that will be featured throughout this book.

First, there is an important difference between points and vectors. Although in a plane both points and vectors are typically represented by pairs of rectangular coordinates, conceptually points and vectors are different types of objects with different kinds of behaviors. Points are affected by translation, vectors are not. Vectors can be rotated and scaled, points cannot. Thus, if the turtle is in state $(P,w)$, then the FORWARD command translates the point $P$, but leaves the vector $w$ unchanged. Similarly, the TURN and RESIZE commands rotate and scale the vector $w$, but leave the point $P$ unchanged.

Second, there are two ways of approaching geometry: conceptual and computational. On the conceptual level, when we think about Euclidean geometry, we think about lengths and angles. Thus, when we program the turtle, we use the FORWARD, TURN, and RESIZE commands to affect distances and directions. Only when we need to communicate with a computer—only when we need to perform actual computations—do we need to descend to the level of coordinates. *Coordinates are a low-level computational tool, not a high-level conceptual device.* Coordinates are akin to an assembly language for geometry. Assembly languages are effective tools for communicating efficiently with a computer, but generally we do not want to think or write programs in an assembly language; rather we want to work in a high-level language. We shall try to keep these two modes—concepts and computations—distinct throughout this book by developing novel ways to think about and to represent geometry. The turtle is only the first of many high-level devices we shall employ for this purpose.

## Exercises

1.1. In classical LOGO, the programmer can control the state of the pen by two commands: PENUP and PENDOWN. When the pen is up, the FORWARD command changes the position of the turtle, but no line is drawn. The default state is pen down. Explain why the MOVE $D$ command is equivalent to the following sequence of turtle commands:

    PENUP
    FORWARD $D$
    PENDOWN

1.2. In classical LOGO, the turtle's state is represented by a pair $(P, A)$, where $P$ is the turtle's position and $A$ is the angle between the turtle's heading and a fixed direction in the plane. The position $P$ is stored by the $xy$ coordinates of the point $P$ relative to a global coordinate system, and the angle $A$ is the angle between the turtle's heading and the $x$-axis of this global coordinate system.

   a. Show how to implement the FORWARD and TURN commands using this representation for the turtle's state.

   Suppose we add to the turtle's state a scalar $S$ representing the turtle's step size.

   b. Show how to implement the FORWARD, TURN, and RESIZE commands using the representation $(P,A,S)$—position, angle, and step size—for the turtle's state.

   c. What are the advantages and disadvantages of the representation $(P,A,S)$ for the turtle's state compared to the representation $(P,w)$ given in the text?

1.3. Write a turtle program that draws a circle circumscribed around:

   a. A polygon with an arbitrary number of sides.

   b. A star with an arbitrary number of vertices.

1.4. Consider the program STAR in Table 1.1:

   a. For what integer values of $N > 5$ does the program STAR fail to draw a star?

   b. What happens if the command TURN $4\pi/N$ is replaced by TURN $8\pi/N$?

1.5. Consider the following program:

NEWSTAR $N$
   REPEAT $N$ TIMES
      FORWARD 1
      TURN $\pi - \pi/N$

   a. How do the stars drawn by this NEWSTAR program differ from the stars drawn by the STAR program in Table 1.1?

   b. How do the stars drawn by this NEWSTAR program differ for even and odd values of $N$? Explain the reason for this curious behavior.

1.6. Consider the following program:

TRISTAR $N$
   REPEAT $N$ TIMES
      FORWARD 1
      TURN $2\pi/3$
      FORWARD 1
      TURN $2\pi/N - 2\pi/3$

   a. How do the stars drawn by this TRISTAR program differ from the stars drawn by the STAR program in Table 1.1?

   b. Suppose that the command TURN $2\pi/3$ is replaced by the command TURN $\alpha$ for an arbitrary angle $\alpha$, and that the command TURN $2\pi/N - 2\pi/3$ is replaced by the command TURN $\beta$. Show that the TRISTAR program still generates a star provided that $\alpha + \beta = 2\pi/N$ and $N\alpha > 2\pi$. What happens when $N\alpha = 2\pi$?

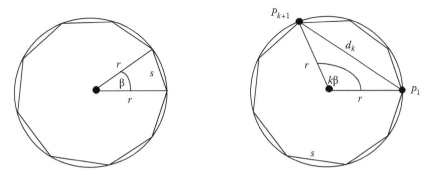

**FIGURE 1.11:** The Law of Cosines provides a relationship between the length of a side and the distance from a vertex to the center of a regular polygon (left). Similarly, the Law of Cosines provides a relationship between the distance from a vertex to the center of a regular polygon and the length of each diagonal of the polygon. Notice that if the polygon has $n$ sides, then $\beta = 2\pi/n$.

1.7. Write a turtle program that for a regular polygon with an arbitrary number of sides draws:

    a. A wheel
    b. A rosette
    (Hint: See Figure 1.11.)

1.8. Apply the SPIN program to the turtle program for a circle to generate the pattern in Figure 1.1, left.

1.9. Write a turtle program that draws an ellipse.

1.10. Prove that the turtle commands TURN and RESIZE commute by showing that if the turtle starts in the state $(P,w)$, then after executing consecutively the commands TURN $A$ and RESIZE $S$ the turtle will arrive at the same state as when executing consecutively the commands RESIZE $S$ and TURN $A$. Explain intuitively why this phenomenon occurs.

1.11. Let $w^\perp$ denote the vector perpendicular to the vector $w$ of the same length as $w$. Show that if $w = (u,v)$, then $w^\perp = (-v,u)$. (Hint: Use the distance formula and the Pythagorean theorem to show that $\triangle AOB$ in Figure 1.12 is a right triangle.)

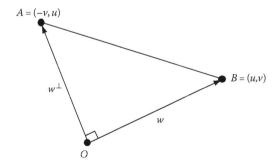

**FIGURE 1.12:** The vectors $w = (u,v)$ and $w^\perp = (-v,u)$ form a right angle at the origin $O$.

# Chapter 2

## Fractals from Recursive Turtle Programs

*use not vain repetitions . . .*

– Matthew 6:7

### 2.1 Fractals

Pictured in Figure 2.1 are four fractal curves. What makes these shapes different from most of the curves we encountered in Chapter 1 is their amazing amount of fine detail. In fact, if we were to magnify a small region of a fractal curve, what we would typically see is the entire fractal in the large. In Chapter 1, we showed how to generate complex shapes like the rosette by applying iteration to repeat over and over again a simple sequence of turtle commands. Fractals, however, by their very nature cannot be generated simply by repeating even an arbitrarily complicated sequence of turtle commands. This observation is a consequence of the Looping Lemmas for Turtle Graphics.

Sierpinski triangle

Fractal swiss flag

Koch snowflake

*C*-curve

**FIGURE 2.1:** Four fractal curves: a Sierpinski gasket, a fractal flag, a Koch snowflake, and a C-curve.

### 2.2 Looping Lemmas

Two of the simplest and most basic turtle programs are the iterative procedures in Table 2.1 for generating polygons and spirals. The Looping Lemmas assert that all iterative turtle programs, no matter how many turtle commands appear inside the loop, generate shapes with the same general symmetries as these basic programs. (There is one caveat here, that the iterating index is not used inside the loop; otherwise any turtle program can be simulated by iteration.)

**TABLE 2.1:**   Basic procedures for generating polygons and spirals via iteration.

| *POLY (Length, Angle)* | *SPIRAL (Length, Angle, Scalefactor)* |
|---|---|
| Repeat Forever | Repeat Forever |
| FORWARD Length | FORWARD Length |
| TURN Angle | TURN Angle |
|  | RESIZE Scalefactor |

**Circle Looping Lemma**

 *Any procedure that is a repetition of the same collection of FORWARD and TURN commands has the structure of POLY(Length, Angle), where*

 *Angle = Total Turtle Turning within the Loop*

 *Length = Distance from Turtle's Initial Position to Turtle's Final Position during the First Iteration of the Loop*

*That is, the two programs have the same boundedness, closing, and symmetry. In particular, if Angle ≠ 2πk for some integer k, then all the vertices generated by the same FORWARD command inside the loop lie on a common circle.*

**Spiral Looping Lemma**

 *Any procedure that is a repetition of the same collection of FORWARD, TURN, and RESIZE commands has the structure of SPIRAL (Length, Angle, Scalefactor), where*

 *Angle = Total Turtle Turning within the Loop*

 *Length = Distance from Turtle's Initial Position to Turtle's Final Position during the First Iteration of the Loop*

 *Scalefactor = Total Turtle Scaling within the Loop*

*That is, the two programs have the same boundedness and symmetry. In particular, if Angle ≠ 2πk for some integer k, and Scalefactor ≠ ± 1, then all the vertices generated by the same FORWARD command inside the loop lie on a common spiral.*

To prove the Circle Looping Lemma, follow the turtle through two iterations of the loop and observe that the initial turtle state undergoes exactly the same change as the turtle traversing two iterations of the POLY procedure (see Figure 2.2). Since we are iterating through exactly the same loop over and over again, after every iteration of the loop the looping turtle must end up in exactly the same state as the turtle executing the POLY procedure. This observation is the essence of the Circle Looping Lemma. The Spiral Looping Lemma can be proved in an exactly analogous manner.

**Examples:** The WALK procedure and the accompanying turtle path (Figure 2.3, left) provide typical embodiments of the Circle Looping Lemma. Notice that vertices generated by the same FORWARD command inside the loop lie on a common circle. Similarly, the PENTARAL procedure and the accompanying turtle path (Figure 2.3, right) are embodiments of the Spiral Looping Lemma. Notice that in the PENTARAL curve vertices generated by the same FORWARD command inside the outer loop of the program lie on a common spiral.

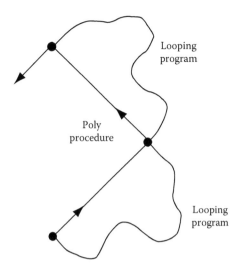

**FIGURE 2.2:** The turtle traversing through two iterations of a loop, both for the Poly procedure and for a Looping program.

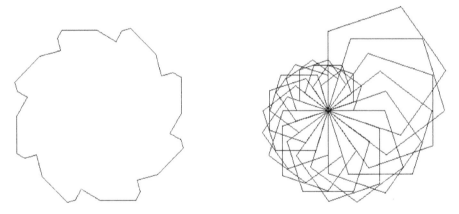

**FIGURE 2.3:** The turtle paths corresponding to the *WALK* procedure (left) and the PENTARAL procedure (right). Notice that corresponding vertices in the WALK curve lie on a common circle and that corresponding vertices in the PENTARAL curve lie on a common spiral.

WALK
  Repeat Forever
    FORWARD 1
    TURN $\pi/3$
    FORWARD 1/4
    TURN $\pi/4$
    FORWARD 1/3
    TURN $-\pi/2$
    FORWARD 3/5
    TURN $\pi/6$

PENTARAL
  Repeat Forever
    Repeat 5 Times
      FORWARD 1
      TURN $2\pi/5$
    TURN $\pi/10$
    RESIZE 1.06

When the POLY procedure generates either a regular polygon or star, the vertices clearly lie on a common circle. But for angles $A$ that are incommensurable with $2\pi$—that is, for angles $A$ for which there are no integers $m, n$ such that $mA = 2n\pi$—the curve does not close and it is not so obvious that the vertices generated by the POLY procedure must necessarily lie on a circle. Yet, as we can see from Figure 2.4, even when $2\pi/A$ is not rational, the vertices do nevertheless seem to lie on a common circle. We shall now give a simple proof of this result based on standard arguments from Euclidean geometry.

**FIGURE 2.4:**   The output of the POLY procedure for different input angles $A$. From left to right: a regular 10-sided polygon ($A = \pi/5$), a 7-pointed star ($A = 6\pi/7$), and a figure that never closes ($A = \pi\sqrt{2}/2$). Notice that in each case, the vertices lie on a common circle.

**Proposition 2.1:** The vertices generated by the procedure POLY(Length, Angle) lie on a common circle for any angle $A \neq 2\pi k$, for some integer $k$, and any length $L \neq 0$.

**Proof:** Consider the circle generated by the first three vertices $D, E, F$ visited by the turtle while executing the POLY procedure (see Figure 2.5). We need to show that the next vertex $G$ visited by the turtle lies on the same circle–that is, we need to show that $x = CG = R$, where $R$ is the radius of the circle circumscribed about $\Delta DEF$ and $C$ is the center of this circle. But

$$\Delta CDE \cong \Delta CEF,$$

since by construction the lengths of the three corresponding sides are equal (SSS). Now

$$A + \beta + \alpha = A + \alpha + \alpha \Rightarrow \beta = \alpha.$$

Therefore,

$$\Delta CEF \cong \Delta CFG,$$

because these triangles agree in two sides and an included angle (SAS). Hence, since $\Delta CEF$ is isosceles, $\Delta CFG$ is also isosceles, so

$$x = CG = R.$$

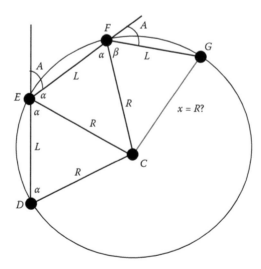

**FIGURE 2.5:**   The circle generated by the first three vertices $D$, $E$, $F$ visited by the turtle while executing the POLY procedure.

Similarly, we will show in Chapter 5, Exercises 5.1–5.3 that the vertices generated by the procedure SPIRAL (Length, Angle, Scalefactor) lie on a common logarithmic spiral for any angle A$\neq 2\pi k$, for some integer $k$ and any Scalefactor $\neq \pm 1$.

The vertices of the fractals in Figure 2.1 do not lie on common circles nor do they lie on common spirals. Thus, as a consequence of the Looping Lemmas, these fractals cannot be generated by iterative turtle procedures. To generate these fractal curves, we must resort instead to recursive turtle programs.

## 2.3  Fractal Curves and Recursive Turtle Programs

There are many different types of fractal curves, including gaskets, snowflakes, terrain, and even plants. Here we are going to concentrate on two representative types of fractals: fractal gaskets and bump fractals. Both of these types of fractal curves can be generated easily by simple recursive turtle programs.

### 2.3.1  Fractal Gaskets

The Sierpinski triangle in Figure 2.1 is an example of a fractal gasket. Because of the Looping Lemmas, we cannot use iteration to generate fractals, so we shall develop a recursive turtle program to generate this gasket. The key observation for building such a recursive program is that the big Sierpinski gasket is made up of three identical scaled down copies of the entire gasket. Thus, a Sierpinski gasket is just three smaller Sierpinski gaskets joined together. This definition would be circular, if we did not have a base case at which to stop the recursion. We shall say, therefore, that a very small Sierpinski gasket is just a very small triangle.

Now the structure of the recursive portion of the turtle program for the Sierpinski gasket must be something like the following:

To make a large Sierpinski gasket:

Make a smaller Sierpinski gasket at one of the corners of the outer triangle.

Then move to another corner of the outer triangle and make another small Sierpinski gasket.

Then move to another corner of the outer triangle and make another small Sierpinski gasket.

Then return the turtle to her initial position and heading.

The moves can be made with MOVE and TURN commands; the rescaling is achieved with RESIZE commands. To make the three smaller gaskets requires three recursive calls. *Form follows function*! By the way, why do you think you have to return the turtle to her initial state? Below is the complete program for generating the Sierpinski gasket. It is short and simple. Different levels of the gasket generated by the SIERPINSKI program are illustrated in Figure 2.6. Notice that all the drawing done by this program is performed exclusively by the FORWARD commands inside the POLY procedure in the base case:

```
SIERPINSKI (Level)
  IF Level = 0, POLY (1, 2π/3)
  OTHERWISE
    REPEAT 3 TIMES
      RESIZE 1/2
      SIERPINSKI (Level −1)
      RESIZE 2
      MOVE 1
      TURN 2π/3
```

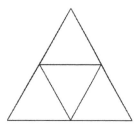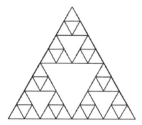

**FIGURE 2.6:**   Levels 1, 3, 6 of a recursive turtle program for the Sierpinski gasket.

Examining the syntax of the SIERPINSKI program, we see that the recursive body has a structure very similar to the program for generating a triangle. Applying this observation, we can easily generalize this program for the Sierpinski triangle to the following program that generates gaskets for arbitrary polygons. Different levels of the pentagonal gasket are illustrated in Figure 2.7. Some additional fractal gaskets are illustrated in Figure 2.8.

*POLYGASKET (N*, Level)
  IF Level $= 0$, POLY $(1, 2\pi/N)$
  OTHERWISE
    REPEAT *N* TIMES
      RESIZE $1/2$
      *POLYGASKET (N*, Level $-1$)
      RESIZE 2
      MOVE 1
      TURN $2\pi/N$

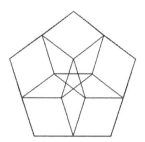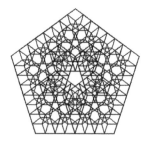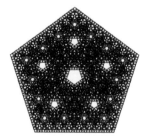

**FIGURE 2.7:**   Levels 1, 3, 5 of a recursive turtle program for a pentagonal gasket.

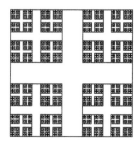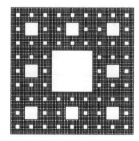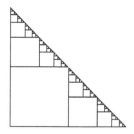

**FIGURE 2.8:**   More fractal gaskets. We leave it as an exercise to the reader to develop recursive turtle programs to generate each of these fractal gaskets (see Exercise 2.4).

### 2.3.2    Bump Fractals

A *bump curve* is just a pulse along a straight line. Figure 2.9 illustrates three examples of bump curves. A *bump fractal* is a fractal curve generated by recursively replacing each straight line of a bump curve by a scaled version of the bump. The *C*-curve in Figure 2.1 is the bump fractal corresponding the rightmost bump curve in Figure 2.9.

**FIGURE 2.9:**   Three examples of bump curves: a triangular pulse, a rectangular pulse, and a triangular bump.

It is actually quite easy to write a recursive turtle program to generate any arbitrary bump fractal. The base case is a straight line, so the code for the base case consists of the single turtle command FORWARD 1. To generate the recursive body of the program, proceed in two steps:

1. Write a turtle program to generate the corresponding bump curve. In this turtle program, the FORWARD commands should all take the parameter 1; the actual distance traveled should be adjusted by using the RESIZE command.

2. Replace all the FORWARD commands in step 1 by recursive calls.

In the recursive turtle program for a bump fractal, the base case draws a straight line. Therefore, since the FORWARD commands in the turtle program for the bump curve are replaced by recursive calls, level 1 of the recursion reproduces the original bump curve, and level $n$ replaces each line on level $n-1$ by a scaled down version of the original bump. Thus, given an arbitrary bump curve, a bump fractal is generated recursively by placing a bump fractal along each straight line in the original bump curve.

We illustrate this approach to generating recursive turtle programs for bump fractals in Table 2.2 for the leftmost triangular bump curve in Figure 2.9. The corresponding bump fractal is called the *Koch curve*. Different levels of the Koch curve are depicted in Figure 2.10. The Koch Snowflake in Figure 2.1 can be generated by spinning the Koch curve three times through the angle $2\pi/3$.

**TABLE 2.2:**   The turtle program for a triangular bump and the recursive turtle program for the corresponding bump fractal (the Koch curve).

| *TriangularBump* | *TriangularBumpFractal* (*Level*) |
|---|---|
|  | If Level $=0$ FORWARD 1 |
|  | Otherwise |
| RESIZE 1/3 | RESIZE 1/3 |
| FORWARD 1 | *TriangularBumpFractal* (*Level* $-1$) |
| TURN $\pi/3$ | TURN $\pi/3$ |
| FORWARD 1 | *TriangularBumpFractal* (*Level* $-1$) |
| TURN $-2\pi/3$ | TURN $-2\pi/3$ |
| FORWARD 1 | *TriangularBumpFractal* (*Level* $-1$) |
| TURN $\pi/3$ | TURN $\pi/3$ |
| FORWARD 1 | *TriangularBumpFractal* (*Level* $-1$) |
| RESIZE 3 | RESIZE 3 |

**FIGURE 2.10:**   Levels 1, 2, 5 of the Koch curve corresponding to the recursive turtle program in Table 2.2.

Notice that the recursive turtle program for the Koch curve restores the original step size of the turtle, but unlike the turtle program for the Sierpinski gasket, the turtle program for the Koch curve does not restore the turtle to her original position. Be sure that you understand why it is not necessary to restore the turtle to her initial position in order to generate the Koch curve or, in fact, to generate any other bump fractal.

---

## 2.4   Summary: Fractals—Recursion Made Visible

We never actually defined the term *fractal*. Nevertheless, from the turtle's point of view, *fractals are just recursion made visible*.

In this chapter we have studied two kinds of fractals: fractal gaskets and bump fractals. But pretty much any recursive turtle program generates a fractal curve. Conversely, many different fractal curves can be generated by recursive turtle programs.

Consider, for example, the fractal curves in Figure 2.11. Each of these fractals is made up of several identical smaller scaled copies of the entire fractal. Curves that consist of several identical scaled copies of themselves are said to be *self-similar*. In the recursive turtle programs that generate self-similar fractals, there must be as many recursive calls as there are equivalent scaled copies. Once again, *form follows function*.

**FIGURE 2.11:**   A few more self-similar fractal curves: a fractal hangman, a fractal bush, and a fractal tree. We leave it as a challenge to the reader to develop recursive turtle programs to generate these fractals, using the techniques described in this chapter (see Exercise 2.11).

When writing a recursive turtle program to generate a specific fractal, you should always keep in mind the answers to following questions:

1. Where does the turtle start?

2. Where does the turtle finish?

3. How many recursive calls must the turtle make?

4. What are the scale factors?

5. What are the turtle transitions (MOVE and TURN commands) between recursive calls?

*The most common mistake students make in writing recursive turtle programs to generate specific fractal curves is failing to return the turtle to her original state.* Not all fractals require that the turtle must return to her original state, but if the curve resulting from your recursive turtle program does not produce the desired result, the first thing you should check is whether you have restored your turtle to the proper state at the end of your recursive procedure.

---

## Exercises

2.1. Prove the Spiral Looping Lemma. (For more about spirals, see Chapter 5, Exercises 5.1 through 5.3.)

2.2. Notice that in Figure 2.4 not only do all the vertices generated by the POLY procedure lie on a circle, but all the edges too seem to be tangent to a common circle, even for angles $A$ incommensurate with $2\pi$.

    a. Prove that for any angle $A \neq 2\pi k$, for some integer $k$, and any length $L \neq 0$, the line segments of the curve generated by the turtle program $POLY(L, A)$ are tangent to a common circle. (Hint: In Figure 2.12, prove that $\beta = \alpha$. Conclude that $\triangle CHG \cong \triangle CFG$, so $x = R$ and $\angle CHG = \pi/2$.)

    b. How does the radius of the inner circle vary as the angle $A$ is increased?

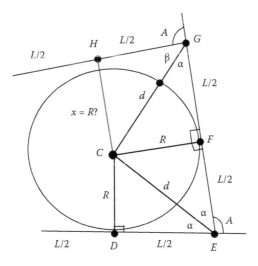

**FIGURE 2.12:** The circle tangent to the first two edges generated by the turtle while executing the POLY procedure.

2.3. Suppose that in either of the Looping Lemmas the total turning angle inside the loop is $2\pi k$ for some integer $k$. On what curve do all the vertices generated by the same

FORWARD command inside the loop necessarily lie? Why? Give some examples to illustrate your answer.

2.4. Write turtle programs to generate each of the fractal gaskets depicted in Figure 2.8.

2.5. Generate new fractal gaskets by changing the scaling parameter inside the body of the recursion for gaskets you have already generated.

2.6. Write turtle programs to generate each of the bump fractals corresponding to the bump curves depicted in Figure 2.9.

2.7. Write turtle programs to generate each of the bump fractals corresponding to the bump curves depicted in Figure 2.13.

**FIGURE 2.13:**   A triangular pulse (left) and a rectangular pulse (right). Compare to Figure 2.9.

2.8. Write turtle programs to generate each of the bump fractals corresponding to the bump curves depicted in Figure 2.14.

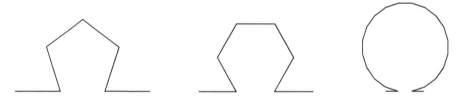

**FIGURE 2.14:**   Polygonal bumps: a pentagonal bump (left), a hexagonal bump (middle), and a polygonal bump with 20 sides (right).

2.9. Write turtle programs to generate each of the bump fractals corresponding to the bump curves depicted in Figure 2.15.

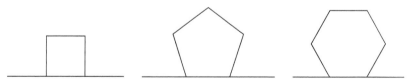

**FIGURE 2.15:**   More polygonal bumps. To generate these bumps, start at the left end point of the line at the base, proceed counterclockwise around the polygon, and finish at the right end point of the line.

2.10. Generate new fractals by applying the SPIN and SCALE procedure to fractals that you have already generated.

2.11. Write turtle programs to generate each of the fractal curves depicted in Figure 2.11.

2.12. Write turtle programs to generate each of the fractal curves depicted in Figure 2.16.

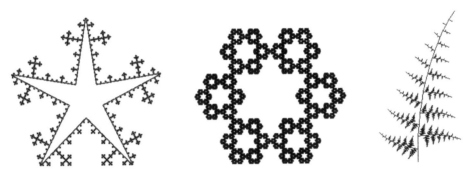

**FIGURE 2.16:** Three more fractal curves: a fractal star, a fractal flower, and a fractal leaf.

2.13. In classical LOGO there is no RESIZE command.

a. Show that the two commands RESIZE $S$, FORWARD $D$ are equivalent to the single command FORWARD $SD$.

b. Conclude from part (a) that the result of any turtle program that includes RESIZE commands can be simulated by a turtle program without any RESIZE commands.

c. Explain how to generate bump fractals without using the RESIZE command.

---

## Programming Projects

2.1. *The Classical Turtle*

Implement LOGO in your favorite programming language using your favorite API.

a. Write LOGO programs for the curves discussed in the text and in the exercises for Chapters 1 and 2.

b. Write a LOGO program to design a novel flag for a new country.

c. Write recursive LOGO programs to create your own novel fractals.

d. In the text we generated fractals using simple recursive turtle programs. But fractals can also be generated using mutually recursive turtle programs—two or more recursive turtle programs that call one another.

i. Build some novel fractals using mutually recursive turtle programs.

ii. The Hilbert curve is a space filling fractal curve (see Figure 2.17). Write two mutually recursive turtle programs to generate the Hilbert curve.

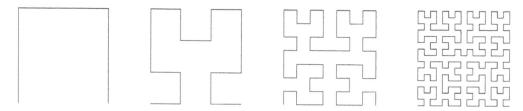

**FIGURE 2.17:** The first four levels of the Hilbert curve. As the number of levels approaches infinity, the Hilbert curve gets arbitrarily close to every point in the square.

2.2. *The Hodograph Turtle*

The classical turtle draws a line from her initial position to her new position after each turtle command. The hodograph turtle draws a line from the initial position of the tip of her direction vector to the new position of the tip of her direction vector after each turtle command.

a. Implement the hodograph turtle in your favorite programming language using your favorite API.

   i. Implement the turtle commands PENUP and PENDOWN, so that, if necessary, the hodograph turtle can TURN and RESIZE without drawing a line.

   ii. Implement two new turtle commands ANCHORUP and ANCHORDOWN. When the anchor is down, the hodograph turtle ignores the FORWARD and MOVE commands.

b. Write turtle programs to generate simple curves like polygons, stars, and spirals using the hodograph turtle.

c. Draw the curves in Figure 2.18 using:

   i. The classical turtle.

   ii. The hodograph turtle.

   Which is easier?

d. Compare and contrast the curves generated by the classical turtle and the hodograph turtle for the same turtle programs when the anchor is down. For example:

   i. What curve does the hodograph turtle draw, when the classical turtle draws a polygon, a star, or a rosette?

   ii. What curve does the hodograph turtle draw, when the classical turtle draws the Sierpinski gasket or the Koch curve?

e. Compare the fractals generated by the classical turtle and the hodograph turtle for the same recursive turtle program. Form a conjecture concerning when the two turtles generate the same fractal curve.

2.3. *The Classical Turtle on a Bounded Domain*

The classical turtle lives on an infinite plane. Suppose, however, that the turtle is restricted to a finite domain bounded by walls. When the turtle hits a wall, she bounces off the wall so that

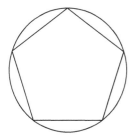
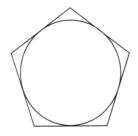
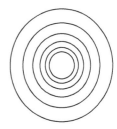

**FIGURE 2.18:** A pentagon inscribed in a circle (left), a pentagon circumscribed about a circle (center), and a collection of concentric circles (right). Are these curves easier to generate with the classical turtle or with the hodograph turtle?

her angle of incidence is equal to her angle of reflection, and then she continues on her way. Thus the turtle behaves like a billiard ball on a table with no friction or like a light beam surrounded by mirrors. To implement the turtle on a bounded domain, the FORWARD command must be replaced by:

NEWFORWARD $D$
   $D_1 = Distance\ from\ Turtle\ to\ Wall\ in\ Direction\ of\ Turtle\ Heading$
   IF $D < D_1$, FORWARD $D$
   OTHERWISE
      FORWARD $D_1$
      TURN $A$ {$A = angle\ of\ reflection$}
      NEWFORWARD $D - D_1$

To compute the angle of reflection $A$, let $\angle(w, N)$ denote the angle between the turtle direction vector $w$ and the normal $N$ to the wall at the point where the turtle hits the wall. By a straightforward geometric argument, you should be able to show that $A = \pi - 2\angle(w, N)$.

a. Implement LOGO on a bounded domain, where the walls form:

   i. A rectangle.

   ii. A circle.

   iii. An ellipse.

b. Consider the turtle program consisting of the single command

   NEWFORWARD $D$.

   Investigate the curves generated by this program for different shape walls, different values of $D$, and different initial positions and headings for the turtle.

c. Investigate how curves drawn by the walled-in turtle differ from curves drawn by the turtle on an infinite plane using the same turtle programs, where FORWARD is replaced by NEWFORWARD. In particular, investigate the curves generated by POLY, SPIRAL, and different recursive turtle procedures for generating fractals.

## 2.4. *The Left-Handed Turtle*

The state of the classical turtle is specified by her position and one vector, her forward facing direction vector. The state of the left-handed turtle is specified by her position and two vectors: one vector specifies her forward facing direction, the other vector specifies the direction of her left hand. Thus the left-handed turtle carries around her own local coordinate system. To manipulate the left-hand vector, the turtle commands TURN and RESIZE are redefined to be functions of two parameters: the first parameter refers to the forward vector, the second parameter refers to the left-hand vector. The RESIZE command is easy to define; just rescale each vector by the specified amount. The TURN command, however, is more subtle. The left-handed turtle measures all angles relative to her own coordinate system, which she assumes is a rectangular coordinate system. Thus when she executes the TURN command, she rotates the forward facing vector and the left-hand vector relative to her own coordinate system rather than relative to a global coordinate system. Let $(P,w,w^*)$ denote the current state of the turtle, where $w$ is the forward facing direction vector and $w^*$ is the vector that specifies the direction of the turtle's left hand. Then the turtle commands have the following effect on the turtle's state:

- *FORWARD D, MOVE D*

$$P_{new} = P + Dw.$$

- *TURN $A_1, A_2$*

$$\begin{aligned}w_{new} &= w \cos(A_1) + w^* \sin(A_1) \\ w^*_{new} &= w^* \cos(A_2) - w \sin(A_2)\end{aligned} \Leftrightarrow \begin{pmatrix} w_{new} \\ w^*_{new} \end{pmatrix} = \begin{pmatrix} \cos(A_1) & \sin(A_1) \\ -\sin(A_2) & \cos(A_2) \end{pmatrix} \begin{pmatrix} w \\ w^* \end{pmatrix}.$$

- *RESIZE $S_1, S_2$*

$$\begin{aligned}w_{new} &= S_1 w \\ w^*_{new} &= S_2 w^*\end{aligned} \Leftrightarrow \begin{pmatrix} w_{new} \\ w^*_{new} \end{pmatrix} = \begin{pmatrix} S_1 & 0 \\ 0 & S_2 \end{pmatrix} \begin{pmatrix} w \\ w^* \end{pmatrix}.$$

In order that the our old LOGO programs will still work for the left-handed turtle, we define

*TURN A = TURN A,A*

*RESIZE S = RESIZE S,S*

a. Implement LOGO for the left-handed turtle in your favorite programming language using your favorite API.

b. Draw an ellipse using the left-handed turtle and then spin the ellipse to generate the pattern in Figure 2.19.

c. Study the shapes generated by the following turtle programs:

| NEWPOLY (*Length, Angle1, Angle2*) | NEWSPIRAL (*Length, Angle, Scale1, Scale2*) |
|---|---|
| Repeat Forever | Repeat Forever |
| FORWARD Length | FORWARD Length |
| TURN Angle1, Angle2 | TURN Angle |
| | RESIZE Scale1, Scale2 |

d. What are the analogs of the Looping Lemmas for the left-handed turtle?

e. Write recursive programs for the left-handed turtle to create your own novel fractals.

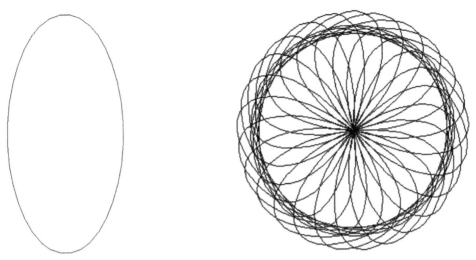

**FIGURE 2.19:** An ellipse (left) and the spin operator applied to the turtle program for an ellipse (right).

### 2.5. *The Turtle on Two-Dimensional Manifolds*

The classical turtle lives on a flat plane. Here we shall investigate turtles that live on curved two-dimensional manifolds, including cylinders, mobius strips, tori, klein bottles, and projective planes. These surfaces can be modeled by rectangles, where one or more pairs of opposite sides are glued together, possibly with a twist (see Figure 2.20). (Notice that by gluing together the appropriate sides, the cylinder, the mobius strip, and the torus can be embedded in three dimensions. However, the klein bottle and the projective plane cannot be embedded in three dimensions due to the twists; these manifolds live naturally in four dimensions, even though they are two-dimensional surfaces.) If a pair of opposite sides is not glued together, then these sides are treated like a wall just as with the classical turtle on a bounded domain (see Programming Project 2.3). If, however, a turtle encounters a wall where opposite sides are identified, then she does not bounce off the wall, but rather emerges on the opposite side of the rectangle heading in the same direction relative to the new wall in which she hit the opposing wall.

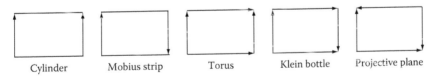

**FIGURE 2.20:** Rectangular representations of five different two-dimensional manifolds. The Klein bottle and the projective plane cannot be embedded in three dimensions, but live naturally in four dimensions.

a. Implement LOGO on one or more of the following two-dimensional manifolds:

   i. Cylinder.

   ii. Mobius strip.

   iii. Torus.

   iv. Klein bottle.

   v. Projective plane.

Note: For each of these manifolds, you will need to define a command MFORWARD to replace the command FORWARD in classical LOGO (see Programming Project 2.3).

b. Consider the turtle program consisting of the single command

MFORWARD *D*.

Investigate the curves generated by this program for different manifolds, different values of *D*, and different initial positions and headings for the turtle.

c. Investigate how curves drawn by turtles on these two-dimensional manifolds differ from curves drawn by the turtle on an infinite plane using the same turtle programs, where FORWARD is replaced by MFORWARD. In particular, investigate the curves generated by POLY, SPIRAL, and different recursive turtle procedures for generating fractals.

# Chapter 3

## Some Strange Properties of Fractal Curves

*I have been a stranger in a strange land*

– Exodus 2:22

### 3.1 Fractal Strangeness

Fractals have a look and feel that is very different from ordinary curves. Unlike commonplace curves such as lines or circles, there are no simple formulas for representing fractals like the Sierpinski gasket or the Koch curve. Therefore it should not be surprising that fractal curves also have geometric features that are unlike the properties of any other curves you have previously encountered. In this chapter we are going to explore some of the strangest peculiarities of fractals, including their dimension, differentiability, and attraction.

### 3.2 Dimension

You may have been wondering about the origin of the term *fractal*. Many fractal curves have a nonintegral dimension, a fractional dimension somewhere between one and two. Fractal refers to this fractional dimension.

Standard curves like the line and the circle are one-dimensional. To say that a curve is one-dimensional means that the curve has no thickness; if the curve is black and the background is white, then when we look at the curve we see white on either side of a thin black band (see Figure 3.1). But fractals are different. Look at the Sierpinski gasket or the $C$-curve in Figure 3.1. There seem to be regions that are neither black nor white, but instead are gray. Such curves typically have dimension greater than one, but less than two; these curves do not completely fill up any region of the plane, so they are not two-dimensional, but neither are these gray curves as thin as one-dimensional curves. To calculate the actual dimensions of fractal curves, we first need to formalize the notion of dimension for some standard geometric shapes.

The dimension of a line segment is one, the dimension of a square is two, and the dimension of a cube is three. There is a formal way to capture these dimensions. Suppose we split these objects by inserting new vertices at the centroids of their edges and faces. Then the line segment splits into 2 line segments, the square into $4 = 2^2$ squares, and the cube into $8 = 2^3$ cubes (see Figure 3.2). In each case the dimension appears in the exponent. There is nothing magical about splitting each edge into two equal parts. If we split each edge into $N$ equal parts, then the line segment splits into $N$ line

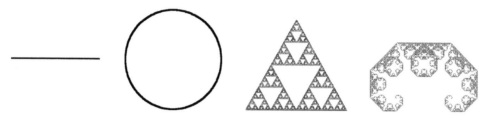

**FIGURE 3.1:** For the line and the circle there is a clear distinction between the curve (black) and the background (white). But for fractals like the Sierpinski gasket and the *C*-curve, the curve (black) and the background (white) interpenetrate and there are gray regions where the distinction between the curve and the background does not appear so clearly defined.

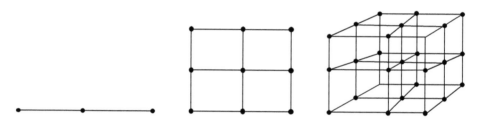

**FIGURE 3.2:** Inserting vertices at the centroids splits a line segment into $2 = 2^1$ line segments, a square into $4 = 2^2$ squares, and a cube into $8 = 2^3$ cubes.

segments, the square into $N^2$ squares, and the cube into $N^3$ cubes. Once again, the dimension appears in the exponent. Another name for an exponent is a logarithm, so we are going to formalize the notion of dimension in terms of logarithms.

Another way of thinking about what we have just done is that we have split the line, the square, and the cube into identical parts, where each part is a scaled down version of the original. This decomposition should remind you of the fractals that you encountered in Chapter 2, where each fractal is composed of several identical scaled down copies of the original fractal. Evidently, in this construction for the line, the square, and the cube, if $N$ is the number of line segments along each edge and $E$ is the number of equal scaled down parts, then if $D$ denotes dimension

$$E = N^D,$$

so

$$D = Log_N(E) = \frac{Log(E)}{Log(N)}. \tag{3.1}$$

But if $N$ is the number of line segments along each edge, then $S = 1/N$ is the scaling along each edge. Since $Log(S) = -Log(N)$, we can rewrite Equation 3.1 by setting

$$D = -\frac{Log(E)}{Log(S)}, \tag{3.2}$$

where
    $S$ is the scale factor
    $E$ is the number of identical scaled down parts.

Equation 3.2 has several important properties. First, notice that $S < 1$, so $Log(S) < 0$. Therefore, the minus sign on the right-hand side of Equation 3.2 insures that the dimension $D$ is positive. Second, since $D$ is defined as the ratio of two logarithms, the base of the logarithm does not matter; dimension is the same in all bases. Finally, Equation 3.2 gives the same result as Equation 3.1 for the line, the square, and the cube, since in these cases $E = N^D$ and $S = 1/N$, so

$$-\frac{Log(E)}{Log(S)} = -\frac{Log(N^D)}{Log(1/N)} = \frac{DLog(N)}{Log(N)} = D.$$

Let us see now what happens when we apply Equation 3.2 to self-similar fractal curves.

### 3.2.1 Fractal Dimension

To apply our dimension formula, we need to consider self-similar curves. Recall that a curve is *self-similar* if the curve can be decomposed into a collection of identical curves each of which is a scaled version of the original curve. Most of the fractal curves we encountered in Chapter 2 such as the Sierpinski gasket and the Koch curve are self-similar curves. In fact, self-similarity is what allows us to write simple recursive turtle programs to generate these shapes. Let us look now at some examples.

**Example 3.1 Sierpinski Gasket**
The Sierpinski gasket consists of three smaller Sierpinski gaskets, where the length of each edge of the smaller gaskets is one-half the length of an edge of the original gasket (see Figure 3.3, left). Thus $E = 3$ and $S = 1/2$, so

$$D = -\frac{Log(E)}{Log(S)} = -\frac{Log(3)}{Log(1/2)} = \frac{Log(3)}{Log(2)} \approx 1.585 \Rightarrow 1 < D < 2.$$

**Example 3.2 Koch Curve**
The Koch curve consists of four smaller Koch curves, each one-third the size of the original curve (see Figure 3.3, right). Thus $E = 4$ and $S = 1/3$, so

$$D = -\frac{Log(E)}{Log(S)} = -\frac{Log(4)}{Log(1/3)} = \frac{Log(4)}{Log(3)} \approx 1.262 \Rightarrow 1 < D < 2.$$

**FIGURE 3.3:** The Sierpinski gasket (left) and the Koch curve (right). The Sierpinski gasket is composed of three self-similar parts; the length of each edge of one of the smaller gaskets is one-half the length of an edge of the original gasket. The Koch curve is composed of four self-similar parts, each one-third the size of the original Koch curve.

### 3.2.2   Computing Fractal Dimension from Recursive Turtle Programs

The fractal dimension of a self-similar fractal curve can often be computed directly from its recursive turtle program: the number of recursive calls corresponds to the number $E$ of equal self-similar parts, and the scale factor in the RESIZE command corresponds to the scale factor $S$. Thus,

$$Fractal\ Dimension = -\frac{Log(\#RecursiveCalls)}{Log(ScaleFactor)}. \tag{3.3}$$

This formula, like Equation 3.2, is valid provided that the self-similar parts do not overlap. To illustrate the validity of this formula, let us revisit the fractal dimension of the Sierpinski gasket and the Koch curve.

**Example 3.3   Sierpinski Gasket**

Recall from Chapter 2 that in the recursive turtle program for the Sierpinski gasket $\#RecursiveCalls = 3$ and $ScaleFactor = 1/2$, so

$$Fractal\ Dimension = -\frac{Log(\#RecursiveCalls)}{Log(ScaleFactor)} = -\frac{Log(3)}{Log(1/2)} = \frac{Log(3)}{Log(2)}.$$

**Example 3.4   Koch Curve**

Again recall from Chapter 2 that in the recursive turtle program for the Koch curve $\#RecursiveCalls = 4$ and $ScaleFactor = 1/3$, so

$$Fractal\ Dimension = -\frac{Log(\#RecursiveCalls)}{Log(ScaleFactor)} = -\frac{Log(4)}{Log(1/3)} = \frac{Log(4)}{Log(3)}.$$

Equations 3.2 and 3.3 may appear somewhat artificial and even contrived, since these formula apply only to self-similar curves, but even many familiar shapes, such as circles and ellipses, are not self-similar curves. The notion of dimension can be defined for these curves as well, but the mathematics required is way beyond the scope of this text. The general notion of dimension is quite deep; we shall not pursue this topic any further here.

---

## 3.3   Differentiability

A curve is said to be *smooth* or *differentiable* if the curve has a well-defined slope or equivalently a well-defined tangent at every point. Lines and circles are smooth curves. Polygons and stars are piecewise smooth curves; polygons and stars have well-defined slopes everywhere except at their vertices. The functions you studied in calculus—polynomials, trigonometric functions, exponentials, and logarithms—are all differentiable functions. What about fractals?

Differentiable functions are continuous everywhere, but continuous functions need not be differentiable everywhere. The function $y = |x|$ represents a continuous curve composed of two lines: the line $y = -x$ for $x \leq 0$ and the line $y = x$ for $x \geq 0$ (see Figure 3.4). Thus $y = |x|$ has slope $-1$ for $x < 0$ and slope $+1$ for $x > 0$, but the slope of $y = |x|$ is not well defined at the origin. A curve is said to be *piecewise linear* if like the function $y = |x|$ it is composed of a sequence of straight lines. A piecewise linear curve is smooth everywhere except where two lines join. Thus it is easy to

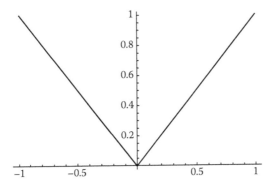

**FIGURE 3.4:** The graph of $y = |x|$. When $x$ is negative, the slope is $-1$; when $x$ is positive, the slope is $+1$. But at the origin the curve comes to a sharp point, so at the origin the slope is not defined.

generate curves that are smooth everywhere except at a finite number of points. The level $n$ Koch curve is a piecewise linear curve that is smooth everywhere except at finitely many points.

Intuitively it seems evident that a continuous curve can have only a finite number of points where the slope is not well defined. But intuition can be misleading. *Many fractal curves are continuous everywhere, but differentiable nowhere.*

Consider the Koch curve (see Figure 3.5). Level 0 is a straight line, which is smooth at every point. Level 1, however, is a piecewise linear curve, and there are three points where the slope is not defined. Similarly, at level 2 there are 15 points where the slope is not defined. At level $n$ there are $4^n - 1$ points where the slope is not defined. Thus in the limit as $n$ approaches infinity, there are infinitely many points on the Koch curve where the slope is not defined.

In fact, in the limit, the slope of the Koch curve is undefined at every point. We can substantiate this assertion in the following manner. Suppose that the length of the line in level 0 is 1. Then the length of the line segments in level 1 is $1/3$; the length of the line segments in level 2 is $1/9$; and, in general, the length of the line segments at level $n$ is $3^{-n}$. Thus the lengths of these line segments approach zero as the number of levels approaches infinity. Now the curve is smooth only at points on the interior of these line segments. But in the limit as the number of levels approaches infinity, there are no points in the interior of line segments because the lengths of the line segments approach zero. Thus in the limit the Koch curve is continuous everywhere, but differentiable nowhere. That is, the Koch curve has a wrinkle at every point!

The Koch curve is not an anomaly. In fact, much the same arguments can be used to show that virtually any bump fractal is continuous everywhere and differentiable nowhere. Fractals are indeed very strange curves.

**FIGURE 3.5:** Levels 0, 1, 2, 3 of the Koch curve. Level 0 is a straight line, which is smooth at every point. Level 1 has three points where the slope is not defined; level 2 has 15 points where the slope is not defined, and level 3 has 63 points where the slope is not defined. Notice too that the lengths of the line segments decrease from level to level by a factor of 3.

## 3.4  Attraction

How precisely does the fractal generated by a recursive turtle program depend on the choice of the base case? Intuitively you might think that changing the base case will alter the fractal in some fundamental way. But as we have just seen when we investigated differentiability, our intuition regarding fractals can often be misleading.

We shall begin then by considering two familiar examples: the Sierpinski gasket and the Koch curve. Based on our experience with these two fractals, we shall draw some highly counterintuitive and rather remarkable conclusions.

### 3.4.1  Base Cases for the Sierpinski Gasket

In the base case, the standard program for the Sierpinski gasket draws a triangle (see Chapter 2). The curves generated by levels 0, 1, 3, 6 of this program are illustrated in Figure 3.6. In Figures 3.7 through 3.9, we have changed the base case to draw a square, a star, and a horizontal line, but we have not altered the recursive body of the program. Each of these programs generates curves that start out quite different at levels 0, 1, but by level 6 they all seem to be converging to the same Sierpinski gasket.

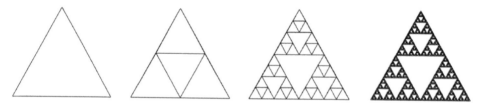

**FIGURE 3.6:**   Levels 0, 1, 3, 6 of the Sierpinski gasket. The base case is a triangle.

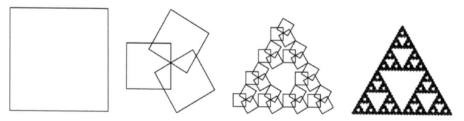

**FIGURE 3.7:**   Levels 0, 1, 3, 6 of the Sierpinski gasket. The base case is a square.

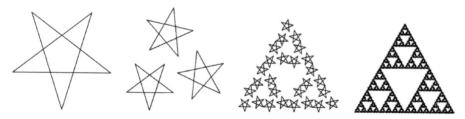

**FIGURE 3.8:**   Levels 0, 1, 3, 6 of the Sierpinski gasket. The base case is a star.

**FIGURE 3.9:** Levels 0, 1, 3, 6 of the Sierpinski gasket. The base case is a horizontal line.

### 3.4.2 Base Cases for the Koch Curve

Let us try another example: the Koch curve. In the base case, the standard program draws a straight line. The curves generated by levels 0, 1, 2, 5 of this program are illustrated in Figure 3.10. In Figures 3.11 through 3.13, we have changed the base case to draw a square bump, a square, and a square where the turtle repeats the first side after completing her path around the square, but we have not altered the recursive body of the program. Again each of the curves generated by these programs starts out quite different at levels 0, 1, but, except for the program where the base case is a square

**FIGURE 3.10:** Levels 0, 1, 2, 5 of the Koch curve. The base case is a horizontal line.

**FIGURE 3.11:** Levels 0, 1, 2, 5 of the Koch curve. The base case is a square bump.

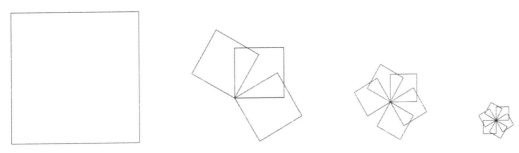

**FIGURE 3.12:** Levels 0, 1, 2, 5 of the Koch program. The base case is a square.

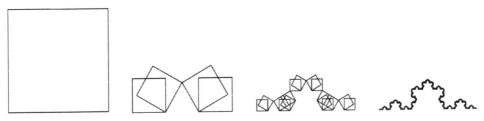

**FIGURE 3.13:** Levels 0, 1, 2, 5 of the Koch curve. The base case is a square, where the turtle repeats the first side after completing her path around the square.

(see Figure 3.12—we will return to a discussion of this anomalous case shortly), by level 5 they all seem to be converging to the same Koch curve.

### 3.4.3   Attractors

When we first encountered recursive turtle programs, our intuition lead us to believe that the precise form of the base case is critical for generating the desired fractal curve. The natural base case for the Sierpinski gasket is a triangle; the natural base case for the Koch curve is either a straight line or a triangular bump. But the examples we have just witnessed both for the Sierpinski gasket and for the Koch curve indicate that in the limit (and as a practical matter after only about five or six levels of recursion) the base case is all but irrelevant. Apparently, a recursive turtle program will always converge to the same fractal curve independent of the turtle program in the base case. When a process converges to the same limit value independent of the choice of the initial value, the limit value is called an *attractor*. Fractals are attractors.

Figure 3.12 seems to challenge this conclusion; here the expected convergence to the Koch curve fails to occur. However, the reason for this failure turns out to be that the recursion in the body of the turtle program for a bump fractal assumes that the turtle advances to the end of the line on which the bump occurs, rather than return to her initial position. This assumption is violated by the base case in Figure 3.12, but is satisfied by the base case in Figure 3.13, where the turtle repeats the first side of the square after completing her path around the square.

In Chapter 8, we shall prove that in the limit as the number of levels approaches infinity the fractals generated by recursive turtle programs are indeed independent of the turtle program in the base case, provided that the final state of the turtle in the base case is the same as the final state of the turtle in the recursion. Usually, as in the Sierpinski gasket, the initial state and the final state of the turtle are identical, but, as illustrated by the Koch curve, one must always be careful to check this assumption.

Thus, with some mild assumptions, the fractal generated in the limit by any recursive turtle program is indeed independent of the turtle program in the base case. Fractals are attractors. Indeed, attraction is the signature property by which we recognize fractals.

But why does such a strange property hold? To understand the reason behind this curious phenomenon, we shall have to forgo for now our study of Turtle Graphics and investigate instead a very different approach to fractals called *iterated function systems*. To understand iterated function systems, we shall first need to take up the study of affine transformations and affine graphics. We shall commence with these topics in Chapter 4.

---

## 3.5   Summary

From the turtle's point of view, fractals are simply *recursion made visible*. But from our perspective, fractals generated from recursive turtle programs can have many strange properties.

1. Fractals can be self-similar; fractals can often be decomposed into a collection of identical curves each of which is a scaled version of the original fractal curve.

2. Fractals can be continuous everywhere, yet differentiable nowhere.

3. Fractals can have fractional dimensions, noninteger dimensions between 1 and 2.

4. Fractals are attractors, independent of the base case in the recursive turtle program that generates the fractal.

These four properties are characteristics of fractal curves that we do not find in common everyday curves; together these features are what set fractals apart from our ordinary experience of geometry.

---

## Exercises

3.1. Let $P_n$ denote the perimeter of the $n$th level of the Koch snowflake.

   a. Show that:

   i. $P_n = \frac{4}{3}P_{n-1}$

   ii. $P_n = \left(\frac{4}{3}\right)^n P_0$.

   b. Conclude that the perimeter of the Koch snowflake is infinite.

3.2. Let $A_n$ denote the area enclosed by the $n$th level of the Koch snowflake.

   a. Show that:

   i. $A_n = A_0 + 3A_0 \left(1/9 + 4/9^2 + 4^2/9^3 + \cdots 4^{n-1}/9^n\right)$.

   ii. $Lim_{n\to\infty}A_n = (8/5)A_0$.

   b. Conclude from Exercise 3.1 and part a that the Koch snowflake has an infinite perimeter but encloses a finite area.

   c. Explain why it is not possible for a curve to have a finite perimeter but to enclose an infinite area.

3.3. Show that the bump fractals corresponding to the bump curves in Chapter 2, Figure 2.13 are continuous everywhere but differentiable nowhere.

3.4. Consider the following recursive turtle program:

```
SQUARE (Level)
   IF Level = 0, POLY (1, π/2)
   OTHERWISE
      REPEAT 4 TIMES
      RESIZE 1/2
      SQUARE (Level − 1)
      RESIZE 2
      MOVE 1
      TURN π/2
```

   a. Show that as the level approaches infinity, this program generates a filled in square.

   b. Compute the dimension of the square from this recursive turtle program.

3.5. Compute the fractal dimension of the *C*-curve.

3.6. Compute the fractal dimension of the gaskets in Chapter 2, Figure 2.8.

3.7. Compute the fractal dimension for each of the fractal curves in Chapter 2, Figure 2.11.

3.8. Compute the fractal dimension for each of the bump fractals corresponding to the bump curves in Chapter 2, Figure 2.13.

3.9. Generate a fractal pentagonal gasket using the same recursive turtle program with three different turtle programs in the base case.

3.10. Generate the fractal *C*-curve using the same recursive turtle program with three different turtle programs in the base case.

3.11. Generate the fractal tree in Chapter 2, Figure 2.11, using the same recursive turtle program with three different turtle programs in the base case.

# Chapter 4

## Affine Transformations

*for Satan himself is transformed into an angel of light.*

– 2 Corinthians 11:14

## 4.1  Transformations

Transformations are the lifeblood of geometry. Euclidean geometry is based on *rigid motions*— translation and rotation—transformations that preserve distances and angles. *Congruent triangles* are triangles where corresponding lengths and angles match.

Transformations generate geometry. The turtle uses translation (FORWARD), rotation (TURN), and uniform scaling (RESIZE) to generate curves by moving about on the plane. We can also apply translation (SHIFT), rotation (SPIN), and uniform scaling (SCALE) to build new shapes from previously defined turtle programs. These three transformations—translation, rotation, and uniform scaling—are called *conformal transformations*. Conformal transformations preserve angles, but not distances. *Similar triangles* are triangles where corresponding angles agree, but the lengths of corresponding sides are scaled. The ability to scale is what allows the turtle to generate self-similar fractals like the Sierpinski gasket.

In Computer Graphics, transformations are employed to position, orient, and scale objects as well as to model shape. Much of elementary Computational Geometry and Computer Graphics is based upon an understanding of the effects of different fundamental transformations.

The transformations that appear most often in two-dimensional Computer Graphics are the affine transformations. *Affine transformations* are composites of four basic types of transformations: translation, rotation, scaling (uniform and nonuniform), and shear. Affine transformations do not necessarily preserve either distances or angles, but affine transformations map straight lines to straight lines and affine transformations preserve ratios of distances along straight lines (see Figure 4.1). For example, affine transformations map midpoints to midpoints. In this chapter, we are going to develop explicit formulas for various affine transformations; in the next chapter, we will use these affine transformations as an alternative to turtle programs to model shapes for Computer Graphics.

A word of warning before we begin. There is a good deal of linear algebra in this chapter. Do not panic. Though there is a lot of algebra, the details are all fairly straightforward. The bulk of this chapter is devoted to deriving matrix representations for the affine transformations, which will be used in subsequent chapters to generate shapes for two-dimensional Computer Graphics. All the matrices that you will need later on are listed at the end of this chapter, so you will not have to slog through these derivations every time you require a specific matrix; you can find these matrices easily whenever you need them. Persevere now because there will be a big payoff shortly.

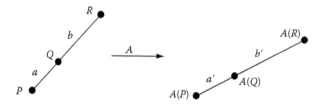

**FIGURE 4.1:**   An affine transformation $A$ maps lines to lines and preserves ratios of distances along straight lines. Thus, $A(Q)$ lies on the line joining $A(P)$ and $A(R)$, and $a'/b' = a/b$.

## 4.2   Conformal Transformations

Among the most important affine transformations are the conformal transformations: translation, rotation, and uniform scaling. We shall begin our study of affine transformations by developing explicit matrix representations for these conformal transformations.

To fix our notation, we will use uppercase letters $P$, $Q$, $R$, ... from the middle of the alphabet to denote points and lowercase letters $u$, $v$, $w$, ... from the end of the alphabet to denote vectors. Lowercase letters with subscripts will denote rectangular coordinates. Thus, for example, we shall write $P = (p_1, p_2)$ to denote that $(p_1, p_2)$ are the rectangular coordinates of the point $P$. Similarly, we shall write $v = (v_1, v_2)$ to denote that $(v_1, v_2)$ are the rectangular coordinates of the vector $v$.

### 4.2.1   Translation

We have already encountered translation in our study of Turtle Graphics (see Figure 4.2). To translate a point $P = (p_1, p_2)$ by a vector $w = (w_1, w_2)$, we set

$$P^{new} = P + w, \tag{4.1}$$

or in terms of coordinates

$$p_1^{new} = p_1 + w_1,$$
$$p_2^{new} = p_2 + w_2.$$

Vectors are unaffected by translation, so

$$v^{new} = v.$$

We can rewrite these equations in matrix form. Let $I$ denote the $2 \times 2$ identity matrix; then

$$P^{new} = P * I + w,$$
$$v^{new} = v * I.$$

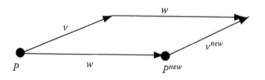

**FIGURE 4.2:**   Translation. Points are affected by translation; vectors are not affected by translation.

## 4.2.2 Rotation

We also encountered rotation in Turtle Graphics. Recall from Chapter 1 that to rotate a vector $v = (v_1, v_2)$ through an angle $\theta$, we introduce the orthogonal vector $v^{\perp} = (-v_2, v_1)$ and set

$$v^{new} = v\cos(\theta) + v^{\perp}\sin(\theta),$$

or equivalently in terms of coordinates

$$v_1^{new} = v_1\cos(\theta) - v_2\sin(\theta),$$
$$v_2^{new} = v_1\sin(\theta) + v_2\cos(\theta). \tag{4.2}$$

Introducing the rotation matrix

$$Rot(\theta) = \begin{pmatrix} \cos(\theta) & \sin(\theta) \\ -\sin(\theta) & \cos(\theta) \end{pmatrix},$$

we can rewrite Equation 4.2 in matrix form as

$$v^{new} = v * Rot(\theta). \tag{4.3}$$

In our study of affine geometry in the plane, we want to rotate not only vectors about their tails, but also points about other points. Since $P = Q + (P - Q)$, rotating the point $P$ about another point $Q$ through the angle $\theta$ is equivalent to rotating the vector $P - Q$ through the angle $\theta$ and then adding the resulting vector to $Q$ (see Figure 4.3). Thus, by Equation 4.3, the formula for rotating the point $P$ about another point $Q$ through the angle $\theta$ is

$$P^{new} = Q + (P - Q) * Rot(\theta) = P * Rot(\theta) + Q * (I - Rot(\theta)), \tag{4.4}$$

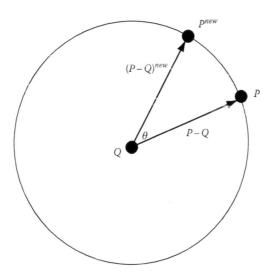

**FIGURE 4.3:** Rotation about the point $Q$. Since the point $Q$ in fixed and since $P = Q + (P - Q)$, rotating a point $P$ about the point $Q$ through the angle $\theta$ is equivalent to rotating the vector $P - Q$ through the angle $\theta$ and then adding the resulting vector to the point $Q$.

where $I$ is again the $2 \times 2$ identity matrix. Notice that if $Q$ is the origin, then this formula reduces to

$$P^{new} = P * Rot(\theta),$$

so $Rot(\theta)$ is also the matrix representing rotation of points about the origin.

### 4.2.3 Uniform Scaling

Uniform scaling also appears in Turtle Graphics. To scale a vector $v = (v_1, v_2)$ by a factor $s$, we simply set

$$v^{new} = sv, \qquad (4.5)$$

or in terms of coordinates

$$v_1^{new} = sv_1,$$
$$v_2^{new} = sv_2.$$

Introducing the scaling matrix

$$Scale(s) = \begin{pmatrix} s & 0 \\ 0 & s \end{pmatrix},$$

we can rewrite Equation 4.5 in matrix form as

$$v^{new} = v * Scale(s).$$

Again, in affine geometry, we want not only to scale the length of vectors, but also to scale uniformly the distance of points from a fixed point. Since $P = Q + (P - Q)$, scaling the distance of an arbitrary point $P$ from a fixed point $Q$ by the factor $s$ is equivalent to scaling the length of the vector $P - Q$ by the factor $s$ and then adding the resulting vector to $Q$ (see Figure 4.4). Thus, the formula for scaling the distance of an arbitrary point $P$ from a fixed point $Q$ by the factor $s$ is

$$P^{new} = Q + (P - Q) * Scale(s) = P * Scale(s) + Q * (I - Scale(s)). \qquad (4.6)$$

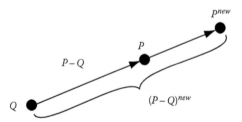

**FIGURE 4.4:** Uniform scaling about the fixed point $Q$. Since the point $Q$ is fixed and since $P = Q + (P - Q)$, scaling the distance of a point $P$ from the point $Q$ by the factor $s$ is equivalent to scaling the length of the vector $P - Q$ by the factor $s$ and then adding the resulting vector to the point $Q$.

Notice that if $Q$ is the origin, then this formula reduces to

$$P^{new} = P * Scale(s),$$

so *Scale(s)* is also the matrix that represents uniformly scaling the distance of points from the origin.

---

## 4.3   The Algebra of Affine Transformations

The three conformal transformations—translation, rotation, and uniform scaling—all have the following form: there exists a matrix $M$ and a vector $w$ such that

$$\begin{aligned} v^{new} &= v * M, \\ P^{new} &= P * M + w. \end{aligned} \tag{4.7}$$

In fact, this form characterizes all affine transformations. That is, a transformation is said to be *affine* if and only if there is a matrix $M$ and a vector $w$ so that Equation 4.7 is satisfied.

The matrix $M$ represents a linear transformation on vectors. Recall that a transformation $L$ on vectors is *linear* if

$$\begin{aligned} L(u + v) &= L(u) + L(v), \\ L(cv) &= cL(v). \end{aligned} \tag{4.8}$$

Matrix multiplication represents a linear transformation because matrix multiplication distributes through vector addition and commutes with scalar multiplication—that is,

$$(u + v) * M = u * M + v * M,$$

$$(cv) * M = c(v * M).$$

The vector $w$ in Equation 4.7 represents translation on the points. Thus, an affine transformation can always be decomposed into a linear transformation followed by a translation. Notice that adding a constant vector to a vector is not a linear transformation, since adding a constant vector does not satisfy Equation 4.8 (see Exercise 4.1). Therefore, translation cannot be represented by a $2 \times 2$ matrix.

Affine transformations can also be characterized abstractly in a manner similar to linear transformations. A transformation $A$ is said to be *affine* if $A$ maps points to points, $A$ maps vectors to vectors, and

$$\begin{aligned} A(u + v) &= A(u) + A(v), \\ A(cv) &= c\,A(v), \\ A(P + v) &= A(P) + A(v). \end{aligned} \tag{4.9}$$

The first two equalities in Equation 4.9 say that an affine transformation is a linear transformation on vectors; the third equality asserts that affine transformations are well behaved with respect to the addition of points and vectors. You should check that with this definition, translation is indeed an affine transformation (see Exercise 4.1).

In terms of coordinates, linear transformations can be written as

$$
\begin{aligned}
x^{new} &= ax + by \\
y^{new} &= cx + dy
\end{aligned}
\quad\Leftrightarrow\quad
(x^{new}, y^{new}) = (x, y) * \begin{pmatrix} a & c \\ b & d \end{pmatrix},
\tag{4.10}
$$

whereas affine transformations have the form:

$$
\begin{aligned}
x^{new} &= ax + by + e \\
y^{new} &= cx + dy + f
\end{aligned}
\quad\Leftrightarrow\quad
(x^{new}, y^{new}) = (x, y) * \begin{pmatrix} a & c \\ b & d \end{pmatrix} + (e, f).
\tag{4.11}
$$

The constant terms $e$ and $f$ that appear in Equation 4.11 are what distinguish the affine transformations of Computer Graphics from the linear transformations of classical linear algebra. These constants represent translation, which, as we have seen, is not a linear transformation.

---

## 4.4   The Geometry of Affine Transformations

There is also a geometric way to characterize both linear and affine transformations. Interpreted geometrically, Equation 4.8 says that linear transformations map triangles into triangles and lines into lines (see Figure 4.5). Moreover, linear transformations preserve ratios of distances along lines, since

$$
\frac{|L(cv)|}{|L(v)|} = \frac{|c\,L(v)|}{|L(v)|} = \frac{|c||L(v)|}{|L(v)|} = |c| = \frac{|c||v|}{|v|} = \frac{|c\,v|}{|v|}.
$$

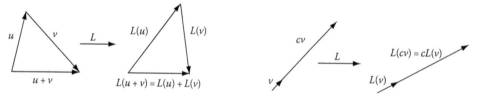

**FIGURE 4.5:**   Linear transformations map triangles into triangles (left) and lines into lines (right).

Similarly, interpreted geometrically, Equation 4.9 makes precisely the same assertions about affine transformations, extending these results to points as well as to vectors (see Figure 4.6).

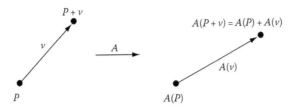

**FIGURE 4.6:** Affine transformations map triangles into triangles and lines into lines (Figure 4.5), and these results are consistent for points as well as for vectors.

## 4.5 Affine Coordinates and Affine Matrices

Linear transformations are typically represented by matrices because composing two linear transformations is equivalent to multiplying the corresponding matrices. We would like to have the same facility with affine transformations—that is, we would like to be able to compose two affine transformations by multiplying their matrix representations. Unfortunately, our current representation of an affine transformation in terms of a transformation matrix $M$ and a translation vector $w$ does not work so well when we want to compose two affine transformations. In order to overcome this deficiency, we shall now introduce a clever device called *affine coordinates*.

Points and vectors are both represented by pairs of rectangular coordinates, but points and vectors are different types of objects with different behaviors for the same affine transformations. For example, points are affected by translation, but vectors are not. We are now going to introduce a third coordinate—an affine coordinate—to distinguish between points and vectors. Affine coordinates will also allow us to represent each affine transformation using a single $3 \times 3$ matrix and to compose any two affine transformations by matrix multiplication.

Since points are affected by translation and vectors are not, the affine coordinate for a point is 1; the affine coordinate for a vector is 0. Thus, from now on, we shall write $P = (p_1, p_2, 1)$ for points and $v = (v_1, v_2, 0)$ for vectors.

Affine transformations can now be represented by $3 \times 3$ affine matrices. An *affine matrix* is a $3 \times 3$ matrix where the third column is $(0\ 0\ 1)^{\mathrm{T}}$. The affine transformation:

$$v^{new} = v * M,$$
$$P^{new} = P * M + w,$$

represented by the $2 \times 2$ transformation matrix $M$ and the translation vector $w$ can be rewritten using affine coordinates in terms of a single $3 \times 3$ affine matrix:

$$\tilde{M} = \begin{pmatrix} M & 0 \\ w & 1 \end{pmatrix}.$$

Now

$$(v^{new}, 0) = (v, 0) * \tilde{M} = (v, 0) * \begin{pmatrix} M & 0 \\ w & 1 \end{pmatrix} = (v * M, 0),$$

$$(P^{new}, 1) = (P, 1) * \tilde{M} = (P, 1) * \begin{pmatrix} M & 0 \\ w & 1 \end{pmatrix} = (P * M + w, 1).$$

Notice how the affine coordinate—0 for vectors and 1 for points—is correctly reproduced by multiplication with the last column of the matrix $\tilde{M}$. Notice too that the 0 in the third coordinate for vectors effectively insures that vectors ignore the translation represented by the vector $w$.

---

## 4.6   Conformal Transformations: Revisited

To illustrate some specific examples of affine matrices, we now show how to represent the standard conformal transformations—translation, rotation, and uniform scaling—in terms of $3 \times 3$ affine matrices. These results follow easily from Equations 4.1, 4.4, and 4.6.

*Translation—by the vector $w = (w_1, w_2)$*

$$Trans(w) = \begin{pmatrix} I & 0 \\ w & 1 \end{pmatrix} = \begin{pmatrix} 1 & 0 & 0 \\ 0 & 1 & 0 \\ w_1 & w_2 & 1 \end{pmatrix}.$$

*Rotation—around the origin through the angle $\theta$*

$$Rot(Origin, \theta) = \begin{pmatrix} Rot(\theta) & 0 \\ 0 & 1 \end{pmatrix} \begin{pmatrix} \cos(\theta) & \sin(\theta) & 0 \\ -\sin(\theta) & \cos(\theta) & 0 \\ 0 & 0 & 1 \end{pmatrix}.$$

*Rotation—around the point $Q = (q_1, q_2)$ through the angle $\theta$*

$$Rot(Q, \theta) = \begin{pmatrix} Rot(\theta) & 0 \\ Q * (I - Rot(\theta)) & 1 \end{pmatrix}$$

$$= \begin{pmatrix} \cos(\theta) & \sin(\theta) & 0 \\ -\sin(\theta) & \cos(\theta) & 0 \\ q_1(1 - \cos(\theta)) + q_2\sin(\theta) & -q_1\sin(\theta) + q_2(1 - \cos(\theta)) & 1 \end{pmatrix}.$$

*Uniform scaling—around the origin by the factor $s$*

$$Scale(Origin, s) = \begin{pmatrix} sI & 0 \\ 0 & 1 \end{pmatrix} = \begin{pmatrix} s & 0 & 0 \\ 0 & s & 0 \\ 0 & 0 & 1 \end{pmatrix}.$$

*Uniform scaling—around the point $Q = (q_1, q_2)$ by the factor $s$*

$$Scale(Q, s) = \begin{pmatrix} sI & 0 \\ Q * (1 - s)I & 1 \end{pmatrix} = \begin{pmatrix} s & 0 & 0 \\ 0 & s & 0 \\ (1 - s)q_1 & (1 - s)q_2 & 1 \end{pmatrix}.$$

---

## 4.7   General Affine Transformations

We shall now develop $3 \times 3$ matrix representations for arbitrary affine transformations. The two most general ways of specifying an affine transformation of the plane are by specifying either the image of one point and two linearly independent vectors or by specifying the image of three noncollinear points. We shall treat each of these cases in turn.

### 4.7.1 Image of One Point and Two Lineary Independent Vectors

Fix a point $Q$ and two linearly independent vectors $u$, $v$. Since two linearly independent vectors $u$, $v$ form a basis for vectors in the plane, for any vector $w$ there are scalars $\sigma$, $\tau$ such that

$$w = \sigma u + \tau v.$$

Hence for any point $P$, there are scalars $s$, $t$ such that

$$P - Q = su + tv,$$

or equivalently

$$P = Q + su + tv.$$

Therefore, if $A$ is an affine transformation, then by Equation 4.9,

$$A(P) = A(Q) + sA(u) + t\,A(v),$$
$$A(w) = \sigma A(u) + \tau A(v). \tag{4.12}$$

Thus, if we know the effect of the affine transformation $A$ on one point $Q$ and on two linearly independent vectors $u$, $v$, then we know the effect of $A$ on every point $P$ and every vector $w$. Hence, in addition to conformal transformations, there is another important way to define affine transformations of the plane: by specifying the image of one point and two linearly independent vectors.

Geometrically, Equation 4.12 says that an affine transformation $A$ maps the parallelogram determined by the point $Q$ and the vectors $u$, $v$ into the parallelogram determined by the point $A(Q)$ and the vectors $A(u)$, $A(v)$ (see Figure 4.7). Thus, affine transformations map parallelograms into parallelograms.

To find the $3 \times 3$ matrix representation $M$ for such an affine transformation $A$, let $Q^*$, $u^*$, $v^*$ be the images of $Q$, $u$, $v$ under the transformation $A$. Then using affine coordinates, we have

$$(u^*, 0) = (u, 0) * M,$$

$$(v^*, 0) = (v, 0) * M,$$

$$(Q^*, 1) = (Q, 1) * M,$$

where

$$M = \begin{pmatrix} a & c & 0 \\ b & d & 0 \\ e & f & 1 \end{pmatrix}.$$

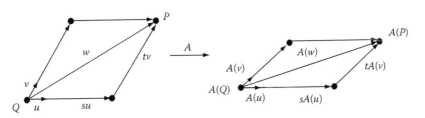

**FIGURE 4.7:** Affine transformations map parallelograms into parallelograms.

Thus,

$$\underbrace{\begin{pmatrix} u^* & 0 \\ v^* & 0 \\ Q^* & 1 \end{pmatrix}}_{S_{new}} = \underbrace{\begin{pmatrix} u & 0 \\ v & 0 \\ Q & 1 \end{pmatrix}}_{S_{old}} * \underbrace{\begin{pmatrix} a & c & 0 \\ b & d & 0 \\ e & f & 1 \end{pmatrix}}_{M}.$$

Solving for $M$ yields

$$M = S_{old}^{-1} * S_{new},$$

or equivalently

$$M = \begin{pmatrix} u & 0 \\ v & 0 \\ Q & 1 \end{pmatrix}^{-1} * \begin{pmatrix} u^* & 0 \\ v^* & 0 \\ Q^* & 1 \end{pmatrix} = \begin{pmatrix} u_1 & u_2 & 0 \\ v_1 & v_2 & 0 \\ q_1 & q_2 & 1 \end{pmatrix}^{-1} * \begin{pmatrix} u_1^* & u_2^* & 0 \\ v_1^* & v_2^* & 0 \\ q_1^* & q_2^* & 1 \end{pmatrix}. \qquad (4.13)$$

To compute $M$ explicitly, we need only invert a $3 \times 3$ affine matrix. But for a nonsingular $3 \times 3$ affine matrix, the inverse can easily be written explicitly; for the formula, see Exercise 4.13a. An alternative explicit formula for the matrix $M$ without inverses is provided in Exercise 4.21.

### 4.7.2   Nonuniform Scaling

One application of our preceding approach to general affine transformations is a formula for nonuniform scaling. Uniform scaling scales distances by the same amount in all directions; nonuniform scaling scales distances by different amounts in different directions. We are interested in nonuniform scaling for many reasons. For example, we can apply nonuniform scaling to generate an ellipse by scaling a circle from its center along a fixed direction (see Figure 4.8). To scale the distance from a fixed point $Q$ along an arbitrary direction $w$ by a scale factor $s$, we shall now apply our method for generating arbitrary affine transformations by specifying the image of one point and two linearly independent vectors (see Figure 4.9).

Let $Scale(Q, w, s)$ denote the matrix that represents the affine transformation that scales the distance from a fixed point $Q$ along an arbitrary direction $w$ by a scale factor $s$. To find the $3 \times 3$ matrix $Scale(Q, w, s)$, we need to know the image of one point and two linearly independent vectors. Consider the point $Q$ and the vectors $w$ and $w^{\perp}$, where $w^{\perp}$ is a vector of the same length as $w$ perpendicular to $w$. It is easy to see how the transformation $Scale(Q, w, s)$ affects $Q, w, w^{\perp}$:

- $Q \rightarrow Q$ because $Q$ is a fixed point of the transformation.

- $w \rightarrow s\,w$ because distances are scaled by the factor $s$ along the direction $w$.

- $w^{\perp} \rightarrow w^{\perp}$ because distances along the vector $w^{\perp}$ are not changed, since $w^{\perp}$ is orthogonal to the scaling direction $w$.

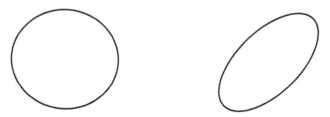

**FIGURE 4.8:**   Scaling a circle from its center along a fixed direction generates an ellipse.

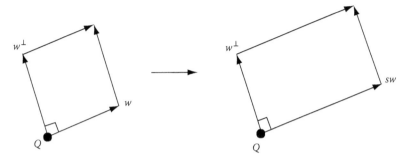

**FIGURE 4.9:** Scaling from the fixed point $Q$ in the direction $w$ by the scale factor $s$. The point $Q$ is fixed, the vector $w$ is scaled by $s$, and the orthogonal vector $w^\perp$ remains fixed. Thus, nonuniform scaling maps a square into a rectangle.

Therefore, by the results of the previous section,

$$Scale(Q, w, s) = \begin{pmatrix} w & 0 \\ w^\perp & 0 \\ Q & 1 \end{pmatrix}^{-1} * \begin{pmatrix} sw & 0 \\ w^\perp & 0 \\ Q & 1 \end{pmatrix}.$$

Now recall that if $w = (w_1, w_2)$, then $w^\perp = (-w_2, w_1)$. Therefore, in terms of coordinates,

$$Scale(Q, w, s) = \begin{pmatrix} w_1 & w_2 & 0 \\ -w_2 & w_1 & 0 \\ q_1 & q_2 & 1 \end{pmatrix}^{-1} * \begin{pmatrix} sw_1 & sw_2 & 0 \\ -w_2 & w_1 & 0 \\ q_1 & q_2 & 1 \end{pmatrix}.$$

An explicit formula for the matrix $Scale(Q, w, s)$ without inverses is provided in Exercise 4.20.

If $Q$ is the origin and $w$ is the unit vector along the $x$-axis, then $w^\perp$ is the unit vector along the $y$-axis, so

$$\begin{pmatrix} w_1 & w_2 & 0 \\ -w_2 & w_1 & 0 \\ q_1 & q_2 & 1 \end{pmatrix} = \begin{pmatrix} 1 & 0 & 0 \\ 0 & 1 & 0 \\ 0 & 0 & 1 \end{pmatrix} = \text{Identity matrix}.$$

Therefore, scaling along the $x$-axis is represented by the $3 \times 3$ matrix:

$$Scale(Origin, i, s) = \begin{pmatrix} s & 0 & 0 \\ 0 & 1 & 0 \\ 0 & 0 & 1 \end{pmatrix}.$$

Similarly, scaling along the $y$-axis is represented by the $3 \times 3$ matrix:

$$Scale(Origin, j, s) = \begin{pmatrix} 1 & 0 & 0 \\ 0 & s & 0 \\ 0 & 0 & 1 \end{pmatrix}.$$

### 4.7.3   Image of Three Noncollinear Points

If we know the image of three noncollinear points under an affine transformation, then we also know the image of one point and two linearly independent vectors. Indeed, suppose that we know the image of the three noncollinear points $P_1$, $P_2$, $P_3$ under the affine transformation $A$. Then $u = P_2 - P_1$ and $v = P_3 - P_1$ are certainly linearly independent vectors, and

$$A(u) = A(P_2 - P_1) = A(P_2) - A(P_1),$$
$$A(v) = A(P_3 - P_1) = A(P_3) - A(P_1).$$

Hence if we know the image of $P_1$, $P_2$, $P_3$ under the affine transformation $A$, then we know the image of the point $P_1$ and the two linearly independent vectors $u$, $v$. Therefore, the image of three noncollinear points specifies a unique affine transformation.

To find the $3 \times 3$ matrix $M$ that represents an affine transformation $A$ defined by the image of three noncollinear points, we can proceed as in Section 4.7.1. Let $P_1^*$, $P_2^*$, $P_3^*$ be the images of the points $P_1$, $P_2$, $P_3$ under the affine transformation $A$. Using affine coordinates, we have

$$(P_1^*, 1) = (P_1, 1) * M,$$

$$(P_2^*, 1) = (P_2, 1) * M,$$

$$(P_3^*, 1) = (P_3, 1) * M,$$

where

$$M = \begin{pmatrix} a & c & 0 \\ b & d & 0 \\ e & f & 1 \end{pmatrix}.$$

Thus,

$$\underbrace{\begin{pmatrix} P_1^* & 1 \\ P_2^* & 1 \\ P_3^* & 1 \end{pmatrix}}_{P_{new}} = \underbrace{\begin{pmatrix} P_1 & 1 \\ P_2 & 1 \\ P_3 & 1 \end{pmatrix}}_{P_{old}} * \underbrace{\begin{pmatrix} a & c & 0 \\ b & d & 0 \\ e & f & 1 \end{pmatrix}}_{M}.$$

Solving for $M$ yields

$$M = P_{old}^{-1} * P_{new},$$

or equivalently if $P_k = (x_k, y_k)$ and $P_k^* = (x_k^*, y_k^*)$ for $k = 1, 2, 3$, then

$$M = \begin{pmatrix} P_1 & 1 \\ P_2 & 1 \\ P_3 & 1 \end{pmatrix}^{-1} * \begin{pmatrix} P_1^* & 1 \\ P_2^* & 1 \\ P_3^* & 1 \end{pmatrix} = \begin{pmatrix} x_1 & y_1 & 1 \\ x_2 & y_2 & 1 \\ x_3 & y_3 & 1 \end{pmatrix}^{-1} * \begin{pmatrix} x_1^* & y_1^* & 1 \\ x_2^* & y_2^* & 1 \\ x_3^* & y_3^* & 1 \end{pmatrix}. \tag{4.14}$$

To compute $M$ explicitly, we need only invert a $3 \times 3$ matrix whose last column is $(1\ 1\ 1)^T$. This inverse can easily be written explicitly; for the formula, see Exercise 4.13b. An alternative explicit formula for the matrix $M$ without inverses is provided in Exercise 4.21.

Notice the similarity between the right-hand side of Equation 4.14 and the right-hand side of Equation 4.13 for the matrix representing the affine transformation defined by the image of one point and two linearly independent vectors. The only differences are the ones appearing in the first two rows of the third columns, indicating that two vectors have been replaced by two points.

## 4.8 Summary

Next we provide is a brief summary of the main high-level concepts that you need to remember from this chapter. Also listed for your convenience are the $3 \times 3$ matrix representations for all the affine transformations that we have discussed in this chapter.

### 4.8.1 Affine Transformations and Affine Coordinates

Affine transformations are the fundamental transformations of two-dimensional Computer Graphics. There are several different ways of characterizing affine transformations:

1. A transformation $A$ is affine if there exists a $2 \times 2$ matrix $M$ and a vector $w$ such that

$$A(v) = v * M,$$
$$A(P) = P * M + w.$$

2. A transformation $A$ is affine if there is a $3 \times 3$ affine matrix $M$ such that

$$(A(v), 0) = v * M,$$
$$(A(P), 1) = (P, 1) * M.$$

3. A transformation $A$ is affine if

$$A(u + v) = A(u) + A(v),$$
$$A(cv) = cA(v),$$
$$A(P + v) = A(P) + A(v).$$

4. A transformation $A$ is affine if $A$ maps triangles to triangles and lines to lines, and $A$ preserves ratios of distances along lines.

Affine transformations include the standard conformal transformations—translation, rotation, and uniform scaling—of Turtle Graphics, but affine transformations also incorporate other transformations such as nonuniform scaling. Every affine transformation can be uniquely specified either by the image of one point and two linearly independent vectors or by the image of three noncollinear points.

Affine coordinates are used to distinguish between points and vectors. The affine coordinate of a point is 1; the affine coordinate of a vector is 0. Affine coordinates can be applied to represent all affine transformations by $3 \times 3$ affine matrices of the form:

$$\widetilde{M} = \begin{pmatrix} M & 0 \\ w & 1 \end{pmatrix},$$

so that

$$(v^{new}, 0) = (v, 0) * M,$$

$$(P^{new}, 1) = (P, 1) * M.$$

These matrix representations allow us to compose affine transformations using matrix multiplication.

The affine transformations of Computer Graphics are closely related to the linear transformations of standard linear algebra. In fact, in terms of matrix representations,

$$\text{Affine transformation} = \begin{pmatrix} Linear & 0 \\ Transformation & 0 \\ Translation & 1 \end{pmatrix}.$$

In Table 4.1, we provide a more detailed comparison of the differences between affine and linear transformations.

**TABLE 4.1:**  Comparison of affine and linear transformations.

| Affine Transformation | Linear Transformation |
|---|---|
| $A(u+v) = A(u) + A(v)$ | $L(u+v) = L(u) + L(v)$ |
| $A(cv) = c\,A(v)$ | $L(cv) = c\,L(v)$ |
| $A(P+v) = A(P) + A(v)$ | |
| Includes translation | Excludes translation |
| Affine coordinates | Rectangular coordinates |
| Coordinate formulas | Coordinate formulas |
| $x_{new} = ax_{old} + by_{old} + e$ | $x_{new} = ax_{old} + by_{old}$ |
| $y_{new} = cx_{old} + dy_{old} + f$ | $y_{new} = cx_{old} + dy_{old}$ |
| $3 \times 3$ matrix representation | $2 \times 2$ matrix representation |
| $(x_{new}, y_{new}, 1) = (x_{old}, y_{old}, 1) * \begin{pmatrix} a & c & 0 \\ b & d & 0 \\ e & f & 1 \end{pmatrix}$ | $(x_{new}, y_{new}) = (x_{old}, y_{old}) * \begin{pmatrix} a & c \\ b & d \end{pmatrix}$ |

*Note:*  For both affine and linear transformations, composition of transformations is equivalent to matrix multiplication.

Finally, a word of caution before we conclude. We use matrices *to represent* transformations, but matrices are not the same things as transformations. Matrix representations for affine transformations are akin to coordinate representations for points and vectors—that is, a low-level computational tool for communicating efficiently with a computer. Matrices are the assembly language of transformations. As usual, we do not want to think or write programs in an assembly language; we want to work in a high-level language. In the next chapter, we shall introduce a high-level language for two-dimensional Computer Graphics based on affine transformations. You will code the matrices for these transformations only once, but you will use the corresponding high-level transformations many, many times.

## 4.8.2 Matrices for Affine Transformations in the Plane

*Translation—by the vector $w = (w_1, w_2)$*

$$Trans(w) = \begin{pmatrix} I & 0 \\ w & 1 \end{pmatrix} = \begin{pmatrix} 1 & 1 & 0 \\ 0 & 1 & 0 \\ w_1 & w_2 & 1 \end{pmatrix}.$$

*Rotation—around the origin through the angle $\theta$*

$$Rot(Origin, \theta) = \begin{pmatrix} Rot(\theta) & 0 \\ 0 & 1 \end{pmatrix} = \begin{pmatrix} \cos(\theta) & \sin(\theta) & 0 \\ -\sin(\theta) & \cos(\theta) & 0 \\ 0 & 0 & 1 \end{pmatrix}.$$

*Rotation—around the point $Q = (q_1, q_2)$ through the angle $\theta$*

$$Rot(Q, \theta) = \begin{pmatrix} Rot(\theta) & 0 \\ Q * (I - Rot(\theta)) & 1 \end{pmatrix}$$

$$= \begin{pmatrix} \cos(\theta) & \sin(\theta) & 0 \\ -\sin(\theta) & \cos(\theta) & 0 \\ q_1(1 - \cos(\theta)) + q_2 \sin(\theta) & -q_1 \sin(\theta) + q_2(1 - \cos(\theta)) & 1 \end{pmatrix}.$$

*Uniform scaling—around the origin by the factor $s$*

$$Scale(Origin, s) = \begin{pmatrix} sI & 0 \\ 0 & 1 \end{pmatrix} = \begin{pmatrix} s & 0 & 0 \\ 0 & s & 0 \\ 0 & 0 & 1 \end{pmatrix}.$$

*Uniform scaling—around the point $Q = (q_1, q_2)$ by the factor $s$*

$$Scale(Origin, s) = \begin{pmatrix} sI & 0 \\ Q * (1 - s)I & 1 \end{pmatrix} = \begin{pmatrix} s & 0 & 0 \\ 0 & s & 0 \\ (1 - s)q_1 & (1 - s)q_2 & 1 \end{pmatrix}.$$

*Nonuniform scaling—around the point $Q = (q_1, q_2)$ in the direction $w = (w_1, w_2)$ by the factor $s$*

$$Scale(Q, w, s) = \begin{pmatrix} w_1 & w_2 & 0 \\ -w_2 & w_1 & 0 \\ q_1 & q_2 & 1 \end{pmatrix}^{-1} * \begin{pmatrix} sw_1 & sw_2 & 0 \\ -w_2 & w_1 & 0 \\ q_1 & q_2 & 1 \end{pmatrix}.$$

*Scaling from the origin along the x-axis by the factor $s$*

$$Scale(Origin, i, s) = \begin{pmatrix} s & 0 & 0 \\ 0 & 1 & 0 \\ 0 & 0 & 1 \end{pmatrix}.$$

*Scaling from the origin along the y-axis by the factor s*

$$Scale(Origin, j, s) = \begin{pmatrix} 1 & 0 & 0 \\ 0 & s & 0 \\ 0 & 0 & 1 \end{pmatrix}.$$

*Image of one point $Q = (x_3, y_3)$ and two linearly independent vectors $v_1 = (x_1, y_1)$, $v_2 = (x_2, y_2)$*

$$Image(v_1, v_2, Q) = \begin{pmatrix} x_1 & y_1 & 0 \\ x_2 & y_2 & 0 \\ x_3 & y_3 & 1 \end{pmatrix}^{-1} * \begin{pmatrix} x_1^* & y_1^* & 0 \\ x_2^* & y_2^* & 0 \\ x_3^* & y_3^* & 1 \end{pmatrix}.$$

*Image of three noncollinear points $P_1 = (x_1, y_1)$, $P_2 = (x_2, y_2)$, $P_3 = (x_3, y_3)$*

$$Image(P_1, P_2, P_3) = \begin{pmatrix} x_1 & y_1 & 1 \\ x_2 & y_2 & 1 \\ x_3 & y_3 & 1 \end{pmatrix}^{-1} * \begin{pmatrix} x_1^* & y_1^* & 1 \\ x_2^* & y_2^* & 1 \\ x_3^* & y_3^* & 1 \end{pmatrix}.$$

*Inverses*

$$\begin{pmatrix} x_1 & y_1 & 0 \\ x_2 & y_2 & 0 \\ x_3 & y_3 & 1 \end{pmatrix}^{-1} = \frac{\begin{pmatrix} y_2 & -y_1 & 0 \\ -x_2 & x_1 & 0 \\ x_2y_3 - y_2x_3 & y_1x_3 - x_1y_3 & x_1y_2 - y_1x_2 \end{pmatrix}}{x_1y_2 - y_1x_2},$$

$$\begin{pmatrix} x_1 & y_1 & 1 \\ x_2 & y_2 & 1 \\ x_3 & y_3 & 1 \end{pmatrix}^{-1} = \frac{\begin{pmatrix} y_2 - y_3 & y_3 - y_1 & y_1 - y_2 \\ x_3 - x_2 & x_1 - x_3 & x_2 - x_1 \\ x_2y_3 - y_2x_3 & y_1x_3 - x_1y_3 & x_1y_2 - y_1x_2 \end{pmatrix}}{x_1(y_2 - y_3) + x_2(y_3 - y_1) + x_3(y_1 - y_2)}.$$

---

## Exercises

4.1. Define $L_w(v) = v + w$, and let $T_w(v) = v$, $T_w(P) = P + w$. Show that $L_w(v)$ is not a linear transformation on vectors, but that $T_w(v)$, $T_w(P)$ is an affine transformation on points and vectors.

4.2. Show that the formula for scaling the distance of a point $P$ from a point $Q$ by a scale factor $s$ is equivalent to

$$P^{new} = (1 - s)Q + sP.$$

4.3. Let $w_Q$ denote the vector from the origin to the point $Q$. Without appealing to coordinates and without explicitly multiplying matrices, explain why:

a. $Rot(Q, \theta) = Trans(-w_Q) * Rot(\theta) * Trans(w_Q)$.

b. $Scale(Q, s) = Trans(-w_Q) * Scale(s) * Trans(w_Q)$.

(Hint: Give a geometric interpretation for formulas a and b.)

4.4. Let $w_Q$ denote the vector from the origin to the point $Q$, $i$ denote the unit vector along the $x$-axis, and $\alpha$ denote the angle between the vectors $i$ and $w_Q$. Without appealing to coordinates and without explicitly multiplying matrices, show that

$$Scale(Q, w, s) = Trans(-w_Q) * Rot(-\alpha) * Scale(Origin, i, s) * Rot(\alpha) * Trans(w_Q).$$

(Hint: Give a geometric interpretation for this formula.)

4.5. Show that rotation and scaling about a fixed point commute. That is, show that

$$Rot(Q, \theta) * Scale(Q, s) = Scale(Q, s) * Rot(Q, \theta).$$

4.6. Verify that if $w = (w_1, w_2)$, then $w^{\perp} = (-w_2, w_1)$ by invoking the formula $w^{\perp} = w * Rot(\pi/2)$.

4.7. What is the geometric effect of the transformation matrix

$$\begin{pmatrix} w & 0 \\ w^{\perp} & 0 \\ Q & 1 \end{pmatrix}^{-1} * \begin{pmatrix} sw & 0 \\ tw^{\perp} & 0 \\ Q & 1 \end{pmatrix}.$$

4.8. Show that when affine transformations are represented by $3 \times 3$ matrices, composition of affine transformations is equivalent to matrix multiplication.

4.9. Let $A$ be an affine transformation. Using Equation 4.9, show that for all points $P, Q$

$$A(Q - P) = A(Q) - A(P).$$

4.10. Let $M$ be the $3 \times 3$ affine matrix that represents the affine transformation $A$. Show that

$$M = \begin{pmatrix} A(i) & 0 \\ A(j) & 0 \\ A(Origin) & 1 \end{pmatrix}.$$

4.11. Show that:

a. The composite of two affine transformations is an affine transformation.

b. The inverse of a nonsingular affine transformation is an affine transformation.

Conclude that the nonsingular affine transformations form a group.

4.12. Let $M, N$ be two matrices whose third column is $(0\ 0\ 1)^{T}$. Show that:

a. The third column of $M * N$ is $(0\ 0\ 1)^{T}$.

b. The third column of $M^{-1}$ is $(0\ 0\ 1)^{T}$.

Conclude that the nonsingular affine matrices form a group.

4.13. Verify that:

a. $M = \begin{pmatrix} x_1 & y_1 & 0 \\ x_2 & y_2 & 0 \\ x_3 & y_3 & 1 \end{pmatrix} \Rightarrow M^{-1} = \dfrac{\begin{pmatrix} y_2 & -y_1 & 0 \\ -x_2 & x_1 & 0 \\ x_2y_3 - y_2x_3 & y_1x_3 - x_1y_3 & x_1y_2 - y_1x_2 \end{pmatrix}}{x_1y_2 - y_1x_2}$.

b. $M = \begin{pmatrix} x_1 & y_1 & 1 \\ x_2 & y_2 & 1 \\ x_3 & y_3 & 1 \end{pmatrix} \Rightarrow M^{-1} = \dfrac{\begin{pmatrix} y_2 - y_3 & y_3 - y_1 & y_1 - y_2 \\ x_3 - x_2 & x_1 - x_3 & x_2 - x_1 \\ x_2y_3 - y_2x_3 & y_1x_3 - x_1y_3 & x_1y_2 - y_1x_2 \end{pmatrix}}{x_1(y_2 - y_3) + x_2(y_3 - y_1) + x_3(y_1 - y_2)}$.

4.14. Here we consider affine transformations determined by the image of two points and one vector.

   a. Show that an affine transformation is uniquely determined by the image of two points and one vector that is not parallel to the line determined by the two points.

   b. Let $P_1^*$, $P_2^*$, $v^*$ be the images of $P_1$, $P_2$, $v$ under the affine transformation $A$. Find the $3 \times 3$ matrix $M$ that represents the affine transformation $A$.

4.15. Show that every conformal transformation is uniquely determined by:

   a. The image of one point and one vector.

   b. The image of two points.

4.16. Show that if $M$ is a conformal transformation, then $ScaleFactor(M) = \sqrt{|\det(M)|}$.

4.17. Let $Mirror(Q, w)$ denote the matrix representing the transformation that mirrors points and vectors in the line determined by the point $Q$ and the direction vector $w$.

   a. Show that

   i. $Mirror(Q, w) = \begin{pmatrix} w & 0 \\ w^{\perp} & 0 \\ Q & 1 \end{pmatrix}^{-1} * \begin{pmatrix} w & 0 \\ -w^{\perp} & 0 \\ Q & 1 \end{pmatrix} = Scale(Q, w^{\perp}, -1)$.

   b. Conclude that the matrices representing mirroring in the $x$ and $y$ axes are given by:

   i. $Mirror(Origin, i) = \begin{pmatrix} 1 & 0 & 0 \\ 0 & -1 & 0 \\ 0 & 0 & 1 \end{pmatrix}$.

   ii. $Mirror(Origin, j) = \begin{pmatrix} -1 & 0 & 0 \\ 0 & 1 & 0 \\ 0 & 0 & 1 \end{pmatrix}$.

4.18. Shear is the affine transformation defined by mapping a unit square with vertex $Q$ and sides $w$, $w^{\perp}$ into a parallelogram by tilting the edge $w^{\perp}$ so that $w^{\perp}_{new}$ makes an angle of $\theta$ with $w^{\perp}$ (see Figure 4.10). Show that:

   a. $Shear(Q, w, w^{\perp}, \theta) = \begin{pmatrix} w & 0 \\ w^{\perp} & 0 \\ Q & 1 \end{pmatrix}^{-1} * \begin{pmatrix} w & 0 \\ \tan(\theta)\, w + w^{\perp} & 0 \\ Q & 1 \end{pmatrix}$.

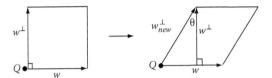

**FIGURE 4.10:** Shear maps the rectangle with vertex $Q$ and sides $w$, $w^{\perp}$ into a parallelogram by tilting the edge $w^{\perp}$ so that $w^{\perp}_{new}$ makes an angle of $\theta$ with $w^{\perp}$.

b. *Shear* $(Origin, i, j, \theta) = \begin{pmatrix} 1 & 0 & 0 \\ \tan(\theta) & 1 & 0 \\ 0 & 0 & 1 \end{pmatrix}$.

c. $\det(Shear(Q, w, w^{\perp}, \theta)) = 1$.

d. Shear preserves area.

4.19. Show that every nonsingular affine transformation is the composition of one translation, one rotation, one shear, and two nonuniform scalings.

4.20. Fix a vector $w$ and a scalar $s$. Define $A(v) = v + (s-1)\dfrac{v \cdot w}{w \cdot w} w$, where $v \cdot w = v_1 w_1 + v_2 w_2$.

a. Show that:

    i. $A(w) = sw$.

    ii. $A(w^{\perp}) = w^{\perp}$.

    iii. $v = \alpha w + \beta w^{\perp} \rightarrow A(v) = s\alpha w + \beta w^{\perp}$.

b. Conclude from part a that

$$v * Scale(Q, w, s) = v + (s-1)\frac{v \cdot w}{w \cdot w} w.$$

c. Using the result of part b, show that for every point $P$

$$P * Scale(Q, w, s) = P + (s-1)\frac{(P-Q) \cdot w}{w \cdot w} w.$$

d. Using the results of parts b and c, show that

$$Scale(Q, w, s) = \begin{pmatrix} 1 + \dfrac{(s-1)w_1^2}{w_1^2 + w_2^2} & \dfrac{(s-1)w_1 w_2}{w_1^2 + w_2^2} & 0 \\[3mm] \dfrac{(s-1)w_1 w_2}{w_1^2 + w_2^2} & 1 + \dfrac{(s-1)w_2^2}{w_1^2 + w_2^2} & 0 \\[3mm] \dfrac{-(s-1)(q_1 w_1^2 + q_2 w_1 w_2)}{w_1^2 + w_2^2} & \dfrac{-(s-1)(q_1 w_1 w_2 + q_2 w_2^2)}{w_1^2 + w_2^2} & 1 \end{pmatrix}.$$

4.21. Fix six points $P_1$, $P_2$, $P_3$ and $P_1^*$, $P_2^*$, $P_3^*$ expressed in affine coordinates, and define

$$A(R) = \frac{\det\left(R^T \; P_2^T \; P_3^T\right)P_1^* + \det\left(P_1^T \; R^T \; P_3^T\right)P_2^* + \det\left(P_1^T \; P_2^T \; R^T\right)P_3^*}{\det\left(P_1^T \; P_2^T \; P_3^T\right)}.$$

a. Show that

$$A(P_i) = P_i^*, \quad i = 1, 2, 3.$$

b. Using part a, show that the $3 \times 3$ matrix representing the affine transformation that maps the points $P_1$, $P_2$, $P_3$ to the points $P_1^*$, $P_2^*$, $P_3^*$ is given by

$$Image(P_1,P_2,P_3) = \frac{(P_2 \times P_3)^T * P_1^* + (P_3 \times P_1)^T * P_2^* + (P_1 \times P_2)^T * P_3^*}{\det\left(P_1^T \; P_2^T \; P_3^T\right)}.$$

c. Explain how the formula for $A(R)$ changes if we replace the points $P_1, P_2$ and $P_1^*, P_2^*$ by linearly independent vectors $v_1, v_2$ and $v_1^*, v_2^*$.

d. Using part c, find a closed form expression analogous to the fomula in part b for $Image(P_1, P_2, P_3)$ for the $3 \times 3$ matrix representing the affine transformation that maps $v_1, v_2, P_3$ to $v_1^*, v_2^*, P_3^*$.

4.22. In this exercise, we are going to prove that for every triangle, up to sign,

$$Area(\Delta P_1 P_2 P_3) = \frac{1}{2} \det\begin{pmatrix} P_1 & 1 \\ P_2 & 1 \\ P_3 & 1 \end{pmatrix} = \frac{1}{2} \det\begin{pmatrix} x_1 & y_1 & 1 \\ x_2 & y_2 & 1 \\ x_3 & y_3 & 1 \end{pmatrix}. \tag{4.15}$$

a. Verify that Equation 4.15 is valid if $P_1$ is at the origin and $P_2$ lies on the $x$-axis.

b. Show that for any $\Delta P_1 P_2 P_3$ there is a rigid motion $R$ consisting of a rotation followed by a translation that maps $\Delta P_1 P_2 P_3$ to $\Delta Q_1 Q_2 Q_3$, where $Q_1$ is the origin and $Q_2$ is a point on the $x$-axis.

c. Show that:

   i. $Area(\Delta P_1 P_2 P_3) = Area(\Delta Q_1 Q_2 Q_3)$.

   ii. $\det\begin{pmatrix} Q_1 & 1 \\ Q_2 & 1 \\ Q_3 & 1 \end{pmatrix} = \det\begin{pmatrix} P_1 & 1 \\ P_2 & 1 \\ P_3 & 1 \end{pmatrix}.$

d. Conclude from parts a through c that Equation 4.15 is valid for every triangle.

4.23. Prove in two ways that nonsingular affine transformations preserve ratios of areas:

a. By using the fact that nonsingular transformations preserve ratios of distances along lines.

b. By invoking Equation 4.15.

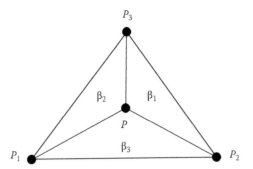

**FIGURE 4.11:** The barycentric coordinates of the point $P$ relative to $\Delta P_1 P_2 P_3$ are given by the ratios of the areas of inner triangles $\Delta PP_iP_j$ to the area of the outer triangle $\Delta P_1 P_2 P_3$.

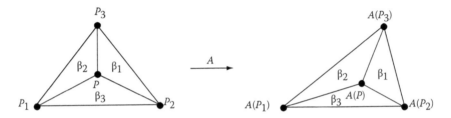

**FIGURE 4.12:** Affine transformations preserve barycentric coordinates. In general, affine transformations do not preserve areas, but affine transformation do preserve ratios of areas.

4.24. Let $P_1$, $P_2$, $P_3$ be three noncollinear points.

 a. Show that for each point $P$ there exists unique scalars $\beta_1, \beta_2, \beta_3$, such that:

 i. $P = \beta_1 P_1 + \beta_2 P_2 + \beta_3 P_3$.

 ii. $\beta_1 + \beta_2 + \beta_3 = 1$.

 The values $\beta_1, \beta_2, \beta_3$ are called the *barycentric coordinates* of $P$ with respect to $\Delta P_1 P_2 P_3$.

 b. Using Equation 4.15 and standard properties of determinants, show that, up to sign,

$$\beta_k = \frac{Area(\Delta PP_iP_j)}{Area(\Delta P_1 P_2 P_3)}, \quad i, j \neq k \text{ (see Figure 4.11)}.$$

4.25. Let $P_1$, $P_2$, $P_3$ be three noncollinear points, and let $\beta_1, \beta_2, \beta_3$ be the barycentric coordinates of $P$ with respect to $\Delta P_1 P_2 P_3$ (see Exercise 4.24).

 a. Show that if $A$ is an affine transformation, then

$$A(P) = \beta_1 A(P_1) + \beta_2 A(P_2) + \beta_3 A(P_3).$$

 b. Conclude that the barycentric coordinates of $A(P)$ with respect to $\Delta A(P_1)\, A(P_2)\, A(P_3)$ are identical to the barycentric coordinates of $P$ with respect to $\Delta P_1 P_2 P_3$ (see Figure 4.12).

# Chapter 5

## Affine Geometry: A Connect-the-Dots Approach to Two-Dimensional Computer Graphics

*The lines are fallen unto me in pleasant places*

– Psalms 16:6

### 5.1  Two Shortcomings of Turtle Graphics

Two points determine a line. In Turtle Graphics we use this simple fact to draw a line joining the two points at which the turtle is located before and after the execution of each FORWARD command. By programming the turtle to move about and to draw lines in this fashion, we are able to generate some remarkable figures in the plane.

Nevertheless, the turtle has two annoying idiosyncrasies. First, the turtle has no memory, so the order in which the turtle encounters points is crucial. Thus, even though the turtle leaves behind a trace of her path, there is no direct command in LOGO to return the turtle to an arbitrary previously encountered location. Second, the turtle is blissfully unaware of the outside universe. The turtle carries her own local coordinate system—her state—but the turtle does not know her position relative to any other point in the plane. Turtle geometry is a local, intrinsic geometry; the turtle knows nothing of the extrinsic, global geometry of the external world. The turtle can draw a circle, but the turtle has no idea where the center of the circle might be or even that there is such a concept as a center, a point outside her path around the circumference.

These two shortcomings—no memory and no knowledge of the outside world—often make the turtle cumbersome to program. For example, even though it is straightforward to program the turtle to generate a regular polygon, it is not so easy to program the turtle to draw a rosette. A rosette is simply the collection of all the lines joining all pairs of vertices in a polygon. Nevertheless, though we can easily program the turtle to visit all the vertices of a polygon, we cannot simply command the turtle to draw the lines joining each pair of vertices because the turtle does not remember where these vertices are located. It is for this reason that to generate the rosette in Turtle Graphics, we need to precompute the distances between each pair of vertices and the angles between adjacent diagonals. Two points determine a line, but Turtle Graphics gives us access to only one pair of points at a time.

## 5.2   Affine Graphics

In Affine Graphics, we will have simultaneous access to many different points in the plane, and we will be able to join any two of these points with a straight line. These two abilities distinguish affine geometry from turtle geometry, and these two properties make Affine Graphics an extremely powerful tool for Computer Graphics.

To implement Affine Graphics, we are going to introduce a new language: CODO (COnnect the DOts). In CODO, as in LOGO, points and vectors are stored internally in terms of coordinates, but the CODO programmer, just like the LOGO programmer, has no direct access to these coordinates. Instead, the main tools in CODO for generating new points and lines are the affine transformations discussed in Chapter 4.

### 5.2.1   The CODO Language

In CODO there are four types of objects: points, vectors, line segments, and affine transformations. Points and vectors are stored internally using affine coordinates; line segments are represented by their end points; affine transformations are stored as $3 \times 3$ matrices.

There are only three types of commands in CODO: commands that create geometry, commands that generate affine transformations, and a command to display geometry. Next we provide a complete description of the CODO language.

*Geometry Creation*

1. VECTOR

   a. *VECTOR(P, Q)*—creates the vector from $P$ to $Q$, usually denoted by $Q - P$.

2. LINE

   a. *LINE(P, Q)*—creates the line segment $PQ$.

   b. *LINE(P_1, \ldots, P_n)*—creates the line segments $P_1 P_2, \ldots, P_{n-1} P_n$.

3. TRANSFORM

   a. *TRANSFORM(X, M)*—applies the affine transformation $M$ to the object $X$, which may be a point, a vector, or a line segment, to create a new point, vector, or line segment. Note that, by definition,

   $$TRANSFORM(LINE(P, Q), M) = LINE(TRANSFORM(P, M), TRANSFORM(Q, M)).$$

   b. *TRANSFORM((X_1, \ldots, X_n), (M_1, \ldots, M_p))*—applies each affine transformation $M_j$ to each object $X_i$ to create the new collection of objects $\{TRANSFORM(X_i, M_j)\}$.

*Affine Transformations*

1. Vectors

   a. *ROT(θ)*—rotates vectors by the angle $\theta$.

   b. *SCALE(s)*—scales the length of vectors by the scale factor $s$.

2. Points

   a. *TRANS(v)*—translates points by the vector $v$.

   b. *ROT(Q, θ)*—rotates points about the fixed point $Q$ by the angle $\theta$.

   c. *SCALE(Q, s)*—scales the distance of points from the fixed point $Q$ by the factor $s$.

   d. *SCALE(Q, w, s)*—scales the distance of points from the fixed point $Q$ in the direction $w$ by the scale factor $s$.

   e. *AFFINETRANS(P$_1$, P$_2$, P$_3$; Q$_1$, Q$_2$, Q$_3$)*—creates the unique affine transformation that maps $P_k \to Q_k$, $k = 1, 2, 3$.

   f. *COMPOSE(M$_1$, M$_2$)*—creates the composite transformation of $M_1$ and $M_2$. We often write $M_1 * M_2$ instead of *Compose(M$_1$, M$_2$)*.

   g. *INVERT(M)*—creates the inverse of the transformation $M$. We often write $M^{-1}$ instead of *INVERT(M)*.

*Rendering*

   *DISPLAY(X$_1$, ..., X$_n$)*—displays the objects $X_1, \ldots, X_n$. The objects displayed may be only either points or lines.

   Notice that CODO does not include a full suite of vector operations. In CODO we can perform scalar multiplication on vectors—$cv = TRANSFORM(v, SCALE(c))$—but there are no commands to compute $u \pm v$, $u \cdot v$, or $\det(u, v)$. The emphasis in CODO is on affine transformations, not on vector operations. Nevertheless, these vector operations could easily be added as a simple enhancement to the version of CODO presented here.

   The commands to create geometry can be used to generate new points, lines, and vectors from existing points, lines, and vectors. But we still need some initial geometry to get started. We begin by fixing a single point $C$ and a single vector *ivec*. In affine coordinates, we set $C = (0, 0, 1)$ and $ivec = (1, 0, 0)$. Thus $C$ is located at the origin of the underlying coordinate system, and *ivec* is the unit vector pointing along the positive $x$-axis.

   We can generate a unit vector along the $y$-axis by the command $TRANSFORM(ivec, ROT(\pi/2))$, and we can generate a new point along the $x$-axis by the command $TRANSFORM(C, TRANS(ivec))$. In Section 5.2.2 we will show how to generate lots of interesting shapes starting from this modest beginning.

## 5.2.2   Sample CODO Programs

   We begin by writing a CODO program to generate the vertices of a polygon. To find the vertices of an $n$-sided polygon, we simply rotate one vertex around the center of the polygon $n$ times through an angle of $2\pi/n$.

*POLYVERTS (Center, Vertex, Number of Vertices)*
   *α = 2π/Number of Vertices*
   *For k = 0, Number of Vertices*
      *PolyVerts [k] = TRANSFORM(Vertex, ROT(Center, kα))*

Another program **STARVERTS** generates the vertices of a star in much the same way that the program **POLYVERTS** generates the vertices of a polygon, except that the command $\alpha = 2\pi/Number\ of\ Vertices$ is replaced by the command $\alpha = 4\pi/Number\ of\ Vertices$.

The program *LINE(POLYVERTS(P, Q, n))* generates the edges of an *n*-sided polygon with center at *P* and a vertex at *Q*; the program DISPLAY(*LINE(POLYVERTS(P, Q, n))*) displays the edges of this polygon. To generate a circle, simply generate a polygon with a large number of sides; to construct an ellipse, apply nonuniform scaling to the circle (see Figure 5.1). Notice how in CODO nonuniform scaling facilitates the construction of the ellipse, a simple curve that is not so easy to generate in LOGO.

Rosettes and wheels are somewhat cumbersome to generate in LOGO, but rosettes and wheels are easy to create in CODO (see Figure 5.2). The lines of the rosette (the diagonals of a polygon as well as the edges) are created by cycling through all pairs of vertices. Notice that in CODO, unlike LOGO, we do not need to know the lengths of the diagonals or the angles between adjacent diagonals to generate the diagonals of a polygon; all we need to know are the vertices of the polygon. Similarly, the spokes for a wheel are generated by cycling through the vertices of the polygon; we do not need to calculate the length of the circumradius. The CODO programs for the lines of a rosette and the spokes of a wheel are presented in Table 5.1.

We can also implement the operators SHIFT, SPIN, and SCALE inside of CODO.

*SHIFT (X, w, n)*
    *For i* = 0, *n*
        *Shift*[*i*] = *TRANSFORM(X, TRANS(TRANSFORM(w, SCALE (i))))*
*(Here n is the number of times we shift the object X in the direction w.)*

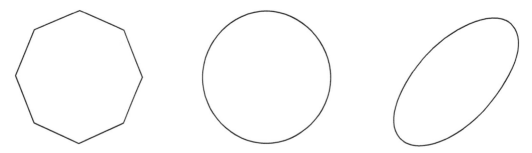

**FIGURE 5.1:**    A regular octagon, a circle, and an ellipse. For the octagon and the circle, the center is at *C* and one vertex is located at *TRANSFORM(C, TRANS(ivec))*. The ellipse is generated by scaling the circle nonuniformly by 0.5 along the minor axis of the ellipse.

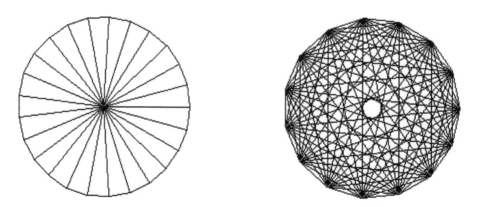

**FIGURE 5.2:**    A wheel with 25 sides and a rosette with 15 sides. The CODO programs for the spokes of the wheel and the lines of the rosette are simple and are presented in Table 5.1.

**TABLE 5.1:** The CODO programs for the lines of a rosette and the spokes of a wheel with vertices $V_1, \ldots, V_n$ and center at $P$.

| *ROSETTE* $(V_1, \ldots, V_n)$ | *SPOKES* $(P, V_1, \ldots, V_n)$ |
|---|---|
| *For* $i = 1, n-1$ | *FOR* $i = 1, n$ |
| *For* $j = i+1, n$ | *Spokes*$[i] = LINE(P, V_i)$ |
| *Rosette*$[i, j] = LINE(V_i, V_j)$ | |

 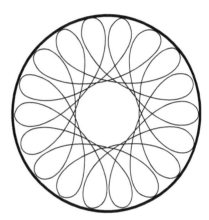

**FIGURE 5.3:** The spin operator applied to the ellipse: Spinning the ellipse around an end point of the major axis (left), and spinning the ellipse around a point outside the ellipse along the minor axis (right).

*SPIN* $(X, Q, \alpha, n)$
    For $i = 0, n$
       *Spin*$[i] = TRANSFORM (X, ROT (Q, i\alpha))$
*(Here n is the number of times we spin the object X around the point Q by the angle $\alpha$.)*

*SCALE* $(X, Q, s, n)$
    For $i = 0, n$
       *Scale*$[i] = TRANSFORM (X, SCALE (Q, s^i))$
*(Here n is the number of times we scale the object X about the point Q by the scale factor s.)*

    Examples of the SPIN operator applied to an ellipse are illustrated in Figure 5.3.

    Thus all the simple shapes we can draw using Turtle Graphics, we can easily recreate using Affine Graphics. In Figure 5.4, we illustrate some additional shapes that are easy to construct in CODO, but are less straightforward to generate in LOGO. In Chapter 6, we shall show how to generate fractals using the affine transformations in CODO.

## 5.3   Summary

    Turtle Graphics is a simple, but powerful, tool for rendering curves. Nevertheless, the lack of memory in LOGO and the absence in the turtle of any knowledge about the outside world make LOGO somewhat cumbersome for generating certain routine geometric shapes.

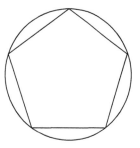 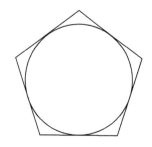 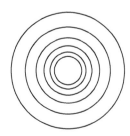

**FIGURE 5.4:** Three shapes involving circles that are easy to construct in CODO, but not so straightforward to generate in LOGO: A pentagon inscribed in a circle (left), a pentagon circumscribed about a circle (center), and a collection of concentric circles (right). We leave it as a simple exercise for the reader to write CODO programs to generate each of these simple shapes (see Exercise 5.5).

Affine Graphics provides an alternative to Turtle Graphics. CODO—the connect-the-dots language embodying Affine Graphics—has both memory and knowledge of how different geometric objects are related in the plane. These features facilitate simple CODO programs for generating shapes that often require more complicated LOGO programs.

Turtle geometry is a local, intrinsic geometry; affine geometry a global, extrinsic geometry. Turtle geometry is based on conformal maps; affine geometry on general affine transformations. Turtle geometry is implemented in rectangular coordinates; affine geometry in affine coordinates. These properties as well as other contrasting features of turtle geometry and affine geometry are summarized in Table 5.2.

Though coordinates are used to implement both turtle geometry and affine geometry, coordinates are not available to LOGO or CODO programmers. Geometry is generated in LOGO by manipulating the motion of the turtle using the commands FORWARD, MOVE, TURN, and RESIZE. These commands represent the conformal transformations translation, rotation, and uniform scaling. Geometry is built in CODO by applying directly to already existing objects the affine transformations translation, rotation, and scaling (uniform and nonuniform), as well as general affine transformations generated from the image of three noncollinear points. The complete LOGO and CODO languages are listed in Table 5.3.

**TABLE 5.2:** Comparison of turtle geometry and affine geometry.

| Turtle Geometry | Affine Geometry |
| --- | --- |
| Local | Global |
| Intrinsic | Extrinsic |
| Order dependent | Order independent |
| No memory | Memory |
| State | Coordinate system |
| Rectangular coordinates | Affine coordinates |
| Conformal transformations | Affine transformations |
| Conformal geometry | Affine geometry |

**TABLE 5.3:** Comparison of the LOGO and CODO languages.

| LOGO | CODO |
|------|------|
| FORWARD/MOVE | TRANS |
| TURN | ROT |
| RESIZE | SCALE |
| | AFFINETRANS |
| | COMPOSE |
| | INVERT |
| | VECTOR |
| | LINE |
| | TRANSFORM |
| | DISPLAY |

## Exercises

5.1. Write a CODO program SPIRALVERTS to generate the vertices of a spiral, where:

$\alpha =$ the angle between the spiral center and adjacent vertices on the spiral arms
$s =$ the ratio of the distances from the spiral center to adjacent vertices on the spiral arms
$Q =$ the center of the spiral
$n =$ the number of spiral vertices
(see Figure 5.5).

5.2. The goal of this exercise is to demonstrate that the spirals generated by the turtle program SPIRAL in Chapters 1 and 2 are identical to the spirals generated by the CODO program SPIRALVERTS in Exercise 5.1. (Both are discrete versions of the logarithmic spiral—see Exercise 5.3.)

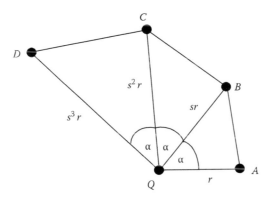

**FIGURE 5.5:** A CODO spiral with center $Q$ and initial vertices $A, B, C, D$. The scalar $s$ is the ratio of the distance from the spiral center to adjacent points on the spiral arms.

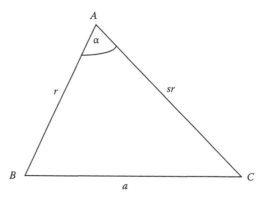

**FIGURE 5.6:** An angle $A$, the length $a$ of the side opposite to the angle $A$, and the ratio $s$ of the two sides adjacent to the angle $A$ determine a unique triangle.

    a. Show that an angle $A$, the length a of the side $BC$ opposite to $A$, and the ratio $s$ between the lengths of the adjacent sides $AB$ and $AC$ containing the angle $A$ (see Figure 5.6) determine a unique triangle $\Delta ABC$ by using the Law of Cosines to find the length $r$ of the side $AB$.

    b. Using the result of part a, show that three points $A$, $B$, $C$ determine a unique CODO spiral by showing that:

        i. the center $Q$ of the spiral is uniquely determined by the points $A$, $B$, $C$ (see Figure 5.7).

        ii. $\alpha = \pi - \angle ABC$ is the angle between the spiral center and adjacent points on the spiral arms.

        iii. $s = \frac{|BC|}{|AB|}$ is the ratio of the distances from the spiral center to adjacent points on the spiral arms.

    (Hint: Show that $\Delta QBC$ is similar to $\Delta QAB$ and that $|QC| = s^2 |QA|$—see Figure 5.7.)

    c. Using the result of part b, show that the spirals generated by the turtle program SPIRAL $(d, \alpha, s)$ are identical to the spirals generated by the CODO program SPIRALVERTS where:

    • $\alpha$ is the angle between the spiral center and adjacent points on the spiral arms,

    • $s$ is the ratio of the distances from the spiral center to adjacent points on the spiral arms,

    • $d$ is the distance between the initial two vertices of the spiral,

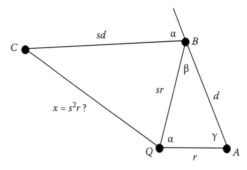

**FIGURE 5.7:** The three points $A$, $B$, $C$ determine a unique CODO spiral. The scalar $d$ is the distance between the two initial vertices of the spiral, and the scalar $s$ is the ratio of the distance from the spiral center to adjacent points on the spiral arms.

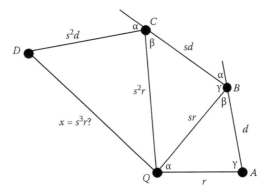

**FIGURE 5.8:** The first four points $A$, $B$, $C$, $D$ generated by the turtle program SPIRAL $(d, \alpha, s)$.

by showing that if $A$, $B$, $C$, $D$ are the first four turtle positions, then:

   i. $\triangle QCD$ is similar to $\triangle QBC$

  ii. $|QD| = s^3 r$
(see Figure 5.8).

5.3 Consider the CODO program SPIRALVERTS in Exercise 5.1. Suppose that the center $Q$ is at the origin and that the initial point on the spiral is located at a unit distance along the $x$-axis.

  a. Show that in this case the $k$th vertex of the spiral lies at the point with coordinates $(s^k \cos(k\alpha), s^k \sin(k\alpha))$.

  b. Conclude that in this case all the vertices of the spiral generated by SPIRALVERTS lie on the logarithmic spiral defined by the parametric equations:

$$x = s^{t/\alpha} \cos(t),$$
$$y = s^{t/\alpha} \sin(t).$$

5.4. Write a CODO program to generate each of the shapes illustrated in Figure 5.3.

5.5. Write a CODO program to generate each of the shapes illustrated in Figure 5.4.

5.6. Write a CODO program to generate the vertices of the same star as the TRISTAR turtle program in Chapter 1, Exercise 1.6.

5.7. Write a CODO program to generate each of the shapes illustrated in Chapter 1, Figure 1.9.

5.8. Write a CODO program to generate each of the shapes illustrated in Chapter 1, Figure 1.10.

5.9. What are the rectangular coordinates of the point generated by the CODO command *TRANS-FORM(C, M₁ * M₂)*, where

$$M_1 = TRANS(TRANSFORM(ivec, SCALE(x))),$$
$$M_2 = TRANS(TRANSFORM(ivec, ROT(\pi/2) * SCALE(y))).$$

# Chapter 6

## Fractals from Iterated Function Systems

*He draweth also the mighty with his power*

— Job 24:22

### 6.1   Generating Fractals by Iterating Transformations

The Sierpinski gasket and the Koch snowflake can both be generated in LOGO using recursive turtle programs. But in CODO there is no recursion. How then can we generate fractals in CODO?

Consider the first few levels of the Sierpinski gasket depicted in Figure 6.1. We begin with a triangle. The next level contains three smaller triangles. (Ignore the central upside-down triangle.) Each of these smaller triangles is a scaled down version of the original triangle. If we denote the vertices of the original triangle by $P_1, P_2, P_3$, then we can generate the three small triangles by scaling by 1/2 around each of the points $P_1, P_2, P_3$ —that is, by applying the three transformations $Scale(P_1, 1/2)$, $Scale(P_2, 1/2)$, and $Scale(P_3, 1/2)$ to $\Delta P_1 P_2 P_3$. Now the key observation is that to go from level 1 to level 2, we can apply these same three transformations to the triangles at level 1. Scaling by 1/2 at any corner of the original triangle maps the three triangles at level 1 to the three smaller triangles at the same corner of level 2. We can go from level 2 to level 3 in the same way, by applying the transformations $Scale(P_1, 1/2)$, $Scale(P_2, 1/2)$, and $Scale(P_3, 1/2)$ to the triangles in level 2. And so on and so on. Thus, starting with $\Delta P_1 P_2 P_3$ and iterating the transformations $Scale(P_1, 1/2)$, $Scale(P_2, 1/2)$, and $Scale(P_3, 1/2)$ $n$ times generates the $n$th level of the Sierpinski gasket.

Let us try another example. Consider the first few levels of the Koch curve illustrated in Figure 6.2. Here we begin with a triangular bump. The next level contains a small triangular bump on each line of the original bump. Each of these smaller bumps is a scaled down version of the original bump. Thus, there are four scaling transformations that map the original bump at level 0 to the four smaller bumps at level 1. (Explicit expressions for these four transformations are provided in Example 6.1.) Once again the key observation is that to go from level 1 to level 2, we can apply these same four transformations. Now we can go from level 2 to level 3 in the same way; and on and on. Thus, starting with a triangular bump and iterating four scaling transformations $n$ times generates the $n$th level of the Koch curve.

Evidently then, if we know the first few levels of a fractal, we can discover transformations that iteratively generate the fractal by finding transformations that generate level 1 from level 0 and then checking that these transformations also generate level 2 from level 1. But suppose we encounter a fractal with which we are not so familiar such as one of the fractals shown in Figure 6.3. If we have not seen these fractals before, we may not know what the first few levels should look like. We could, of course, write a turtle program to generate these fractals and then observe the first few levels of the turtle program, but this approach is rather tortuous. Fortunately, there is a better, more direct approach.

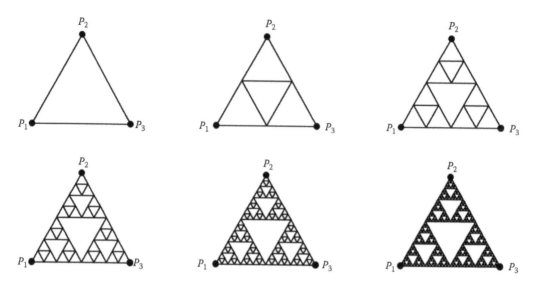

**FIGURE 6.1:** Levels 0–5 of the Sierpinski gasket, generated by scaling by $1/2$ from each of the vertices of $\Delta P_1 P_2 P_3$.

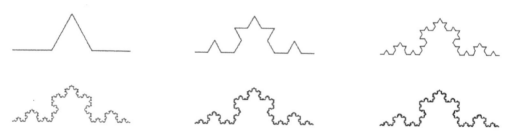

**FIGURE 6.2:** Levels 0–5 of the Koch curve, generated by the four affine transformations that replace each line segment of the original bump by a scaled down version of the original bump.

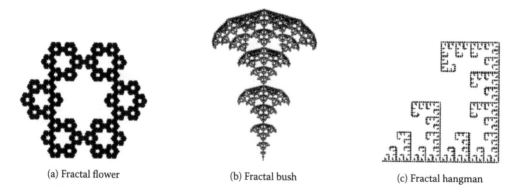

(a) Fractal flower    (b) Fractal bush    (c) Fractal hangman

**FIGURE 6.3:** Three fractals for which the first few levels are not evident: (a) a fractal flower, (b) a fractal bush, and (c) a fractal hangman.

## 6.2 Fractals as Fixed Points of Iterated Function Systems

There is a way to find the transformations that generate a given fractal without knowing the first few levels of the fractal. The Sierpinski gasket consists of three scaled down versions of the Sierpinski gasket. The transformations $Scale(P_1, 1/2)$, $Scale(P_2, 1/2)$, and $Scale(P_3, 1/2)$ that generate the Sierpinski gasket map the Sierpinski gasket into these three scaled down copies. Similarly, the Koch curve consists of four scaled down versions of the Koch curve. The four transformations that map the $n$th level of the Koch curve to the $(n + 1)$st level of the Koch curve also map the original Koch curve into these four scaled down copies.

These observations are not a coincidence. Suppose we have a collection of transformations that for all $n$ map the $n$th level of a fractal to the $(n + 1)$st level. Then in the limit as $n$ goes to infinity, these transformations would necessarily map the entire fractal onto itself. The converse also turns out to be true. If we can find a set of transformations that map a fractal onto itself as a collection of smaller self-similar copies, then iterating these transformations necessarily generates the fractal. *Note that the requirement of smaller self-similar copies excludes the identity transformation.*

A set of transformations that generates a fractal by iteration is called an *iterated function system* (IFS). An iterated function system maps the corresponding fractal onto itself as a collection of smaller self-similar copies. Fractals are often defined as fixed points of iterated function systems because when applied to the fractal the transformations that generate a fractal do not alter the fractal.

The iterated function systems that generate the fractals in Figure 6.3 are not difficult to construct. The fractal flower consists of six smaller self-similar parts, so six transformations are required to generate this fractal. The fractal bush has four smaller self-similar parts, so four transformations are required to generate this fractal. Finally, the fractal hangman consists of three smaller self-similar parts, so three transformations are needed to generate this fractal. We leave it as a straightforward exercise for the reader to find explicit expressions for these transformations. The required transformations are all affine, and can be constructed either by composing translation, rotation, and uniform scaling or by specifying the images of three noncollinear points in the plane (see Exercise 6.2).

### Example 6.1   The Koch Curve

We shall illustrate two ways to construct the four affine transformations $\{M_1, M_2, M_3, M_4\}$ for the IFS that generates the Koch curve. For the labeling of the points on the Koch curve, refer to Figure 6.4.

Method 1: Building each transformation by composing translation, rotation, and uniform scaling.

$$M_1 = Scale(A, 1/3),$$
$$M_2 = Scale(A, 1/3) * Trans(Vector(A, D)) * Rot(D, \pi/3),$$
$$M_3 = Scale(C, 1/3) * Trans(Vector(C, E)) * Rot(E, -\pi/3),$$
$$M_4 = Scale(C, 1/3).$$

Method 2: Building each transformation by specifying the image of three noncollinear points.

$$M_1 = AffineTrans(A, B, C; A, F, D),$$
$$M_2 = AffineTrans(A, B, C; D, H, B),$$
$$M_3 = AffineTrans(A, B, C; B, I, E),$$
$$M_4 = AffineTrans(A, B, C; E, G, C).$$

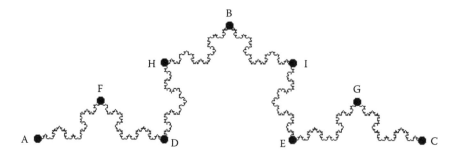

**FIGURE 6.4:** The Koch curve. Key points along the curve are labeled to help specify the four transformations in the IFS that generates this curve.

Notice that in this example, both methods generate the same set of transformations. The first method may seem more natural, but the second method is often much simpler to apply. For yet another approach to deriving these same transformations, see Exercise 6.6.

## 6.3    Fractals as Attractors

We have explained how to find an IFS to generate a given fractal curve, but we still do not know on what shape to start the iteration. That is, we still do not know how, in general, to find level 0 of a given fractal curve. For the Sierpinski triangle, we know to begin with an ordinary triangle, and for the Koch curve, we know to begin with a triangular bump. But how do we begin for fractals with which we are not so familiar such as the fractals shown in Figure 6.3?

Fractals generated by recursive turtle programs are independent of the base case. Could the same independence property hold for fractals generated by iterated function systems? Yes! Figure 6.5 illustrates this independence for the Sierpinski gasket. The same IFS is applied to three different base cases: a square, a horizontal line, and a single point. In each case, after a few iterations, the figures all converge to the Sierpinski gasket.

*Fractals are attractors.* No matter what geometry we choose for the initial shape, the subsequent shapes are inexorably attracted to the fractal generated by the IFS. We shall explain the reason for this attraction in the next chapter. For now, it is enough to know that it does not matter how we choose the initial shape. The base case is irrelevant. If we choose the right IFS, the fractal we want will magically emerge no matter what geometry we choose at the start.

## 6.4    Fractals with Condensation Sets

There is another type of fractal that is relatively common: fractals with condensation sets. Consider the fractals in Figure 6.6. These fractals are not quite self-similar. The fractal staircase contains two smaller fractal staircases, but the largest square in the figure does not belong to either one of the two smaller staircases. Similarly, the fractal tree contains two smaller trees, but the trunk of the tree does not belong to either of these smaller trees. The square in the fractal staircase and the trunk of the fractal tree are called *condensation sets*. If a fractal consists of self-similar parts together

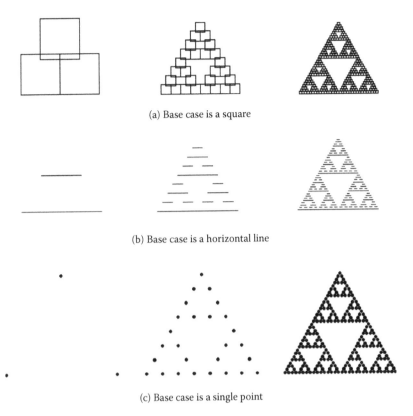

(a) Base case is a square

(b) Base case is a horizontal line

(c) Base case is a single point

**FIGURE 6.5:** Levels 1, 3, and 5 of the Sierpinski gasket: (a) the base case is a square, (b) the base case is a horizontal line, and (c) the base case is a single point.

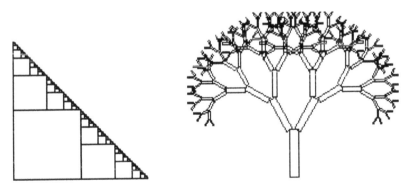

**FIGURE 6.6:** Fractals with condensation sets: a fractal staircase (left) and a fractal tree (right).

with an additional set, the additional set is called a *condensation set*. Thus, in general, to generate a fractal, we need to specify three things: a set of transformations (an IFS), a base case (at which to start the iteration), and a condensation set (possibly the empty set).

There are actually two ways to deal with condensation sets. One approach is to take the base case to be the condensation set and include the identity transformation in the IFS. For the fractal staircase, the IFS consists of the two transformations that scale by 1/2 about the top and bottom of the staircase. If we add the identity to this IFS and take the large square as our base case, then

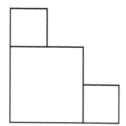 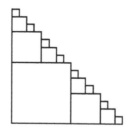 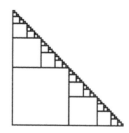

**FIGURE 6.7:**   Levels 1, 3, and 5 of the fractal staircase. The base case is the large square, which is also the condensation set. Here the identity transformation is included in the IFS.

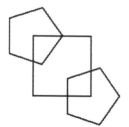 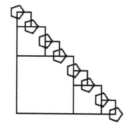 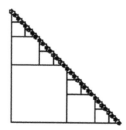

**FIGURE 6.8:**   Levels 1, 3, and 5 of the fractal staircase. The base case is a pentagon and the condensation set is the large square. Notice that the pentagons converge to the diagonal of the staircase, and that the square generates the main part of the fractal. Here the identity transformation is not included in the IFS, but at each level after applying the transformations in the IFS, the large square is added back into the curve.

iterating the IFS will generate a sequence of smaller and smaller squares that converge to the fractal staircase. Notice that because of the identity function in the IFS, the large square is included at every level of the iteration (see Figure 6.7).

The advantage of this technique is that it is simple to implement, but the disadvantage is that the identity function must be applied at every level to every square. Moreover, in this approach, the fractal is no longer independent of the base case. An alternative method is simply to add the condensation set back into the curve at every stage after applying the transformations. This approach has the advantage that we do not waste time redundantly applying the identity function at every level to every square. We can also use any base case; we are no longer restricted to starting with the condensation set. This approach is illustrated for the fractal staircase in Figure 6.8.

## 6.5   Summary

*Fractals are fixed points of iterated function systems.* This observation allows us to find transformations that generate a given fractal curve by finding a set of transformations that map the fractal onto itself as the union of a collection of smaller self-similar copies.

*Fractals are attractors.* No matter what base case we choose to initiate the iteration for a fixed IFS, the shapes we generate will automatically converge to the identical fractal.

*Fractals may have condensation sets.* We can incorporate these condensation sets into our scheme for generating fractals from an IFS by adding the condensation set back into the curve at every stage of the iteration after applying the transformations in the IFS.

---

## Exercises

6.1. Find an IFS to generate a polygonal gasket with $n$ sides.

6.2. Find an IFS to generate each of the fractals in Figure 6.3 by:

    a. Composing translations, rotations, and uniform scalings.

    b. Specifying the images of three noncollinear points in the plane.

6.3. Find an IFS to generate each of the fractals in Chapter 2, Figure 2.8.

6.4. Find an IFS to generate each of the fractals in Chapter 2, Figure 2.16.

6.5. Find explicit expressions for the IFS and the condensation set of the fractal tree in Figure 6.6.

6.6. Consider the Koch curve in Figure 6.4. Let $M$ denote the point midway between $D$ and $E$, and let $v = E - D$.

    a. Show that the Koch curve is generated by the IFS consisting of the four transformations:

$$T_1 = Scale(M, 1/3) * Trans(v), \quad T_2 = Scale(M, 1/3) * Trans(-v),$$
$$T_3 = Scale(M, 1/3) * Rot(D, \pi/3), \quad T_4 = Scale(M, 1/3) * Rot(E, -\pi/3).$$

    b. Verify that these transformations are identical to the transformations in Example 6.1.

6.7. Prove that for any self-similar fractal generated by an IFS,

$$Fractal\ dimension = -\frac{Log\ (Number\ of\ transformations\ in\ the\ IFS)}{Log\ (Scale\ factor)}.$$

6.8. Using the result of Exercise 6.7, compute the fractal dimension of the fractals in Figure 6.3.

6.9. The *Cantor set* is generated by starting with the unit interval and recursively removing the middle third of every line segment.

    a. Show that the Cantor set is the fractal generated by the IFS consisting of the two conformal transformations $Scale(Origin, 1/3)$ and $Scale(P_1, 1/3)$, where $P_1 = (1, 0)$.

    b. Using part a and the result in Exercise 6.7, show that $0 < Dimension(\text{Cantor set}) < 1$.

    c. Show that the length of the Cantor set is zero.

    d. Consider the set generated by starting with the unit interval and recursively removing the middle half of every line segment.

        i. Find the transformations in the IFS that generates this fractal.

        ii. Using the result in Exercise 6.7, show that the dimension of this fractal is $1/2$.

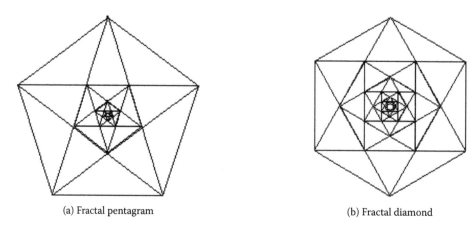

(a) Fractal pentagram                              (b) Fractal diamond

**FIGURE 6.9:**   Two fractals with condensation sets: (a) A fractal pentagram and (b) a fractal diamond.

6.10. Consider the fractal pentagram and the fractal diamond in Figure 6.9.

   a. Find a condensation set and an IFS to generate the fractal pentagram.

   b. Find a condensation set and an IFS to generate the fractal diamond.

6.11. Consider the fractal tower and the fractal star in Figure 6.10.

   a. Find a condensation set and an IFS to generate the fractal tower.

   b. Apply the SPIN operator to the fractal tower to generate the fractal star.

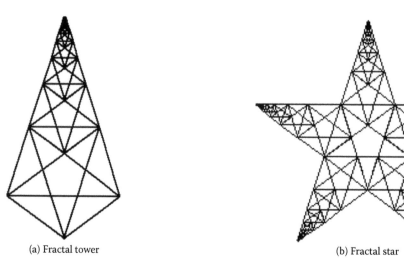

(a) Fractal tower                              (b) Fractal star

**FIGURE 6.10:**   Two fractals: (a) A fractal tower and (b) a fractal star. The size of successive stars in both figures varies by the golden ratio.

## Programming Project

6.1. CODO

Implement CODO in your favorite programming language using your favorite API.

  a. Write CODO programs for the curves discussed in the text and in the exercises for Chapters 5 and 6.

  b. Write a CODO program to design a novel flag for a new country.

  c. Develop some iterated function systems to create your own novel fractals with and without condensation sets.

# Chapter 7

## The Fixed-Point Theorem and Its Consequences

*My heart is fixed...*

– Psalm 57:7

### 7.1 Fixed Points and Iteration

We are now going to solve the central mystery concerning fractals: why is it that the fractals generated both by recursive turtle programs and by iterated function systems are independent of the base case? The heart of the answer lies in the *trivial fixed point theorem*.

A *fixed point* of a function $F$ is a point $P$ such that $F(P) = P$. That is, $P$ is a fixed point of $F$ if $P$ is unchanged by $F$. For example, if $f(x) = x^2$, then $f(0) = 0$ and $f(1) = 1$, so 0 and 1 are fixed points of $f$. We are interested in fixed points of transformations because fractals are fixed points of iterated function systems.

Fixed points are often generated by iteration. Consider the sequence,

$$P_1 = F(P_0),$$
$$P_2 = F^2(P_0) = F(P_1),$$
$$\vdots \qquad \vdots$$
$$P_\infty = F^\infty(P_0),$$

where $F^\infty(P_0)$ denotes the limit of $F^n(P_0) = F(P_{n-1})$ as $n$ approaches infinity (assuming this limit exists). Suppose that we try to apply $F$ to $P_\infty$; informally,

$$F(P_\infty) = F(F^\infty(P_0)) = F^{\infty+1}(P_0) = F^\infty(P_0) = P_\infty.$$

Thus the limit point $P_\infty$ is a fixed point of the function $F$. The point $P_0$ with which the iteration begins does not matter; as long as the iteration converges, iteration converges to a fixed point.

There are two important caveats here. First, there is no guarantee that the iteration must converge. For some starting values—perhaps even for all starting values—the iteration may diverge. Second, even if the iteration always converges, the fixed point need not be unique; different starting values may converge to different fixed points.

In the next section, we are going to establish conditions which guarantee that these two problems never arise. That is, we shall provide conditions which guarantee that the iteration always converges and that the iteration converges to a unique fixed point independent of the initial value. If this result sounds a lot like fractals, this similarity is no accident. Indeed, this general result about iteration is at the heart of our observations that fractals generated from iterated function systems are independent of the base case.

A word of caution before we proceed. Be forewarned: this chapter is by far the most mathematically challenging chapter in this book. We are going to prove a powerful mathematical theorem with applications to many different areas of Pure and Applied Mathematics, including root finding for transcendental functions, relaxation methods for solving large systems of linear equations, solution techniques for ordinary differential equations, and fractals. Therefore, we shall require some general mathematical machinery that we can apply in a wide range of diverse situations. *The power of Mathematics is the ability to reduce many seemingly different problems to their common mathematical essence.* The price we pay for this power is abstraction: a collection of mathematical terms, techniques, and theorems, the significance of which may, at first, be difficult to understand. *Cauchy sequences, complete spaces, compact sets*, and *Haussdorf metrics* will all appear in this discussion. For each of these terms, I will provide some informal intuition as well as a formal definition. For fractals, I will explain only the key ideas behind the proofs of the main theorems, providing references for those readers interested in all the fine details. Understanding this material may be difficult at first, and you may need to read this chapter more than once. It may be hard going, but hang in there; the final results will be well worth all the effort.

---

## 7.2   The Trivial Fixed Point Theorem

We begin with a rigorous formulation of our informal argument in Section 7.1 that if iteration converges, iteration necessarily converges to a fixed point.

**Proposition 7.1:** *Let $T$ be a continuous function and let $P_{n+1} = T(P_n)$ for all $n \geq 0$. If $P_\infty = \text{Lim}_{n \to \infty} P_n$ exists, then $P_\infty$ is a fixed point of T.*

**Proof:** Since, by assumption, $T$ is continuous, we can push $T$ past the limit sign. Therefore,

$$T(P_\infty) = T(\text{Lim}_{n \to \infty} P_n) = \text{Lim}_{n \to \infty} T(P_n) = \text{Lim}_{n \to \infty} P_{n+1} = P_\infty.$$

For our applications, the most important continuous functions are *contractive transformations*. A transformation $T$ is said to be *contractive* if there is a constant $s$, $0 < s < 1$, such that for all points $P, Q$

$$Dist(T(P), T(Q)) \leq s \, Dist(P, Q).$$

That is, contractive maps bring points closer together. The canonical example of a contractive transformation is uniform scaling, where the absolute value of the scale factor is less than one. It follows from Proposition 7.1 that if we iterate a contractive transformation and if the iteration converges, then the iteration necessarily converges to a fixed point. Our next result asserts that for contractive maps, this fixed point is unique.

**Proposition 7.2:** *If T is a contractive map, then T can have at most one fixed point.*

**Proof:** Suppose that $P$ and $Q$ are both fixed points of $T$. Then $T(P) = P$ and $T(Q) = Q$. Therefore,

$$Dist(T(P), T(Q)) = Dist(P, Q).$$

Hence, $T$ is not a contractive map. Contradiction.

Proposition 7.2 guarantees the uniqueness, but not the existence, of fixed points for contractive maps. In order to establish existence, we need to introduce the notion of a *cauchy sequence*.

Informally, a sequence of points $\{P_n\}$ is a *cauchy sequence* if the points get closer and closer as $n$ gets larger and larger. For example, the sequence $\{2^{-n}\}$ is cauchy, but the sequence $\{2^n\}$ is not cauchy. Formally, a sequence of points $\{P_n\}$ is said to be *cauchy* if for every $\varepsilon > 0$ there is an integer $N$ such that

$$Dist(P_{n+m}, P_n) < \varepsilon \text{ for all } n > N \text{ and all } m > 0.$$

Our next result shows that one way to generate a cauchy sequence is by iterating a contractive map.

**Proposition 7.3:** *Suppose that T is a contractive transformation, and that $P_{n+1} = T(P_n)$ for all $n \geq 0$. Then $\{P_n\}$ is a cauchy sequence for any choice of $P_0$.*

**Proof:** Since $T$ is a contractive transformation, there is a constant $s$, $0 < s < 1$, such that

$$Dist(P_{n+1}, P_n) = Dist(T(P_n), T(P_{n-1}))$$
$$\leq s\, Dist(P_n, P_{n-1}) = s\, Dist(T(P_{n-1}), T(P_{n-2}))$$
$$\vdots$$
$$\leq s^n\, Dist(P_1, P_0).$$

Therefore, by the triangle inequality, for $n$ sufficiently large,

$$Dist(P_{n+m+1}, P_n) \leq Dist(P_{n+m+1}, P_{n+m}) + \cdots + Dist(P_{n+1}, P_n)$$
$$\leq (s^{n+m} + \cdots + s^n)Dist(P_1, P_0)$$
$$\leq \frac{s^n}{1-s}Dist(P_1, P_0)$$
$$< \varepsilon.$$

The main fact about cauchy sequences is that for many well-known spaces every cauchy sequences converges to some limiting value. A space in which every cauchy sequence converges is said to be *complete*. Intuitively, a space is *complete* if it has no holes because if the points in a sequence are getting closer and closer to each other, they should be getting closer and closer to some limiting value. The standard Euclidean spaces in $n$ dimensions are examples of complete spaces.

We are now ready to establish the main result of this section: the existence and uniqueness of fixed points for contractive transformations on complete spaces.

**Trivial Fixed-Point Theorem:** *Suppose that:*

    a. *T is a contractive transformation on a complete metric space.*

    b. $P_{n+1} = T(P_n)$ *for all* $n \geq 0$.

*Then* $P_\infty = \text{Lim}_{n \to \infty} P_n$ *exists, and* $P_\infty$ *is the unique fixed point of T for any choice of* $P_0$.

**Proof:** This result follows from our previous propositions because:

1. Since $T$ is a contractive transformation, the sequence $\{P_n\}$ is cauchy (Proposition 7.3).

2. Therefore, since by assumption the space is complete, $\text{Lim}_{n \to \infty} P_n$ exists.

3. Hence, since $T$ is continuous, $P_\infty = \text{Lim}_{n \to \infty} P$ is a fixed point of $T$ (Proposition 7.1).

4. Moreover, since $T$ is a contractive transformation, this fixed point is unique (Proposition 7.2).

## 7.3    Consequences of the Trivial Fixed-Point Theorem

The trivial fixed-point theorem has many important applications in Mathematics and Computer Science. Here we will look at three of these applications: root finding for transcendental functions, relaxation methods for solving large systems of linear equations, and algorithms for generating fractals from iterated function systems.

### 7.3.1    Root Finding Methods

To find a root of a function $F(x)$ means to find a solution of the equation $F(x) = 0$. For low-degree polynomials, such as quadratic equations, there are explicit formulas, such as the quadratic formula, for computing the roots. But for polynomials of degree greater than 4 and for transcendental functions such as trigonometric and exponential functions, simple explicit formulas for the roots do not exist. In these cases, iterative techniques are typically used to locate the roots.

One well-known iterative method that you probably encountered in calculus is Newton's method. To find a solution of the equation $F(x) = 0$ using Newton's method, we first make some initial guess, $x_0$. Once we have produced the $n$th guess, $x_n$, the $(n+1)$st guess is given by

$$x_{n+1} = x_n - \frac{F(x_n)}{F'(x_n)}.$$

The idea behind Newton's method is that the value of $x_{n+1}$ is the location along the $x$-axis where the line tangent to the curve $y = F(x)$ at $x = x_n$ intersects the $x$-axis (see Exercise 7.7). Thus, $x_{n+1}$ is often closer to the root of $F(x) = 0$ than $x_n$ (see Figure 7.1, left). If the initial guess $x_0$ is close to a root of $F(x)$ and if $F'(x_n)$ is not close to zero, then the sequence $x_0, x_1, \ldots$ typically converges rapidly to a solution of the equation $F(x) = 0$. Thus, Newton's method can be summarized in the following fashion.

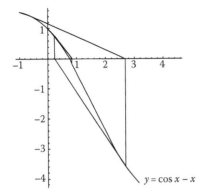
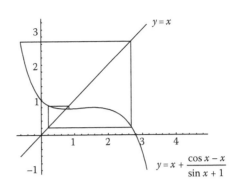

**FIGURE 7.1:** Finding a root of $F(x) = \cos x - x$ by using Newton iteration to find a fixed point of $T(x) = x - \frac{F(x)}{F'(x)} = x + \frac{\cos x - x}{\sin x + 1}$. Here the initial guess is at $x_0 = -0.6$. On the left is the traditional view of Newton's method: the next guess is where the line tangent to the curve $y = F(x) = \cos x - x$ at the current guess intersects the $x$-axis. On the right is the fixed point perspective for Newton's method: here the next guess is where the horizontal line $y = T(x_n)$ at the current guess intersects the line $y = x$. Notice how in both figures the guesses spiral into the root at $x \approx 0.739086$.

*Newton's Method*

1. Select any(!) initial guess $x_0$.

2. Compute $x_{n+1} = x_n - \dfrac{F(x_n)}{F'(x_n)}$ for $n = 1, 2, \ldots$.

3. Stop when $|x_{k+1} - x_k| < \varepsilon$.

Newton's method is actually a fixed-point method. Let

$$T(x) = x - \frac{F(x)}{F'(x)}.$$

Then $F(x^*) = 0$ if and only if $T(x^*) = x^*$. Thus, to find a root of $F(x)$, we seek a fixed point of $T(x)$. Hence, we iterate $T(x)$ on some initial guess $x_0$. But iterating $T(x)$ is precisely the second step of Newton's method. We illustrate Newton's method in Figures 7.1 and 7.2. Notice that Newton's method is not guaranteed to converge (see Figure 7.2) because $T(x)$ need not be a contractive map; moreover, if $F(x)$ has more than one root, different initial guesses may converge to different roots.

Newton's formula for the $(n+1)$st guess $x_{n+1}$ is expensive; we need to compute both the value of the function and the value of the derivative at the current guess $x_n$. An easier approach that requires only evaluation and no differentiation is to use a simpler fixed-point method.

Suppose that we want to find a solution of the equation $F(x) = 0$. Let $G(x) = x + F(x)$. Then $F(x)$ has a root if and only if $G(x)$ has a fixed point—that is, $F(x^*) = 0$ if and only if $G(x^*) = x^*$. But by the *trivial fixed-point theorem*, we can often find a fixed point by iteration. This observation leads to the following simple root finding algorithm.

*Fixed-Point Root Finding Algorithm*

1. Replace $F(x)$ by $G(x) = x + F(x)$.

2. Select any(!) initial guess $x_0$.

3. Compute $x_{k+1} = G(x_k)$ for $k = 1, \ldots, n$.

4. Stop when $|x_{k+1} - x_k| < \varepsilon$.

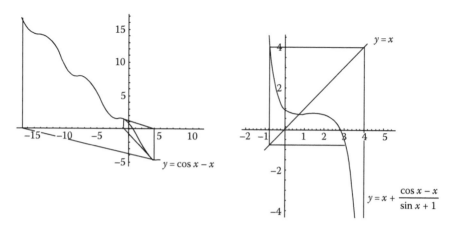

**FIGURE 7.2:**  Again we look at Newton's method for the function $F(x) = \cos x - x$. Here the initial guess is at $x_0 = 3.08$. As in Figure 7.1, on the left is the traditional view of Newton's method; on the right is the fixed-point perspective for Newton's method. Notice how in this case the guesses diverge, spiraling away from the root at $x \approx 0.739086$.

Notice that unlike Newton's method, this algorithm requires only evaluation; no differentiation is necessary. By the *trivial fixed-point theorem*, the fixed-point root finding algorithm is guaranteed to work whenever $G(x)$ is a contractive map. Moreover, any initial guess will do.

How can we tell if a map is contractive? One way is to invoke the mean value theorem:

$$G(b) - G(a) \le G'(c)(b - a) \quad a \le c \le b.$$

By the mean value theorem, if $|G'(c)| < 1$, then $G(x)$ is necessarily a contractive map. For example, $G(x) = \cos x$ is contractive in the interval $[0, a]$ for any $a < \pi/2$, and $G(x) = e^{-x}$ is contractive in the interval $[0, a]$ for any $a > 0$.

We shall illustrate the fixed-point algorithm for finding the roots of two transcendental equations. In Figure 7.3, we find a root of $F(x) = \cos x - x = 0$ by finding a fixed point of $G(x) = x + F(x) = \cos x$;

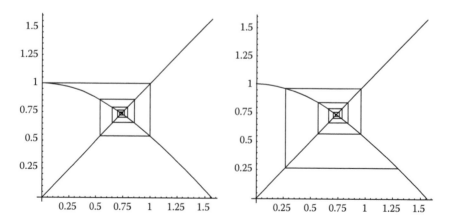

**FIGURE 7.3:**  Finding a root of $F(x) = \cos x - x$ by finding a fixed point of $G(x) = F(x) + x = \cos x$ using iteration. On the left, the initial guess is $x_0 = 0$; on the right, the initial guess is $x_0 = 1.3$. In both cases, the guesses converge to the intersection of the curves $y = \cos x$ and $y = x$—that is, to a fixed point of $\cos x$.

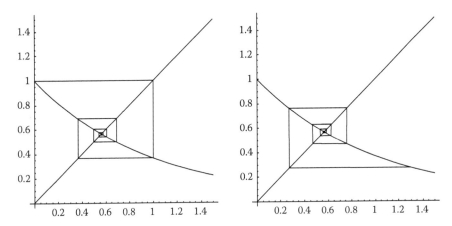

**FIGURE 7.4:** Finding a root of $F(x) = e^{-x} - x$ by finding a fixed point of $G(x) = F(x) + x = e^{-x}$ using iteration. On the left, the initial guess is $x_0 = 0$; on the right, the initial guess is $x_0 = 1.3$. In both cases, the guesses converge to the intersection of the curves $y = e^{-x}$ and $y = x$—that is, to a fixed point of $e^{-x}$.

in Figure 7.4, we find a root of $F(x) = e^{-x} - x = 0$ by finding a fixed point of $G(x) = x + F(x) = e^{-x}$. In both cases, we show that the method converges to the same root, independent of the initial guess. Notice that this fixed-point method often requires more iterations than Newton's method to get close to a root, but the reader should check that this fixed-point method can converge to a root even in cases where Newton's method diverges (see Exercise 7.8).

### 7.3.2 Relaxation Methods

Consider a large system of linear equations:

$$
\begin{aligned}
M_{11}x_1 + M_{12}x_2 + \cdots + M_{1n}x_n &= b_1, \\
M_{21}x_1 + M_{22}x_2 + \cdots + M_{2n}x_n &= b_2, \\
&\;\;\vdots \\
M_{n1}x_1 + M_{n2}x_2 + \cdots + M_{nn}x_n &= b_n.
\end{aligned}
\tag{7.1}
$$

We can rewrite these linear equations in matrix form as

$$
\underbrace{\begin{pmatrix} M_{11} & M_{12} & \cdots & M_{1n} \\ M_{21} & M_{22} & \cdots & M_{2n} \\ \vdots & \vdots & \ddots & \vdots \\ M_{n1} & M_{n2} & \cdots & M_{nn} \end{pmatrix}}_{M} \underbrace{\begin{pmatrix} x_1 \\ x_2 \\ \vdots \\ x_n \end{pmatrix}}_{X} = \underbrace{\begin{pmatrix} b_1 \\ b_2 \\ \vdots \\ b_n \end{pmatrix}}_{B} \Leftrightarrow M * X = B,
\tag{7.2}
$$

and we can solve this linear system for the unknowns $X = (x_1, \ldots, x_n)^{\mathrm{T}}$ using either Gaussian elimination, or Cramer's rule, or simply by inverting the matrix $M$. But when $n$, the number of equations, is large, Cramer's rule and matrix inversion are numerically unstable, and Gaussian elimination is slow; iterative methods often work best. Iterative methods for solving systems of linear equations are called *relaxation methods*. Here we shall look at two such relaxation methods: one due to Jacobi and the other due to Gauss–Seidel.

In *Jacobi relaxation*, the $p$th guess $X^p = (x_1^p, \ldots, x_n^p)^T$ is generated from the $(p-1)$st guess $X^{p-1} = (x_1^{p-1}, \ldots, x_n^{p-1})^T$ by solving the equations

$$
\begin{aligned}
M_{11}x_1^p &= b_1 - M_{12}x_2^{p-1} - \cdots - M_{1n}x_n^{p-1}, \\
M_{22}x_2^p &= b_2 - M_{21}x_1^{p-1} - \cdots - M_{2n}x_n^{p-1}, \\
&\ \ \vdots \qquad\qquad \vdots \\
M_{nn}x_n^p &= b_n - M_{n1}x_1^{p-1} - \cdots - M_{n\,n-1}x_{n-1}^{p-1},
\end{aligned}
\tag{7.3}
$$

which we can write in matrix form as

$$
\underbrace{\begin{pmatrix} M_{11} & 0 & \cdots & 0 \\ 0 & M_{22} & \cdots & 0 \\ \vdots & \vdots & \ddots & \vdots \\ 0 & 0 & \cdots & M_{nn} \end{pmatrix}}_{D}
\underbrace{\begin{pmatrix} x_1^p \\ x_2^p \\ \vdots \\ x_n^p \end{pmatrix}}_{X^p}
=
\underbrace{\begin{pmatrix} b_1 \\ b_2 \\ \vdots \\ b_n \end{pmatrix}}_{B}
-
\underbrace{\begin{pmatrix} 0 & M_{12} & \cdots & M_{1n} \\ M_{21} & 0 & \cdots & M_{2n} \\ \vdots & \vdots & \ddots & \vdots \\ M_{n1} & M_{n2} & \cdots & 0 \end{pmatrix}}_{M-D}
\underbrace{\begin{pmatrix} x_1^{p-1} \\ x_2^{p-1} \\ \vdots \\ x_n^{p-1} \end{pmatrix}}_{X^{p-1}}.
\tag{7.4}
$$

Thus, from Equation 7.3, in Jacobi relaxation,

$$
x_i^p = \frac{b_i}{M_{ii}} - \sum_{j \neq i} \frac{M_{ij}}{M_{ii}} x_j^{p-1}, \quad 1 \leq i \leq n.
\tag{7.5}
$$

On the other hand, in *Gauss–Seidel relaxation*, the $p$th guess $X^p = (x_1^p, \ldots, x_n^p)^T$ is generated from the $(p-1)$st guess $X^{p-1} = (x_1^{p-1}, \ldots, x_n^{p-1})^T$ by solving the diagonal system of equations

$$
\begin{aligned}
M_{11}x_1^p &= b_1 - M_{12}x_2^{p-1} - \cdots - M_{1n}x_n^{p-1}, \\
M_{21}x_1^p + M_{22}x_2^p &= b_2 - M_{23}x_3^{p-1} - \cdots - M_{2n}x_n^{p-1}, \\
&\ \ \vdots \qquad\qquad \vdots \\
M_{n1}x_1^p + M_{n2}x_2^p + \cdots + M_{nn}x_n^p &= b_n,
\end{aligned}
\tag{7.6}
$$

which can be written in matrix form as

$$
\underbrace{\begin{pmatrix} M_{11} & 0 & \cdots & 0 \\ M_{21} & M_{22} & \cdots & 0 \\ \vdots & \vdots & \ddots & \vdots \\ M_{n1} & M_{n2} & \cdots & M_{nn} \end{pmatrix}}_{L}
\underbrace{\begin{pmatrix} x_1^p \\ x_2^p \\ \vdots \\ x_n^p \end{pmatrix}}_{X^p}
=
\underbrace{\begin{pmatrix} b_1 \\ b_2 \\ \vdots \\ b_n \end{pmatrix}}_{B}
-
\underbrace{\begin{pmatrix} 0 & M_{12} & \cdots & M_{1n} \\ 0 & 0 & \cdots & M_{2n} \\ \vdots & \vdots & \ddots & \vdots \\ 0 & 0 & \cdots & 0 \end{pmatrix}}_{M-L}
\underbrace{\begin{pmatrix} x_1^{p-1} \\ x_2^{p-1} \\ \vdots \\ x_n^{p-1} \end{pmatrix}}_{X^{p-1}}.
\tag{7.7}
$$

Thus, from Equation 7.6, in Gauss–Seidel relaxation,

$$
x_i^p = \frac{b_i}{M_{ii}} - \sum_{j=1}^{i-1} \frac{M_{ij}}{M_{ii}} x_j^p - \sum_{j=i+1}^{n} \frac{M_{ij}}{M_{ii}} x_j^{p-1}, \quad 1 \leq i \leq n.
\tag{7.8}
$$

Notice from Equation 7.8 that in Gauss–Seidel relaxation, earlier values of the $p$th guess are used to generate later values of the $p$th guess. For this reason, Gauss–Seidel relaxation is usually somewhat faster than Jacobi relaxation.

Typically, the initial guess for both methods is either $X^0 = 0 = (0, \ldots, 0)^T$ or $X^0 = (b_1, \ldots, b_n)^T$. Both relaxation methods converge for any initial guess, when $M$ is *diagonally dominant*—this is, when

$$|M_{ii}| \geq \sum_{j \neq i} |M_{ij}|, \qquad 1 \leq i \leq n,$$

because both relaxation methods are really fixed-point methods.

To understand why relaxation methods are fixed-point methods, observe that if we multiply Equation 7.4 by $D^{-1}$, then

$$X^p = D^{-1}B - (D^{-1}M - I) * X^{p-1} \qquad \text{(Jacobi relaxation)}. \tag{7.9}$$

Similarly, if we multiply Equation 7.7 by $L^{-1}$, then

$$X^p = L^{-1}B - (L^{-1}M - I) * X^{p-1} \qquad \text{(Gauss–Seidel relaxation)}. \tag{7.10}$$

Let

$$T(X) = Q^{-1}B - (Q^{-1}M - I) * X \tag{7.11}$$

where $Q = D$ for Jacobi relaxation and $Q = L$ for Gauss–Seidel relaxation. Now it is easy to see that the following statements are equivalent:

1. $T(X^*) = X^*$

2. $Q^{-1}B - (Q^{-1}M - I) * X^* = X^*$

3. $B - (M - Q) * X^* = Q * X^*$

4. $B - M * X^* = 0$

5. $M * X^* = B$

Therefore, $X^*$ is a solution of the original system of equations $M * X = B$ if and only if $T(X^*) = X^*$— that is, if and only if $X^*$ is a fixed point of the transformation $T(X)$. Now it turns out that the transformation $T$ is a contractive map whenever $M$ is diagonally dominant and either $Q = D$ is the diagonal part of $M$ (Jacobi relaxation) or $Q = L$ is the lower triangular part of $M$ (Gauss–Seidel relaxation).

We shall use these relaxation methods when we study *radiosity* in Chapter 23. The radiosity equations are a large system of linear equations, so relaxation methods are typically the fastest, most robust methods for solving the radiosity equations.

### 7.3.3 Fractals

Finally fractals. The *trivial fixed-point theorem* asserts that iterating a contractive map on a point from a complete space is guaranteed to converge to the unique fixed point of the transformation independent of the point at which we start the iteration. For fractals, the contractive maps are iterated function systems—collections of contractive transformations on Euclidean space. But, what exactly is the complete space that provides the necessary scaffolding for the *trivial fixed-point theorem*?

### 7.3.3.1　Compact Sets and the Haussdorf Metric

To generate fractals, the points at which the iteration starts are actually sets of points: line segments, polygons, stars, almost any subset of the plane seems to work. However, we do not use unbounded subsets of the plane, since unbounded sets remain unbounded even after applying contractive affine transformations (e.g., uniform scaling), whereas the fractals we want to generate are typically bounded sets of points. Therefore, we shall restrict our attention to compact sets.

A set is *compact* if it is both closed and bounded. *Bounded sets* are sets that lie within a circle of finite radius. Thus, line segments, polygons, and stars are bounded; lines, infinite spirals, and half planes are unbounded. A set is *closed* if the set contains its boundary. A line segment is closed, but if we remove the end points of a line segment, the segment is no longer closed. Similarly, a disk is closed, but if we remove the bounding circle from the disk, then the disk is no longer closed. We shall see shortly why we want to consider sets that are closed as well as bounded.

To study convergence, we need a notion of distance. We say that a sequence $s_0, s_1, \ldots$ converges to a limit $s_\infty$ if and only if the distance between $s_n$ and $s_\infty$ gets smaller and smaller as $n$ gets larger and larger. For fractals, we are dealing with sets. If we want to say that some sequence of sets converges to a fractal, we need some way to measure the distance between two sets. The standard notion of distance for compact sets is called the *Haussdorf metric*.

To define the Haussdorf metric, we first define the distance between a single point $p$ and a compact set $S$ by setting

$$Dist(p, S) = \min_{q \in S}\{Dist(p, q)\}.$$

Now we could try to define the distance between two sets $R$, $S$ by setting

$$Dist(R, S) = \max_{p \in R}\{Dist(p, S)\}.$$

Unfortunately, with this definition, $Dist(R, S) \neq Dist(S, R)$. For example, if $L_1$, $L_2$ are two parallel lines of unequal length, then $Dist(L_1, L_2) \neq Dist(L_2, L_1)$ (see Figure 7.5, left). Similarly, if $R$ is a proper subset of $S$, then $Dist(R, S) = 0$ but $Dist(S, R) > 0$ (see Figure 7.5, right). But clearly, any reasonable notion of distance should be symmetric—that is, for any reasonable notion of distance, we want $Dist(R, S) = Dist(S, R)$.

To fix our definition of distance, we define the *Haussdorf metric* $H(R, S)$ between two compact sets $R$, $S$ by setting

$$H(R, S) = \max\{Dist(R, S), Dist(S, R)\}.$$

**FIGURE 7.5:** *Dist(R, S)* is not symmetric. On the left, two unequal parallel lines $L_1$, $L_2$: $Dist(L_1, L_2) = d$, but $Dist(L_2, L_1) = Dist(B, A) > d$. On the right, two sets $R$, $S$, where $R$ is a proper subset of $S$: $Dist(R, S) = 0$, but $Dist(S, R) > 0$.

Clearly, this definition is symmetric. Moreover, the Haussdorf metric on compact sets has the following properties:

1. $H(R, S) \geq 0$

2. $H(R, S) = 0 \Leftrightarrow R = S$

3. $H(R, S) = H(S, R)$

4. $H(R, T) \leq H(R, S) + H(S, T)$

Properties 1–3 are immediate from the definition. To verify Property 4 is straightforward but somewhat tedious, so we shall not present the proof of Property 4 here. The interested reader can find the proof in Barnsley's book, *Fractals Everywhere* (p. 33).

Note that we need to restrict to closed sets, otherwise Property 2 would be false. For example, the distance between a line segment containing its end points and the same line segment without its end points would be zero, even though the two sets are clearly not identical. If we want the distance function to distinguish between different sets, we cannot have infinitely close sets; therefore, we are forced to restrict our attention to closed sets. Similarly, if we want to consider only finite distances, then we must restrict our attention to bounded sets. For these reasons, we consider only compact sets—that is, sets that are both closed and bounded.

Property 4 is called the *triangle inequality*. We invoked the triangle inequality in the proof of Proposition 7.3 and we used Proposition 7.3 in the proof of the *trivial fixed-point theorem*. Since we are going to invoke the *trivial fixed-point theorem* to prove that fractals generated by iterated function systems are independent of the base case, we need the triangle inequality.

The space that provides the fractal framework for the *trivial fixed-point theorem* is the space of compact sets where distance is measured by the Haussdorf metric. To invoke the *trivial fixed-point theorem*, we need to know that this space is complete.

Recall that a space is *complete* if every cauchy sequence converges. Informally, a cauchy sequence of compact sets is a sequence of compact sets $\{S_n\}$ that get closer and closer as measured by the Haussdorf metric as $n$ gets larger and larger. Formally, a sequence $\{S_n\}$ of compact sets is said to be *cauchy* if for every $\varepsilon > 0$ there is an integer $N$ such that

$$H(S_{n+m}, S_n) < \varepsilon \quad \text{for all } n > N \text{ and all } m > 0.$$

For cauchy sequences of compact sets, we have the following theorem.

**Completeness Theorem:** *Every cauchy sequence of compact sets converges to a compact set.*

The main idea behind the proof of this theorem is that we can find the set $S$ to which a cauchy sequence of compact sets $\{S_n\}$ converges by studying cauchy sequences of points $\{x_n\}$ where $x_n \in S_n$. In fact,

$$x \in S = \text{Lim}_{n \to \infty} S_n \Leftrightarrow x = \text{Lim}_{n \to \infty} x_n, \quad x_n \in S_n,$$

where $x = \text{Lim}_{n \to \infty} x_n$ is guaranteed to exist because ordinary Euclidean space is complete. We omit the gory details of this proof. The interested reader can find these details in Barnsley's book, *Fractals Everywhere* (pp. 35–37).

### 7.3.3.2 Contractive Transformations and Iterated Function Systems

The *trivial fixed-point theorem* is a result about contractive transformations on complete spaces. The space of compact sets with the Haussdorf metric is complete, but what are the contractive transformations on this space? Fractals are generated by iterated function systems, so if we are to invoke the *trivial fixed-point theorem*, iterated function systems must be the contractive transformations that we seek. Let us investigate.

Suppose that $w$ is a map of points in Euclidean space—that is, $w$ maps points to points. Then we can extend $w$ to a map on sets of points $S$ by setting

$$w(S) = \{w(x)|x \in S\}.$$

Now consider a collection of such maps $W = \{w_1, \ldots, w_l\}$. Again we can extend $W$ to a map on sets $S$ by setting

$$W(S) = w_1(S) \cup \cdots \cup w_l(S).$$

When the maps $\{w_1, \ldots, w_l\}$ are contractive, $W$ is called an *iterated function system*. For the *trivial fixed point theorem* to apply, the map $W$ must also be contractive.

**Lemma 7.1:** *If $w_1, \ldots, w_l$ are each contractive transformations on Euclidean space, then $W = \{w_1, \ldots, w_l\}$ is a contractive transformation on the space of compact sets.*

The essential idea behind the proof of this lemma is that since the transformations $w_1, \ldots, w_l$ bring points closer together, the map $W = \{w_1, \ldots, w_l\}$ must bring sets closer together since sets are collections of points. Again the details of the proof are straightforward but somewhat tedious, so we shall not present the proof here. The interested reader can find the proof in Barnsley's book, *Fractals Everywhere* (pp. 79–81).

### 7.3.3.3    Fractal Theorem, Fractal Algorithm, and Fractal Strategy

Now, finally, we have everything we need to invoke the *trivial fixed-point theorem*. The complete space is the set of compact sets with the Haussdorf metric, and the contractive transformations are iterated function systems—collections of contractive transformations on Euclidean space.

**Fractal Theorem:** *Suppose that:*

- $W = \{w_1, \ldots, w_l\}$ *is an iterated function system.*

- $S_0$ *is a compact set.*

- $S_{n+1} = W(S_n)$ *for all $n \geq 0$.*

*Then $S_\infty = \text{Lim}_{n \to \infty} S_n$ exists, and $S_\infty$ is the unique fixed point of $W$ for any choice of $S_0$.*

**Proof:** This result follows immediately from the *trivial fixed-point theorem* because $W$ is a contractive map on a complete space.

*Fractals are fixed points of iterated function systems.* Therefore, given an iterated function system $W = \{w_1, \ldots, w_l\}$, we can generate the associated fractal $A$ using the following algorithm.

*Fractal Algorithm*

- $A_0 = B$ (pick any compact set $B$)

- $A_{n+1} = W(A_n) = w_1(A_n) \cup \cdots \cup w_l(A_n)$

- $A_n$ converges to the fractal (fixed point) $A$

The *fractal algorithm* is a consequence of the *fractal theorem*—really then a consequence of the *trivial fixed-point theorem*. Notice that this fractal algorithm can easily be modified to generate a fractal with a condensation set $C$ simply by setting

$$A_{n+1} = W(A_n) \cup C = w_1(A_n) \cup \cdots \cup w_l(A_n) \cup C.$$

The fractal algorithm generates a fractal from an iterated function system. Conversely, if we know the fractal, but want to find the corresponding iterated function system so that we can generate the fractal, we can proceed in the following manner.

*Fractal Strategy*
    Given a fractal $A$:

- Find a collection of contractive transformations $W = \{w_1, \ldots, w_l\}$ that map the fractal $A$ onto itself—that is, find $W$ so that $W(A) = w_1(A) \cup \cdots \cup w_l(A) = A$.

- Then $A$ is the fractal generated by the iterated function system $W$ starting from any compact set $S$.

Recall that we used this fractal strategy in Chapter 6 to find iterated functions systems for a variety of different fractals.

---

## 7.4 Summary

    The theme of this chapter—the one idea you should carry away for the rest of your life—is that if you want to find a stable point of any process, iterate! If you are lucky—if the process actually converges to some equilibrium—then by iterating you will arrive at an equilibrium point—a fixed point of the process. Moreover, it does not matter how you start the process, just keep iterating. If you are unlucky, the iteration might diverge or there might be more than one equilibrium point. The *trivial fixed point theorem* guarantees that if the space of possible states is complete and the transformation defined by the process is contractive, then the iteration must converge to a unique equilibrium point of the process independent of the initial state of the system.

    We provided three examples of fixed point methods: root finding, relaxation techniques, and fractals. Since the theme of this chapter is that equilibrium values are generated by iteration, we have solved the mystery of why fractals generated by iterated function systems are independent of the base case: contractive transformations always converge to a unique fixed point independent of the starting value.

    Concerning fractals, the main things to take away from this chapter are the following three items, which summarize what we know about fractals generated from iterated function systems:

*Fractal Theorem*
    *Suppose that:*

- $W = \{w_1, \ldots, w_l\}$ *is an iterated function system.*

- $S_0$ *is a compact set.*

- $S_{n+1} = W(S_n)$ *for all $n \geq 0$.*

*Then $S_\infty = \mathrm{Lim}_{n \to \infty} S_n$ exists, and $S_\infty$ is the unique fixed point of $W$ for any choice of $S_0$.*

**TABLE 7.1:** Some analogies between Euclidean space and the fractal space of compact sets.

| Euclidian Space | Fractal Space |
|---|---|
| Points | Compact sets |
| Euclidean distance | Haussdorf metric |
| Completeness | Completeness |
|    Cauchy sequences of points converge |    Cauchy sequences of sets converge |
| Contractive maps | Iterated function systems |
| Fixed points | Fractals |
| Trivial fixed point theorem | Fractal theorem |
| Fixed point algorithm | Fractal algorithm |
|    Start with any point and iterate |    Start with any set and iterate |

*Fractal Algorithm*

- $A_0 = B$ (pick any compact set $B$)

- $A_{n+1} = W(A_n) = w_1(A_n) \cup \cdots \cup w_l(A_n)$

- $A_n$ converges to the fractal (fixed point) $A$ determined by the transformations $W$

*Fractal Strategy*
   Given a fractal $A$:

- Find a collection of contractive transformations $W = \{w_1, \ldots, w_l\}$ that map the fractal $A$ onto itself—that is, find $W$ so that $W(A) = w_1(A) \cup \cdots \cup w_l(A_n) = A$.

- Then $A$ is the fractal generated by the iterated function system $W$ starting from any compact set $S$.

For easy comparison, we have assembled in Table 7.1 some analogies between ordinary Euclidean space and the fractal space of compact sets.

The term *fractal* has two distinct definitions in this book:

1. Recursion made visible.

2. Fixed point of an iterated function system.

Each of these definitions reflects a different aspect of fractal curves. The first captures the spirit of recursive turtle programs; the second captures the essence of contractive transformations. We shall make the connection between iterated function systems and recursive turtle programs more precise in the next chapter. When we understand this connection, we shall finally understand why fractals generated by recursive turtle programs are also independent of the base case of the recursion.

---

## Exercises

7.1. Let $A$ be an affine transformation. Then there is a matrix $M$ and a vector $w$ such that

$$A(\nu) = \nu * M,$$
$$A(P) = P * M + w.$$

a. Show that $A$ has a unique fixed point if and only if $\det(I - M) \neq 0$, where $I$ is the identity matrix.

b. Show that when the condition in part a is satisfied, the unique fixed point of $A$ is given by $P = w * (I - M)^{-1}$.

7.2. Let $A, B$ be two transformations and let $A \circ B$ denote the composite transformation. That is, $(A \circ B)(P) = A(B(P))$. Show that if $A$ and $B$ are contractive transformations, then $A \circ B$ is a contractive transformation.

7.3. A transformation $A$ is said to be a *rigid motion* if for every pair of points $P, Q$

$$Dist(A(P), A(Q)) = Dist(P, Q).$$

Let $A \circ B$ denote the composite of $A$ and $B$ (see Exercise 7.2). Show that:

a. If $A$ and $B$ are rigid motions, then $A \circ B$ is also a rigid motion.

b. If $A$ is a rigid motion and $B$ is a contractive transformation, then $A \circ B$ and $B \circ A$ are both contractive transformations.

7.4. For each of the affine transformations $Trans(w)$, $Rot(Q, \theta)$, $Scale(Q, s)$, $Scale(Q, w, s)$:

a. State whether or not the transformation is a contractive transformation.

b. Find all the fixed points of the transformation.

7.5. What is the result of iterating the function $f(x) = x^2$ on the three initial values: $x_0 = 2$, $x_0 = 1$, and $x_0 = 1/2$.

7.6. Let $T(x) = mx + b$, and suppose that $|m| < 1$. Define $x_{n+1} = T(x_n)$ for all $n \geq 0$. Without appealing to the *trivial fixed-point theorem*, show that:

a. $\mathrm{Lim}_{n \to \infty} x_n$ does not depend on the choice of $x_0$.

b. $\mathrm{Lim}_{n \to \infty} x_n = \dfrac{b}{1 - m}$.

c. $\dfrac{b}{1 - m}$ is a fixed point of $T(x)$ and this fixed point of $T(x)$ is unique.

7.7. Show that the line tangent to the curve $y = F(x)$ at $x = x_n$ intersects the $x$-axis at

$$x_{n+1} = x_n - \frac{F(x_n)}{F'(x_n)}.$$

7.8. Let $F(x) = \cos x - x$. In Figure 7.2, we observed that Newton's method applied to this function diverges for the initial guess $x_0 = 3.08$. Show that the fixed-point method applied to this function converges for the initial value $x_0 = 3.08$.

7.9. For two linear equations in two unknowns:

a. Prove that when the coefficient matrix is diagonally dominant, the transformation matrices associated with the Jacobi and Gauss–Seidel relaxation methods are contractive maps.

b. Conclude that both Jacobi and Gauss–Seidel relaxation always converge to a unique fixed point when the coefficient matrix is diagonally dominant.

7.10. For Gauss–Seidel relaxation:

    a. Show that for two linear equations in two unknowns all the points generated after the initial guess satisfy the second equation—that is, all the points lie on the second line.

    b. Generalize the result in part a to $n$ linear equations in $n$ unknowns by showing that all the points generated after the initial guess satisfy the last equation—that is, all the points lie on the last hyperplane.

7.11. Show that if an iterated function system $W$ consists of a single contractive transformation $w$, then the fractal corresponding to $W$ consists of a single point $P$.

7.12. Let $W = \{w_1, \ldots, w_l\}$ be an iterated function system, and let $F$ be the fractal generated by $W$. Show that if $P_k$ is a fixed point of $w_k$, then $P_k \in F$.

7.13. Show that a fixed point of a contractive affine transformation $w$ is equivalent to an eigenvector of $w$ corresponding to the eigenvalue 1.

7.14. Let $F_i$ be the fractal generated by the iterated function system $W_i$, $i = 1, 2$. Show that if $W_1 \subset W_2$ then $F_1 \subset F_2$.

7.15. Let $F$ be the fractal generated by the iterated function system $W = \{w_1, \ldots, w_l\}$, and let $T$ be a nonsingular affine transformation.

    a. Show that $T(F)$ is the fractal generated by the set of transformations $T^{-1} * W * T = \{T^{-1} * w_1 * T, \ldots, T^{-1} * w_l * T\}$.

    b. Conclude that an affine transformation applied to a fractal generates another fractal.

7.16. Consider two iterated functions systems $F = \{f_1, \ldots, f_m\}$ and $G = \{g_1, \ldots, g_m\}$.

    a. Show that

$$H(t) = (1 - t)F + tG = \{(1 - t)f_1 + tg_1, \ldots, (1 - t)f_m + tg_m\},$$

    is an iterated function system for all $0 \leq t \leq 1$.

    b. Using the result in part a, build an animation that maps:

        i. The Sierpinski gasket into the fractal hangman.

        ii. The Koch curve into a fractal bush.

7.17. Let $f, g$ be continuous functions on the interval $[0, 1]$. Define

$$Dist(f, g) = \max_{0 \leq x \leq 1} |f(x) - g(x)|,$$

$$T(f)(x) = 1 + \int_0^x \int_0^t -f(s)\,ds\,dt.$$

    a. Show that:

        i. $Dist(f, g) \geq 0$.

        ii. $Dist(f, g) = 0 \Leftrightarrow f = g$.

        iii. $Dist(f, g) = Dist(g, f)$.

        iv. $Dist(f, h) \leq Dist(f, g) + Dist(g, h)$.

b. Show that $T(f)$ is a contractive transformation. That is, show that:

$$Dist(T(f), T(g)) \leq Dist(f, g).$$

c. Show that:

$$T(f)(0) = 1 \text{ and } T(f)'(0) = 0.$$

d. Verify that $f(x) = \cos x$ is a fixed point of $T(f)$.

e. Verify that $f(x) = \cos x$ is the unique solution of the ordinary differential equation $f''(x) = -f(x)$ satisfying the initial conditions $f(0) = 1, f'(0) = 0$.

7.18. In this exercise, we shall show how to use iterative fixed-point methods to solve the second-order ordinary differential equation:

$$y'' = a_1(x)y' + a_0(x)y + b(x) \tag{*}$$

subject to the initial conditions

$$y(0) = c_0 \quad \text{and} \quad y'(0) = c_1. \tag{**}$$

Let $f, g$ be continuous functions on the interval $[0, 1]$. Define

$$Dist(f, g) = \max_{0 \leq x \leq 1} |f(x) - g(x)|,$$

$$T(f)(x) = c_0 + c_1 x + \int_0^x \int_0^t (a_1(s)f'(s) + a_0(s)f(s) + b(s))ds \, dt.$$

a. Show that:

    i. $Dist(f, g) \geq 0$.

    ii. $Dist(f, g) = 0 \Leftrightarrow f = g$.

    iii. $Dist(f, g) = Dist(g, f)$.

    iv. $Dist(f, h) \leq Dist(f, g) + Dist(g, h)$.

b. Show that:

    i. $Dist(T(f), T(g)) \leq Dist(f, g)$.

    ii. $T(f)(0) = c_0$ and $T(f)'(0) = c_1$.

    iii. $T(f)(x) = f \Leftrightarrow f''(x) = a_1(x) f'(x) + a_0(x) f(x) + b(x)$.

c. Consider the sequence of continuous functions defined by setting:

- $f_0 = g$.

- $f_{n+1} = T(f_n)$.

By invoking the *trivial fixed-point theorem*, explain why $\{f_n\}$ converges to a solution of the differential equation (*) that satisfies the initial conditions (**) independent of the choice of $g$.

  d. Use the result of part c to develop an iterative algorithm for solving second-order ordinary differential equations.

  e. Use the method in part d to solve the ordinary differential equation $y'' = -y$ subject to the initial conditions $y(0) = 1$ and $y'(0) = 0$, starting with the initial guess $y = 0$.

---

## Programming Projects

7.1. *Fractal Tennis*

In this project, we provide an alternative approach to rendering fractals $F$ corresponding to iterated function systems $W = \{w_1, \ldots, w_l\}$.

*Fractal Tennis*

  1. Select any point $P_0$ in $F$—we shall explain shortly how to find such a point.

  2. Choose a point $P_{n+1} \in \{w_1(P_n), \ldots, w_l(P_n)\}$—pick a point at random by assigning probabilities to the maps $\{w_1, \ldots, w_l\}$.

  3. Display the points $P = \{P_0, P_1, \ldots\}$—these points form a dense subset of the fractal $F$ generated by the IFS $W$.

There are two ways to select a point $P_0$ in the fractal $F$:

  1. Find an eigenvector corresponding to the eigenvalue 1 for one of the transformations $w_k$ (see Exercises 7.12 and 7.13).

  2. Pick any point $Q$ in the plane and apply Step 2 $n$ times for $n$ sufficiently large.

Note that *Fractal Tennis* is more efficient than the standard *Fractal Algorithm* because only one additional point is computed and stored at each stage of the algorithm.

  a. Implement *Fractal Tennis* in your favorite programming language using your favorite API.

  b. Use *Fractal Tennis* to generate the following fractals:

   i. The Sierpinski gasket and the Koch curve.

   ii. The fractals in Chapter 2, Figure 2.8 left and center.

   iii. The fractals in Chapter 2, Figure 2.11.

   iv. The fractals in Chapter 2, Figure 2.16.

   v. Some novel fractals of your own design.

7.2. *Root Finding*

  a. Implement Newton's method and the fixed-point method in your favorite programming language, using your favorite API to make pictures like those in Figures 7.1 through 7.4 of the resulting iterations.

  b. Compare the speed and accuracy of these two root finding methods for different functions and different initial guesses.

### 7.3. Relaxation Methods

a. Implement the Jacob and Gauss–Seidel relaxation methods in your favorite programming language.

b. Compare the speed and accuracy of these two methods to standard methods such as Gaussian elimination for solving large systems of linear equations.

### 7.4. Second-Order Ordinary Differential Equations

a. Implement in your favorite programming language the iterative fixed-point method described in Exercise 7.18 for solving second-order ordinary differential equations

$$y'' = a_1(x)y' + a_0(x)y + b(x) \qquad (*)$$

subject to the initial conditions

$$y(0) = c_0 \quad \text{and} \quad y'(0) = c_1. \qquad (**)$$

b. Using your favorite API, make pictures of the functions generated by successive iterations of this algorithm.

c. Whenever closed form solutions exist, compare your solutions to these closed form expressions.

# Chapter 8

## Recursive Turtle Programs and Conformal Iterated Function Systems

*convert, and be healed*

– Isaiah 6:10

## 8.1 Motivating Questions

We have studied two methods for generating fractals: recursive turtle programs and iterated function systems. We have also examined many fractal curves that can be generated by both methods. Whenever there are two seemingly dissimilar approaches to accomplishing the same task, questions naturally arise concerning how the two techniques are related. For fractals we ask the following questions:

1. Is it always true that a fractal generated by one method can be generated by other technique?

2. Do there exist algorithms that convert recursive turtle programs into iterated function systems?

3. Do there exist algorithms that convert iterated function systems into recursive turtle programs?

The turtle commands FORWARD, MOVE, TURN, and RESIZE embody the conformal transformations TRANSLATE, ROTATE, and SCALE. Since there exist affine transformations such as nonuniform scaling that are not conformal, there exist fractals that can be generated by iterated function systems that cannot be generated by recursive turtle programs. But as long as we restrict our attention to iterated function systems consisting solely of conformal transformations, then the answer to all three questions is yes.

The purpose of this chapter is to establish these results by investigating the connections between recursive turtle programs and conformal iterated function systems. Using the equivalence of recursive turtle programs and conformal iterated function systems, we shall also establish that the fractals generated by recursive turtle programs are independent of the base case.

## 8.2 The Effect of Changing the Turtle's Initial State

A (non-recursive) turtle program is simply a finite sequence of FORWARD, MOVE, TURN, and RESIZE commands. In this section, we are going to study the effect of changing the turtle's initial state on the geometry generated by a turtle program. The insights we shall develop here are

101

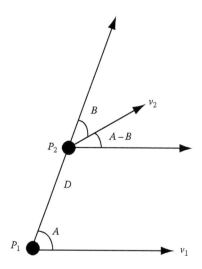

**FIGURE 8.1:**   A pair of turtle states: $S_1 = (P_1, v_1)$ and $S_2 = (P_2, v_2)$. The turtle can get from $S_1$ to $S_2$ by executing the commands: TURN(A), MOVE(D), TURN(−B), RESIZE(R), where $R =$ *Length*$(v_2)$/*Length*$(v_1)$. Similarly, executing consecutively the three conformal transformations— TRANS$(P_2 − P_1)$, ROT$(P_2, A − B)$, SCALE$(P_2, R)$—maps $S_1$ to $S_2$.

actually the keys to understanding the relationship between recursive turtle programs and conformal iterated function systems.

The turtle knows only her state $(P, v)$, where $P$ is her position and $v$ her direction vector. Besides sometimes drawing a line, all that the turtle commands really do is change the state of the turtle. Now consider two possible turtle states: $S_1 = (P_1, v_1)$ and $S_2 = (P_2, v_2)$. Is it always possible for the turtle to get from one state to the other? Sure. In fact, we can see from Figure 8.1 that there is always a sequence of four turtle commands:

TURN(A), MOVE(D), TURN(−B), RESIZE(R)

that maps the turtle from $S_1$ to $S_2$. Similarly, there is a sequence of three conformal transformations consisting of translation, rotation, and uniform scaling that maps $S_1$ to $S_2$.

Now we are going to make two key observations that will help guide us to an understanding of the relationships between recursive turtle programs and conformal iterated function systems.

**Lemma 8.1:** *Suppose that*

   $S_1, S_2 =$ *two turtle states*

   $T_1, T_2 =$ *two turtle programs that differ only by their initial states* $S_1, S_2$

   $G_1, G_2 =$ *geometry generated by* $T_1, T_2$

   $M =$ *conformal transformation that maps* $S_1$ *to* $S_2$

*Then*

$$G_2 = M(G_1).$$

**Proof:** Clearly, if we change the turtle's initial state by, for example, a TURN(A) command, the turtle will still draw exactly the same shape, but rotated by $A$ radians. Similarly, MOVE(D) will translate the shape and RESIZE(R) will scale the shape. But we have just seen that we can get from any turtle state to any other turtle state by a sequence of four such turtle commands. Thus, if $M$ is the affine transformation corresponding to these four turtle commands and $G_1$, $G_2$ are the geometry generated by the turtle programs $T_1$, $T_2$, then $G_2 = M(G_1)$.

The effects of the turtle programs SHIFT, SPIN, and SCALE are examples of Lemma 8.1 in action.

**Lemma 8.2:** *Suppose that*

$S_1 = initial\ turtle\ state$

$T_1 = turtle\ program$

$P = sequence\ of\ MOVE,\ TURN,\ and\ RESIZE\ commands\ (Turtle\ Program)$

$T_2 = (P,\ T_1)\ (do\ P\ first,\ then\ do\ T_1)$

$G_i = geometry\ generated\ by\ T_i\ \ i = 1,\ 2$

*If*

$$P: S_1 \rightarrow S_2,$$
$$M: S_1 \rightarrow S_2\ (conformal\ transformation),$$

*then*

$$G_2 = M(G_1).$$

**Proof:** The only effect of the turtle program $P$ is to change the state of the turtle from state $S_1$ to state $S_2$. Therefore, this result is a consequence of Lemma 8.1.

---

## 8.3  Equivalence Theorems

We are now ready to establish our main results concerning the equivalence of recursive turtle programs and conformal iterated function systems.

**Theorem 8.1:** *For every recursive turtle program, there exists a conformal iterated function system such that starting with the same base case, level for level, the recursive turtle program and the conformal iterated function system generate exactly the same geometry.*

**Proof:** Consider a recursive turtle program:

- Base Case: Turtle Program ($T_B$)

- Recursion:

  *Turtle Commands($T_1$), Turtle Recursion*

  $$\vdots$$

  *Turtle Commands($T_p$), Turtle Recursion*
  *Turtle Commands($T_{p+1}$)*

Let $G_R$ be the geometry drawn by the recursion without the recursive calls—that is, $G_R$ is the geometry drawn by the turtle program $T = T_1, \ldots, T_{p+1}$. Suppose that

$S_0 = $ the initial turtle state;

$S_k = $ the turtle state immediately after executing the commands $T_k$.

Let $M_k$ be the conformal transformation that maps $S_0$ to $S_k$—that is,

$$M_k: S_0 \rightarrow S_k.$$

Define a conformal iterated function system by setting

$$CIFS = \{M_1, \ldots, M_p\},$$

$$\text{Condensation Set} = G_R.$$

We shall show that starting with the same base case, level for level, the recursive turtle program and this conformal iterated function system generate exactly the same geometry.

Let $G_B$ be the geometry drawn in base case—that is, the geometry drawn by $T_B$—and let $L_n$ be the geometry drawn by the $n$th level of the recursive turtle program. The key insight is that at the $(n+1)$st level, every recursive call draws the geometry $L_n$, but starting from a different initial state. Thus, by Lemma 8.2,

$$L_0 = G_B,$$

$$L_{n+1} = M_1(L_n) \cup \cdots \cup M_p(L_n) \cup G_R.$$

On the other hand, let $F_n$ be the geometry drawn by the $n$th iteration of the iterated function system. If we start the iteration with the base case $G_B$, then by the definition of an iterated function system with a condensation set,

$$F_0 = G_B,$$

$$F_{n+1} = M_1(F_n) \cup \cdots \cup M_p(F_n) \cup G_R.$$

Therefore, by induction, level for level, the recursive turtle program and the conformal iterated function system generate exactly the same geometry.

**Corollary 8.1:** *Every fractal generated by a recursive turtle program can be generated by an iterated function system consisting solely of conformal transformations. Moreover, the number of conformal transformations in the conformal iterated function system is equal to the number of recursive calls in the recursive turtle program.*

**Proof:** This result is an immediate consequence of the statement and proof of Theorem 8.1.

**Corollary 8.2:** *The fractal drawn by a recursive turtle program is independent of the geometry drawn in the base case.*

**Proof:** This result follows from Theorem 8.1 because by the Fractal Theorem in Chapter 7, Section 7.3.3.3, the fractal generated by an iterated function system is independent of the geometry in the base case.

**Theorem 8.2:** *For every iterated function system consisting solely of conformal transformations, there exists a recursive turtle program such that starting with the same base case, level for level, the conformal iterated function system and the recursive turtle program generate exactly the same geometry.*

**Proof:** Given a conformal iterated function system

$$CIFS = \{M_1, \ldots, M_p\},$$

$$\text{Condensation Set} = G_R,$$

we seek a recursive turtle program that generates, level for level, the same curves as this iterated function system. Fix an initial turtle state $S_0$, and let $S_k$ be the turtle state into which $S_0$ is mapped by the conformal transformation $M_k$. From our observations in Section 8.2, we know that for every $k$ there is a sequence of turtle commands $T_k$ consisting of MOVE, TURN, and RESIZE commands that map $S_{k-1}$ to $S_k$. We will use these turtle commands to transition between recursive calls. To finish the recursive turtle program, let $T_{p+1}$ be the turtle commands that draw the condensation set $G_R$ and return the turtle to her initial state. For the base case of the recursive turtle program, we can choose any turtle program $T_B$ that draws some geometry $G_B$ and returns the turtle to her initial state $S_0$. We claim that the following recursive turtle program generates, level for level, the same geometry as the given conformal iterated function system:

- Base Case: Turtle Program ($T_B$)

- Recursion:

    *Turtle Commands*($T_1$), *Turtle Recursion*

         ⋮

    *Turtle Commands*($T_p$), *Turtle Recursion*

    *Turtle Commands*($T_{p+1}$)

Let $L_n$ be the geometry drawn by the $n$th level of this recursive turtle program. Again the key insight is that at the $(n+1)$st level, the $k$th recursive call draws the geometry $L_n$, but starting from the initial state $S_k$. Since $M_k: S_0 \rightarrow S_k$, it follows by Lemma 8.2 that

$$L_0 = G_B,$$
$$L_{n+1} = M_1(L_n) \cup \cdots \cup M_p(L_n) \cup G_R.$$

On the other hand, let $F_n$ be the geometry drawn by the $n$th iteration of the conformal iterated function system. If we start the iteration with the base case $G_B$, then by the definition of an iterated function system,

$$F_0 = G_B,$$
$$F_{n+1} = M_1(F_n) \cup \cdots \cup M_p(F_n) \cup G_R.$$

Therefore, by induction, level for level, the recursive turtle program and the conformal iterated function system generate exactly the same geometry.

**Corollary 8.3:** *Every fractal generated by a conformal iterated function system can be generated by a recursive turtle program whose initial and final states are identical both in the base case and in the recursion.*

**Proof:** This result is an immediate consequence of the statement and proof of Theorem 8.2.

**Corollary 8.4:** *Every fractal generated by a recursive turtle program can be generated by a recursive turtle program whose initial and final states are identical both in the base case and in the recursion.*

**Proof:** By Corollary 8.1, we know that every fractal generated by a recursive turtle program can be generated by a conformal iterated function system, and by Corollary 8.3, we know that every fractal generated by a conformal iterated function system can be generated by a recursive turtle program whose initial and final states are identical.

## 8.4   Conversion Algorithms

Although we have just proved that every recursive turtle program is equivalent to a conformal iterated function system, we would like to have an explicit algorithm for finding this conformal iterated function system from a given recursive turtle program. Next we present two such explicit conversion algorithms: Ron's Algorithm and Tao's Algorithm.

### 8.4.1   Ron's Algorithm

Ron's Algorithm takes an exocentric point of view, a perspective from outside the turtle's domain. Suppose that the turtle is in some state $S = (P, v)$. Viewed from outside the turtle, each turtle command corresponds to a conformal transformation:

$$FORWARD\ D \leftrightarrow TRANS(v * SCALE(D)),$$
$$MOVE\ D \leftrightarrow TRANS(v * SCALE(D)),$$
$$TURN\ A \leftrightarrow ROT(P, A),$$
$$RESIZE\ S \leftrightarrow SCALE(P, S).$$

Ron's Algorithm proceeds in the following fashion.

*Ron's Algorithm: Recursive Turtle Program → Conformal Iterated Function System*

1. *Preprocessing*

   a. For each turtle command, compute the corresponding conformal transformation.

   b. If the turtle changes state during the recursion, compute the conformal transformation corresponding to the state change caused by the recursion.

2. *Finding the Conformal Iterated Function System*

   a. $N_k$ = product of the transformations between the $(k-1)$st and $k$th recursive call.

   b. $M_1 = N_1$ = product of the transformations *before* the first recursive call.

c. $M_k = M_{k-1} * N_k$.

d. $CIFS = \{M_1, \ldots M_p\}$ $\{p = \text{number of recursive calls}\}$.

## Example 8.1   Sierpinski Gasket—Ron's Method

*Outer Triangle*
　　Vertices $= (P_1, P_2, P_3)$
　　Edges $= (w_1, w_2, w_3)$

| *Recursive Turtle Program* | | *Transformations* |
|---|---|---|
| BASE CASE: TRIANGLE | | |
| RECURSION | | |
| 　REPEAT 3 TIMES | | |
| 　　RESIZE 1/2 | $\leftrightarrow$ | $SCALE(P_k, 1/2)$ |
| 　　RECUR | | |
| 　　RESIZE 2 | $\leftrightarrow$ | $SCALE(P_k, 2)$ |
| 　　MOVE 1 | $\leftrightarrow$ | $TRANS(w_k)$ |
| 　　TURN $2\pi/3$ | $\leftrightarrow$ | $ROT(P_{k+1}, 2\pi/3)$ |

*Conformal Iterated Function System*

$$M_1 = SCALE(P_1, 1/2),$$
$$M_2 = M_1 * SCALE(P_1, 2) * TRANS(w_1) * ROT(P_2, 2\pi/3) * SCALE(P_2, 1/2)$$
$$= TRANS(w_1) * ROT(P_2, 2\pi/3) * SCALE(P_2, 1/2),$$
$$M_3 = M_2 * SCALE(P_2, 2) * TRANS(w_2) * ROT(P_3, 2\pi/3) * SCALE(P_3, 1/2)$$
$$= TRANS(w_1) * ROT(P_2, 2\pi/3) * TRANS(w_2) * ROT(P_3, 2\pi/3) * SCALE(P_3, 1/2).$$

Notice that in Ron's Algorithm, the transformations depend on the turtle's current state. Thus, Ron's Algorithm has the annoying property that it must keep track of the turtle's state after each turtle command. For the Sierpinski gasket, keeping track of the turtle's state is easy, but for many other fractals, it is not so straightforward to keep track of the turtle's state.

## 8.4.2   Tao's Algorithm

Tao's Algorithm takes a turtle-centric point of view, a perspective from inside the turtle's domain. From the turtle's viewpoint, the turtle is always at the center of the universe (the origin) facing in the direction of the x-axis (*ivec*), so there is no need to keep track of the turtle's state. Therefore, viewed from inside the turtle, the turtle commands correspond to the following conformal transformations:

$$\text{FORWARD } D \leftrightarrow TRANS(ivec * SCALE(D)),$$
$$\text{MOVE } D \leftrightarrow TRANS(ivec * SCALE(D)),$$
$$\text{TURN } A \leftrightarrow ROT(Origin, A) = ROT(A),$$
$$\text{RESIZE } s \leftrightarrow SCALE(Origin, s) = SCALE(s).$$

There is one novel detail here. When we apply transformation matrices to compute the new coordinates of a point or vector, we multiply on the right: $Q_{new} = Q_{old} * M$. But when we apply these transformations to change the state of the turtle—the turtle's local coordinate system—then we must multiply on the left. Indeed, suppose that the turtle is in the state $S = (P, v)$. Then:

$$\begin{matrix} \text{FORWARD } D \\ \text{MOVE } D \end{matrix} \Leftrightarrow \begin{matrix} v_{new} = v \\ v_{new}^{\perp} = v^{\perp} \\ P_{new} = P + Dv \end{matrix} \Leftrightarrow \begin{pmatrix} v_{new} \\ v_{new}^{\perp} \\ P_{new} \end{pmatrix} = \begin{pmatrix} 1 & 0 & 0 \\ 0 & 1 & 0 \\ D & 0 & 1 \end{pmatrix} * \begin{pmatrix} v \\ v^{\perp} \\ P \end{pmatrix},$$

$$\text{RESIZE } s \Leftrightarrow \begin{matrix} v_{new} = s\,v \\ v_{new}^{\perp} = sv^{\perp} \\ P_{new} = P \end{matrix} \Leftrightarrow \begin{pmatrix} v_{new} \\ v_{new}^{\perp} \\ P_{new} \end{pmatrix} = \begin{pmatrix} s & 0 & 0 \\ 0 & s & 0 \\ 0 & 0 & 1 \end{pmatrix} * \begin{pmatrix} v \\ v^{\perp} \\ P \end{pmatrix},$$

$$\text{TURN } A \Leftrightarrow \begin{matrix} v_{new} = \cos(A)v + \sin(A)v^{\perp} \\ v_{new}^{\perp} = -\sin(A)v + \cos(A)v^{\perp} \\ P_{new} = P \end{matrix} \Leftrightarrow \begin{pmatrix} v_{new} \\ v_{new}^{\perp} \\ P_{new} \end{pmatrix} = \begin{pmatrix} \cos(A) & \sin(A) & 0 \\ -\sin(A) & \cos(A) & 0 \\ 0 & 0 & 1 \end{pmatrix} * \begin{pmatrix} v \\ v^{\perp} \\ P \end{pmatrix}.$$

These observations lead to the following result.

**Proposition 8.1:** *If $L_1, \ldots, L_n$ are the transformation matrices corresponding to the turtle commands $C_1, \ldots, C_n$, then $L = L_n * \cdots * L_1$ is the transformation matrix corresponding to the turtle program $C_1, \ldots, C_n$.*

**Proof:** Let $S_k$ denote the state of the turtle after executing the command $C_k$. Since $L_k * S_{k-1} = S_k$, it follows that $L * S_0 = (L_n * \cdots * L_1) * S_0 = S_n$.

*Tao's Algorithm: Recursive Turtle Program $\rightarrow$ Conformal Iterated Function System*

1. *Preprocessing*

   a. For each turtle command, compute the corresponding conformal transformation.

   b. If the turtle changes state during the recursion, compute the conformal transformation corresponding to the state change caused by the recursion.

2. *Finding the Conformal Iterated Function System*

   a. $N_k =$ product of the transformations between the $(k-1)$st and $k$th recursive call in reverse order (right to left, not left to right).

   b. $M_1 = N_1 =$ product of the transformations *before* the first recursive call.

   c. $M_k = N_k * M_{k-1}$.

   d. $CIFS = \{M_1, \ldots, M_p\}$   $\{p =$ number of recursive calls$\}$.

Note that all rotation and scaling is about the origin and all translation is in the $x$-direction.

**Example 8.2   Sierpinski Gasket—Tao's Method**

| Recursive Turtle Program | Transformations |
|---|---|
| BASE CASE: TRIANGLE | |
| RECURSION | |
| REPEAT 3 TIMES | |
| RESIZE 1/2     $\leftrightarrow$ | SCALE(1/2) |
| RECUR | |
| RESIZE 2     $\leftrightarrow$ | SCALE(2) |

$$\begin{array}{lcl} \text{MOVE } 1 & \leftrightarrow & \textit{TRANS(ivec)} \\ \text{TURN } 2\pi/3 & \leftrightarrow & \textit{ROT}(2\pi \, / \, 3) \end{array}$$

*Conformal Iterated Function System*

$$M_1 = SCALE(1/2),$$

$$M_2 = SCALE(1/2) * ROT(2\pi/3) * TRANS(ivec) * SCALE(2) * M_1$$

$$= SCALE(1/2) * ROT(2\pi/3) * TRANS(ivec),$$

$$M_3 = SCALE(1/2) * ROT(2\pi/3) * TRANS(ivec) * SCALE(2) * M_2$$

$$= SCALE(1/2) * ROT(2\pi/3) * TRANS(ivec) * ROT(2\pi/3) * TRANS(ivec).$$

Note that in Tao's method, unlike Ron's method, there is no need to keep track of the turtle's state! Nevertheless, both Tao's method and Ron's method generate the identical iterated function system.

## 8.5   Bump Fractals

The easiest way to describe a bump is with a turtle program. When we studied bump fractals, we observed that we could generate the recursive turtle program for a bump fractal directly from the turtle program for the bump. We would like to do the same for iterated function systems: that is,

**TABLE 8.1:**   Bump $\rightarrow$ recursive turtle program $\rightarrow$ conformal iterated function system. All the rotation and scaling commands are around the origin. The last three bump commands—FORWARD 1, TURN $A_{n+1}$, and RESIZE $S_{n+1}$—are ignored by the iterated function system, since the only purpose of these commands in the bump program is to conduct the turtle to her final state. To read off the iterated function system from the transformations in the last column, simply write down the transformations in reverse order and group them between successive appearances of *Trans(ivec)* (see Example 8.3).

| Bump | | Recursive Turtle Program (RTP) | | Conformal Iterated Function System |
|------|---|--------------------------------|---|-----------------------------------|
| | | *Base Case*: FORWARD 1 | | |
| | | *Recursion*: | | |
| TURN $A_1$ | $\rightarrow$ | TURN $A_1$ | $\rightarrow$ | $ROT(A_1)$ |
| RESIZE $S_1$ | $\rightarrow$ | RESIZE $S_1$ | $\rightarrow$ | $SCALE(S_1)$ |
| FORWARD 1 | $\rightarrow$ | RTP (*Level*-1) | $\rightarrow$ | $TRANS(ivec)$ |
| TURN $A_2$ | $\rightarrow$ | TURN $A_2$ | $\rightarrow$ | $ROT(A_2)$ |
| RESIZE $S_2$ | $\rightarrow$ | RESIZE $S_2$ | $\rightarrow$ | $SCALE(S_2)$ |
| FORWARD 1 | $\rightarrow$ | RTP (*Level*-1) | $\rightarrow$ | $TRANS(ivec)$ |
| . | | . | | . |
| . | | . | | . |
| . | | . | | . |
| TURN $A_n$ | $\rightarrow$ | TURN $A_n$ | $\rightarrow$ | $ROT(A_n)$ |
| RESIZE $S_n$ | $\rightarrow$ | RESIZE $S_n$ | $\rightarrow$ | $SCALE(S_n)$ |
| FORWARD 1 | $\rightarrow$ | RTP (*Level*-1) | | |
| TURN $A_{n+1}$ | $\rightarrow$ | TURN $A_{n+1}$ | | |
| RESIZE $S_{n+1}$ | $\rightarrow$ | RESIZE $S_{n+1}$ | | |

**Example 8.3   Koch Curve**

| *Koch Iterated Function System* | | *Koch Bump* | | *Koch (Level)* |
|---|---|---|---|---|
| | | | | Base Case: FORWARD 1 |
| | | | | Recursion: |
| *SCALE*(1/3) | ← | RESIZE 1/3 | → | RESIZE 1/3 |
| *TRANS(ivec)* | ← | FORWARD 1 | → | Koch (*Level*-1) |
| *ROT*(π/3) | ← | TURN π/3 | → | TURN π/3 |
| *TRANS(ivec)* | ← | FORWARD 1 | → | Koch (*Level*-1) |
| *ROT*(−2π/3) | ← | TURN −2π/3 | → | TURN −2π/3 |
| *TRANS(ivec)* | ← | FORWARD 1 | → | Koch (*Level*-1) |
| *ROT*(π/3) | ← | TURN π/3 | → | TURN π/3 |
| | | FORWARD 1 | → | Koch (*Level*-1) |
| | | RESIZE 3 | → | RESIZE 3 |

$$Rot(\pi/3) * Trans(ivec) * Rot(-2\pi/3) * Trans(ivec) * Rot(\pi/3) * Trans(ivec) * \underbrace{Scale(1/3)}_{M_1}$$

$$\underbrace{\phantom{Rot(\pi/3) * Trans(ivec) * Rot(-2\pi/3) * Trans(ivec) * Rot(\pi/3)}}_{M_2}$$

$$\underbrace{\phantom{Rot(\pi/3) * Trans(ivec) * Rot(-2\pi/3)}}_{M_3}$$

$$\underbrace{\phantom{Rot(\pi/3)}}_{M_4}$$

A conformal iterated function system for the Koch curve is given by the transformations $\{M_1, M_2, M_3, M_4\}$. To find these transformations, we list in column 1 the transformations corresponding to the turtle program in column 2 for the Koch bump. We then write these transformations horizontally in reverse order and group these products between successive appearance of *Trans(ivec)*.

we would like to extract the iterated function system for the bump fractal directly from a simple turtle description of the bump. Since we know how to get from a turtle description of a bump to a recursive turtle program for the bump fractal and since we know how to convert from a recursive turtle program to an iterated function system, all we need to do to get from a turtle description of a bump to an iterated function system for the bump fractal is to combine these two procedures. We exhibit this algorithm in Table 8.1. In order to avoid keeping track of the turtle's state, we use Tao's Algorithm to convert from a recursive turtle program to an iterated function system. In Example 8.3, we apply this algorithm to the Koch curve.

## 8.6   Summary

The main result of this chapter is the equivalence between recursive turtle programs and conformal iterated function systems. The conformal transformations in the conformal iterated function system correspond to changes of state between successive recursive calls in the recursive turtle program; the condensation set of the conformal iterated function system is the geometry drawn during highest level of the recursive phase of the recursive turtle program. These and other explicit correspondences between recursive turtle programs and conformal iterated function systems are summarized in Table 8.2. Since, by the trivial fixed point theorem, the fractal generated by an iterated function system is independent of the base case, it follows from our equivalence theorems that the fractal generated by a recursive turtle program is also independent of the base case.

We presented two explicit algorithms to convert from a recursive turtle program to an equivalent conformal iterated function system: Ron's Algorithm and Tao's Algorithm. Tao's Algorithm was used

**TABLE 8.2:** Correspondences between the conformal iterated function system and the recursive turtle program that generate the same fractal.

| Conformal Iterated Function System | Recursive Turtle Program |
|---|---|
| Number of transformations | Number of recursive calls |
| Transformations | Changes of state between successive recursive calls |
| Base case | Geometry drawn in the base case |
| Condensation set | Geometry drawn at the highest level of recursion |

to generate the iterated function system for a bump fractal directly from a turtle program for the bump. To conclude, we compare and contrast the main features of these two algorithms.

*Ron's Algorithm (Exo-Centric)*

- Matrix products are taken in order from left to right.

- Needs to keep track of the turtle's state.

- All transformations are rooted at the turtle's current state.

*Tao's Algorithm (Turtle-Centric)*

- Matrix products are taken in reverse order from right to left.

- Does not need to keep track of the turtle's state.

- All transformations are rooted at the origin and along the *x*-axis.

# Exercises

8.1. Develop an explicit algorithm to convert a conformal iterated function system to an equivalent recursive turtle program.

8.2. The text presents two algorithms to convert a recursive turtle program to an equivalent conformal iterated function system: Ron's Algorithm and Tao's Algorithm.

    a. Verify that Ron's method and Tao's method generate the same conformal iterated function system for the Sierpinski gasket.

    b. Does Ron's method always give the same conformal iterated function system as Tao's method? Prove or give a counterexample.

8.3. Is the conformal iterated function system for the Koch curve presented in Example 8.3 the same as the conformal iterated function system for the Koch curve presented in Chapter 6, Exercise 6.6?

8.4. Consider the bumps in Chapter 2, Figure 2.9.

    a. Write a turtle program to generate each of these bump curves.

    b. Write a recursive turtle program to generate the bump fractals corresponding to each of these bump curves.

    c. Describe the transformations in the conformal iterated function systems that generate each of these bump fractals.

8.5. Consider the bumps in Chapter 2, Figure 2.13.

    a. Write a turtle program to generate each of these bump curves.

    b. Write a recursive turtle program to generate the bump fractals corresponding to each of these bump curves.

    c. Describe the transformations in the conformal iterated function systems that generate each of these bump fractals.

8.6. Consider the bumps in Chapter 2, Figure 2.14.

    a. Write a turtle program to generate each of these bump curves.

    b. Write a recursive turtle program to generate the bump fractals corresponding to each of these bump curves.

    c. Describe the transformations in the conformal iterated function systems that generate each of these bump fractals.

8.7. Consider the bumps in Chapter 2, Figure 2.15.

    a. Write a turtle program to generate each of these bump curves.

    b. Write a recursive turtle program to generate the bump fractals corresponding to each of these bump curves.

    c. Describe the transformations in the conformal iterated function systems that generate each of these bump fractals.

8.8. Consider the iterated function systems for the Sierpinski gasket generated from the standard recursive turtle program for the Sierpinski gasket by Ron's Algorithm (Example 8.1) and by Tao's Algorithm (Example 8.2).

    a. Show that these two iterated function systems are identical.

    b. Show that these two iterated function systems are not identical to the standard iterated function system for the Sierpinski gasket given by the three scaling transformations: *Scale* $(P_1, 1/2)$, *Scale*$(P_2, 1/2)$, and *Scale*$(P_3, 1/2)$.

    c. Conclude from part b that the iterated function system for generating a fractal is not unique.

    d. Construct a recursive turtle program that corresponds to the standard iterated function system for the Sierpinski gasket given in part b.

8.9. Prove that the classical turtle and the hodograph turtle (see Chapter 2, Programming Project 2.2) generate the same fractal when both of the following conditions are satisfied:

   a. The pen is down and the anchor is up during the base case.

   b. Both the pen and the anchor are up during the recursive body of the program.

8.10. Consider the left-handed turtle described in Chapter 2, Programming Project 2.4.

   a. Show that if we replace the classical turtle by the left-handed turtle and conformal transformations by arbitrary affine transformations, then:

      i. Lemmas 8.1 and 8.2 remain valid.

      ii. Theorems 8.1, 8.2 and Corollaries 8.1 through 8.4 remain valid.

   b. Explain how to extend Tao's Algorithm to convert recursive turtle programs for the left-handed turtle to iterated function systems consisting of arbitrary affine transformations.

8.11. Using an iterated function system with affine transformations:

   a. Build a fractal that cannot be generated by a recursive turtle program for the classical turtle.

   b. Construct a recursive turtle program for the left-handed turtle that generates the fractal in part a.

---

## Programming Projects

8.1. *Parsers*

   a. Write a parser to convert recursive programs for the classical turtle into conformal iterated function systems that generate the same fractal.

   b. Write a parser to convert recursive turtle programs for the left-handed turtle into iterated function systems that generate the same fractal (see Exercise 8.10).

8.2. *Open Problems* (see Exercise 8.8)

   a. Develop an algorithm that given two arbitrary iterated function systems determines whether the two systems generate the same fractal or show that no such algorithm exists.

   b. Develop an algorithm that given two arbitrary recursive turtle programs determines whether the two programs generate the same fractal or show that no such algorithm exists.

   Note that by the results in this chapter, the preceding two open problems are equivalent.

# Part II

# Mathematical Methods
# for Three-Dimensional
# Computer Graphics

# Chapter 9

## Vector Geometry: A Coordinate-Free Approach

*And ye shall know the truth, and the truth shall make you free*

*– John 8:32*

### 9.1   Coordinate-Free Methods

We are going to ascend now from the flat world of two dimensions into the real world of three dimensions. When working in three dimensions, we shall insist predominantly on coordinate-free methods. Even in two dimensions we adopted coordinate-free techniques to describe shapes because it is easier to represent geometry without troubling about coordinates. For example, it is simpler to describe a bump using a turtle program rather than trying to delineate the coordinates of all the vertices of the bump. Similarly, we invoke affine transformations—translation, rotation, scaling, and shear—to move and reshape geometry without worrying about the entries—the coordinates—of the corresponding matrices. Coordinates are useful for computations, but conceptually we prefer to work at a higher level of abstraction. Turtle programs and affine transformations were our entry to coordinate-free methods in two dimensions.

In three dimensions coordinate-free methods are even more crucial. You may be comfortable with coordinate techniques in two dimensions, but in three dimensions it is much harder to conceptualize geometry using coordinates. In addition to allowing us to work at a higher level of abstraction, coordinate-free methods provide a more concise notation. Typically we will need to deal with only one equation for a point or a vector, rather than with three equations for their coordinates. Also we shall see that coordinate-free methods lead naturally to geometrically meaningful expressions. With coordinates we can perform senseless computations that have no intrinsic geometric significance—for example, we could add the coordinates of two points. Coordinate-free methods will help us to avoid the pitfalls of such mindless computations.

Finally, and most importantly, coordinate-free techniques capture the geometric meaning behind our computational methods. You may know that the dot product of two vectors $u, v$ can be calculated from the formula:

$$u \cdot v = u_1 v_1 + u_2 v_2 + u_3 v_3,$$

but why would anyone ever want to compute the expression involving the rectangular coordinates on the right-hand side? The geometric meaning of the dot product is captured by the coordinate-free expression:

$$u \cdot v = |u||v| \cos(\vartheta),$$

where:

$|u|$ and $|v|$ are the lengths of vectors $u$ and $v$

$\vartheta$ is the angle between $u$ and $v$.

Angle and length (as well as projection), these quantities are the geometric content of the dot product, content that is completely hidden in the coordinate computation.

Our approach to Computer Graphics is to simplify geometry as much as possible by invoking the Mathematics most appropriate to the problem at hand. In two dimensions we employ Turtle Graphics (LOGO) rooted in conformal transformations and Affine Graphics (CODO) based on affine transformations. In three dimensions, affine transformations along with a new transformation, perspective projection, play an even more central role. These transformations are applied to manipulate shapes in three dimensions, but at bottom most shapes in three dimensions are represented in terms of points and vectors. Therefore we begin our study of three-dimensional Computer Graphics by introducing a coordinate-free approach to the algebra and geometry of points and vectors.

## 9.2    Vectors and Vector Spaces

Vectors and vector spaces should be familiar to you from standard courses on Linear Algebra. Vectors can be added, subtracted, and multiplied by scalars, and these vector operations all have coordinate-free definitions (see Figure 9.1).

These vector operations obey the usual laws of arithmetic: addition is associative (Figure 9.2a) and commutative (Figure 9.2b) and scalar multiplication distributes through addition (Figure 9.2c).

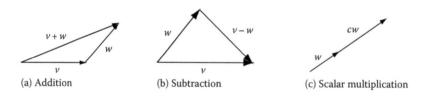

(a) Addition          (b) Subtraction          (c) Scalar multiplication

**FIGURE 9.1:**    Coordinate-free geometric definitions of (a) addition, (b) subtraction, and (c) scalar multiplication for vectors.

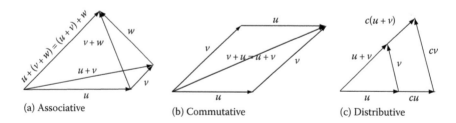

(a) Associative          (b) Commutative          (c) Distributive

**FIGURE 9.2:**    The associative, commutative, and distributive properties of vector addition and scalar multiplication. Vector addition is (a) associative because no matter how we group $u$, $v$, $w$, if we place these vectors head to tail, the vector $u+v+w$ goes from the tail of $u$ to the head of $w$. Vector addition is (b) commutative because $v+u$, $u+v$ both represent the same diagonal of the parallelogram with sides $u$, $v$. Finally, by similar triangles, (c) scalar multiplication distributes through vector addition.

Nevertheless, although vectors and vector operations are useful and familiar, vectors are not the primary focus of Computer Graphics. On the graphics terminal we see points, not vectors, so it is to points that we next turn our attention.

---

## 9.3   Points and Affine Spaces

Points are not vectors. Points have a fixed position, but no direction or length; vectors have direction and length, but no fixed location. Vectors can be added, subtracted, and multiplied by scalars and the result is always a vector. Points can be subtracted from one another, but the result is a vector, not a point (see Figure 9.3). A vector can be added to a point, and the result is a point (see Figure 9.3), but there is no coordinate-free way to add two points or to multiply a point by a scalar.

Expressions of the form $\sum_k c_k v_k$ are called *linear combinations*. For vectors, linear combinations always make sense because addition and scalar multiplication are always defined for vectors. Although we cannot, in general, add two points or multiply points by scalars, nevertheless some linear combinations of points also make sense. For example, the expression:

$$\frac{P+Q}{2} = \frac{P}{2} + \frac{Q}{2}$$

represents the midpoint of the line segment joining the points $P$ and $Q$, even though none of the expressions $P+Q$, $P/2$, $Q/2$ are well defined. We would like to have some analogue of linear combinations for points to account for expressions like the midpoint $(P+Q)/2$.

Although, in general, scalar multiplication for points cannot be defined in a coordinate-free manner, we can define a coordinate-free version of scalar multiplication in two special cases:

$$1 \cdot P = P,$$

$$0 \cdot P = 0,$$

$$c \cdot P = \text{undefined}, \quad c \neq 0, 1.$$

By the way, the 0 on the right-hand side of the second equation is the zero vector, not the origin of the coordinate system.

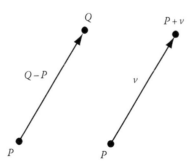

**FIGURE 9.3:**   Subtracting a point from a point (left), and adding a vector to a point (right). Notice that $P+(Q-P)=Q$, so the usual cancellation law of addition applies.

Now for a crucial observation. Notice that formally

$$\sum_{k=0}^{n} c_k P_k = \left(\sum_{k=0}^{n} c_k\right) P_0 + \sum_{k=1}^{n} c_k(P_k - P_0).$$

That is, if the usual rules of arithmetic are to apply, then the terms $c_k P_0$ and $-c_k P_0$ for $k \geq 1$ should cancel. The expression:

$$\sum_{k=1}^{n} c_k(P_k - P_0),$$

is well defined, since this expression is just a linear combination of vectors. Moreover, the expression:

$$\left(\sum_{k=0}^{n} c_k\right) P_0,$$

is well defined whenever the constants sum to zero or one. Based on these observations, we introduce the following definitions:

$$\sum_{k=0}^{n} c_k P_k = P_0 + \sum_{k=1}^{n} c_k(P_k - P_0) \quad \text{if } \sum_{k=0}^{n} c_k = 1$$

$$= \sum_{k=1}^{n} c_k(P_k - P_0) \quad \text{if } \sum_{k=0}^{n} c_k = 0$$

$$= \text{undefined} \quad \text{if } \sum_{k=0}^{n} c_k \neq 0, 1.$$

Expressions of the form $\Sigma_k c_k P_k$, where $\Sigma_k c_k \equiv 1$, are called *affine combinations*. Affine combinations are the analogues for points of linear combinations for vectors.

Vectors form a vector space; points form an *affine space*. Vectors are closed under linear combinations; points are closed under affine combinations. This distinction between points and vectors, between affine spaces and vector spaces, is what makes the algebra underlying Computer Graphics just a bit different from the standard algebra of vector spaces that you learn in courses on Linear Algebra.

---

## 9.4   Vector Products

There is more to vector algebra than just addition, subtraction, and scalar multiplication. For vectors, there are also several distinct notions of multiplication: dot product, cross product, and determinant. These products are related geometrically to length, area, and volume, so these products show up in many geometric applications. Here we review the coordinate-free definitions of these three products, emphasizing their major algebraic and geometric properties.

### 9.4.1 Dot Product

The dot product of two vectors $u$, $v$ is the scalar defined by

$$u \cdot v = |u||v| \cos(\vartheta),$$

where:
$|u|$ and $|v|$ are the lengths of vectors $u$ and $v$
$\vartheta$ is the angle between the vectors $u$ and $v$.

Dot product is commutative and distributes through addition (see Exercise 9.6a).

We are interested in dot product because dot product can be used to compute several important geometric quantities. The following properties follow easily from the definition of the dot product.

1. *Length*

$$|u|^2 = u \cdot u.$$

2. *Cosine (angle between two vectors)*

$$\cos(\vartheta) = \frac{u \cdot v}{|u||v|}.$$

3. *Orthogonality*

$$u \cdot v = 0 \Leftrightarrow u \perp v.$$

4. *Projections* (see Figure 9.4)

$$u_{\parallel} = \left(\frac{u \cdot v}{v \cdot v}\right)v, \qquad u_{\parallel} = (u \cdot v)v \quad \text{if} \quad |v| = 1,$$

$$u_{\perp} = u - u_{\parallel} = u - \left(\frac{u \cdot v}{v \cdot v}\right)v, \quad u_{\perp} = u - (u \cdot v)v \quad \text{if} \quad |v| = 1.$$

The first three properties follow easily from the definition of the dot product. To derive the formulas for projection, notice that the second formula follows from the first formula because $u_{\parallel} + u_{\perp} = u$ (see Figure 9.4). To prove the first formula, observe from Figure 9.4 that

$$|u_{\parallel}| = |u| \cos(\vartheta) = \frac{|u||v| \cos(\vartheta)}{|v|} = \frac{u \cdot v}{|v|}.$$

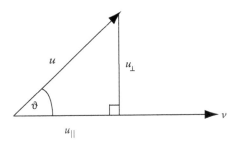

**FIGURE 9.4:** Parallel and perpendicular projection. Think dot product!

Therefore, since $u_{\parallel}$ is parallel to $v$,

$$u_{\parallel} = |u_{\parallel}|\frac{v}{|v|} = \left(\frac{u \cdot v}{|v|}\right)\frac{v}{|v|} = \left(\frac{u \cdot v}{v \cdot v}\right)v.$$

### 9.4.2   Cross Product

The cross product of two vectors $u$, $v$ is the unique vector with the following three properties (see Figure 9.5):

$$|u \times v| = |u||v| \sin(\vartheta),$$

$$u \times v \perp u, v,$$

$$\text{sgn}(u, v, u \times v) > 0.$$

The third condition means that the vectors $u$, $v$, $u \times v$ have positive orientation—that is, these three vectors obey the right-hand rule: if you curl the fingers of your right hand from $u$ to $v$, then your thumb will point in the direction of $u \times v$.

Cross product distributes through addition (see Exercise 9.6b), but cross product is not associative and cross product is anticommutative. That is,

$$u \times (v + w) = u \times v + u \times w \quad \text{(distributive)},$$

$$v \times u = -u \times v \quad \text{(anticommutative)},$$

$$(u \times v) \times w \neq u \times (v \times w) \quad \text{(nonassociative)}.$$

The anticommutativity follows because by the right-hand rule: $\text{sgn}(u, v, u \times v) = -\text{sgn}(v, u, v \times u)$. The nonassociativity follows because $(u \times v) \times w \perp u \times v, w$ whereas $u \times (v \times w) \perp u, v \times w$. A useful formula to remember is that two cross products can be reduced to two dot products:

$$(v \times w) \times u = (u \cdot v)w - (u \cdot w)v.$$

We shall derive this identity in Appendix A.

We are interested in the cross product because, just like the dot product, cross product can be used to compute several important geometric quantities. The following properties follow easily from the definition of the cross product.

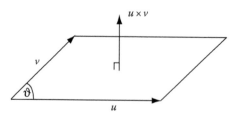

**FIGURE 9.5:**   The cross product of two vectors is the vector perpendicular to the two vectors with positive orientation, and has length equal to the area of the associated parallelogram.

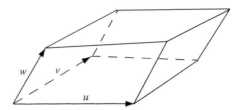

**FIGURE 9.6:** The determinant of three vectors is the signed volume of the associated parallel-epiped because $|\det(u, v, w)| = |u \times v||w||\cos(\vartheta)| = \text{area}(u, v) \cdot \text{height}(u, v, w) = \text{vol}(u, v, w)$.

1. *Area*

$$\text{area}(u, v) = |u \times v|.$$

2. *Sine (angle between two vectors)*

$$|\sin(\vartheta)| = \frac{|u \times v|}{|u||v|}.$$

3. *Parallelism*

$$u\|v \Leftrightarrow u \times v = 0.$$

### 9.4.3 Determinant

The determinant is the scalar defined by the triple product:

$$\det(u, v, w) = (u \times v) \cdot w.$$

The determinant inherits the properties of the dot and cross product: determinant is multilinear (linear in each variable) and skew symmetric (interchanging the order of the two vectors changes the sign of the determinant). The main geometric property of the determinant is the relation between determinant and volume (see Figure 9.6).

1. *Volume*

$$\text{vol}(u, v, w) = |\det(u, v, w)|.$$

## 9.5 Summary

Linear Algebra provides the mathematical foundation for Computer Graphics. However, the Linear Algebra underlying Computer Graphics differs somewhat from standard Linear Algebra because the fundamental constituents of Computer Graphics—what one actually sees on a graphics terminal—are points, not vectors. Points, unlike vectors, cannot be combined using arbitrary linear combinations; points are restricted to affine combinations. Thus the points form an affine space, rather than a vector space. This distinction between affine spaces (spaces of points) and vector spaces (spaces of vectors) is the main difference between the mathematical foundation of three-dimensional Computer Graphics and the standard Linear Algebra of three-dimensional vector spaces.

Nevertheless, vectors play a fundamental role in Computer Graphics because most of the information in an affine space is stored in the vectors. Fix a single point $Q$ in affine space. Then any other point $P$ can be written as the sum of the fixed point $Q$ and the vector $v = P - Q$ from $Q$ to $P$—that is, $P = Q + v = Q + (P - Q)$. Thus, for example, if we know how an affine transformation $T$ affects the vectors $v$ and we want to know how this transformation behaves on the points $P$, all we need to do is to compute the transformation on the single point $Q$; the rest of the transformation is known from its effect on the vectors because $T(P) = T(Q) + T(P - Q)$.

To emphasize concepts over computation, we adopt a coordinate-free approach to vector geometry. Addition, subtraction, scalar multiplication, dot product, cross product, and determinant all have coordinate-free, geometric interpretations. These geometric properties are the main conceptual tools we will apply in our analysis of geometry for Computer Graphics. Coordinate computations are discussed in Chapter 10. These coordinate computations can be implemented once and forgotten, but the geometric concepts behind these computations will be reused again and again.

---

## Appendix A: The Nonassociativity of the Cross Product

The purpose of this appendix is to derive the handy identities:

$$(v \times w) \times u = (u \cdot v)w - (u \cdot w)v,$$

$$u \times (v \times w) = (u \cdot w)v - (u \cdot v)w.$$

We begin with an important special case from which the general results will follow.

**Lemma 9.A.1:** $(u \times w) \times u = (u \cdot u)w - (u \cdot w)u$

**Proof:** To prove that two vectors are equal, we need to check that they have the same direction, orientation, and length.

a. *Direction*

$$((u \cdot u)w - (u \cdot w)u) \cdot u = (u \cdot u)(w \cdot u) - (u \cdot w)(u \cdot u) = 0,$$
$$((u \cdot u)w - (u \cdot w)u) \cdot (u \times w) = (u \cdot u)(w \cdot (u \times w)) - (u \cdot w)(u \cdot (u \times w)) = 0.$$

Therefore $(u \cdot u)w - (u \cdot w)u$ is perpendicular both to $u$ and to $u \times w$, so $(u \cdot u)w - (u \cdot w)u$ is either parallel or antiparallel to $(u \times w) \times u$.

b. *Orientation*

$$\begin{aligned}
\det(u \times w, u, (u \cdot u)w - (u \cdot w)u) &= \det(u, (u \cdot u)w - (u \cdot w)u, u \times w) \\
&= (u \times \{(u \cdot u)w - (u \cdot w)u\}) \cdot (u \times w) \\
&= \{(u \cdot u)(u \times w)\} \cdot (u \times w) \\
&= (u \cdot u)((u \times w) \cdot (u \times w)) \\
&= |u|^2 |u \times w|^2 > 0.
\end{aligned}$$

Therefore the vectors $u \times w$, $u$, $(u \cdot u)w - (u \cdot w)u$ form a right-handed system, so $(u \cdot u)w - (u \cdot w)u$ is parallel to $(u \times w) \times u$.

c. *Length*

Let $\theta$ be the angle between $u$ and $w$. Then,

$$|(u \cdot u)w - (u \cdot w)u|^2 = ((u \cdot u)w - (u \cdot w)u) \cdot ((u \cdot u)w - (u \cdot w)u)$$
$$= (u \cdot u)^2(w \cdot w) - (u \cdot w)^2(u \cdot u)$$
$$= (u \cdot u)\big((u \cdot u)(w \cdot w) - (u \cdot w)^2\big)$$
$$= |u|^2 \left(|u|^2|w|^2 - |u|^2|w|^2 \cos^2 \theta\right)$$
$$= |u|^4|w|^2 \sin^2 \theta$$
$$= |u|^2|u \times w|^2.$$

Since $(u \cdot u)w - (u \cdot w)u$ and $(u \times w) \times u$ have the same direction, orientation, and length, it follows that $(u \times w) \times u = (u \cdot u)w - (u \cdot w)u$.

**Theorem 9.A.1:** $(v \times w) \times u = (u \cdot v)w - (u \cdot w)v.$

**Proof:** If $v$ is parallel to $w$, then both sides are zero. Hence we can assume that $v$, $w$, $v \times w$ form a basis. Let us first check that this result is true when $u$ is an element of this basis. If $u = v$ or $u = w$, then the result follows from the Lemma 9.A.1. If $u = v \times w$, then,

$$(v \times w) \times u = (v \times w) \times (v \times w) = 0,$$

and

$$(u \cdot v)w - (u \cdot w)v = ((v \times w) \cdot v)w - ((v \times w) \cdot w)v = 0.$$

Hence the result is valid when $u$ is an element of the basis $v$, $w$, $v \times w$. Now for an arbitrary vector $u$, there are constants $\lambda, \mu, \nu$ such that

$$u = \lambda v + \mu w + \nu v \times w.$$

Therefore the result follows in the general case by linearity, since dot product and cross product both distribute through addition.

**Corollary 9.A.1:** $u \times (v \times w) = (u \cdot w)v - (u \cdot v)w.$

**Proof:** $u \times (v \times w) = -(v \times w) \times u = (u \cdot w)v - (u \cdot v)w.$

## Appendix B: The Algebra of Points and Vectors

In this appendix we collect for easy reference the main algebraic identities for addition, subtraction, scalar multiplication, dot product, cross product, and determinant. Most of these formulas either follow easily from the definitions or are derived in the text; the remainder are proved in the exercises at the end of this chapter.

*Addition, Subtraction, and Scalar Multiplication for Vectors*

$$u + (v + w) = (u + v) + w \qquad \text{(associative)}$$
$$u + v = v + u \qquad \text{(commutative)}$$
$$c(u + v) = cu + cv \qquad \text{(distributive)}$$
$$|u + v| \le |u| + |v| \qquad \text{(triangle inequality)}$$

*Addition, Subtraction, and Affine Combinations for Points*

$$P + (v + w) = (P + v) + w \qquad \text{(associative)}$$

$$P + (Q - P) = Q$$
$$(R - Q) + (Q - P) = R - P \qquad \text{(cancellation)}$$

$$\sum_{k=0}^{n} c_k P_k = P_0 + \sum_{k=1}^{n} c_k (P_k - P_0) \quad \text{if } \sum_{k=0}^{n} c_k = 1 \quad \text{(affine combinations)}$$

$$= \sum_{k=1}^{n} c_k (P_k - P_0) \qquad \text{if } \sum_{k=0}^{n} c_k = 0$$

$$= \text{undefined} \qquad \text{if } \sum_{k=0}^{n} c_k \ne 0, 1$$

*Dimensional Analysis for Points and Vectors*

$$\text{Point} + \text{Point} = \text{Undefined}$$
$$\text{Point} - \text{Point} = \text{Vector}$$
$$\text{Point} \pm \text{Vector} = \text{Point}$$
$$\text{Vector} \pm \text{Vector} = \text{Vector}$$
$$\text{Scalar} \cdot \text{Vector} = \text{Vector}$$
$$\Sigma \text{ Scalar} \cdot \text{Vector} = \text{Vector}$$
$$\Sigma \text{ Scalar} \cdot \text{Point} = \text{Point} \qquad \Sigma \text{ Scalar} = 1$$
$$= \text{Vector} \qquad \Sigma \text{ Scalar} = 0$$
$$= \text{Undefined} \qquad \text{Otherwise}$$

*Dot Product*

$$u \cdot v = |u||v| \cos(\theta) \qquad \text{(definition)}$$
$$u \cdot v = v \cdot u \qquad \text{(commutative)}$$

$$u \cdot (v + w) = u \cdot v + u \cdot w$$
$$(v + w) \cdot u = v \cdot u + w \cdot u \qquad \text{(distributive)}$$

$$|v|^2 = v \cdot v \qquad \text{(length)}$$

$$\cos(\theta) = \frac{u \cdot v}{|u||v|} \qquad \text{(angle)}$$

$$u_\| = \left(\frac{u \cdot v}{v \cdot v}\right) v$$
$$u_\| = (u \cdot v)v \quad \text{if} \quad |v| = 1 \qquad \text{(parallel projection)}$$

$$u_\perp = u - u_\| = u - \left(\frac{u \cdot v}{v \cdot v}\right) v$$
$$u_\perp = u - (u \cdot v)v \quad \text{if} \quad |v| = 1 \qquad \text{(perpendicular projection)}$$

$$u \cdot v = 0 \Leftrightarrow u \perp v \qquad \text{(orthogonality)}$$

*Cross Product*

$$|u \times v| = |u||v| \sin(\theta)$$
$$u \times v \perp u, v \qquad \text{(definition)}$$
$$\mathrm{sgn}(u, v, u \times v) > 0$$

$$\mathrm{area}(u, v) = |u \times v| \qquad \text{(area)}$$

$$(u \times v) \cdot u = 0$$
$$(u \times v) \cdot v = 0 \qquad \text{(orthogonality)}$$

$$u \times u = 0$$
$$u \times v = 0 \Leftrightarrow v \| \pm u \qquad \text{(parallelism)}$$

$$u \times (v + w) = u \times v + u \times w$$
$$(v + w) \times u = v \times u + w \times u \qquad \text{(distributive)}$$

$$u \times v = -v \times u \qquad \text{(anticommutative)}$$

$$u \times (v \times w) = (u \cdot w)v - (u \cdot v)w$$
$$(v \times w) \times u = (u \cdot v)w - (u \cdot w)v \qquad \text{(nonassociative)}$$

$$|u \times v|^2 = |u|^2 |v|^2 - (u \cdot v)^2 \qquad \text{(length)}$$

$$(u_1 \times u_2) \cdot (v_1 \times v_2) = (u_1 \cdot v_1)(u_2 \cdot v_2) - (u_1 \cdot v_2)(u_2 \cdot v_1) \qquad \text{(Lagrange identity)}$$

*Determinant*

$$\det(u, v, w) = (u \times v) \cdot w \qquad \text{(definition)}$$

$$vol(u, v, w) = |\det(u, v, w)| \qquad \text{(volume)}$$

$$\det(u, v, w) > 0 \Leftrightarrow \mathrm{sgn}(u, v, w) > 0 \qquad \text{(orientation)}$$

$$\det(u, v, w) \neq 0 \Leftrightarrow u, v, w \text{ are linearly independent}$$
$$\det(u, u, w) = \det(u, v, u) = \det(u, v, v) = 0 \qquad \text{(linear independence)}$$

$$\det(u, v, w) = \det(v, w, u) = \det(w, u, v)$$
$$\det(v, u, w) = -\det(u, v, w) \qquad \text{(skew symmetry)}$$

$$\det(u_1 + c\ u_2, v, w) = \det(u_1, v, w) + c\ \det(u_2, v, w)$$
$$\det(u, v_1 + c\ v_2, v, w) = \det(u, v_1, w) + c\ \det(u, v_2, w) \qquad \text{(multilinearity)}$$
$$\det(u, v, w_1 + c\ w_2) = \det(u, v, w_1) + c\ \det(u, v, w_2)$$

---

## Exercises

9.1. Prove that for any point $R$:

   a. $\displaystyle\sum_{k=0}^{n} c_k P_k = R + \sum_{k=0}^{n} c_k(P_k - R)$   whenever   $\displaystyle\sum_{k=0}^{n} c_k = 1$.

   b. $\displaystyle\sum_{k=0}^{n} c_k P_k = \sum_{k=0}^{n} c_k(P_k - R)$   whenever   $\displaystyle\sum_{k=0}^{n} c_k = 0$.

Interpret these results geometrically.
(Hint: You may not use the equality $c_k(P_k - R) = c_k P_k - c_k R$ because the right-hand side has no intrinsic meaning.)

9.2. Prove that:

$$|u \times v|^2 = |u|^2 |v|^2 - (u \cdot v)^2.$$

9.3. The purpose of this exercise is to prove the Lagrange identity:

$$(u_1 \times u_2) \cdot (v_1 \times v_2) = (u_1 \cdot v_1)(u_2 \cdot v_2) - (u_1 \cdot v_2)(u_2 \cdot v_1).$$

   a. If $u_2$ is parallel to $u_1$, show that both sides of the Lagrange identity are zero.

   b. If $u_2$ is not parallel to $u_1$, then $u_1, u_2, u_1 \times u_2$ forms a basis for vectors in three-space. Therefore there are constants $c_1, c_2, c_3, d_1, d_2, d_3$, such that:

$$v_1 = c_1 u_1 + c_2 u_2 + c_3 u_1 \times u_2,$$
$$v_2 = d_1 u_1 + d_2 u_2 + d_3 u_1 \times u_2.$$

Now using the distributive law and other standard rules for the cross product together with the result of Exercise 9.2, prove the Lagrange identity.

9.4. Prove that:

   a. $v \cdot w = v_{||} \cdot w$.

   b. $v \times w = v_{\perp} \times w$.

9.5. Fix a vector $w$. Let $u_{||}$, $v_{||}$ denote the parallel projections and let $u_{\perp}$, $v_{\perp}$ denote the perpendicular projections of $u$ and $v$ on $w$. Prove that:

   a. $(u + v)_{||} = u_{||} + v_{||}$.

   b. $(u + v)_{\perp} = u_{\perp} + v_{\perp}$.

9.6. Using the results of Exercises 9.4 and 9.5, prove that dot product and cross product both distribute through addition. That is, prove that:

a. $(u+v) \cdot w = u \cdot w + v \cdot w$.

b. $(u+v) \times w = u \times w + v \times w$.

9.7. By drawing the appropriate figures, show that:

a. $(P+v)+w = P+(v+w)$   (associativity).

b. $(R-Q)+(Q-P) = R-P$   (cancellation).

c. $u+(-1)v = u-v$   (negation).

9.8. Show that:

a. $\det(u, v, w) \neq 0 \Leftrightarrow u, v, w$ are linearly independent.

b. $\det(u, u, w) = \det(u, v, u) = \det(u, v, v) = 0$.

9.9. Show that:

a. $\mathrm{sgn}(u, v, w) = \mathrm{sgn}(v, w, u) = \mathrm{sgn}(w, u, v)$.

b. $\mathrm{sgn}(v, u, w) = -\mathrm{sgn}(u, v, w)$.

9.10. Using the results of Exercise 9.9, show that the determinant function is skew symmetric. That is, show that:

a. $\det(u, v, w) = \det(v, w, u) = \det(w, u, v)$.

b. $\det(v, u, w) = -\det(u, v, w)$.

9.11. Show that the determinant function is multilinear. That is, show that:

a. $\det(u_1 + c\,u_2, v, w) = \det(u_1, v, w) + c\,\det(u_2, v, w)$.

b. $\det(u, v_1 + c\,v_2, w) = \det(u, v_1, w) + c\,\det(u, v_2, w)$.

c. $\det(u, v, w_1 + c\,w_2) = \det(u, v, w_1) + c\,\det(u, v, w_2)$.

9.12. Using the results of Exercises 9.8 and 9.11, show that:

a. $\det(u + c\,v, v, w) = \det(u, v, w)$.

b. $\det(u, v + c\,w, w) = \det(u, v, w)$.

c. $\det(u + c\,w, v, w) = \det(u, v, w)$.

9.13. Show that:

a. $(u_1 \times u_2) \times (v_1 \times v_2) = \det(u_1, u_2, v_2)v_1 - \det(u_1, u_2, v_1)v_2$.

b. $(u_1 \times u_2) \times (v_1 \times v_2) = \det(u_1, v_1, v_2)u_2 - \det(u_2, v_1, v_2)u_1$.

9.14. Show that:

$u \times (v \times w) + v \times (w \times u) + w \times (u \times v) = 0$.

# Chapter 10

## Coordinate Algebra

*I am not come to destroy, but to fulfill*

– Matthew 5:17

### 10.1  Rectangular Coordinates

To introduce coordinate computations for the algebra of points and vectors, we begin by fixing an origin and three orthogonal axes. Denote the origin by $O$, and the three unit vectors along the three orthogonal axes by $i, j, k$ (see Figure 10.1).

Any vector $v$ can be written in a unique way as a linear combination of the three basis vectors $i, j, k$. Thus, for any vector $v$, there are three scalars $v_1, v_2, v_3$ such that

$$v = v_1 i + v_2 j + v_3 k.$$

The scalars $v_1, v_2, v_3$ are called the *rectangular coordinates* of the vector $v$, and the notation $v = (v_1, v_2, v_3)$ is simply shorthand for the equation $v = v_1 i + v_2 j + v_3 k$.

Similarly, any point $P$ can be written as $P = O + (P - O)$. Since $P - O$ is a vector, there are three scalars $p_1, p_2, p_3$ such that

$$P - O = p_1 i + p_2 j + p_3 k.$$

The scalars $p_1, p_2, p_3$ are called the *rectangular coordinates* of the point $P$, and the notation $P = (p_1, p_2, p_3)$ is once again simply shorthand for the equation $P = O + p_1 i + p_2 j + p_3 k$.

With these definitions and this notation in hand, we are now ready to introduce coordinate computations for the linear algebra of points and vectors.

### 10.2  Addition, Subtraction, and Scalar Multiplication

The rules for computing sums, differences, and scalar products in terms of coordinates follow directly from the associative, commutative, and distributive properties of addition, subtraction, and scalar multiplication. Thus,

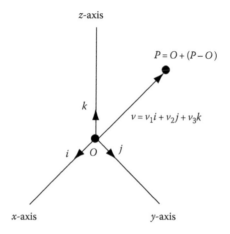

**FIGURE 10.1:**   A rectangular coordinate system consists of an origin and three orthogonal axes. Every vector can be written as a unique linear combination of the three orthogonal unit vectors along the coordinate axes, and every point is the sum of the origin and the vector from the origin to the point.

$$u \pm v = (u_1 i + u_2 j + u_3 k) \pm (v_1 i + v_2 j + v_3 k) = (u_1 \pm v_1)i + (u_2 \pm v_2)j + (u_3 \pm v_3)k,$$
$$\therefore \ u \pm v = (u_1 \pm v_1, u_2 \pm v_2, u_3 \pm v_3).$$

Similarly,

$$P \pm v = (O + p_1 i + p_2 j + p_3 k) \pm (v_1 i + v_2 j + v_3 k) = O + (p_1 \pm v_1)i + (p_2 \pm v_2)j + (p_3 \pm v_3)k,$$
$$\therefore \ P \pm v = (p_1 \pm v_1, p_2 \pm v_2, p_3 \pm v_3).$$

and

$$Q - P = (O + q_1 i + q_2 j + q_3 k) - (O + p_1 i + p_2 j + p_3 k) = (q_1 - p_1)i + (q_2 - p_2)j + (q_3 - p_3)k,$$
$$\therefore \ Q - P = (q_1 - p_1, q_2 - p_2, q_3 - p_3).$$

Also, since scalar multiplication distributes through vector addition,

$$cu = c(u_1 i + u_2 j + u_3 k) = cu_1 i + cu_2 j + cu_3 k,$$
$$\therefore \ cu = (cu_1, cu_2, cu_3).$$

## 10.3   Vector Products

The rules for computing vector products in terms of rectangular coordinates are derived in two stages:

1. Rules for products of the basis vectors are derived directly from the coordinate-free definitions.

2. Rules for the products of arbitrary vectors are derived from the rules for the basis vectors and the distributive property of multiplication.

### 10.3.1 Dot Product

Recall that, by definition,

$$u \cdot v = |u||v| \cos \theta,$$

where $\theta$ is the angle between the vectors $u$ and $v$. Hence, since the basis vectors $i, j, k$ are orthogonal unit vectors, we have the following rules:

$$i \cdot i = j \cdot j = k \cdot k = 1,$$
$$i \cdot j = j \cdot i = j \cdot k = k \cdot j = k \cdot i = i \cdot k = 0.$$

Therefore, by the distributive property of the dot product,

$$u \cdot v = (u_1 i + u_2 j + u_3 k) \cdot (v_1 i + v_2 j + v_3 k) = u_1 v_1 + u_2 v_2 + u_3 v_3.$$

### 10.3.2 Cross Product

Recall that, by definition,

$$|u \times v| = |u||v| \sin \theta,$$

$$u \times v \perp u, v,$$

$$\text{sgn}(u, v, u \times v) = 1$$

where $\theta$ is the angle between the vectors $u$ and $v$. Hence, since the basis vectors $i, j, k$ are orthogonal unit vectors, we have the following rules:

$$i \times i = j \times j = k \times k = 0,$$
$$i \times j = k, \ j \times k = i, \ k \times i = j,$$
$$j \times i = -k, \ k \times j = -i, \ i \times k = -j.$$

Therefore, by the distributive property of the cross product,

$$u \times v = (u_1 i + u_2 j + u_3 k) \times (v_1 i + v_2 j + v_3 k)$$
$$= (u_2 v_3 - u_3 v_2)i + (u_3 v_1 - u_1 v_3)j + (u_1 v_2 - u_2 v_1)k,$$

so

$$u \times v = (u_2 v_3 - u_3 v_2, u_3 v_1 - u_1 v_3, u_1 v_2 - u_2 v_1).$$

Equivalently, expanding by cofactors of the first row yields

$$u \times v = \det \begin{pmatrix} i & j & k \\ u_1 & u_2 & u_3 \\ v_1 & v_2 & v_3 \end{pmatrix}.$$

### 10.3.3   Determinant

We can derive an expression for computing the determinant of three vectors from the rules for computing the dot product and cross product. Recall that

$$\det(u, v, w) = (u \times v) \cdot w.$$

Therefore,

$$\det(u, v, w) = (u_2 v_3 - u_3 v_2, u_3 v_1 - u_1 v_3, u_1 v_2 - u_2 v_1) \cdot (w_1, w_2, w_3),$$

so

$$\det(u, v, w) = (u_2 v_3 - u_3 v_2) w_1 + (u_3 v_1 - u_1 v_3) w_2 + (u_1 v_2 - u_2 v_1) w_3.$$

Equivalently, expanding by cofactors of the third column yields

$$\det(u, v, w) = \det \begin{pmatrix} u_1 & v_1 & w_1 \\ u_2 & v_2 & w_2 \\ u_3 & v_3 & w_3 \end{pmatrix}.$$

---

## 10.4   Summary

Although we do not need coordinates to construct an algebra for points and vectors, we do need coordinates to communicate this algebra to a computer. For easy reference, we collect here the low-level coordinate computations corresponding to the high-level algebraic operations.

*Addition and Subtraction*

$$u \pm v = (u_1 \pm v_1, u_2 \pm v_2, u_3 \pm v_3),$$
$$P \pm v = (p_1 \pm v_1, p_2 \pm v_2, p_3 \pm v_3),$$
$$Q - P = (q_1 - p_1, q_2 - p_2, q_3 - p_3).$$

*Scalar Multiplication*

$$cu = (cu_1, cu_2, cu_3).$$

*Dot Product*

$$u \cdot v = u_1 v_1 + u_2 v_2 + u_3 v_3.$$

*Cross Product*

$$u \times v = (u_2 v_3 - u_3 v_2, u_3 v_1 - u_1 v_3, u_1 v_2 - u_2 v_1) = \det \begin{pmatrix} i & j & k \\ u_1 & u_2 & u_3 \\ v_1 & v_2 & v_3 \end{pmatrix}.$$

*Determinant*

$$\det(u, v, w) = \det \begin{pmatrix} u_1 & v_1 & w_1 \\ u_2 & v_2 & w_2 \\ u_3 & v_3 & w_3 \end{pmatrix}.$$

Notice that the cross product is much more expensive to compute than the dot product: cross product requires six multiplications and three subtractions, whereas dot product uses only three multiplications and two additions. Thus, the cross product is about twice as expensive to compute as the dot product. For this reason given a choice, we usually prefer formulas that invoke dot product rather than cross product. The Lagrange identity from Chapter 9 allows us to replace two cross products and a dot product by four dot products.

## Exercises

10.1. Explain the computational significance of Chapter 9, Exercise 9.1.

10.2. Using a computer algebra system such as Mathematica or Maple to simplify the coordinate computations, verify symbolically the following identities:

a. $(v \times w) \times u = (u \cdot v)\, w - (u \cdot w)v.$

b. $(u_1 \times u_2) \cdot (v_1 \times v_2) = (u_1 \cdot u_2)\,(v_1 \cdot v_2) - (u_1 \cdot v_2)(u_2 \cdot v_1).$

10.3. Show that if $v_1 = (r_1, \theta_1, z_1)$ and $v_2 = (r_2, \theta_2, z_2)$ are cylindrical coordinates for $v_1$, $v_2$, then

$$v_1 \cdot v_2 = r_1 r_2 \cos(\theta_1 - \theta_2) + z_1 z_2.$$

10.4. Using the formula $u \cdot v = u_1 v_1 + u_2 v_2 + u_3 v_3$, verify the commutative and distributive properties of the dot product.

10.5. Using the formulas

$$u \cdot v = u_1 v_1 + u_2 v_2 + u_3 v_3,$$

$$u \times v = \det \begin{pmatrix} i & j & k \\ u_1 & u_2 & u_3 \\ v_1 & v_2 & v_3 \end{pmatrix},$$

verify the following identities:

a. $u \times (v + w) = u \times v + u \times w.$

b. $(v + w) \times u = v \times u + w \times u.$

c. $u \times u = 0.$

d. $u \times v = -v \times u.$

e. $(u \times v) \cdot u = 0.$

f. $(u \times v) \cdot v = 0$.

g. $|u \times v|^2 = |u|^2 \, |v|^2 - (u \cdot v)^2$.

10.6. Using the formula

$$\det(u, v, w) = \det \begin{pmatrix} u_1 & v_1 & w_1 \\ u_2 & v_2 & w_2 \\ u_3 & v_3 & w_3 \end{pmatrix},$$

verify the following identities:

a. $\det(u, u, w) = \det(u, v, u) = \det(u, v, v) = 0$.

b. $\det(u, v, w) = \det(v, w, u) = \det(w, u, v)$.

c. $\det(v, u, w) = -\det(u, v, w)$.

d. $\det(u_1 + c\, u_2, v, w) = \det(u_1, v, w) + c\, \det(u_2, v, w)$.

e. $\det(u, v_1 + c\, v_2, w) = \det(u, v_1, w) + c\, \det(u, v_2, w)$.

f. $\det(u, v, w_1 + c\, w_2) = \det(u, v, w_1) + c\, \det(u, v, w_2)$.

In the following exercises, we will develop an alternative approach to deriving expressions for the dot and cross products in terms of rectangular coordinates by a method that does not invoke the distributive law (Chapter 9, Exercise 9.6).

10.7. Recall from Chapter 4, Section 4.2.2 that in two dimensions the matrix representing rotation of vectors around the origin by the angle $\theta$ is given by:

$$Rot(\theta) = \begin{pmatrix} \cos\theta & \sin\theta \\ -\sin\theta & \cos\theta \end{pmatrix}.$$

a. Show that in three dimensions, the matrices representing rotation of vectors around the coordinate axes are given by:

i. $Rot(k, \theta) = \begin{pmatrix} \cos\theta & \sin\theta & 0 \\ -\sin\theta & \cos\theta & 0 \\ 0 & 0 & 1 \end{pmatrix}$    rotation around the $z$-axis.

ii. $Rot(j, \theta) = \begin{pmatrix} \cos\theta & 0 & \sin\theta \\ 0 & 1 & 0 \\ -\sin\theta & 0 & \cos\theta \end{pmatrix}$    rotation around the $y$-axis.

iii. $Rot(i, \theta) = \begin{pmatrix} 1 & 0 & 0 \\ 0 & \cos\theta & \sin\theta \\ 0 & -\sin\theta & \cos\theta \end{pmatrix}$    rotation around the $x$-axis.

b. Show that:

i. $Rot(k, \theta)^T = Rot(k, \theta)^{-1}$.

ii. $Rot(j, \theta)^T = Rot(j, \theta)^{-1}$.

iii. $Rot(i, \theta)^T = Rot(i, \theta)^{-1}$.

10.8. Let $R$ be a matrix representing a rotation about an arbitrary vector in three dimensions.

    a. Show that $R$ is equivalent to a product of rotations around the coordinate axes.

    b. Using part a and the results of Exercise 10.7, conclude that $R^T = R^{-1}$.

10.9. Let $u = (u_1, u_2, u_3)$ and $v = (v_1, v_2, v_3)$, and let $\theta$ be the angle between $u$ and $v$.

    a. Show that:

      i. For any two vectors $u, v$,

$$u * v^T = u_1\, v_1 + u_2\, v_2 + u_3\, v_3.$$

      ii. For any rotation matrix $R$,

$$(u * R) * (v * R)^T = u * v^T.$$

      iii. If $u$ lies along the $x$-axis and $v$ lies in the $xy$-plane, then

$$u * v^T = |u||v| \cos \theta.$$

      iv. There exists a rotation matrix $R$ such that $u * R$ lies along the $x$-axis and $v * R$ lies in the $xy$-plane.

    b. Conclude from part a that:

$$u_1 v_1 + u_2 v_2 + u_3 v_3 = |u||v| \cos \theta = u \cdot v.$$

10.10. Let $u = (u_1, u_2, u_3)$ and $v = (v_1, v_2, v_3)$, and let $\theta$ be the angle between $u$ and $v$. Set

$$u \cdot v = u_1 v_1 + u_2 v_2 + u_3 v_3,$$
$$u \times v = (u_2 v_3 - u_3 v_2, u_3 v_1 - u_1 v_3, u_1 v_2 - u_2 v_1).$$

    a. Using these definitions, show that:

      i. $(u \times v) \cdot (u \times v) = (u \cdot u)(v \cdot v) - (u \cdot v)^2.$

      ii. $(u \times v) \cdot u = (u \times v) \cdot v = 0.$

      iii. $\det(u, v, u \times v) = (u \times v) \cdot (u \times v).$

    b. Using the results of Exercise 10.9 and part a, show that:

      i. $|u \times v| = |u|\,|v|\, \sin \theta.$

      ii. $u \times v \perp u, v.$

      iii. $\mathrm{sgn}(u, v, u \times v) > 0.$

    c. Conclude from part b that $(u_2 v_3 - u_3 v_2, u_3 v_1 - u_1 v_3, u_1 v_2 - u_2 v_1)$ represents the cross product in terms of rectangular coordinates.

# Chapter 11

## Some Applications of Vector Geometry

*apply thine heart unto my knowledge.*

– Proverbs 22:18

## 11.1 Introduction

Vector geometry and vector algebra are powerful tools for deriving geometric relationships and algebraic identities. In this chapter, we shall employ vector methods to derive trigonometric laws, metric expressions, intersection formulas, interpolation techniques, and inside–outside tests that have wide-ranging applications in Computer Graphics and Geometric Modeling.

## 11.2 Trigonometric Laws

Trigonometric laws are often required for the analysis of geometry. As early as Chapter 1, we employed the Law of Cosines to compute the length of the radius in a wheel and the lengths of the diagonals in a rosette. Here, we shall provide simple derivations for the Law of Cosines and the Law of Sines using dot product and cross product.

### 11.2.1 Law of Cosines

Consider $\triangle ABC$ with corresponding sides $a, b, c$ (see Figure 11.1). The Law of Cosines asserts that

$$c^2 = a^2 + b^2 - 2ab\cos(C). \tag{11.1}$$

The Law of Cosines is a generalization of the Pythagorean Theorem; indeed, when $C = \pi/2$, the Law of Cosines is the Pythagorean Theorem, since $\cos(\pi/2) = 0$.

*Whenever you see cosine, think dot product.* To derive the Law of Cosines, recall that

$$c^2 = |B - A|^2 = (B - A) \cdot (B - A). \tag{11.2}$$

Since we need to introduce the sides $a, b$ into the formula for $c^2$, observe that

$$B - A = (B - C) + (C - A). \tag{11.3}$$

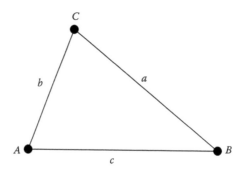

**FIGURE 11.1:**   Law of Cosines:   $c^2 = a^2 + b^2 - 2ab \cos(C)$.

Now the rest of the derivation is mechanical. Substituting Equation 11.3 into Equation 11.2, expanding by the distributive law, and invoking the definition of the dot product yields

$$
\begin{aligned}
c^2 &= (B-A) \cdot (B-A) \\
&= ((B-C)+(C-A)) \cdot ((B-C)+(C-A)) \\
&= (B-C) \cdot (B-C) + 2(B-C) \cdot (C-A) + (C-A) \cdot (C-A) \\
&= |B-C|^2 - 2(B-C) \cdot (A-C) + |C-A|^2 \\
&= a^2 - 2ab \cos(C) + b^2.
\end{aligned}
$$

### 11.2.2   Law of Sines

Consider $\triangle ABC$ in Figure 11.2. The Law of Sines asserts that

$$\frac{a}{\sin(A)} = \frac{b}{\sin(B)} = \frac{c}{\sin(C)}. \tag{11.4}$$

*Whenever you see sine, think cross product.* To derive the Law of Sines, observe that by the definition of the cross product,

$$
\begin{aligned}
2Area(\triangle ABC) &= |(A-B) \times (C-B)| = ca \sin(B), \\
2Area(\triangle ABC) &= |(C-A) \times (B-A)| = bc \sin(A), \\
2Area(\triangle ABC) &= |(B-C) \times (A-C)| = ab \sin(C).
\end{aligned}
$$

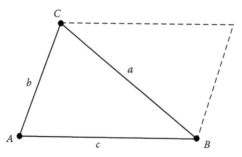

**FIGURE 11.2:**   Law of Sines:   $\dfrac{a}{\sin(A)} = \dfrac{b}{\sin(B)} = \dfrac{c}{\sin(C)}$.

Therefore,

$$ca \sin (B) = bc \sin (A) = ab \sin (C),$$

so

$$\frac{a}{\sin (A)} = \frac{b}{\sin (B)} = \frac{c}{\sin (C)}.$$

## 11.3 Representations for Lines and Planes

We are going to derive formulas for distances and intersections between simple geometric objects. We shall focus primarily on lines and planes, so a word about representations for lines and planes is in order here before we proceed any further.

### 11.3.1 Lines

Two points determine a line, but a point and a vector also determine a line (see Figure 11.3, left). Generally, it will be more convenient to represent a line $L$ in terms of a point $P$ on the line and a vector $v$ parallel to the line, since the formulas for distances and intersections are readily represented in terms of dot and cross products, which apply to vectors but not to points. The line $L$ determined by the point $P$ and the vector $v$ can be expressed parametrically by the linear equation:

$$L(t) = P + tv, \tag{11.5}$$

which is equivalent to the three linear parametric equations for the coordinates

$$x(t) = p_1 + tv_1, \quad y(t) = p_2 + tv_2, \quad z(t) = p_3 + tv_3.$$

### 11.3.2 Planes

There are two convenient ways to represent a plane in three dimensions: parametrically and implicitly. Just like a line $L$ can be represented by a point $P$ and a nonzero vector $v$, a plane $S$ can be represented by a point $P$ and two linearly independent vectors $u, v$ (see Figure 11.3, right). The plane $S$ determined by the point $P$ and the vectors $u, v$ can be expressed parametrically by the linear equation:

$$S(s, t) = P + su + tv, \tag{11.6}$$

**FIGURE 11.3:** A line $L$ determined by a point $P$ and a direction vector $v$ (left). A plane $S$ determined either by a point $P$ and two linearly independent vectors $u, v$ or by a point $P$ and a normal vector $N$ (right). Notice that $N \parallel u \times v$.

which is equivalent to the three linear parametric equations for the coordinates

$$x(s,t) = p_1 + su_1 + tv_1, \quad y(s,t) = p_2 + su_2 + tv_2, \quad z(s,t) = p_3 + su_3 + tv_3.$$

Alternatively, a plane $S$ can be represented by a point $P$ on $S$ and a vector $N$ normal to $S$. A point $Q$ lies on the plane $S$ if and only if the vector $Q - P$ is perpendicular to the normal vector $N$—that is, if and only if

$$N \cdot (Q - P) = 0. \tag{11.7}$$

Equation 11.7 is called the *implicit equation* of the plane $S$. Notice that the implicit equation is also a linear equation. Indeed, if $N = (a, b, c)$, $Q = (x, y, z)$, and $d = -N \cdot P$, then

$$N \cdot (Q - P) = 0 \iff ax + by + cz + d = 0. \tag{11.8}$$

Equation 11.8 is called an *implicit equation* because there are no explicit expressions for the coordinates $x, y, z$ as there are in the parametric representation. Notice that if we have a parametric representation for the plane $S$ in terms of a point $P$ and two linearly independent vectors $u, v$, then we can easily generate an implicit representation, since $N = u \times v$ is normal to the plane $S$.

---

## 11.4   Metric Formulas

Here we collect some simple formulas for distance, area, and volume, which will be useful in a variety of future applications.

### 11.4.1   Distance

We are interested mainly in the distance between points, lines, and planes. Distance is often related to projection and projection is computed using dot product, so distance formulas are typically expressed in terms of dot products.

#### 11.4.1.1   Distance between Two Points

The distance between two points $P, Q$ is the same as the length of the vector from $P$ to $Q$ (see Figure 11.4). Therefore,

$$Dist^2(P, Q) = |Q - P|^2 = (Q - P) \cdot (Q - P). \tag{11.9}$$

**FIGURE 11.4:**   $Dist^2(P, Q) = |Q - P|^2 = (Q - P) \cdot (Q - P).$

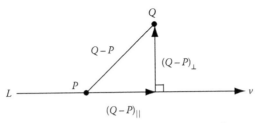

**FIGURE 11.5:** $Dist^2(P,L) = (Q-P) \cdot (Q-P) - \dfrac{((Q-P) \cdot v)^2}{v \cdot v}$.

### 11.4.1.2 Distance between a Point and a Line

Consider a point $Q$ and a line $L$ determined by a point $P$ and a direction vector $v$ (see Figure 11.5). The distance between the point $Q$ and the line $L$ is the length of the perpendicular component of the vector $Q - P$ relative to the vector $v$. Therefore, by the Pythagorean Theorem,

$$Dist^2(Q,L) = |(Q-P)_\perp|^2 = |Q-P|^2 - |(Q-P)_\parallel|^2,$$

so

$$Dist^2(Q,L) = (Q-P) \cdot (Q-P) - \frac{((Q-P) \cdot v)^2}{v \cdot v}. \qquad (11.10)$$

Notice that if $v$ is a unit vector, then $v \cdot v = 1$, so in this case,

$$Dist^2(Q,L) = (Q-P) \cdot (Q-P) - ((Q-P) \cdot v)^2. \qquad (11.11)$$

### 11.4.1.3 Distance between a Point and a Plane

Consider a point $Q$ and a plane $S$ determined by a point $P$ and a normal vector $N$ (see Figure 11.6). The distance between the point $Q$ and the plane $S$ is the length of the parallel component of the vector $Q - P$ relative to the vector $N$. Therefore, by the definition of the dot product:

$$Dist(Q,S) = |(Q-P)_\parallel| = |Q-P|\cos(\theta) = \frac{|Q-P||N|\cos(\theta)}{|N|} = \frac{|(Q-P) \cdot N|}{|N|}. \qquad (11.12)$$

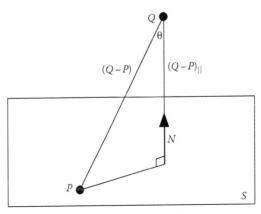

**FIGURE 11.6:** $Dist(Q,S) = |(Q-P)_\parallel| = \dfrac{|(Q-P) \cdot N|}{|N|}$.

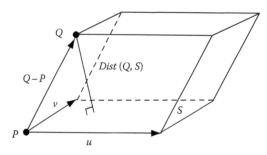

**FIGURE 11.7:** $Dist(Q, S) = \dfrac{Vol(u, v, Q - P)}{Area(u, v)} = \dfrac{|Det(u, v, Q - P)|}{|u \times v|}$.

Notice that if $N$ is a unit vector, then $|N| = 1$, so in this case,

$$Dist(Q, S) = |(Q - P) \cdot N|. \tag{11.13}$$

Alternatively, instead of representing the plane $S$ by a point $P$ and a normal vector $N$, suppose $S$ is represented by a point $P$ and a pair of vectors $u,v$ (see Figure 11.7). Since

$$\textit{Volume of a parallelepiped} = \textit{Area of the base} \times \textit{Height of the parallelepiped},$$

it follows that

$$Vol(u, v, Q - P) = Area(u, v) \times Dist(Q, S).$$

Therefore, solving for $Dist(Q, S)$ and invoking the definitions of cross product and determinant, we find that

$$Dist(Q, S) = \frac{Vol(u, v, Q - P)}{Area(u, v)} = \frac{|Det(u, v, Q - P)|}{|u \times v|}. \tag{11.14}$$

### 11.4.1.4   Distance between Two Lines

If two lines fail to intersect, then either the lines are parallel or the lines are skew. We shall consider each of these cases in turn.

**11.4.1.4.1   Distance between Two Parallel Lines**     Consider a pair of lines, $L_1, L_2$, parallel to a vector $v$. Let $P_1$ be a point on $L_1$ and let $P_2$ be a point on $L_2$ (see Figure 11.8). Then,

$$Dist(L_1, L_2) = Dist(P_2, L_1).$$

Therefore, by Equation 11.10,

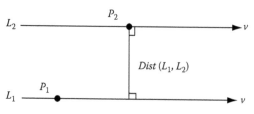

**FIGURE 11.8:**  $Dist^2(L_1, L_2) = (P_2 - P_1) \cdot (P_2 - P_1) - \dfrac{((P_2 - P_1) \cdot v)^2)}{v \cdot v}$.

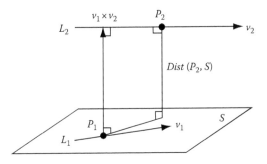

**FIGURE 11.9:** $Dist(L_1, L_2) = \dfrac{\left|Det((P_2 - P_1), v_1, v_2)\right|}{|v_1 \times v_2|}.$

$$Dist^2(L_1, L_2) = (P_2 - P_1) \cdot (P_2 - P_1) - \frac{((P_2 - P_1) \cdot v)^2}{v \cdot v}. \qquad (11.15)$$

**11.4.1.4.2   Distance between Two Skew Lines**   Consider a pair of skew lines, $L_1, L_2$, where $L_1$ is parallel to the vector $v_1$ and $L_2$ is parallel to the vector $v_2$. Let $P_1$ be a point on $L_1$, let $P_2$ be a point on $L_2$, and let $S$ be the plane determined by the point $P_1$ and the normal vector $v_1 \times v_2$ (see Figure 11.9). Then since $v_1 \times v_2$ is perpendicular to both $L_1$ and $L_2$,

$$Dist(L_1, L_2) = Dist(P_2, S).$$

Therefore, by Equation 11.12,

$$Dist(L_1, L_2) = \frac{|(P_2 - P_1) \cdot v_1 \times v_2|}{|v_1 \times v_2|} = \frac{|Det(P_2 - P_1, v_1, v_2)|}{|v_1 \times v_2|}. \qquad (11.16)$$

## 11.4.2   Area

We are interested mainly in the areas of triangles, parallelograms, and planar polygons. Since the cross product is defined in terms of area, formulas for area are typically expressed in terms of cross products.

### 11.4.2.1   Triangles and Parallelograms

Consider a parallelogram determined by two vectors $u, v$ (see Figure 11.10, left). Then, by the definition of the cross product,

$$Area(u, v) = |u \times v|. \qquad (11.17)$$

Two triangles make a parallelogram (see Figure 11.10, right). Therefore,

$$Area(\Delta PQR) = \frac{Area(Q - P, R - P)}{2} = \frac{|(Q - P) \times (R - P)|}{2}. \qquad (11.18)$$

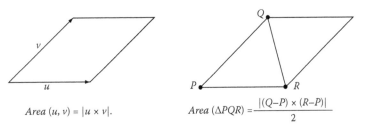

**FIGURE 11.10:**   Area formulas for the parallelogram (left) and the triangle (right).

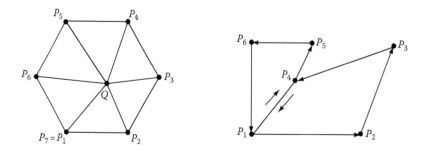

**FIGURE 11.11:** Convex polygon (left) and non-convex polygon (right). The non-convex polygon on the right can be split into two convex polygons by inserting the edge $P_1P_4$. Notice that the new edge $P_1P_4$ has opposite orientations in the polygons $P_1P_2P_3P_4$ and $P_1P_4P_5P_6$.

### 11.4.2.2   Polygons: Newell's Formula

To extend our area formula from the triangles to polygons, consider a closed convex planar polygon $P$ with vertices $P_1, \ldots, P_{n+1}$, where $P_{n+1} = P_1$ (see Figure 11.11, left). Let $Q$ be any point in the interior of the polygon $P$, and define

$$N(P) = \frac{1}{2} \sum_{k=1}^{n} (P_k - Q) \times (P_{k+1} - Q). \tag{11.19}$$

Since the vectors $(P_k - Q)$, $k = 1, \ldots, n+1$, all lie in the plane of the polygon $P$, the cross products $(P_k - Q) \times (P_{k+1} - Q)$, $k = 1, \ldots, n$, all point in the same direction normal to the plane of the polygon $P$. Moreover, by Equation 11.18,

$$Area(\Delta P_k P_{k+1} \, Q) = \frac{|(P_{k+1} - Q) \times (P_k - Q)|}{2}.$$

Therefore,

$$N(P) \perp P$$

$$|N(P)| = \frac{1}{2} \left| \sum_{k=1}^{n} (P_k - Q) \times (P_{k+1} - Q) \right| = \frac{1}{2} \sum_{k=1}^{n} |(P_k - Q) \times (P_{k+1} - Q)| = Area(P).$$

$$\tag{11.20}$$

The expression on the right-hand side of Equation 11.19 for the vector $N(P)$ is independent of the choice of the point $Q$ in the polygon $P$, since the direction and length of $N(P)$ in Equation 11.20 are independent of the choice of $Q$. In fact, the point $Q$ need not lie inside the polygon $P$ or even in the plane of the polygon $P$; the expression on the right-hand side of Equation 11.19 represents the same vector $N(P)$ for any point $Q$. Indeed, let $Q, R$ be two arbitrary points. Then,

$$\sum_{k=1}^{n} (P_k - R) \times (P_{k+1} - R) = \sum_{k=1}^{n} ((P_k - Q) + (Q - R)) \times ((P_{k+1} - Q) + (Q - R)).$$

Expanding the right-hand side by the distributive property and using the identity $(Q - R) \times (Q - R) = 0$ yields

$$\sum_{k=1}^{n} (P_k - R) \times (P_{k+1} - R) = \sum_{k=1}^{n} (P_k - Q) \times (P_{k+1} - Q) + \sum_{k=1}^{n} (P_k - Q) \times (Q - R)$$
$$+ \sum_{k=1}^{n} (Q - R) \times (P_{k+1} - Q).$$

Now observe that

$$\sum_{k=1}^{n} (P_k - Q) \times (Q - R) = (P_1 - Q) \times (Q - R) + (P_2 - Q) \times (Q - R) + \cdots + (P_n - Q) \times (Q - R),$$

$$\sum_{k=1}^{n} (Q - R) \times (P_{k+1} - Q) = (Q - R) \times (P_2 - Q) + \cdots + (Q - R) \times (P_{n+1} - Q).$$

Since the cross product is anticommutative and since $P_{n+1} = P_1$, these sums cancel, and we are left with

$$\sum_{k=1}^{n} (P_k - R) \times (P_{k+1} - R) = \sum_{k=1}^{n} (P_k - Q) \times (P_{k+1} - Q).$$

Choosing $R$ to be the origin, the expression for computing the vector $N(P)$ simplifies to Newell's formula:

$$N(P) = \frac{1}{2} \sum_{k=1}^{n} P_k \times P_{k+1}, \tag{11.21}$$

- $N(P) \perp P$

- $|N(P)| = Area(P)$

We derived Equation 11.21 for convex polygons, but this formula remains valid even for non-convex polygons. If $P$ is not convex, we can split $P$ into a collection of convex polygons by inserting some extra edges $P_j P_k$ (see Figure 11.11, right). Applying Equation 11.21 to each of the convex polygons and summing, we find that the terms involving the new edges $P_j P_k$ appear twice with opposite orientations. Thus, the terms $P_j \times P_k$ and $P_k \times P_j$ cancel and we arrive again at Equation 11.21.

### 11.4.3 Volume

We are interested here mainly in two formulas: one for the volume of tetrahedra and one for the volume of parallelepipeds. A formula for the volume of arbitrary polyhedra is provided in Exercise 11.5. Since the determinant is defined in terms of volume, formulas for volume are typically expressed in terms of determinants.

Consider a parallelepiped determined by three vectors $u$, $v$, $w$ (see Figure 11.12, left). By the definition of the determinant function,

$$Vol(u, v, w) = |Det(u, v, w)|. \tag{11.22}$$

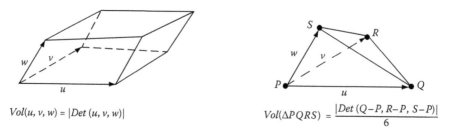

$$Vol(u, v, w) = |Det\ (u, v, w)| \qquad Vol(\Delta PQRS) = \frac{|Det\ (Q{-}P, R{-}P, S{-}P)|}{6}$$

**FIGURE 11.12:**   Volume formulas for the parallelepiped (left) and the tetrahedron (right).

Now consider a tetrahedron with vertices $P, Q, R, S$, and let $u = Q - P$, $v = R - P$, $w = S - P$. A tetrahedron is a triangular pyramid (see Figure 11.12, right). Since

$$Volume\ of\ a\ pyramid = \frac{1}{3}Area\ of\ the\ base\ of\ the\ pyramid \times Height\ of\ the\ pyramid,$$

the volume of a tetrahedron with vertices $P, Q, R, S$ is

$$Vol(\Delta PQRS) = \frac{1}{3}Base \times Height = \frac{1}{3}Area(\Delta PQR) \times Height(\Delta PQRS)$$

$$= \frac{1}{6}Area(u, v) \times Height(u, v, w)$$

$$= \frac{1}{6}Vol(u, v, w)$$

$$= \frac{1}{6}|Det(u, v, w)|.$$

Therefore,

$$Vol(\Delta PQRS) = \frac{1}{6}|Det(Q - P, R - P, S - P)|. \qquad (11.23)$$

## 11.5   Intersection Formulas for Lines and Planes

Formulas for intersecting lines and planes are important in Computer Graphics both for hidden surface procedures and for ray casting algorithms. We shall study hidden surface procedures and ray casting algorithms later in this book. Here we prepare the way by deriving closed formulas for the intersections of lines and planes.

### 11.5.1   Intersecting Two Lines

Consider a pair of intersecting lines, $L_1, L_2$, where $L_1$ is parallel to the vector $v_1$ and $L_2$ is parallel to the vector $v_2$. Let $P_1$ be a point on $L_1$ and let $P_2$ be a point on $L_2$. We seek the intersection point $P$ lying on both $L_1$ and $L_2$ (see Figure 11.13).

At the intersection point $P$, there are parameters $s, t$ such that

$$P_1 + sv_1 = P_2 + tv_2, \qquad (11.24)$$

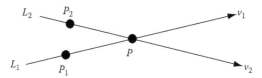

**FIGURE 11.13:** A pair of intersecting lines $L_1$ and $L_2$. The line $L_1$ passes through the point $P_1$ in the direction $v_1$; the line $L_2$ passes through the point $P_2$ in the direction $v_2$.

or equivalently

$$sv_1 - tv_2 = P_2 - P_1. \tag{11.25}$$

Dotting both sides of Equation 11.25 with both $v_1$ and $v_2$ yields two linear equations in the two unknown parameters $s, t$:

$$\begin{aligned} (v_1 \cdot v_1)s - (v_1 \cdot v_2)t &= v_1 \cdot (P_2 - P_1), \\ (v_1 \cdot v_2)s - (v_2 \cdot v_2)t &= v_2 \cdot (P_2 - P_1). \end{aligned} \tag{11.26}$$

Solving for $s, t$ by Cramer's rule, we find that at the intersection point

$$s^* = \frac{Det\begin{pmatrix} v_1 \cdot (P_2 - P_1) & -v_1 \cdot v_2 \\ v_2 \cdot (P_2 - P_1) & -v_2 \cdot v_2 \end{pmatrix}}{Det\begin{pmatrix} v_1 \cdot v_1 & -v_1 \cdot v_2 \\ v_1 \cdot v_2 & -v_2 \cdot v_2 \end{pmatrix}} \quad \text{and} \quad t^* = \frac{Det\begin{pmatrix} v_1 \cdot v_1 & v_1 \cdot (P_2 - P_1) \\ v_1 \cdot v_2 & v_2 \cdot (P_2 - P_1) \end{pmatrix}}{Det\begin{pmatrix} v_1 \cdot v_1 & -v_1 \cdot v_2 \\ v_1 \cdot v_2 & -v_2 \cdot v_2 \end{pmatrix}}. \tag{11.27}$$

Therefore,

$$P = P_1 + s^* v_1 = P_2 + t^* v_2. \tag{11.28}$$

### 11.5.2 Intersecting Three Planes

Consider a plane $S$ determined by a point $P$ on the plane and a vector $N$ normal to the plane. Recall that another point $Q$ lies on the plane $S$ if and only if the vector $Q - P$ is perpendicular to the normal vector $N$ (see Figure 11.14)—that is, if and only if

$$N \cdot (Q - P) = 0. \tag{11.29}$$

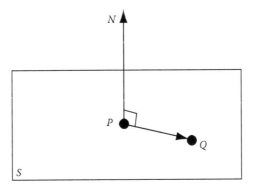

**FIGURE 11.14:** A plane $S$ determined by a point $P$ and a normal vector $N$. A point $Q$ lies on the plane $S$ if and only if the vector $Q - P$ is perpendicular to the normal vector $N$.

Now consider three planes $S_1, S_2, S_3$ intersecting in a common point $Q$. Let the plane $S_k$ be determined by the point $P_k$ and the normal vector $N_k$, $k = 1, 2, 3$. Since the intersection point $Q$ lies on each of the three planes, the point $Q$ must satisfy the equation of each of these planes. Therefore,

$$N_k \cdot (Q - P_k) = 0, \quad k = 1, 2, 3. \tag{11.30}$$

Equation 11.30 represents three linear equations in three unknowns, the three coordinates of the point $Q$. We could solve these equations by linear algebra, using either Cramer's rule or Gaussian elimination. But there is a better way; we can just write down the answer! In fact,

$$Q = R + \frac{(N_1 \cdot (P_1 - R))N_2 \times N_3 + (N_2 \cdot (P_2 - R))N_3 \times N_1 + (N_3 \cdot (P_3 - R))N_1 \times N_2}{Det(N_1\ N_2\ N_3)}, \tag{11.31}$$

where $R$ is any point whatsoever. We can check that this expression for $Q$ is valid by verifying that $Q$ satisfies Equation 11.30. For example, since

$$N_1 \cdot N_1 \times N_2 = N_1 \cdot N_3 \times N_1 = 0 \quad \text{and} \quad N_1 \cdot N_2 \times N_3 = Det(N_1\ N_2\ N_3),$$

it follows that

$$N_1 \cdot (Q - P_1) \ = N_1 \cdot (R - P_1) + N_1 \cdot (P_1 - R) = 0.$$

Similarly, we can show that

$$N_2 \cdot (Q - P_2) \ = N_3 \cdot (Q - P_3) = 0.$$

Therefore, $Q$ lies on all three planes. Moreover, since $R$ can be any point whatsoever, we can choose $R$ to be the origin. In this case, the expression for computing the intersection point $Q$ simplifies to

$$Q = \frac{(N_1 \cdot P_1)N_2 \times N_3 \ + \ (N_2 \cdot P_2)N_3 \times N_1 + (N_3 \cdot P_3)N_1 \times N_2}{Det(N_1\ N_2\ N_3)}. \tag{11.32}$$

An alternative derivation of Equation 11.32 based on solving the linear system in Equation 11.30 by inverting the coefficient matrix is provided in Exercise 11.13.

### 11.5.3   Intersecting Two Planes

Three planes intersect in a point; two planes intersect in a line. Let the plane $S_k$ be determined by the point $P_k$ and the normal vector $N_k$, $k = 1, 2$. To find the line $L$ in which the planes $S_1$ and $S_2$ intersect, we need to find a point $Q$ on $L$ and a direction vector $v$ parallel to $L$.

The direction vector $v$ is easy. Since the line $L$ lies in both planes, the direction vector $v$ must be perpendicular to the normal to both planes. Therefore, we can choose

$$v = N_1 \times N_2. \tag{11.33}$$

To find a point $Q$ on the line $L$, we need to solve two linear equations in three unknowns:

$$\begin{aligned} N_1 \cdot (Q - P_1) &= 0, \\ N_2 \cdot (Q - P_2) &= 0. \end{aligned} \tag{11.34}$$

There are infinitely many solutions to these two equations, since there are infinitely many points $Q$ on the line $L$. To find a unique solution $Q$ and to make the problem deterministic, we can add the equation of any plane that intersects the line $L$. Since the vector $N_1 \times N_2$ is parallel to the direction of the line $L$, any plane with normal vector $N_1 \times N_2$ will certainly intersect the line $L$. Therefore, we can choose any point $P_3$, and find a unique point $Q$ on the line $L$ by solving the three linear equations in three unknowns:

$$N_1 \cdot (Q - P_1) = 0,$$
$$N_2 \cdot (Q - P_2) = 0,$$
$$(N_1 \times N_2) \cdot (Q - P_3) = 0, \tag{11.35}$$

for the coordinates of the point $Q$.

We already know how to solve for the intersection of three planes, since we solved this problem in the previous section. Setting $N_3 = N_1 \times N_2$ in Equation 11.31 and invoking the identities

$$N_2 \times (N_1 \times N_2) = (N_2 \cdot N_2)N_1 - (N_1 \cdot N_2)N_2,$$
$$(N_1 \times N_2) \times N_1 = (N_1 \cdot N_1)N_2 - (N_1 \cdot N_2)N_1,$$
$$Det(N_1, N_2, N_1 \times N_2) = |N_1 \times N_2|^2,$$

yields

$$Q = R + \frac{(N_1 \cdot (P_1 - R))((N_2 \cdot N_2)N_1 - (N_1 \cdot N_2)N_2)}{|N_1 \times N_2|^2}$$
$$+ \frac{(N_2 \cdot (P_2 - R))((N_1 \cdot N_1)N_2 - (N_1 \cdot N_2)N_1) + ((N_1 \times N_2) \cdot (P_3 - R))N_1 \times N_2}{|N_1 \times N_2|^2}, \tag{11.36}$$

where $R$ and $P_3$ are any points whatsoever. Since $R$ and $P_3$ can be any points whatsoever, we can choose both $R$ and $P_3$ to be the origin. In this case, the expression for the point $Q$ on the intersection line $L$ simplifies to

$$Q = \frac{(N_1 \cdot P_1)((N_2 \cdot N_2)N_1 - (N_1 \cdot N_2)N_2) + (N_2 \cdot P_2)((N_1 \cdot N_1)N_2 - (N_1 \cdot N_2)N_1)}{|N_1 \times N_2|^2}. \tag{11.37}$$

Finally, notice that if $N_1$ and $N_2$ are unit vectors, then,

$$N_1 \cdot N_1 = N_2 \cdot N_2 = 1,$$
$$|N_1 \times N_2|^2 = 1 - (N_1 \cdot N_2)^2,$$

so in this case,

$$Q = \frac{(N_1 \cdot P_1)(N_1 - (N_1 \cdot N_2)N_2) + (N_2 \cdot P_2)(N_2 - (N_1 \cdot N_2)N_1)}{1 - (N_1 \cdot N_2)^2}. \tag{11.38}$$

## 11.6 Spherical Linear Interpolation

Interpolation plays a key role in many algorithms in Computer Graphics. The simplest kind of interpolation is *linear interpolation*. Let $P_0, P_1$ be two points in affine space. Then $v = P_1 - P_0$ is the vector from $P_0$ to $P_1$. Therefore, the equation of the straight line $L$ joining the points $P_0, P_1$ is

$$L(t) = P_0 + tv = P_0 + t(P_1 - P_0),$$

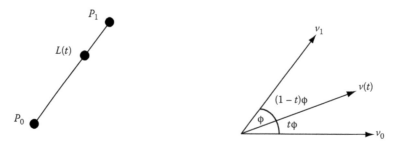

**FIGURE 11.15:** Linear interpolation between two points along a straight line (left), and spherical linear interpolation between two vectors along a circular arc (right).

or equivalently

$$L(t) = (1 - t)P_0 + tP_1. \tag{11.39}$$

Notice that

$$|L(t) - P_0| = t|P_1 - P_0|.$$

Thus, if $d$ is the distance from $P_0$ to $P_1$, then $td$ is the distance from $P_0$ to $L(t)$. Equation 11.39 is called *linear interpolation* because the line $L(t)$ interpolates the points $P_0$ and $P_1$, and distance varies linearly along the line $L(t)$.

Suppose, however, that instead of starting with two points along a straight line, we start with two vectors along a circular arc (see Figure 11.15) and we want to interpolate these vectors uniformly by vectors along this arc. That is, given two vectors $v_0, v_1$ of the same length, we seek an equation for the vectors $v(t)$ along the arc joining the vectors $v_0, v_1$ so that if $\phi$ is the angle from $v_0$ to $v_1$, then $t\phi$ is the angle from $v_0$ to $v(t)$. The following theorem gives a formula for the vectors $v(t)$.

**Theorem 11.1:** $v(t) = \frac{\sin((1-t)\phi)}{\sin(\phi)}v_0 + \frac{\sin(t\phi)}{\sin(\phi)}v_1$, *where $\phi$ is the angle between $v_0$ and $v_1$.*

**Proof:** The vector $v(t)$ lies in the plane of the vectors $v_0, v_1$. Therefore, there are constants $\alpha, \beta$ such that

$$v(t) = \alpha v_0 + \beta v_1.$$

*Whenever you see sine, think cross product.* Crossing both sides of this equation with $v_0$ and $v_1$ yields:

$$v_0 \times v(t) = \beta v_0 \times v_1,$$

$$v(t) \times v_1 = \alpha v_0 \times v_1.$$

Taking the lengths of both sides of these two equations, we get

$$|v_0||v(t)| \sin(t\phi) = \beta|v_0||v_1| \sin(\phi),$$
$$|v_1||v(t)| \sin((1 - t)\phi) = \alpha|v_0||v_1| \sin(\phi).$$

But $|v_0| = |v_1| = |v(t)|$, so solving these two equations for $\alpha, \beta$, we find that

$$\alpha = \frac{\sin((1 - t)\phi)}{\sin(\phi)},$$

$$\beta = \frac{\sin(t\phi)}{\sin(\phi)}.$$

Hence,

$$v(t) = \frac{\sin((1-t)\phi)}{\sin(\phi)} v_0 + \frac{\sin(t\phi)}{\sin(\phi)} v_1.$$

Notice, in particular, that $v(0) = v_0$ and $v(1) = v_1$.

The formula for $v(t)$ in Theorem 11.1 is called *spherical linear interpolation* or SLERP and we write

$$slerp(v_0, v_1, t) = \frac{\sin((1-t)\phi)}{\sin(\phi)} v_0 + \frac{\sin(t\phi)}{\sin(\phi)} v_1, \tag{11.40}$$

where $\phi$ is the angle between $v_0$ and $v_1$, because $slerp(v_0, v_1, t)$ interpolates the vectors $v_0$ and $v_1$, and the angle varies linearly along the arc joining $v_0$ and $slerp(v_0, v_1, t)$. The term *spherical* refers to the fact that since the vectors $v_0$ and $v_1$ have the same length, we can think of these vectors as representing points on a sphere. Equation 11.40 then interpolates two points on a sphere by moving uniformly along a great circle, a geodesic, on the sphere.

Notice that spherical linear interpolation is valid for vectors in any dimension greater than or equal to two, even though our proof using cross product assumes that $v_0$ and $v_1$ lie in three dimensions. In fact, since $v_0$ and $v_1$ lie in a plane, we can always embed these vectors in a three-dimensional subspace and apply the argument in Theorem 11.1. Since the result is coordinate free, spherical linear interpolation is valid in any dimension (see also Exercise 11.17). We shall have occasion to use spherical linear interpolation in dimensions three and four, when we study shading algorithms (Chapter 21) and refraction techniques (Chapter 17) in three dimensions and key frame animation (Chapter 15) with quaternions in four dimensions.

## 11.7 Inside–Outside Tests

For many algorithms in Computer Graphics such as clipping and shading, it is important to know if a given point lies inside or outside of a fixed polygon. There are two standard ways to solve this problem: ray casting and winding number.

### 11.7.1 Ray Casting

In ray casting, we fire a ray from the given point $P$ to the fixed polygon $S$. If the point $P$ lies outside the polygon $S$, the number of intersections between the ray and the polygon is even, since the ray must alternate entering and exiting the polygon; if the point $P$ lies inside the polygon $S$ the number of intersections between the ray and the polygon is odd, since the line must alternate exiting and entering the polygon before exiting the polygon one final time (see Figure 11.16).

Ray casting works fine in most cases, but ray casting can get into trouble if the ray intersects the polygon at or near one of the vertices of the polygon. Should a vertex count as one or two intersections? If we count a vertex as one intersection, we may get the wrong result for points

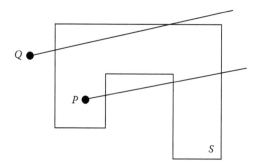

**FIGURE 11.16:**   Ray casting. A ray emanating from a point $P$ inside the polygon $S$ will intersect the polygon in an odd number of points; a ray emanating from a point $Q$ outside the polygon $S$ will intersect the polygon in an even number of points.

outside the polygon; if we count a vertex as two intersections, we may get the wrong result for points inside the polygon. There is no consistent way to handle vertices that always gives the correct answer. Fortunately, if we fire rays in random directions, we are not likely to hit a vertex, so usually ray casting works well, but there is another approach that avoids this problem altogether.

## 11.7.2   Winding Number

The *winding number* relative to a point $P$ of an oriented closed planar curve $C$ with no self intersection points is the number of times the curve $C$ circles around the point $P$. Let $Wind(C, P)$ denote the winding number of $C$ relative to $P$. Then

$$Wind(C, P) = 0 \quad \Leftrightarrow \quad P \text{ lies outside of } C,$$

$$Wind(C, P) = \pm 1 \quad \Leftrightarrow \quad P \text{ lies inside of } C.$$

When the point $P$ lies inside the curve $C$, the orientation of the curve determines the sign of the winding number; counterclockwise is positive orientation, clockwise is negative orientation.

For polygons $S$ there is a closed formula for computing the winding number for any point $P$. Let $P_1, \ldots P_n$ be the vertices of the polygon $S$, and let $P_{n+1} = P_1$. Then

$$Wind(S, P) = \frac{1}{2\pi} \sum_{k=1}^{n} \arcsin \left( \frac{|(P_{k+1} - P) \times (P_k - P)|}{|P_{k+1} - P||P_k - P|} \right). \tag{11.41}$$

This formula is valid for any polygon; the polygon need not be regular or even convex.

To understand why this formula works, think about a turtle located at the point $P$ facing initially in the direction of the vertex $P_1$. To measure how much the polygon $S$ winds around $P$, the turtle measures the angles $A_k = \angle P_k P P_{k+1}$ for $k = 1, \ldots, n$. For each angle $A_k$, the turtle executes the command TURN $A_k$. After executing TURN $A_1$, the turtle is facing the vertex $P_2$; after executing TURN $A_k$, the turtle is facing the vertex $P_{k+1}$. Thus, the turtle starts and ends facing the vertex $P_1$, so the total amount the turtle turns must be an integer multiple of $2\pi$. Some angles may be positive (counterclockwise) and some negative (clockwise), but the total turning must always be an integer multiple of $2\pi$.

To calculate the angles $A_k$ for $k = 1, \ldots, n$, we can use the cross product. The angle $A_k = \angle P_k P P_{k+1}$ is the angle between the vectors $P_k - P$ and $P_{k+1} - P$ (see Figure 11.17), and from the definition of the cross product

$$\sin(\angle P_k P P_{k+1}) = \frac{|(P_{k+1} - P) \times (P_k - P)|}{|P_{k+1} - P||P_k - P|}.$$

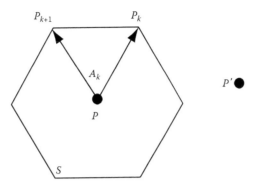

**FIGURE 11.17:** Winding number. If $S$ is a regular polygon and the point $P$ is at the center, then the sum of the central angles is $\pm 2\pi$, so the winding number is $\pm 1$. If the point $P'$ is far away from the polygon $S$, then the turtle will never face in a direction $180°$ away from the centroid of the polygon, so the winding number is 0.

Therefore,

$$A_k = \angle P_k P P_{k+1} = \arcsin\left(\frac{|(P_{k+1} - P) \times (P_k - P)|}{|P_{k+1} - P||P_k - P|}\right),$$

so

$$\sum_{k=1}^{n} A_k = \sum_{k=1}^{n} \arcsin\left(\frac{|(P_{k+1} - P) \times (P_k - P)|}{|P_{k+1} - P||P_k - P|}\right).$$

Since the total turtle turning must be an integer multiple of $2\pi$, the winding number is given by

$$Wind(S, P) = \frac{1}{2\pi} \sum_{k=1}^{n} A_k = \frac{1}{2\pi} \sum_{k=1}^{n} \arcsin\left(\frac{|(P_{k+1} - P) \times (P_k - P)|}{|P_{k+1} - P||P_k - P|}\right).$$

When the point $P$ lies outside of the polygon $S$, the winding number $Wind(S, P) = 0$. We can verify this result in the following fashion. If the point $P$ is far away from the polygon $S$, then the turtle will never face in a direction $180°$ away from the centroid of the polygon (see Figure 11.17). Therefore, the turtle never turns through more than a total of $\pi$ radians, so since the total turtle turning is an integer multiple of $2\pi$, the total turtle turning must be zero. Now as long as we do not cross an edge of the polygon, the winding number is a continuous function, so if we move the turtle a little bit, the winding number can only change by a little bit. But the winding number is an integer, so if the winding number can only change by a little bit, then the winding number cannot change at all. Therefore, as we move the point $P$ outside the polygon $S$, the winding number never changes; hence, if $P$ lies outside of $S$, then $Wind(S, P) = 0$.

Conversely, when the point $P$ lies inside of the polygon $S$, the winding number $Wind(S, P) = \pm 1$. We can verify this result in the following fashion. If $S$ is a regular polygon and the point $P$ is at the center, then the sum of the central angles is $\pm 2\pi$, so the winding number is $\pm 1$ (see Figure 11.17). Again, as long as we do not cross an edge of the polygon, the winding number is a continuous function, so if we move the turtle a little bit, the winding number can only change by a little bit. Hence, once again, since the winding number is an integer, the winding number cannot change at all inside the polygon. Thus, if the point $P$ lies inside a regular polygon $S$, then $Wind(S, P) = \pm 1$. Moreover, this result remains valid even if $S$ is not a regular polygon because we can deform a regular polygon into an arbitrary polygon. Since the winding number is a continuous function, the winding number is unchanged under small deformations of the polygon; therefore, no matter how we deform that polygon, if the point $P$ lies inside the polygon $S$, then $Wind(S, P) = \pm 1$.

In addition to correctly distinguishing the difference between inside and outside, the winding number has one additional advantage: Equation 11.41 for the winding number is numerically stable. If $\angle P_k P P_{k+1}$ is small, then,

$$\angle P_k P P_{k+1} \approx \sin(\angle P_k P P_{k+1}) = \frac{|(P_{k+1} - P) \times (P_k - P)|}{|P_{k+1} - P||P_k - P|},$$

so we do not lose precision for small angles when we calculate the cross product.

## 11.8 Summary

Vector algebra is a potent tool for analyzing geometry. Trigonometric laws, metric expressions, intersection formulas, and inside–outside tests have all been derived here using vector techniques. Vector techniques are cleaner and more powerful than coordinate methods. From now on, whenever you encounter a geometric problem, you should eschew coordinate methods in favor of vector techniques.

We have encountered two helpful analytic rules that you should remember forever:

1. *Whenever you see cosine, think dot product.*

2. *Whenever you see sine, think cross product.*

Similarly, there are three useful geometric heuristics that you should always keep in mind:

3. *Whenever you see distance, think dot product.*

4. *Whenever you see area, think cross product.*

5. *Whenever you see volume, think determinant.*

Next, for easy reference, we list all the key formulas that we have derived in this chapter.

## 11.8.1 Trigonometric Laws

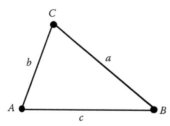

Law of Cosines: $c^2 = a^2 + b^2 - 2ab \cos(C)$.

Law of Sines: $\dfrac{a}{\sin(A)} = \dfrac{b}{\sin(B)} = \dfrac{c}{\sin(C)}$.

## 11.8.2  Metric Formulas

### 11.8.2.1  Distance

*Distance between two points*

$$Dist^2(P, Q) = |Q - P|^2 = (Q - P) \cdot (Q - P).$$

*Distance between a point and a line*

$$Dist^2(P, L) = Dist^2(P, Q + tv) = (Q - P) \cdot (Q - P) - \frac{((Q - P) \cdot v)^2}{v \cdot v}.$$

*Distance between a point and a plane*

$$Dist(Q, S) = Dist(Q, P + su + tv) = \frac{Vol(u, v, Q - P)}{Area(u, v)} = \frac{|Det(u, v, Q - P)|}{|u \times v|},$$

$$Dist(Q, S) = \frac{|(Q - P) \cdot N|}{|N|} \quad N \| u \times v.$$

*Distance between parallel lines*

$$Dist^2(L_1, L_2) = Dist^2(P_1 + sv, P_2 + tv) = (P_2 - P_1) \cdot (P_2 - P_1) - \frac{((P_2 - P_1) \cdot v)^2}{v \cdot v}.$$

*Distance between skew lines*

$$Dist(L_1, L_2) = Dist(P_1 + sv_1, P_2 + tv_2) = \frac{|Det((P_2 - P_1), v_1, v_2)|}{|v_1 \times v_2|}.$$

### 11.8.2.2  Area

*Area of a parallelogram*

$$Area(u, v) = |u \times v|.$$

*Area of a triangle*

$$Area(\Delta PQR) = \frac{|(Q - P) \times (R - P)|}{2}.$$

*Area of a planar polygon*

$$\text{Vertices } P = \{P_1, \ldots, P_n\},$$

$$Area(P) = \frac{1}{2} \left| \sum_{k=1}^{n} P_k \times P_{k+1} \right|.$$

### 11.8.2.3    Volume

*Volume of a parallelepiped*

$$Vol(u, v, w) = |Det(u, v, w)|.$$

*Volume of a tetrahedron*

$$Vol(\Delta PQRS) = \frac{|Det(Q - P, R - P, S - P)|}{6}.$$

### 11.8.3    Intersections

*Two lines:* $L_1(s) = P_1 + sv_1$, $L_2(t) = P_2 + tv_2$

$$s^* = \frac{Det\begin{pmatrix} v_1 \cdot (P_2 - P_1) & -v_1 \cdot v_2 \\ v_2 \cdot (P_2 - P_1) & -v_2 \cdot v_2 \end{pmatrix}}{Det\begin{pmatrix} v_1 \cdot v_1 & -v_1 \cdot v_2 \\ v_1 \cdot v_2 & -v_2 \cdot v_2 \end{pmatrix}} \quad \text{and} \quad t^* = \frac{Det\begin{pmatrix} v_1 \cdot v_1 & v_1 \cdot (P_2 - P_1) \\ v_1 \cdot v_2 & v_2 \cdot (P_2 - P_1) \end{pmatrix}}{Det\begin{pmatrix} v_1 \cdot v_1 & -v_1 \cdot v_2 \\ v_1 \cdot v_2 & -v_2 \cdot v_2 \end{pmatrix}},$$

$$P = P_1 + s^* v_1 = P_2 + t^* v_2.$$

*Three planes:* $S_1 : N_1 \cdot (Q - P_1) = 0$, $S_2 : N_2 \cdot (Q - P_2) = 0$, $S_3 : N_3 \cdot (Q - P_3) = 0$

$$Q = \frac{(N_1 \cdot P_1)N_2 \times N_3 + (N_2 \cdot P_2)N_3 \times N_1 + (N_3 \cdot P_3)N_1 \times N_2}{Det(N_1 \, N_2 \, N_3)}.$$

*Two planes:* $S_1 : N_1 \cdot (Q - P_1) = 0$, $S_2 : N_2 \cdot (Q - P_2) = 0$

$$v = N_1 \times N_2,$$
$$Q = \frac{(N_1 \cdot P_1)((N_2 \cdot N_2)N_1 - (N_1 \cdot N_2)N_2) + (N_2 \cdot P_2)((N_1 \cdot N_1)N_2 - (N_1 \cdot N_2)N_1)}{|N_1 \times N_2|^2},$$
$$L(t) = Q + tv.$$

### 11.8.4    Interpolation

*Linear interpolation*

$$L(t) = (1 - t)P_0 + tP_1.$$

*Spherical linear interpolation (SLERP)*

$$slerp(v_0, v_1, t) = \frac{\sin((1 - t)\phi)}{\sin(\phi)} v_0 + \frac{\sin(t\phi)}{\sin(\phi)} v_1,$$

where $\phi$ is the angle between $v_0$ and $v_1$.

### 11.8.5    Winding Number

$$\text{Vertices } S = \{P_1, \ldots, P_n\},$$

$$Wind(S, P) = \frac{1}{2\pi} \sum_{k=1}^{n} \arcsin\left( \frac{|(P_{k+1} - P) \times (P_k - P)|}{|P_{k+1} - P||P_k - P|} \right).$$

## Exercises

11.1. Consider a point $Q$ and a line $L$ determined by a point $P$ and a direction vector $v$.

a. Show that

$$Dist^2(Q, L) = \frac{((Q - P) \times v)^2}{v \cdot v}.$$

b. Why is Equation 11.10 preferable to the formula in part a?

11.2. Consider a circle inscribed in a triangle (see Figure 11.18). Let $C_{in}$ denote the center of this circle and let $r_{in}$ denote the radius of this circle. Verify that:

a. $C_{in} = \dfrac{|P_3 - P_2|P_1 + |P_1 - P_3|P_2 + |P_2 - P_1|P_3}{|P_3 - P_2| + |P_1 - P_3| + |P_2 - P_1|}.$

b. $r_{in} = \dfrac{2Area(\Delta P_1 P_2 P_3)}{Perimeter(\Delta P_1 P_2 P_3)}.$

11.3. Show that:

a. $Area(\Delta PQR) = \dfrac{1}{2}\left| Det\begin{pmatrix} P & 1 \\ Q & 1 \\ R & 1 \end{pmatrix} \right| = \dfrac{1}{2}\left| Det\begin{pmatrix} p_1 & p_2 & 1 \\ q_1 & q_2 & 1 \\ r_1 & r_2 & 1 \end{pmatrix} \right|.$

b. $Vol(\Delta PQRS) = \dfrac{1}{6}\left| Det\begin{pmatrix} P & 1 \\ Q & 1 \\ R & 1 \\ S & 1 \end{pmatrix} \right| = \dfrac{1}{6}\left| Det\begin{pmatrix} p_1 & p_2 & p_3 & 1 \\ q_1 & q_2 & q_3 & 1 \\ r_1 & r_2 & r_3 & 1 \\ s_1 & s_2 & s_3 & 1 \end{pmatrix} \right|.$

11.4. Consider a pyramid $V$ with apex $A$ and polygonal base $P$ with vertices $P_1, \ldots, P_m$, and let $P_{m+1} = P_1$. Show that

$$Vol(V) = \frac{1}{6}\left| \sum_{j=1}^{m} Det(A - Q, P_j - Q, P_{j+1} - Q) \right|,$$

where $Q$ is any point in the plane of the polygon $P$.

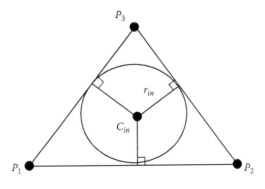

**FIGURE 11.18:** A circle with center $C_{in}$ and radius $r_{in}$ inscribed in $\Delta P_1 P_2 P_3$.

11.5. Consider a convex polyhedron $V$ with polygonal faces $S_1, \ldots, S_n$ oriented consistently either clockwise or counterclockwise with respect to the outward pointing normals to the faces. With respect to this orientation, let $P_{k,1}, \ldots, P_{k,m(k)}$ be the vertices of $S_k$, and let $P_{k,m(k)+1} = P_{k,1}, k = 1, \ldots, n$. Finally, let $A$ be a point in the interior of $V$, and let $Q_k$ be any point in the plane of the polygon $S_k, k = 1, \ldots, n$.

a. Using Exercise 11.4, show that

$$Vol(V) = \frac{1}{6} \left| \sum_{k=1}^{n} \sum_{j=1}^{m(k)} Det(A - Q_k, P_{k,j} - Q_k, P_{k,j+1} - Q_k) \right|.$$

b. Show that the formula in part a is valid for any point $A$, even for points outside the polyhedron $V$. That is, show that for any two points $A$, $B$

$$\frac{1}{6} \left| \sum_{k=1}^{n} \sum_{j=1}^{m(k)} Det(A - Q_k, P_{k,j} - Q_k, P_{k,j+1} - Q_k) \right|$$

$$= \frac{1}{6} \left| \sum_{k=1}^{n} \sum_{j=1}^{m(k)} Det(B - Q_k, P_{k,j} - Q_k, P_{k,j+1} - Q_k) \right|.$$

c. Choosing $A$ to be the origin, conclude from parts a and b that the expression for computing the volume of a polyhedron $V$ simplifies to

$$Vol(V) = \frac{1}{6} \left| \sum_{k=1}^{n} \sum_{j=1}^{m(k)} Det(Q_k, P_{k,j}, P_{k,j+1}) \right|.$$

d. Show that the formula in part c is valid even for nonconvex polyhedra.

11.6. Let $Q$ be the point defined in Equation 11.36. Verify that:

$$N_1 \cdot (Q - P_1) = 0,$$
$$N_2 \cdot (Q - P_2) = 0.$$

11.7. What does it mean if $Det(N_1 \; N_2 \; N_3) = 0$ in Equation 11.31 or 11.32?

11.8. What does it mean if $|N_1 \times N_2|^2 = 0$ in Equation 11.36 or 11.37?

11.9. Consider a line $L$ through a point $P$ in the direction $v$, and a plane $S$ through a point $Q$ with normal vector $N$. Show that the line $L$ and the plane $S$ intersect in the point $R$, where

$$R = P + \frac{N \cdot (Q - P)}{N \cdot v} v.$$

11.10. Consider a pair of intersecting lines, $L_1, L_2$, where $L_1$ is parallel to the vector $v_1$ and $L_2$ is parallel to the vector $v_2$. Let $P_1$ be a point on $L_1$ and let $P_2$ be a point on $L_2$.

a. Show that the point $P$ on the intersection of the two line $L_1$ and $L_2$ is the same as the point on the intersection of the three planes defined by the equations:

$$(v_1 \times v_2) \cdot (P - P_1) = 0,$$
$$((v_1 \times v_2) \times v_1) \cdot (P - P_1) = 0,$$
$$((v_1 \times v_2) \times v_2) \cdot (P - P_2) = 0.$$

b. Using part a and Equation 11.31, find a closed form expression for the intersection point $P$.

c. Show that the expression you derived in part b is equivalent to the result in Equation 11.28.

11.11. Consider a pair of skew lines, $L_1, L_2$, where $L_1$ is parallel to the vector $v_1$ and $L_2$ is parallel to the vector $v_2$. Let $P_1$ be a point on $L_1$, and let $P_2$ be a point on $L_2$. Show that Equations 11.27 and 11.28 provide the points of closest approach of the two lines.

11.12. Verify that:

a. $M = \begin{pmatrix} u^T & v^T & w^T \end{pmatrix} = \begin{pmatrix} u_1 & v_1 & w_1 \\ u_2 & v_2 & w_2 \\ u_3 & v_3 & w_3 \end{pmatrix} \Rightarrow$

$$M^{-1} = \frac{\begin{pmatrix} v \times w \\ w \times u \\ u \times v \end{pmatrix}}{\mathrm{Det}\begin{pmatrix} u^T & v^T & w^T \end{pmatrix}} = \frac{\begin{pmatrix} v_2w_3 - v_3w_2 & v_3w_1 - v_1w_3 & v_1w_2 - v_2w_1 \\ w_2u_3 - w_3u_2 & w_3u_1 - w_1u_3 & w_1u_2 - w_2u_1 \\ u_2v_3 - u_3v_2 & u_3v_1 - u_1v_3 & u_1v_2 - u_2v_1 \end{pmatrix}}{\mathrm{Det}\begin{pmatrix} u^T & v^T & w^T \end{pmatrix}}.$$

b. $M = \begin{pmatrix} u \\ v \\ w \end{pmatrix} = \begin{pmatrix} u_1 & u_2 & u_3 \\ v_1 & v_2 & v_3 \\ w_1 & w_2 & w_3 \end{pmatrix} \Rightarrow$

$$M^{-1} = \frac{\begin{pmatrix} (v \times w)^T & (w \times u)^T & (u \times v)^T \end{pmatrix}}{\mathrm{Det}\begin{pmatrix} u \\ v \\ w \end{pmatrix}} = \frac{\begin{pmatrix} v_2w_3 - v_3w_2 & w_2u_3 - w_3u_2 & u_2v_3 - u_3v_2 \\ v_3w_1 - v_1w_3 & w_3u_1 - w_1u_3 & u_3v_1 - u_1v_3 \\ v_1w_2 - v_2w_1 & w_1u_2 - w_2u_1 & u_1v_2 - u_2v_1 \end{pmatrix}}{\mathrm{Det}\begin{pmatrix} u \\ v \\ w \end{pmatrix}}.$$

11.13. Let $S_k$ be the plane determined by the point $P_k$ and the normal vector $N_k$, $k = 1, 2, 3$, and let $Q$ be the point of intersection of the three planes $S_1, S_2, S_3$.

a. Show that

$$\begin{aligned} N_1 \cdot (Q - P_1) &= 0 \\ N_2 \cdot (Q - P_2) &= 0 \\ N_3 \cdot (Q - P_3) &= 0 \end{aligned} \quad \Leftrightarrow \quad Q * \begin{pmatrix} N_1^T & N_2^T & N_3^T \end{pmatrix} = \begin{pmatrix} N_1 \cdot P_1 & N_2 \cdot P_2 & N_3 \cdot P_3 \end{pmatrix}.$$

b. Using part a and Exercise 11.12a, show that

$$Q = \frac{(N_1 \cdot P_1)N_2 \times N_3 + (N_2 \cdot P_2)N_3 \times N_1 + (N_3 \cdot P_3)N_1 \times N_2}{\mathrm{Det}\begin{pmatrix} N_1^T & N_2^T & N_3^T \end{pmatrix}}.$$

11.14. What is the value of the winding number

$$Wind(S, P) = \frac{1}{2\pi} \sum_{k=1}^{n} \arcsin\left(\frac{|(P_{k+1} - P) \times (P_k - P)|}{|P_{k+1} - P| \, |P_k - P|}\right),$$

when:

a. $P$ is a point on an edge of the polygon $S$.

b. $P$ is a vertex of the polygon $S$.

11.15. Prove in two ways that nonsingular affine transformations preserve ratios of volumes:

a. By using the fact that nonsingular affine transformations preserve ratios of distances along lines.

b. By invoking Exercise 11.3b.

11.16. Let $v_0, v_1$ be unit vectors. Show that the unit vector that bisects the angle between $v_0$ and $v_1$ is

$$v = \sec(\phi/2)\left(\frac{v_0 + v_1}{2}\right),$$

where $\phi$ is the angle between $v_0$ and $v_1$.

11.17. In this exercise, we provide an alternative derivation of the formula for $slerp(v_0, v_1, t)$ using only the dot product.

a. Derive a formula for $slerp(v_0, v_1, t)$ by dotting both sides of the equation:

$$v(t) = \alpha v_0 + \beta v_1,$$

with $v_0, v_1$ and solving the resulting linear equations for $\alpha, \beta$.

b. Using the trigonometric identity,

$$\cos(\theta - \phi) = \cos(\theta)\cos(\phi) + \sin(\theta)\sin(\phi),$$

show that the formula for $slerp(v_0, v_1, t)$ derived in part a is equivalent to the formula for $slerp(v_0, v_1, t)$ provided in Equation 11.40.

c. Conclude that Equation 11.40 for $slerp(v_0, v_1, t)$ is independent of the dimension of the vectors $v_0, v_1$.

# Chapter 12

## Coordinate-Free Formulas for Affine and Projective Transformations

*be ye transformed by the renewing of your mind*

– Romans 12:2

### 12.1 Transformations for Three-Dimensional Computer Graphics

Computer Graphics in two dimensions is grounded on simple transformations of the plane. Turtle Graphics and LOGO are based on conformal transformations; Affine Graphics and CODO on affine transformations. The turtle uses translation, rotation, and uniform scaling to navigate on the plane, and traces of the turtle's path generate a wide variety of shapes, ranging from simple polygons to complex fractals. Affine Graphics uses affine transformations to position, orient, and scale objects as well as to model shape. The vertices of simple figures like polygons and stars are typically positioned using translation and rotation; ellipses are generated by scaling circles nonuniformly along preferred directions; fractals are built using iterated function systems.

Computer Graphics in three dimensions is grounded in transformations of three-space. But in three dimensions Computer Graphics employs not only nonsingular conformal and affine transformations, but also affine projections and projective transformations. Projections play a key role in three-dimensional Computer Graphics because we need to display three-dimensional geometry on a two-dimensional screen. Orthogonal projections of three-dimensional models from several different directions are traditional in many engineering applications. Perspective projection plays a central role in Computer Graphics because, as in classical art, we often want to display three-dimensional shapes in perspective just as they appear to the human eye. Visual realism requires projective transformations.

Here we are going to study three-dimensional affine and projective transformation, using the coordinate-free techniques of vector geometry.

### 12.2 Affine and Projective Transformations

An *affine transformation* is a transformation that maps points to points, maps vectors to vectors, and preserves linear and affine combinations. In particular, $A$ is an *affine transformation* if

$$
\begin{aligned}
A(u + v) &= A(u) + A(v), \\
A(c\,v) &= c\,A(v), \\
A(P + v) &= A(P) + A(v).
\end{aligned}
\tag{12.1}
$$

163

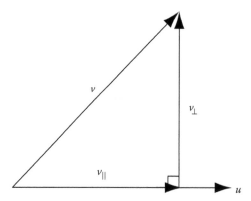

**FIGURE 12.1:**   Decomposing a vector $v$ into components parallel and perpendicular to a fixed vector $u$.

As a consequence of Equation 12.1, nondegenerate affine transformations in three dimensions map lines to lines, triangles to triangles, and parallelograms to parallelograms just as in two dimensions. The most important affine transformations in three-dimensional Computer Graphics are the rigid motions (translation, rotation, and mirror image), scaling (both uniform and nonuniform), and orthogonal projection.

Perspective projection is not an affine transformation. Indeed, as we shall see shortly, perspective projection is not defined at every point nor is perspective well defined for any vector. Nevertheless, in Section 12.5.2 we will derive a simple, straightforward formula for computing perspective projection on any point where perspective is well defined.

In this chapter we are going to derive explicit formulas for each of the most important affine and projective transformations in three-dimensional Computer Graphics—translation, rotation, mirror image, uniform and nonuniform scaling, and orthogonal and perspective projection—using coordinate-free vector algebra. In Chapter 13, we will apply these formulas to construct matrices to represent each of these transformations.

Before we proceed, we need to recall the following results. In most of our derivations we shall be required to decompose an arbitrary vector $v$ into components $v_\parallel$ and $v_\perp$ parallel and perpendicular to a fixed unit vector $u$ (see Figure 12.1). Recall from Chapter 9 that:

$$
\begin{aligned}
v &= v_\parallel + v_\perp, \\
v_\parallel &= (v \cdot u)u, \\
v_\perp &= v - (v \cdot u)u.
\end{aligned}
\tag{12.2}
$$

Be sure that you are comfortable with these formulas before you continue because we shall use these equations again and again throughout this chapter.

## 12.3   Rigid Motions

*Rigid motions* are transformations that preserve lengths and angles. The rigid motions in three dimensions are composites of three basic transformations: translation, rotation, and mirror image.

### 12.3.1   Translation

Translation is defined by specifying a distance and a direction. Since a distance and direction designate a vector, a translation is defined by specifying a translation vector. The formulas for translation in three dimensions are similar to the formulas for translation in two dimensions: points are affected by translation, vectors are unaffected by translation (see Figure 12.2). Let $w$ be a translation vector. If $P$ is a point and $v$ is a vector, then

$$v^{new} = v,$$
$$P^{new} = P + w. \tag{12.3}$$

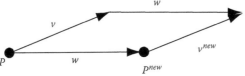

**FIGURE 12.2:** Translation. Points are affected by translation; vectors are not affected by translation.

### 12.3.2   Rotation

Rotation in three dimensions is defined by specifying an angle and an axis of rotation. The axis line $L$ is typically described by specifying a point $Q$ on $L$ and a unit direction vector $u$ parallel to $L$. We shall denote the angle of rotation by $\theta$ (see Figure 12.3).

The *Formula of Rodrigues* computes rotation around the axis $L$ through the angle $\theta$ by setting

$$v^{new} = (\cos \theta)v + (1 - \cos \theta)(v \cdot u)u + (\sin \theta)u \times v$$
$$P^{new} = Q + (\cos \theta)(P - Q) + (1 - \cos \theta)((P - Q) \cdot u)u + (\sin \theta)u \times (P - Q). \tag{12.4}$$

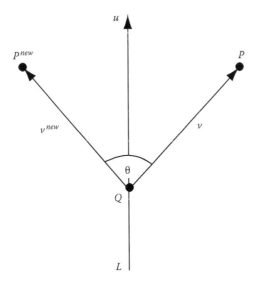

**FIGURE 12.3:** Rotation through the angle $\theta$ around the axis line $L$, determined by the point $Q$ and the unit direction vector $u$.

Notice that since

$$P = Q + (P - Q),$$

it follows that

$$P^{new} = Q^{new} + (P - Q)^{new} = Q + (P - Q)^{new}.$$

So the rotation formula for points follows immediately from the rotation formula for vectors. Therefore it is enough to prove Equation 12.4 for vectors.

To derive the *Formula of Rodrigues* for vectors, split $v$ into two components: the component $v_{\parallel}$ parallel to $u$ and the component $v_{\perp}$ perpendicular to $u$ (see Figure 12.4, left). Then

$$v = v_{\parallel} + v_{\perp},$$

so by linearity

$$v^{new} = v_{\parallel}^{new} + v_{\perp}^{new}.$$

Since $v_{\parallel}$ is parallel to the axis of rotation,

$$v_{\parallel}^{new} = v_{\parallel}.$$

Hence we need only compute the effect of rotation on $v_{\perp}$. Now $v_{\perp}$ lies in the plane perpendicular to $u$; moreover, $u \times v_{\perp}$ is perpendicular to $v_{\perp}$ and lies in the plane perpendicular to $u$ (see Figure 12.4, right). Therefore using our standard approach to rotation in the plane, we find that

$$v_{\perp}^{new} = (\cos \theta)v_{\perp} + (\sin \theta)u \times v_{\perp}.$$

Now we have

$$v^{new} = v_{\parallel}^{new} + v_{\perp}^{new},$$
$$v_{\parallel}^{new} = v_{\parallel},$$
$$v_{\perp}^{new} = (\cos \theta)v_{\perp} + (\sin \theta)u \times v_{\perp}.$$

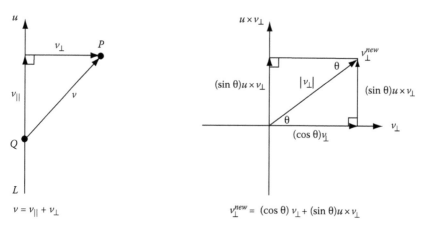

**FIGURE 12.4:**  Decomposing $v$ into components parallel and perpendicular to the axis of rotation (left), and rotation in the plane perpendicular to the axis of rotation (right).

Therefore,

$$v^{new} = v_{\parallel} + (\cos \theta)v_{\perp} + (\sin \theta)u \times v_{\perp}. \tag{12.5}$$

But since $u$ is a unit vector,

$$v_{\parallel} = (v \cdot u)u,$$
$$v_{\perp} = v - v_{\parallel} = v - (v \cdot u)u.$$

Substituting these two identities into Equation (12.5) yields

$$v^{new} = (v \cdot u)u + (\cos \theta)(v - (v \cdot u)u) + (\sin \theta)u \times (v - (v \cdot u)u)$$

Thus, since cross product distributes through addition and $u \times u = 0$,

$$v^{new} = (\cos \theta)v + (1 - \cos \theta)(v \cdot u)u + (\sin \theta)u \times v,$$

which is the *Formula of Rodrigues* for vectors.

### 12.3.3   Mirror Image

A mirror in three dimensions is a plane $S$, typically specified by a point $Q$ on $S$ and a unit vector $n$ normal to $S$ (see Figure 12.5, left). If $P$ is a point and $v$ is a vector, then the formulas for mirror image are:

$$v^{new} = v - 2(v \cdot n)n,$$
$$P^{new} = P + 2((P - Q) \cdot n)n. \tag{12.6}$$

Notice that since

$$P = Q + (P - Q),$$

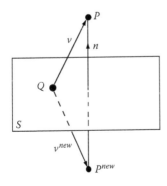
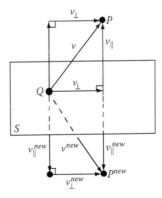

**FIGURE 12.5:**   A mirror plane $S$ specified by a point $Q$ and a unit normal vector $n$ (left), and the decomposition of a vector into components parallel and perpendicular to the normal vector (right).

it follows that

$$P^{new} = Q^{new} + (P - Q)^{new} = Q + (P - Q)^{new},$$

so the formula for mirror image for points follows immediately from the formula for mirror image for vectors. Therefore once again it is enough to prove the mirror image formula for vectors.

To derive the mirror image formula for vectors, we again split $v$ into two components: the component $v_{\parallel}$ parallel to $n$ and the component $v_{\perp}$ perpendicular to $n$ (see Figure 12.5, right). Then

$$v = v_{\perp} + v_{\parallel},$$

so by linearity

$$v^{new} = v_{\perp}^{new} + v_{\parallel}^{new}.$$

Since $v_{\perp}$ lies in the plane $S$,

$$v_{\perp}^{new} = v_{\perp}.$$

Moreover, since $v_{\parallel}$ is parallel to the normal vector $n$,

$$v_{\parallel}^{new} = -v_{\parallel}.$$

Therefore,

$$v^{new} = v_{\perp}^{new} + v_{\parallel}^{new} = v_{\perp} - v_{\parallel}. \tag{12.7}$$

But,

$$v_{\perp} = v - v_{\parallel},$$
$$v_{\parallel} = (v \cdot n)n.$$

Substituting these identities into Equation (12.7) yields

$$v^{new} = v - 2(v \cdot n)n,$$

which is the mirror image formula for vectors.

---

## 12.4   Scaling

Both uniform and nonuniform scaling are important in three-dimensional Computer Graphics. Next we treat each of these transformations in turn.

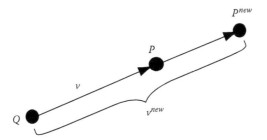

**FIGURE 12.6:** Uniform scaling around the fixed point $Q$ for arbitrary points $P$ and arbitrary vectors $v$.

### 12.4.1 Uniform Scaling

Uniform scaling is defined by specifying a point $Q$ from which to scale distance and a scale factor $s$. The formulas for uniform scaling in three dimensions are identical to the formulas for uniform scaling in two dimensions (see Figure 12.6): If $P$ is a point and $v$ is a vector, then

$$v^{new} = sv,$$
$$P^{new} = Q + s(P - Q) = sP + (1 - s)Q. \tag{12.8}$$

### 12.4.2 Nonuniform Scaling

For nonuniform scaling, we need to designate a unit vector $w$ specifying the scaling direction, as well as a point $Q$ and a scale factor $s$ (see Figure 12.7, left). If $P$ is a point and $v$ is a vector, then the formulas for nonuniform scaling are

$$v^{new} = v + (s - 1)(v \cdot w)w,$$
$$P^{new} = P + (s - 1)((P - Q) \cdot w)w. \tag{12.9}$$

As usual nonuniform scaling for points follows easily from nonuniform scaling for vectors, since

$$P = Q + (P - Q),$$

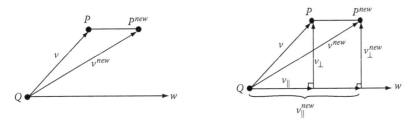

**FIGURE 12.7:** Nonuniform scaling is specified around a point $Q$ in a direction $w$ (left). The decomposition of a vector into components parallel and perpendicular to the direction vector (right).

so

$$P^{new} = Q^{new} + (P - Q)^{new} = Q + (P - Q)^{new}.$$

Therefore once again it is enough to prove the nonuniform scaling formula for vectors.

  To derive the nonuniform scaling formula for vectors, we again split $v$ into two components: the component $v_{\parallel}$ parallel to $w$ and the component $v_{\perp}$ perpendicular to $w$ (see Figure 12.7, right). Then, as usual,

$$v = v_{\perp} + v_{\parallel},$$

so by linearity

$$v^{new} = v_{\perp}^{new} + v_{\parallel}^{new}.$$

Since $v_{\perp}$ is perpendicular to the scaling direction $w$,

$$v_{\perp}^{new} = v_{\perp},$$

and since $v_{\parallel}$ is parallel to the scaling direction $w$,

$$v_{\parallel}^{new} = sv_{\parallel}.$$

Therefore,

$$v^{new} = v_{\perp}^{new} + v_{\parallel}^{new} = v_{\perp} + sv_{\parallel}. \tag{12.10}$$

But once again

$$v_{\perp} = v - v_{\parallel},$$
$$v_{\parallel} = (v \cdot w)w.$$

Substituting these identities into Equation (12.10) yields

$$v^{new} = v + (s - 1)(v \cdot w)w,$$

which is the nonuniform scaling formula for vectors. Notice, by the way, that mirror image is just nonuniform scaling along the normal direction to the mirror plane, with scale factor $s = -1$.

## 12.5 Projections

  To display three-dimensional geometry on a two-dimensional screen, we need to project from three dimensions to two dimensions. The two most important types of projections are orthogonal projection and perspective projection.

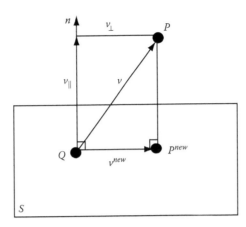

**FIGURE 12.8:** Orthogonal projection into the plane $S$, determined by a point $Q$ and a unit normal vector $n$.

## 12.5.1 Orthogonal Projection

The simplest projection is orthogonal projection (see Figure 12.8).

The screen is represented by a plane $S$ defined by a point $Q$ and a unit normal vector $n$. If $P$ is a point and $v$ is a vector, then the formulas for orthogonal projection are

$$v^{new} = v - (v \cdot n)n,$$
$$P^{new} = P - ((P - Q) \cdot n)n. \tag{12.11}$$

These formulas are easy to derive. The formula for vectors is immediate from the observation that $v^{new}$ is just the perpendicular component of $v$ relative to the normal vector $n$, so

$$v^{new} = v_\perp = v - v_\| = v - (v \cdot n)n.$$

Moreover, for points, we simply observe, as usual, that orthogonal projection for points follows easily from orthogonal projection for vectors. Indeed, since

$$P = Q + (P - Q),$$

it follows that

$$P^{new} = Q^{new} + (P - Q)^{new} = Q + (P - Q)^{new} = P - ((P - Q) \cdot n)n.$$

## 12.5.2 Perspective

Perspective projection is required in Computer Graphics to simulate visual realism. Perspective projection is defined by specifying an eye point $E$ and a plane $S$ into which to project the image. The plane $S$ is specified, as usual, by a point $Q$ on $S$ and a unit vector $n$ normal to $S$. The image of a point $P$ under perspective projection from the eye point $E$ to the plane $S$ is the intersection of the line $EP$ with the plane $S$ (see Figure 12.9, left). Notice that unlike all of the other transformations that we have developed in this chapter, perspective projection is not an affine transformation because perspective projection is not a well-defined transformation at every point nor is perspective well-defined for any vector (see Figure 12.9, right).

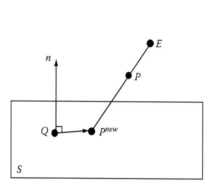 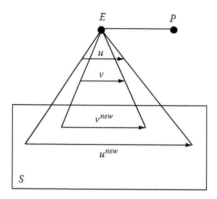

**FIGURE 12.9:** Perspective projection from the eye point $E$ through the point $P$ onto the plane $S$ is defined by the intersection of the line $EP$ with the viewing plane $S$ (left). Notice that perspective projection is not well defined on points $P$ where the line $EP$ is parallel to the plane $S$ nor is perspective defined on any vector, since $u = v$ but $u^{new} \neq v^{new}$ (right).

The image of a point $P$ under perspective projection from the eye point $E$ to the plane $S$ is given by

$$P^{new} = \frac{((E - Q) \cdot n)P + ((Q - P) \cdot n)E}{(E - P) \cdot n}. \tag{12.12}$$

To derive this formula, we simply compute the intersection of the line $EP$ with the plane $S$. Since the point $P^{new}$ lies on the line $EP$, there is a constant $c$ such that

$$P^{new} = P + c(E - P).$$

Moreover, since the point $P^{new}$ also lies on the plane $S$, the vector $P^{new} - Q$ must be perpendicular to the normal vector $n$; therefore,

$$(P^{new} - Q) \cdot n = 0.$$

Substituting the first equation for $P^{new}$ into the second equation for $P^{new}$ gives

$$(P - Q) \cdot n + c(E - P) \cdot n = 0.$$

Solving for $c$, we find that

$$c = \frac{(Q - P) \cdot n}{(E - P) \cdot n}.$$

Therefore,

$$P^{new} = P + c(E - P) = P + \frac{(Q - P) \cdot n}{(E - P) \cdot n}(E - P) = \frac{((E - Q) \cdot n)P + ((Q - P) \cdot n)E}{(E - P) \cdot n}.$$

Notice that if the line $EP$ is parallel to the plane $S$, then the denominator $(E - P) \cdot n = 0$. In this case the expression for $P^{new}$ is not defined, so perspective projection is not well defined for any point

$P$ for which the line $EP$ is parallel to the plane $S$. This problem arises because unlike any of our previous transformations, the expression for $P^{new}$ has $P$ in the denominator, so perspective projection is a rational function. We shall have more to say about rational transformations in Chapter 13.

## 12.6   Summary

We close with a summary of the highlights of this chapter. Also listed here for your convenience are the coordinate-free formulas from vector algebra for all the transformations that we have studied in this chapter.

### 12.6.1   Affine and Projective Transformations without Matrices

One of the main themes of this chapter is that we do not need matrices to compute the effects of affine and projective transformations on points and vectors. Instead we can use vector algebra to derive coordinate-free expressions for all of the important affine and projective transformations in three-dimensional Computer Graphics. Moreover, except for rotation, these formulas from vector algebra are actually more efficient than the corresponding matrix formulations. Nevertheless, formulas from vector algebra are not convenient for composing transformations, whereas matrix representations are easily composed by matrix multiplication. Therefore, in Chapter 13 we shall use these formulas from vector algebra to derive matrix representations for each of the corresponding affine and projective transformations.

Our main strategy, used over and over again, for deriving explicit formulas for transformations on vectors is to decompose an arbitrary vector into components parallel and perpendicular to a particular direction vector in which it is especially easy to compute the transformation. We then analyze the transformation on each of these components and sum the transformed components to find the effect of the transformation on the original vector.

This strategy works because most of the transformations that we have investigated are specified in terms of either a fixed line or a fixed plane: rotations revolve around an axis line; projections collapse into a projection plane. Both lines and planes can be specified by a point $Q$ and a unit vector $u$. For lines $L$, the vector $u$ is parallel to $L$; for planes $S$, the vector $u$ is normal to $S$. Our strategy for computing transformations on a vector $v$ is to decompose $v$ into components $v_{\parallel}$ and $v_{\perp}$ parallel and perpendicular to the unit vector $u$. Typically one of these components is unaffected by the transformation and the effect of the transformation on the other component is easy to compute. We then reassemble the transformation on these components to derive the effect of the transformation on the original vector.

Formulas for transforming points follow easily from formulas for transforming vectors, since a point $P$ can typically be expressed as the sum of a vector $P - Q$ and a fixed point $Q$ of the transformation. This fixed point is typically the point $Q$ that defines the line or plane that specifies the transformation. Thus once we know the effect of the transformation on arbitrary vectors, we can easily find the effect of the transformation on arbitrary points.

The general approach outlined here is important and powerful; you should incorporate this strategy into your analytic toolkit. We shall have occasion to use this strategy again in other applications later in this book.

### 12.6.2    Formulas for Affine and Projective Transformations

*Translation—by the vector w*

$$v^{new} = v,$$
$$P^{new} = P + w.$$

*Rotation—by the angle θ around the line L determined by the point Q and the unit direction vector w (Formula of Rodrigues)*

$$v^{new} = (\cos\theta)v + (1 - \cos\theta)(v \cdot w)w + (\sin\theta)w \times v,$$
$$P^{new} = Q + (\cos\theta)(P - Q) + (1 - \cos\theta)((P - Q) \cdot w)w + (\sin\theta)w \times (P - Q).$$

*Mirror image—in the plane S determined by the point Q and the unit normal n*

$$v^{new} = v - 2(v \cdot n)n,$$
$$P^{new} = P + 2((P - Q) \cdot n)n.$$

*Uniform scaling—around the point Q by the scale factor s*

$$v^{new} = sv,$$
$$P^{new} = Q + s(P - Q) = sP + (1 - s)Q.$$

*Nonuniform scaling—around the point Q by the scale factor s in the direction of the unit vector w*

$$v^{new} = v + (s - 1)(v \cdot w)w,$$
$$P^{new} = P + (s - 1)((P - Q) \cdot w)w.$$

*Orthogonal projection—into the plane S determined by the point Q and the unit normal n*

$$v^{new} = v - (v \cdot n)n,$$
$$P^{new} = P - ((P - Q) \cdot n)n.$$

*Perspective projection—from the eye point E into the plane S determined by the point Q and the unit normal vector n*

$$P^{new} = \frac{((E - Q) \cdot n)P + ((Q - P) \cdot n)E}{(E - P) \cdot n}.$$

---

### Exercises

12.1. Let $A$ be an affine transformation. Show that

$$A(\Sigma_k \lambda_k P_k) = \Sigma_k \lambda_k A(P_k) \quad \text{whenever } \Sigma_k \lambda_k \equiv 0, 1.$$

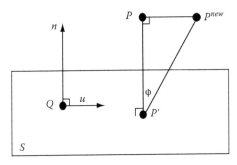

**FIGURE 12.10:** Shear. Fix a plane $S$, determined by a point $Q$ and a unit normal vector $n$, and let $u$ be a unit vector perpendicular to $n$. An arbitrary point $P$ is moved parallel to the direction $u$ to the point $P^{new}$ so that $\angle PP'P^{new} = \tan(\phi)$, where is $P'$ the orthogonal projection of the point $P$ onto the plane $S$ and $\phi$ is a fixed shearing angle.

12.2. A *shear transformation* is defined in terms of a shearing plane $S$, a unit vector $u$ in the plane $S$, and an angle $\phi$ in the following fashion. Given any point $P$, project $P$ orthogonally onto a point $P'$ in the shearing plane $S$. Now slide $P$ parallel to $u$ to a point $P^{new}$ so that $\angle P^{new} P'P = \phi$. The point $P^{new}$ is the result of applying the shearing transformation to the point $P$ (see Figure 12.10).

Let

$S = $ Shearing plane
$n = $ Unit vector perpendicular to $S$
$Q = $ Point on $S$
$u = $ Unit vector in $S$ (i.e., a unit vector perpendicular to $n$)
$\phi = $ Shear angle

a. For any point $P$, show that:

$$P^{new} = P + \tan(\phi)((P - Q) \cdot n)u.$$

b. For each vector $v = P - R$, define $v^{new} = P^{new} - R^{new}$. Show that:

$$v^{new} = v + \tan(\phi)(v \cdot n)u.$$

c. Conclude that:

    i. $v^{new}$ is independent of the choices of the points $P$ and $R$ that represent the vector $v$.

    ii. $v^{new} = v \Leftrightarrow v \perp n$.

d. Interpret the result of part b geometrically.

12.3. *Parallel projection* is projection onto a plane $S$ along the direction of a unit vector $u$. As usual the plane $S$ is defined by a point $Q$ and a unit normal vector $n$ (see Figure 12.11). Show that:

a. $v^{new} = v - \dfrac{v \cdot n}{u \cdot n} u.$

b. $P^{new} = P - \dfrac{(P - Q) \cdot n}{u \cdot n} u.$

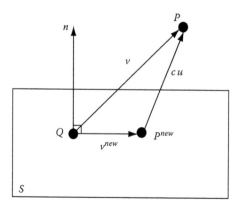

**FIGURE 12.11:** Parallel projection along a direction $u$ into a plane $S$, determined by a point $Q$ and a unit normal vector $n$.

12.4. Let $v_1, v_2$ be unit vectors, and let $u = v_1 \times v_2$.

    a. Show that the transformation

$$R(v) = (v_1 \cdot v_2)v + \frac{v \cdot u}{1 + v_1 \cdot v_2}u + u \times v,$$

    rotates the vector $v_1$ into the vector $v_2$.

    b. Interpret this result geometrically.

12.5. Recall from Chapter 4, Section 4.7.1 that the image of one point and two linearly independent vectors determines a unique affine transformation in the plane.

    a. Show that the image of one point and three linearly independent vectors determines a unique affine transformation in three-space.

    b. Fix two sets of three vectors $v_1, v_2, v_3$ and $w_1, w_2, w_3$, and define

$$A(u) = \frac{Det(u \ v_2 \ v_3)w_1 + Det(v_1 \ u \ v_3)w_2 + Det(v_1 \ v_2 \ u)w_3}{Det(v_1 \ v_2 \ v_3)}.$$

    Show that

$$A(v_i) = w_i, \quad i = 1, 2, 3.$$

12.6. Recall from Chapter 4, Section 4.7.3 that the image of three noncollinear points determines a unique affine transformation in the plane.

    a. Show that the image of four noncoplanar points determines a unique affine transformation in three-space.

    b Fix two sets of four points $P_1, P_2, P_3, P_4$ and $Q_1, Q_2, Q_3, Q_4$, and define

$$A(R) = \frac{Det(P_2 - R \ P_3 - R \ P_4 - R)Q_1 - Det(P_1 - R \ P_3 - R \ P_4 - R)Q_2}{Det(P_2 - P_1 \ P_3 - P_1 \ P_4 - P_1)}$$
$$+ \frac{Det(P_1 - R \ P_2 - R \ P_4 - R)Q_3 - Det(P_1 - R \ P_2 - R \ P_3 - R)Q_4}{Det(P_2 - P_1 \ P_3 - P_1 \ P_4 - P_1)}.$$

Show that

$$A(P_i) = Q_i, \quad i = 1, 2, 3, 4.$$

12.7. Recall from Chapter 4, Exercise 4.19 that every nonsingular affine transformation in the plane is the composition of one translation, one rotation, one shear, and two nonuniform scalings. Show that every nonsingular affine transformation in three dimensions is the composition of one translation, two rotations, two shears, and three nonuniform scalings.

12.8. Let $u, w$ be unit vectors and define the map $w \otimes u$ by setting

$$(w \otimes u)(v) = (v \cdot w)u.$$

Show that:

a. $w \otimes u$ is a linear transformation.

b. $(u \otimes u)(v) = v_{\parallel}$, where $v_{\parallel}$ is the parallel projection of $v$ on $u$.

12.9. Let $u$ be a unit vector and define the map $u \times \_$ by setting

$$(u \times \_)(v) = u \times v.$$

Show that $u \times \_$ is a linear transformation.

12.10. Let $R(v)$ denote the vector generated by rotating the vector $v$ by the angle $\theta$ around the vector $w$. Using the Formula of Rodrigues, show that:

a. $R(u) \cdot R(v) = u \cdot v$.

b. $R(u \times v) = R(u) \times R(v)$.

12.11. Let $M(v)$ denote the mirror image of the vector $v$ in a plane normal to the vector $n$. Show that:

a. $M(u) \cdot M(v) = u \cdot v$.

b. $M(u \times v) = -M(u) \times M(v)$.

12.12. Let $S(v)$ denote the transformation that scales each vector $v$ by the scale factor $s$. Show that:

a. $S(u) \cdot S(v) = s^2(u \cdot v)$.

b. $S(u \times v) = (S(u) \times S(v))/s$.

12.13. Derive the formula for spherical linear interpolation (SLERP, see Chapter 11, Section 11.6) in three dimensions

$$v(t) = \frac{\sin((1-t)\phi)}{\sin(\phi)} v_0 + \frac{\sin(t\phi)}{\sin(\phi)} v_1,$$

where $\phi$ is the angle between the vectors $v_0$ and $v_1$, by rotating the vector $v_0$ by the angle $t\phi$ around the unit vector $w = \dfrac{v_0 \times v_1}{|v_0 \times v_1|}$.

12.14. Let $T_1, T_2$ be transformations from $\mathbf{R}^3$ to $\mathbf{R}^3$, and let $T(\lambda) = (1-\lambda)T_1 + \lambda T_2$. Show that:

a. If $T_1, T_2$ are affine transformations, then $T(\lambda)$ is also an affine transformation.

b. If $T_1, T_2$ are rotations, then $T(\lambda)$ is not necessarily a rotation.

# Chapter 13

## Matrix Representations for Affine and Projective Transformations

*thou shalt set apart unto the LORD all that openeth the matrix*

– Exodus 13:12

### 13.1 Matrix Representations for Affine Transformations

In Chapter 12, we derived formulas for affine and projective transformations using the techniques of vector geometry and vector algebra. Formulas expressed in vector algebra are effective for computing individual transformations, but these formulas are not efficient for composing transformations. Therefore in this chapter, we are going to derive matrix representations for each of the classical transformations of Computer Graphics so that we will be able to compose these transformations by matrix multiplication.

To fix our notation, denote the origin by $O$ and the three unit vectors along the $x, y, z$ coordinate axes by $i, j, k$. Then in three dimensions for any vector $v$ and any point $P$, there are constants $(v_1, v_2, v_3)$ and $(p_1, p_2, p_3)$ such that

$$v = v_1 i + v_2 j + v_3 k,$$
$$P = O + p_1 i + p_2 j + p_3 k.$$

The constants $(v_1, v_2, v_3)$ and $(p_1, p_2, p_3)$ are the rectangular coordinates of $v$ and $P$, and we often abuse notation and write $v = (v_1, v_2, v_3)$ and $P = (p_1, p_2, p_3)$.

Now let $A$ be an affine transformation. Then since $A$ preserves linear and affine combinations

$$A(v) = v_1 A(i) + v_2 A(j) + v_3 A(k),$$
$$A(P) = A(O) + p_1 A(i) + p_2 A(j) + p_3 A(k).$$

Thus if we know the value of $A$ on the three vectors $i, j, k$ and on the single point $O$, then we know the value of $A$ on every vector $v$ and every point $P$—that is, any affine transformation $A$ is completely determined by the four values $A(O), A(i), A(j), A(k)$.

Since affine transformations map points to points and vectors to vectors, there are constants $(a_{ij})$ such that

$$A(i) = a_{11} i + a_{12} j + a_{13} k = (a_{11}, a_{12}, a_{13}),$$
$$A(j) = a_{21} i + a_{22} j + a_{23} k = (a_{21}, a_{22}, a_{23}),$$
$$A(k) = a_{31} i + a_{32} j + a_{33} k = (a_{31}, a_{32}, a_{33}),$$
$$A(O) = O + a_{41} i + a_{42} j + a_{43} k = (a_{41}, a_{42}, a_{43}).$$

179

Therefore the 12 numbers $(a_{ij})$ completely characterize the affine transformation $A$. It seems natural then that we should represent $A$ by a $4 \times 3$ matrix and write

$$A = \begin{pmatrix} A(i) \\ A(j) \\ A(k) \\ A(O) \end{pmatrix} = \begin{pmatrix} a_{11} & a_{12} & a_{13} \\ a_{21} & a_{22} & a_{23} \\ a_{31} & a_{32} & a_{33} \\ a_{41} & a_{42} & a_{43} \end{pmatrix}.$$

But there is something not quite right about this $4 \times 3$ matrix. If we represent a transformation by a matrix, then we expect that if we multiply the coordinates of a point or a vector by the matrix we should get the result of the transformation applied to the point or vector. However, points and vectors in three dimensions have only three coordinates, whereas this matrix has four rows.

To resolve this inconsistency, we shall introduce affine coordinates for points and vectors, and $4 \times 4$ matrices to represent affine transformations. Recall that in two dimensions we insert a third coordinate, an affine coordinate, for points and vectors: the affine coordinate for points is 1, the affine coordinate for vectors is 0. This affine coordinate allows us to represent affine transformations in the plane by $3 \times 3$ matrices. We shall follow similar conventions in three-space; in three dimensions we insert a fourth, affine coordinate for points and for vectors. Again the affine coordinate for points is 1, the affine coordinate for vectors is 0. Following this convention,

$$A = \begin{pmatrix} A(i) & 0 \\ A(j) & 0 \\ A(k) & 0 \\ A(O) & 1 \end{pmatrix} = \begin{pmatrix} a_{11} & a_{12} & a_{13} & 0 \\ a_{21} & a_{22} & a_{23} & 0 \\ a_{31} & a_{32} & a_{33} & 0 \\ a_{41} & a_{42} & a_{43} & 1 \end{pmatrix}. \tag{13.1}$$

Now for vectors we have

$$A(v) = v_1 A(i) + v_2 A(j) + v_3 A(k) = (v_1, v_2, v_3, 0) * \begin{pmatrix} A(i) & 0 \\ A(j) & 0 \\ A(k) & 0 \\ A(O) & 1 \end{pmatrix},$$

or equivalently

$$A(v) = (v_1, v_2, v_3, 0) * \begin{pmatrix} a_{11} & a_{12} & a_{13} & 0 \\ a_{21} & a_{22} & a_{23} & 0 \\ a_{31} & a_{32} & a_{33} & 0 \\ a_{41} & a_{42} & a_{43} & 1 \end{pmatrix} = (v, 0) * A.$$

Similarly, for points

$$A(P) = A(O) + p_1 A(i) + p_2 A(j) + p_3 A(k) = (p_1, p_2, p_3, 1) * \begin{pmatrix} A(i) & 0 \\ A(j) & 0 \\ A(k) & 0 \\ A(O) & 1 \end{pmatrix},$$

or equivalently

$$A(P) = (p_1, p_2, p_3, 1) * \begin{pmatrix} a_{11} & a_{12} & a_{13} & 0 \\ a_{21} & a_{22} & a_{23} & 0 \\ a_{31} & a_{32} & a_{33} & 0 \\ a_{41} & a_{42} & a_{43} & 1 \end{pmatrix} = (P, 1) * A.$$

In this way affine transformations are represented by multiplying the affine coordinates of points and vectors by special $4 \times 4$ matrices, where the last column is always $(0, 0, 0, 1)^{\mathrm{T}}$.

## 13.2 Linear Transformation Matrices and Translation Vectors

To derive explicit $4 \times 4$ matrix representations for all the standard transformations of three-dimensional Computer Graphics—translation, rotation, mirror image, uniform and nonuniform scaling, and orthogonal and perspective projection—we need to convert the corresponding formulas from vector algebra into matrix form.

To achieve this goal, notice that the $4 \times 4$ matrix representing an affine transformation $A$ always has the following form:

$$A = \begin{pmatrix} L_A & 0 \\ w_A & 1 \end{pmatrix}, \tag{13.2}$$

where
  $L_A$ is a $3 \times 3$ matrix
  $w_A$ is a three-dimensional vector.

Moreover, for any vector $v$,

$$A(v) = (v, 0) * \begin{pmatrix} L_A & 0 \\ w_A & 1 \end{pmatrix} = (v * L_A, 0). \tag{13.3}$$

Thus the $3 \times 3$ matrix $L_A$ represents a linear transformation on vectors. Similarly, for any point $P$,

$$A(P) = (P, 1) * \begin{pmatrix} L_A & 0 \\ w_A & 1 \end{pmatrix} = (P * L_A + w_A, 1). \tag{13.4}$$

Thus the vector $w_A$ represents a translation vector. Hence we can decompose $A$ so that

$$A = \begin{pmatrix} L_A & 0 \\ w_A & 1 \end{pmatrix} = \begin{pmatrix} 3 \times 3 \text{ Linear transformation matrix} & 0 \\ \text{Translation vector} & 1 \end{pmatrix}. \tag{13.5}$$

Our goal is to find explicit expressions for the matrix $L_A$ and the vector $w_A$ for each of the standard affine transformations of three-dimensional Computer Graphics.

### 13.2.1   Linear Transformation Matrices

Let us begin with the $3 \times 3$ matrix $L_A$. Looking at the formulas for affine transformations from Chapter 12, we see that on vectors these formulas all have the following form:

$$A(v) = v^{new} = c_1 v + c_2 (v \cdot u)\tilde{u} + c_3 u \times v,$$

where
  $u, \tilde{u}$ are fixed unit direction vectors
  $c_1, c_2, c_3$ are constants.

To represent these linear transformations using matrices, we introduce the following three $3 \times 3$ matrices:

$$I = \begin{pmatrix} 1 & 0 & 0 \\ 0 & 1 & 0 \\ 0 & 0 & 1 \end{pmatrix} = 3 \times 3 \text{ Identity Matrix,} \qquad (13.6)$$

$$u^T * \tilde{u} = \begin{pmatrix} u_1 \\ u_2 \\ u_3 \end{pmatrix} * (\tilde{u}_1 \quad \tilde{u}_2 \quad \tilde{u}_3) = \begin{pmatrix} u_1\tilde{u}_1 & u_1\tilde{u}_2 & u_1\tilde{u}_3 \\ u_2\tilde{u}_1 & u_2\tilde{u}_2 & u_2\tilde{u}_3 \\ u_3\tilde{u}_1 & u_3\tilde{u}_2 & u_3\tilde{u}_3 \end{pmatrix}, \qquad (13.7)$$

$$u \times \_ = \begin{pmatrix} 0 & u_3 & -u_2 \\ -u_3 & 0 & u_1 \\ u_2 & -u_1 & 0 \end{pmatrix}. \qquad (13.8)$$

It follows easily by direct computation from Equations 13.6 through 13.8 that

$$v * I = v,$$
$$v * (u^T * \tilde{u}) = (v * u^T) * \tilde{u} = (v \cdot u)\tilde{u},$$
$$v * (u \times \_) = u \times v.$$

Thus
  $I$ represents the identity transformation,
  $u^T * \tilde{u}$ represents the dot product with $u$ (or parallel projection onto $u$ when $\tilde{u} = u$),
  $u \times \_$ represents the cross product with $u$.

Hence

$$c_1 v + c_2 (v \cdot u)\tilde{u} + c_3 u \times v = v * \left( c_1 I + c_2 u^T * \tilde{u} + c_3 u \times \_ \right). \qquad (13.9)$$

Therefore using Equation 13.9, we can easily find the matrix representing the linear transformation $L_A$ mechanically in the following manner:

$$A(v) = c_1 v + c_2 (v \cdot u)\tilde{u} + c_3 u \times v$$
$$\downarrow \qquad \downarrow \qquad \downarrow \qquad \downarrow \qquad\qquad (13.10)$$
$$L_A = c_1 I + c_2 u^T * \tilde{u} + c_3 u \times \_ .$$

### 13.2.2  Translation Vectors

To find the translation vector $w_A$, observe that each of the affine transformations $A$ from Chapter 12, except for translation, has a fixed point $Q$—that is, a point $Q$ for which

$$A(Q) = Q.$$

Now for any point $P$,

$$P = Q + (P - Q).$$

Therefore since $A$ is an affine transformation,

$$A(P) = A(Q) + A(P - Q) = Q + A(P - Q). \tag{13.11}$$

Moreover, since $P - Q$ is a vector, we know from Section 13.2.1 that

$$A(P - Q) = (P - Q) * L_A = P * L_A - Q * L_A. \tag{13.12}$$

Hence by Equations 13.11 and 13.12,

$$A(P) = Q + P * L_A - Q * L_A = P * L_A + Q(I - L_A). \tag{13.13}$$

But by Equation 13.4

$$A(P) = P * L_A + w_A. \tag{13.14}$$

Therefore, we conclude from Equations 13.13 and 13.14 that

$$w_A = Q * (I - L_A), \tag{13.15}$$

where $Q$ is any fixed point of the affine transformation $A$.

With Equations 13.10 and 13.15 for $L_A$ and $w_A$ in hand, we are now ready to derive matrix representations for each of the standard transformations of three-dimensional Computer Graphics.

---

## 13.3  Rigid Motions

Rigid motions in three dimensions are composites of three basic transformations: translation, rotation, and mirror image.

### 13.3.1  Translation

Let $w$ be a translation vector. For any point $P$ and any vector $v$,

$$v^{new} = v = v * I,$$
$$p^{new} = P + w = P * I + w.$$

Therefore the matrix *Trans(w)* representing translation is simply

$$Trans(w) = \begin{pmatrix} I & 0 \\ w & 1 \end{pmatrix}. \tag{13.16}$$

**Example 13.1 Translation along the Coordinate Axes**
The matrix representing translation parallel to the *x*-axis by the distance *d* is given by

$$Trans(di) = \begin{pmatrix} I & 0 \\ di & 1 \end{pmatrix} = \begin{pmatrix} 1 & 0 & 0 & 0 \\ 0 & 1 & 0 & 0 \\ 0 & 0 & 1 & 0 \\ d & 0 & 0 & 1 \end{pmatrix}.$$

Similarly, the matrices representing translation parallel to the *y*- or *z*-axes by the distance *d* are given by

$$Trans(dj) = \begin{pmatrix} 1 & 0 & 0 & 0 \\ 0 & 1 & 0 & 0 \\ 0 & 0 & 1 & 0 \\ 0 & d & 0 & 1 \end{pmatrix} \quad \text{and} \quad Trans(dk) = \begin{pmatrix} 1 & 0 & 0 & 0 \\ 0 & 1 & 0 & 0 \\ 0 & 0 & 1 & 0 \\ 0 & 0 & d & 1 \end{pmatrix}.$$

### 13.3.2   Rotation

Let *L* be an axis of rotation defined by specifying a point *Q* on *L* and a unit direction vector *u* parallel to *L*, and let $\theta$ be the angle of rotation. Denote by $Rot(Q, u, \theta)$ the $4 \times 4$ matrix representing rotation of points around the line *L* through the angle $\theta$, and denote by $Rot(u, \theta)$ the upper $3 \times 3$ submatrix of $Rot(Q, u, \theta)$. Then by Equation 13.15 since *Q* remains fixed under rotation,

$$Rot(Q, u, \theta) = \begin{pmatrix} Rot(u, \theta) & 0 \\ Q * (I - Rot(u, \theta)) & 1 \end{pmatrix}. \tag{13.17}$$

Moreover by the Formula of Rodrigues (see Chapter 12, Section 12.3.2) for any vector *v*,

$$v^{new} = (\cos \theta)v + (1 - \cos \theta)(v \cdot u)u + (\sin \theta)u \times v.$$

Therefore we can find the matrix $Rot(u, \theta)$ by invoking Equation 13.10:

$$v^{new} = (\cos \theta)v + (1 - \cos \theta)(v \cdot u)u + (\sin \theta)(u \times v)$$
$$\downarrow \qquad \qquad \downarrow \qquad \qquad \downarrow \qquad \qquad \downarrow \tag{13.18}$$
$$Rot(u, \theta) = (\cos \theta)I + (1 - \cos \theta)(u^{\mathrm{T}} * u) + (\sin \theta)(u \times \_ \,).$$

Thus by Equations 13.6 through 13.8,

$$Rot(u, \theta) = \begin{pmatrix} \cos \theta + (1 - \cos \theta)u_1^2 & (1 - \cos \theta)u_1 u_2 + (\sin \theta)u_3 & (1 - \cos \theta)u_1 u_3 - (\sin \theta)u_2 \\ (1 - \cos \theta)u_1 u_2 - (\sin \theta)u_3 & \cos \theta + (1 - \cos \theta)u_2^2 & (1 - \cos \theta)u_2 u_3 + (\sin \theta)u_1 \\ (1 - \cos \theta)u_1 u_3 + (\sin \theta)u_2 & (1 - \cos \theta)u_2 u_3 - (\sin \theta)u_1 & \cos \theta + (1 - \cos \theta)u_3^2 \end{pmatrix}. \tag{13.19}$$

**Example 13.2  Rotation around the Coordinate Axes**
Suppose that the axis of rotation is the $z$-axis. Then we can choose $Q = Origin = (0,0,0)$ and $u = k = (0,0,1)$. Now by Equation 13.18

$$Rot(k, \theta) = (\cos \theta)I + (1 - \cos \theta)(k^{\mathrm{T}} * k) + (\sin \theta)(k \times \_).\qquad(13.20)$$

But by Equations 13.7 and 13.8

$$k^{\mathrm{T}} * k = \begin{pmatrix} 0 & 0 & 0 \\ 0 & 0 & 0 \\ 0 & 0 & 1 \end{pmatrix} \quad \text{and} \quad k \times \_ = \begin{pmatrix} 0 & 1 & 0 \\ -1 & 0 & 0 \\ 0 & 0 & 0 \end{pmatrix}.$$

Therefore

$$Rot(k, \theta) = \begin{pmatrix} \cos \theta & \sin \theta & 0 \\ -\sin \theta & \cos \theta & 0 \\ 0 & 0 & 1 \end{pmatrix}.$$

Hence the matrix $Rot(Origin, k, \theta)$ representing rotation around the $z$-axis by the angle $\theta$ is given by

$$Rot(Origin, k, \theta) = \begin{pmatrix} Rot(k, \theta) & 0 \\ Origin * (I - Rot(k, \theta)) & 1 \end{pmatrix} = \begin{pmatrix} \cos \theta & \sin \theta & 0 & 0 \\ -\sin \theta & \cos \theta & 0 & 0 \\ 0 & 0 & 1 & 0 \\ 0 & 0 & 0 & 1 \end{pmatrix}.$$

Similarly, the matrices representing rotation around the $x$- and $y$-axes by the angle $\theta$ are given by

$$Rot(Origin, i, \theta) = \begin{pmatrix} 1 & 0 & 0 & 0 \\ 0 & \cos \theta & \sin \theta & 0 \\ 0 & -\sin \theta & \cos \theta & 0 \\ 0 & 0 & 0 & 1 \end{pmatrix},$$

and

$$Rot(Origin, j, \theta) = \begin{pmatrix} \cos \theta & 0 & -\sin \theta & 0 \\ 0 & 1 & 0 & 0 \\ \sin \theta & 0 & \cos \theta & 0 \\ 0 & 0 & 0 & 1 \end{pmatrix}.$$

### 13.3.3  Mirror Image

Let $S$ be a mirror plane specified by a point $Q$ on $S$ and a unit vector $n$ normal to $S$. Denote by $Mir(Q, n)$ the $4 \times 4$ matrix representing the mirror image of points in the plane $S$, and denote by $Mir(n)$ the upper $3 \times 3$ submatrix of $Mir(Q, n)$. Then by Equation 13.15 since $Q$ remains fixed under mirror image,

$$Mir(Q, n) = \begin{pmatrix} Mir(n) & 0 \\ Q * (I - Mir(n)) & 1 \end{pmatrix}.\qquad(13.21)$$

Moreover, for mirror image (see Chapter 12, Section 12.3.3),

$$v^{new} = v - 2(v \cdot n)n.$$

Therefore we can find the matrix $Mir(n)$ by invoking Equation 13.10

$$
\begin{array}{ccc}
v^{new} & = & v - 2(v \cdot n)n \\
\downarrow & & \downarrow \quad\quad \downarrow \\
Mir(n) & = & I - 2n^T * n.
\end{array}
\tag{13.22}
$$

Now notice that

$$I - Mir(n) = 2n^T * n,$$

so

$$Q * (I - Mir(n)) = Q * (2n^T * n) = 2(Q \cdot n)n.$$

Therefore by Equations 13.21 and 13.22

$$Mir(Q, n) = \begin{pmatrix} 1 - 2n^T * n & 0 \\ 2(Q \cdot n)n & 1 \end{pmatrix}.
\tag{13.23}$$

### Example 13.3   Mirror Image in the Coordinate Planes

Suppose that the mirror plane is the $xy$-plane. Then we can choose $Q = Origin = (0, 0, 0)$ and $n = k = (0, 0, 1)$. By Equations 13.7

$$k^T * k = \begin{pmatrix} 0 & 0 & 0 \\ 0 & 0 & 0 \\ 0 & 0 & 1 \end{pmatrix}.$$

Therefore

$$Mir(Origin, k) = \begin{pmatrix} I - 2k^T * k & 0 \\ 2(Origin \cdot n)n & 1 \end{pmatrix} = \begin{pmatrix} 1 & 0 & 0 & 0 \\ 0 & 1 & 0 & 0 \\ 0 & 0 & -1 & 0 \\ 0 & 0 & 0 & 1 \end{pmatrix}.$$

Similarly, the matrices representing mirror images in the $xz$-plane and $yz$-plane are

$$Mir(Origin, j) = \begin{pmatrix} 1 & 0 & 0 & 0 \\ 0 & -1 & 0 & 0 \\ 0 & 0 & 1 & 0 \\ 0 & 0 & 0 & 1 \end{pmatrix} \quad \text{and} \quad Mir(Origin, i) = \begin{pmatrix} -1 & 0 & 0 & 0 \\ 0 & 1 & 0 & 0 \\ 0 & 0 & 1 & 0 \\ 0 & 0 & 0 & 1 \end{pmatrix}.$$

## 13.4   Scaling

We are interested in both uniform and nonuniform scaling.

### 13.4.1   Uniform Scaling

Uniform scaling is defined by specifying a point $Q$ from which to scale distance and a scale factor $s$. For any point $P$ and any vector $v$ (see Chapter 12, Section 12.4.1),

$$v^{new} = sv = v * (sI),$$
$$p^{new} = sP + (1 - s)Q = P * (sI) + (1 - s)Q.$$

Therefore the matrix $Scale(Q, s)$ representing uniform scaling about the point $Q$ by the scale factor $s$ is given by

$$Scale(Q, s) = \begin{pmatrix} sI & 0 \\ (1 - s)Q & 1 \end{pmatrix}. \tag{13.24}$$

**Example 13.4   Uniform Scaling about the Origin**
   To scale uniformly about the origin by the scale factor $s$, set $Q = Origin = (0, 0, 0)$. Then by Equation 13.24

$$Scale(Origin, s) = \begin{pmatrix} sI & 0 \\ (1 - s)Origin & 1 \end{pmatrix} = \begin{pmatrix} s & 0 & 0 & 0 \\ 0 & s & 0 & 0 \\ 0 & 0 & s & 0 \\ 0 & 0 & 0 & 1 \end{pmatrix}.$$

### 13.4.2   Nonuniform Scaling

Nonuniform scaling is defined by designating a unit vector $u$ specifying the scaling direction, as well as a point $Q$ about which to scale and a scale factor $s$. Denote by $Scale(Q, u, s)$ the $4 \times 4$ matrix representing nonuniform scaling from the point $Q$ in the direction $u$ by the scale factor $s$, and denote by $Scale(u, s)$ the upper $3 \times 3$ submatrix of $Scale(Q, u, s)$. Then by Equation 13.15 since $Q$ remains fixed under this scaling,

$$Scale(Q, u, s) = \begin{pmatrix} Scale(u, s) & 0 \\ Q * (I - Scale(u, s)) & 1 \end{pmatrix}. \tag{13.25}$$

Moreover, for nonuniform scaling (see Chapter 12, Section 12.4.2),

$$v^{new} = v + (s - 1)(v \cdot u)u.$$

Therefore we can find $Scale(u, s)$ by invoking Equation 13.10

$$
\begin{array}{ccc}
v^{new} & = & v + (s - 1)(v \cdot u)u \\
\downarrow & & \downarrow \qquad \downarrow
\end{array}
\tag{13.26}
$$
$$Scale(u, s) = I + (s - 1)u^{T} * u.$$

Now notice that

$$I - Scale(u, s) = (1 - s)u^T * u,$$

so

$$Q * (I - Scale(u, s)) = (1 - s)Q * (u^T * u) = (1 - s)(Q \cdot u)u.$$

Therefore by Equations 13.25 and 13.26

$$Scale(Q, u, s) = \begin{pmatrix} I + (s - 1)u^T * u & 0 \\ (1 - s)(Q \cdot u)u & 1 \end{pmatrix}. \tag{13.27}$$

**Example 13.5   Nonuniform Scaling from the Origin along the Coordinate Axes**

To scale about the origin in the direction of the $z$-axis by the scale factor $s$, we can choose $Q = Origin = (0, 0, 0)$ and $u = k = (0, 0, 1)$. By Equation 13.7

$$k^T * k = \begin{pmatrix} 0 & 0 & 0 \\ 0 & 0 & 0 \\ 0 & 0 & 1 \end{pmatrix}.$$

Therefore by Equation 13.27

$$Scale(Origin, k, s) = \begin{pmatrix} I + (s - 1)k^T * k & 0 \\ (1 - s)(Origin \cdot k)k & 1 \end{pmatrix} = \begin{pmatrix} 1 & 0 & 0 & 0 \\ 0 & 1 & 0 & 0 \\ 0 & 0 & s & 0 \\ 0 & 0 & 0 & 1 \end{pmatrix}.$$

Similarly, the matrices representing scaling about the origin along the $x$- and $y$-axes are

$$Scale(Origin, i, s) = \begin{pmatrix} s & 0 & 0 & 0 \\ 0 & 1 & 0 & 0 \\ 0 & 0 & 1 & 0 \\ 0 & 0 & 0 & 1 \end{pmatrix} \quad \text{and} \quad Scale(Origin, j, s) = \begin{pmatrix} 1 & 0 & 0 & 0 \\ 0 & s & 0 & 0 \\ 0 & 0 & 1 & 0 \\ 0 & 0 & 0 & 1 \end{pmatrix}.$$

## 13.5   Projections

Orthogonal projection is an affine transformation, but perspective projection is not affine. Therefore in this section we will be concerned only with orthogonal projections. Parallel projections—projections into a fixed plane parallel to a fixed direction—are also affine. These projections are covered in Exercise 13.2. Perspective projections will be investigated in detail in Section 13.6.

### 13.5.1 Orthogonal Projection

Let the plane $S$ into which points are projected be defined by a point $Q$ and a unit normal vector $n$, and denote by $Orth(Q, n)$ the $4 \times 4$ matrix representing orthogonal projection into the plane $S$. The formulas for orthogonal projection are (see Chapter 12, Section 12.5.1):

$$v^{new} = v - (v \cdot n)n,$$
$$p^{new} = P - ((P - Q) \cdot n)n.$$

Notice that these formulas are almost exactly the same as the formulas for mirror image, except for the missing factors of 2. Therefore using the same analysis we used for mirror image, we arrive at

$$Ortho(Q, n) = \begin{pmatrix} I - n^{\mathrm{T}} * n & 0 \\ (Q \cdot n)n & 1 \end{pmatrix}, \tag{13.28}$$

which is the same matrix as $Mir(Q, n)$ in Equation 13.23, except for the missing factors of 2.

**Example 13.6   Orthogonal Projection into the Coordinate Planes**
Suppose that the projection plane is the $xy$-plane. Then we can choose $Q = Origin = (0, 0, 0)$ and $n = k = (0, 0, 1)$. By Equation 13.7

$$k^{\mathrm{T}} * k = \begin{pmatrix} 0 & 0 & 0 \\ 0 & 0 & 0 \\ 0 & 0 & 1 \end{pmatrix}.$$

Therefore

$$Ortho(Origin, k) = \begin{pmatrix} I - k^{\mathrm{T}} * k & 0 \\ (Origin \cdot n)n & 1 \end{pmatrix} = \begin{pmatrix} 1 & 0 & 0 & 0 \\ 0 & 1 & 0 & 0 \\ 0 & 0 & 0 & 0 \\ 0 & 0 & 0 & 1 \end{pmatrix}.$$

Similarly, the matrices representing orthogonal projections into the $xz$-plane and $yz$-plane are

$$Ortho(Origin, j) = \begin{pmatrix} 1 & 0 & 0 & 0 \\ 0 & 0 & 0 & 0 \\ 0 & 0 & 1 & 0 \\ 0 & 0 & 0 & 1 \end{pmatrix} \quad \text{and} \quad Ortho(Origin, i) = \begin{pmatrix} 0 & 0 & 0 & 0 \\ 0 & 1 & 0 & 0 \\ 0 & 0 & 1 & 0 \\ 0 & 0 & 0 & 1 \end{pmatrix}.$$

## 13.6   Perspective

Unlike all of the other transformations that we have investigated so far, perspective projection is not an affine transformation because perspective projection is not a well-defined transformation at every point nor is perspective well defined for any vector. Nevertheless, we shall see shortly that we can still represent perspective projections by $4 \times 4$ matrices.

### 13.6.1   Projective Transformations and Homogeneous Coordinates

Perspective projection is defined by specifying an eye point $E = (e_1, e_2, e_3)$ and a plane $S$ into which to project the image. The plane $S$ is specified, as usual, by a point $Q = (q_1, q_2, q_3)$ on $S$ and a unit vector $n = (n_1, n_2, n_3)$ normal to $S$.

The image of an arbitrary point $P = (x, y, z)$ under perspective projection from the eye point $E = (e_1, e_2, e_3)$ to the plane $S$ is given by the expression (see Chapter 12, Section 12.5.2):

$$P^{\text{new}} = \frac{((E - Q) \cdot n)P + ((Q - P) \cdot n)E}{(E - P) \cdot n}, \tag{13.29}$$

or in terms of coordinates

$$x^{new} = \frac{((e_1 - q_1)n_1 + (e_2 - q_2)n_2 + (e_3 - q_3)n_3)x + ((q_1 - p_1)n_1 + (q_2 - p_2)n_2 + (q_3 - p_3)n_3)e_1}{-n_1 x - n_2 y - n_3 z + e_1 n_1 + e_2 n_2 + e_3 n_3},$$

$$y^{new} = \frac{((e_1 - q_1)n_1 + (e_2 - q_2)n_2 + (e_3 - q_3)n_3)y + ((q_1 - p_1)n_1 + (q_2 - p_2)n_2 + (q_3 - p_3)n_3)e_2}{-n_1 x - n_2 y - n_3 z + e_1 n_1 + e_2 n_2 + e_3 n_3},$$

$$z^{new} = \frac{((e_1 - q_1)n_1 + (e_2 - q_2)n_2 + (e_3 - q_3)n_3)z + ((q_1 - p_1)n_1 + (q_2 - p_2)n_2 + (q_3 - p_3)n_3)e_3}{-n_1 x - n_2 y - n_3 z + e_1 n_1 + e_2 n_2 + e_3 n_3}.$$

Notice that since the point $P$ appears in both the numerator and the denominator on the right-hand side of Equation 13.29, the variables $x, y, z$ appear in both the numerator and the denominator of the expressions for the coordinates $x^{new}, y^{new}, z^{new}$. Transformations where $x, y, z$ appear in both the numerator and the denominator are called *projective transformations*. In order to represent perspective projection by matrix multiplication, we need to be able to represent projective transformations using matrix multiplication. So far we can only invoke matrices to represent linear and affine transformations.

*Linear transformations* are transformations of the form:

$$
\begin{aligned}
x^{new} &= a_{11}x + a_{21}y + a_{31}z \\
y^{new} &= a_{12}x + a_{22}y + a_{32}z \\
z^{new} &= a_{13}x + a_{23}y + a_{33}z
\end{aligned}
\quad \Leftrightarrow \quad
(x \quad y \quad z) *
\begin{pmatrix}
a_{11} & a_{12} & a_{13} \\
a_{21} & a_{22} & a_{23} \\
a_{31} & a_{32} & a_{33}
\end{pmatrix}.
$$

Linear transformations are readily represented by $3 \times 3$ matrices because matrix multiplication with rectangular coordinates generates three linear functions in the coordinates.

*Affine transformations* are transformations of the form:

$$
\begin{aligned}
x^{new} &= a_{11}x + a_{21}y + a_{31}z + a_{41} \\
y^{new} &= a_{12}x + a_{22}y + a_{32}z + a_{42} \\
z^{new} &= a_{13}x + a_{23}y + a_{33}z + a_{43}
\end{aligned}
\quad \Leftrightarrow \quad
(x \quad y \quad z \quad 1) *
\begin{pmatrix}
a_{11} & a_{12} & a_{13} & 0 \\
a_{21} & a_{22} & a_{23} & 0 \\
a_{31} & a_{32} & a_{33} & 0 \\
a_{41} & a_{42} & a_{43} & 1
\end{pmatrix}.
$$

Affine transformations cannot be represented by $3 \times 3$ matrices because matrix multiplication with rectangular coordinates cannot capture the constant (translation) terms. To represent affine transformations by matrices, we introduced affine coordinates for points and special $4 \times 4$ matrices, where the last column is $(0 \quad 0 \quad 0 \quad 1)^{\mathrm{T}}$, in order to capture the constant (translation) terms.

*Projective transformations* are transformations of the form:

$$x^{new} = \frac{a_{11}x + a_{21}y + a_{31}z + a_{41}}{a_{14}x + a_{24}y + a_{34}z + a_{44}},$$

$$y^{new} = \frac{a_{12}x + a_{22}y + a_{32}z + a_{42}}{a_{14}x + a_{24}y + a_{34}z + a_{44}},$$

$$z^{new} = \frac{a_{13}x + a_{23}y + a_{33}z + a_{43}}{a_{14}x + a_{24}y + a_{34}z + a_{44}}.$$

Projective transformations cannot be represented by either $3 \times 3$ or $4 \times 4$ matrices because matrix multiplication with either rectangular or affine coordinates cannot capture denominators. To represent projective transformations using matrices, we need to introduce a new device in order to express rational linear functions using matrix multiplication.

The device we shall adopt is called *homogeneous coordinates*, an extension of affine coordinates designed specifically to handle common denominators. So far we have permitted the fourth coordinate to be only 0 or 1; now we shall permit the fourth coordinate, the homogenous coordinate, to take on arbitrary values. By convention, if $w \neq 0$, then the four homogeneous coordinates $(x, y, z, w)$ represent the point with rectangular coordinates $(x/w, y/w, z/w)$—that is, denominators are stored in the fourth coordinate.

We can now express projective transformations by four equations in the homogeneous coordinates $(x, y, z, w)$:

$$
\begin{aligned}
x^{new} &= a_{11}x + a_{21}y + a_{31}z + a_{41} \\
y^{new} &= a_{12}x + a_{22}y + a_{32}z + a_{42} \\
z^{new} &= a_{13}x + a_{23}y + a_{33}z + a_{43} \\
w^{new} &= a_{14}x + a_{24}y + a_{34}z + a_{44}
\end{aligned}
\quad \Leftrightarrow \quad
(x \quad y \quad z \quad 1) *
\begin{pmatrix}
a_{11} & a_{12} & a_{13} & a_{14} \\
a_{21} & a_{22} & a_{23} & a_{24} \\
a_{31} & a_{32} & a_{33} & a_{34} \\
a_{41} & a_{42} & a_{43} & a_{44}
\end{pmatrix}.
$$

Thus, using homogeneous coordinates, we can represent projective transformations by arbitrary $4 \times 4$ matrices, where the last column captures the denominator. Another advantage of homogeneous coordinates is that we can postpone division until we actually need a numerical result.

### 13.6.2 Matrices for Perspective Projections

Let us return now to perspective projection. Recall that the formula for perspective is

$$P^{new} = \frac{((E - Q) \cdot n)P + ((Q - P) \cdot n)E}{(E - P) \cdot n},$$

where
$E$ is the eye point
$S$ is the projection plane, specified by a point $Q$ on $S$ and a unit vector $n$ normal to $S$.

Using homogenous coordinates, we can place the denominator in the fourth coordinate and write

$$P^{new} = (((E - Q) \cdot n)P + ((Q - P) \cdot n)E, (E - P) \cdot n). \qquad (13.30)$$

Expanding the terms on the right-hand side of Equation 13.30 gives

$$P^{new} = (((E - Q) \cdot n)P - (P \cdot n)E + (Q \cdot n)E, \; -P \cdot n + E \cdot n). \qquad (13.31)$$

Now observe that

$$\begin{aligned} ((E - Q) \cdot n)P &= P * ((E - Q) \cdot n)I, \\ (P \cdot n)E &= (P * n^T) * E = P * (n^T * E). \end{aligned} \qquad (13.32)$$

Therefore, by Equations 13.31 and 13.32, the numerator of $P^{new}$ is

$$((E - Q) \cdot n)P - (P \cdot n)E + (Q \cdot n)E = P * \big(((E - Q) \cdot n)I - (n^T * E)\big) + (Q \cdot n)E.$$

Moreover, the denominator of $P^{new}$ is

$$-P \cdot n + E \cdot n = P * (-n^T) + E \cdot n.$$

Therefore it follows that

$$(P^{new}, 1) = (P, 1) * \begin{pmatrix} ((E - Q) \cdot n)I - n^T * E & -n^T \\ (Q \cdot n)E & E \cdot n \end{pmatrix}.$$

Thus, with the convention that the fourth coordinate represents the denominator, the $4 \times 4$ matrix $Persp(E, Q, n)$ representing perspective projection from the eye point $E$ to the plane $S$ is given by

$$Persp(E, Q, n) = \begin{pmatrix} ((E - Q) \cdot n)I - n^T * E & -n^T \\ (Q \cdot n)E & E \cdot n \end{pmatrix}. \qquad (13.33)$$

### Example 13.7   Perspective Projection into the *xy*-Plane

Suppose that the eye is located along the $z$-axis at $E = (0, 0, 1)$ and the perspective plane is the $xy$-plane. Then we can choose $Q = Origin = (0, 0, 0)$ and $n = k = (0, 0, 1)$. By Equations 13.7

$$k^T * E = \begin{pmatrix} 0 & 0 & 0 \\ 0 & 0 & 0 \\ 0 & 0 & 1 \end{pmatrix}.$$

Therefore by Equation 13.33

$$Persp(E, Origin, k) = \begin{pmatrix} ((E - Origin) \cdot k)I - k^T * E & -k^T \\ (Origin \cdot k)E & E \cdot k \end{pmatrix} = \begin{pmatrix} 1 & 0 & 0 & 0 \\ 0 & 1 & 0 & 0 \\ 0 & 0 & 0 & -1 \\ 0 & 0 & 0 & 1 \end{pmatrix}.$$

Thus the projective transformation representing perspective projection into the $xy$-plane is

$$x^{new} = x, \; y^{new} = y, \; z^{new} = 0, \; w^{new} = 1 - z \Leftrightarrow x^{new} = x/(1 - z), \; y^{new} = y/(1 - z), \; z^{new} = 0.$$

## 13.7  Summary

In this section we summarize the main themes of this chapter. Also listed here for your convenience are the matrix representations for each of the transformations that we have studied in this chapter.

### 13.7.1  Matrix Representations for Affine and Projective Transformations

The core content of this chapter are $4 \times 4$ matrix representations for each of the basic affine and projective transformations—translation, rotation, mirror image, uniform and nonuniform scaling, and orthogonal and perspective projection—of three-dimensional Computer Graphics.

Every affine transformation $A$ is completely determined by the four values $A(O), A(i), A(j), A(k)$, and these values form the rows of the matrix representation for $A$—that is,

$$A = \begin{pmatrix} A(i) & 0 \\ A(j) & 0 \\ A(k) & 0 \\ A(O) & 1 \end{pmatrix}.$$

Thus, matrix representations for affine transformations all have the form:

$$A = \begin{pmatrix} L_A & 0 \\ w_A & 1 \end{pmatrix} = \begin{pmatrix} 3 \times 3 \text{ Linear transformation matrix} & 0 \\ \text{Translation vector} & 1 \end{pmatrix}.$$

The upper $3 \times 3$ matrix

$$L_A = \begin{pmatrix} A(i) & 0 \\ A(j) & 0 \\ A(k) & 0 \end{pmatrix},$$

can be constructed from three basic $3 \times 3$ matrices: $I, u^{\mathrm{T}} * \tilde{u}, u \times_-$ , and the vector $w_A$ can typically be expressed in terms of a fixed point $Q$ of the transformation $A$. In particular,

$$v^{new} = c_1 v + c_2(v \cdot u)\tilde{u} + c_3 u \times v \Rightarrow L_A = c_1 I + c_2 u^{\mathrm{T}} * \tilde{u} + c_3 u \times_-$$

and

$$w_A = Q * (I - L_A).$$

Perspective projection is a projective transformation, not an affine transformation. Affine transformations have the linear form:

$$
\begin{aligned}
x^{new} &= a_{11}x + a_{21}y + a_{31}z + a_{41} \\
y^{new} &= a_{12}x + a_{22}y + a_{32}z + a_{42} \\
z^{new} &= a_{13}x + a_{23}y + a_{33}z + a_{43}
\end{aligned}
\quad \Leftrightarrow \quad (x \quad y \quad z \quad 1) * \begin{pmatrix} a_{11} & a_{12} & a_{13} & 0 \\ a_{21} & a_{22} & a_{23} & 0 \\ a_{31} & a_{32} & a_{33} & 0 \\ a_{41} & a_{42} & a_{43} & 1 \end{pmatrix},
$$

whereas projective transformations have the rational linear form:

$$x^{new} = \frac{a_{11}x + a_{21}y + a_{31}z + a_{41}}{a_{14}x + a_{24}y + a_{34}z + a_{44}}$$

$$y^{new} = \frac{a_{12}x + a_{22}y + a_{32}z + a_{42}}{a_{14}x + a_{24}y + a_{34}z + a_{44}} \quad \Leftrightarrow \quad (x \quad y \quad z \quad 1) * \begin{pmatrix} a_{11} & a_{12} & a_{13} & a_{14} \\ a_{21} & a_{22} & a_{23} & a_{24} \\ a_{31} & a_{32} & a_{33} & a_{34} \\ a_{41} & a_{42} & a_{43} & a_{44} \end{pmatrix}.$$

$$z^{new} = \frac{a_{13}x + a_{23}y + a_{33}z + a_{43}}{a_{14}x + a_{24}y + a_{34}z + a_{44}}$$

Perspective projections can still be represented by $4 \times 4$ matrices, but the last column is no longer $(0, 0, 0, 1)^T$; instead the last column represents the common denominator. To realize this representation, we adopt a new convention called *homogeneous coordinates* for representing points in three dimensions. By convention, if $w \neq 0$, then,

$$(x, y, z, w) \Leftrightarrow \left( \frac{x}{w}, \frac{y}{w}, \frac{z}{w}, 1 \right).$$

Thus the fourth coordinate, the homogeneous coordinate, represents a common denominator. In Chapter 14 we shall look more closely at the intrinsic geometric meaning behind this formal algebraic convention.

### 13.7.2 Matrices for Affine and Projective Transformations

*Translation—by the vector w*

$$Trans(w) = \begin{pmatrix} 1 & 0 \\ w & 1 \end{pmatrix}.$$

*Rotation—by the angle $\theta$ around the line L determined by the point Q and the unit direction vector u*

$$Rot(Q, u, \theta) = \begin{pmatrix} Rot(u, \theta) & 0 \\ Q * (I - Rot(u, \theta)) & 1 \end{pmatrix},$$

$$Rot(u, \theta) = (\cos \theta)I + (1 - \cos \theta)(u^T * u) + (\sin \theta)(u \times_{-}).$$

*Mirror image—in the plane S determined by the point Q and the unit normal n*

$$Mir(Q, n) = \begin{pmatrix} I - 2n^T * n & 0 \\ 2(Q \cdot n)n & 1 \end{pmatrix}.$$

*Uniform scaling—around the point Q by the scale factor s*

$$Scale(Q, s) = \begin{pmatrix} sI & 0 \\ (1 - s)Q & 1 \end{pmatrix}.$$

*Nonuniform scaling—around the point Q by the scale factor s in the direction u*

$$Scale(Q, u, s) = \begin{pmatrix} I + (s - 1)u^T * u & 0 \\ (1 - s)(Q \cdot u)u & 1 \end{pmatrix}.$$

*Orthogonal projection—into the plane S determined by the point Q and the unit normal n*

$$Ortho(Q, n) = \begin{pmatrix} I - n^{\mathrm{T}} * n & 0 \\ (Q \cdot n)n & 1 \end{pmatrix}.$$

*Perspective projection—from the eye point E into the plane S determined by the point Q and the unit normal vector n*

$$Persp(E, Q, n) = \begin{pmatrix} ((E - Q) \cdot n)I - n^{\mathrm{T}} * E & -n^{\mathrm{T}} \\ (Q \cdot n)E & E \cdot n \end{pmatrix}.$$

---

## Exercises

13.1. Consider the shear transformation introduced in Chapter 12, Exercise 12.2. Let

$S =$ Shearing plane

$n =$ Unit vector perpendicular to $S$

$Q =$ Point on $S$

$u =$ Unit vector in $S$ (i.e., a unit vector perpendicular to $n$)

$\phi =$ Shear angle

Recall that the formulas for shear are:

$$\begin{aligned} v^{new} &= v + \tan(\phi)(v \cdot n)u, \\ P^{new} &= P + \tan(\phi)((P - Q) \cdot n)u. \end{aligned}$$

a. Show that the $4 \times 4$ matrix representing shear is given by

$$Shear(Q, n, u, \phi) = \begin{pmatrix} I + (\tan \phi)(n^{\mathrm{T}} * u) & 0 \\ -\tan \phi(Q \cdot n)u & 1 \end{pmatrix}.$$

b. Suppose that:

    i. $S$ is the $xy$-plane.

    ii. $u$ is the unit vector along the $x$-axis.

    iii. $\phi = 45°$.

    Write down the explicit entries of the $4 \times 4$ matrix $Shear(Q, n, u, \phi)$.

13.2. Recall from Chapter 12, Exercise 12.3 that parallel projection is projection onto a plane $S$ along a direction $u$. As usual the plane $S$ is defined by a point $Q$ and a unit normal vector $n$. Recall that the formulas for parallel projection are:

$$v^{new} = v - \frac{v \cdot n}{u \cdot n} u,$$

$$P^{new} = P - \frac{(P - Q) \cdot n}{u \cdot n} u.$$

a. Show that the $4 \times 4$ matrix representing parallel projection is given by

$$PProj(Q, n, u) = \begin{pmatrix} I - \dfrac{n^{\mathrm{T}} * u}{u \cdot n} & 0 \\ \dfrac{Q \cdot n}{u \cdot n} u & 1 \end{pmatrix}.$$

b. Suppose that

   i. $S$ is the $xy$-plane.

   ii. $u$ is the unit vector parallel to the vector $(1,1,1)$.

   Write down the explicit entries of the $4 \times 4$ matrix $PProj(Q, n, u)$.

13.3. Verify that:

   a. $Rot(u, \theta)^{-1} = Rot(u, -\theta) = Rot(u, \theta)^{\mathrm{T}}$.

   b. $Mir(n)^{-1} = Mir(n) = Mir(n)^{\mathrm{T}}$.

13.4. Verify that:

   a. $Det(Rot(u, \theta)) = 1$.

   b. $Det(Mir(n)) = -1$.

13.5. Verify that:

   a. $u$ is an eigenvector of $Rot(u, \theta)$ corresponding the eigenvalue 1.

   b. $n$ is an eigenvector of $Mir(n)$ corresponding the eigenvalue $-1$.

   c. Interpret the results in parts a and b geometrically.

13.6. The trace of a matrix is the sum of the diagonal entries. Thus, for a $3 \times 3$ matrix $M$:

$$Trace(M) = m_{11} + m_{22} + m_{33}.$$

a. Show that:

   i. $Trace(M + N) = Trace(M) + Trace(N)$.

   ii. $Trace(cM) = cTrace(M)$.

b. Using parts a, b and Equation 13.18, conclude that

$$\cos(\theta) = \frac{Trace(Rot(u, \theta)) - 1}{2}.$$

13.7. Consider a rotation matrix $Rot(u, \theta)$.

a. Using Equation 13.18 or 13.19, show that we can recover the unit vector $u$ parallel to the axis of rotation and the sine of the angle of rotation from the rotation matrix $Rot(u, \theta)$ by the formula:

$$2\sin(\theta)u = \left(Rot(u,\theta)_{2,3} - Rot(u,\theta)_{3,2}, Rot(u,\theta)_{3,1} - Rot(u,\theta)_{1,3}, Rot(u,\theta)_{1,2} - Rot(u,\theta)_{2,1}\right).$$

b. Explain why we cannot solve independently for the axis of rotation $u$ and the angle of rotation $\theta$.

13.8. Let $M$ be a $3 \times 3$ matrix, and let $v, w$ be arbitrary vectors.

a. Show that

$$(v * M) \cdot w = v \cdot \left(w * M^{\mathrm{T}}\right).$$

b. Conclude from part a and Exercise 13.3 that:

i. $(v * Rot(u, \theta)) \cdot (w * Rot(u, \theta)) = v \cdot w$.

ii. $(v * Mir(n)) \cdot (w * Mir(n)) = v \cdot w$.

c. Interpret the results in part b geometrically.

d. Find an example of a matrix $M$ such that

$$(u * M) \cdot (v * M) \neq u \cdot v.$$

13.9. Let $M$ be a $3 \times 3$ matrix, and let $v, w$ be arbitrary vectors.

a. Show that

$$(u * M) \times (v * M) = Det(M)(u \times v) * M^{-\mathrm{T}}.$$

(Hint: Consider the dot product of both sides with the vectors $(u * M), (v * M), (u \times v) * M$.)

b. Conclude from part a and Exercises 13.3 and 13.4 that:

i. $(v * Rot(u, \theta)) \times (w * Rot(u, \theta)) = v \times w$.

ii. $(v * Mir(n)) \times (w * Mir(n)) = -v \times w$.

c. Interpret the results in part b geometrically.

d. Find an example of a matrix $M$ such that

$$(u * M) \times (v * M) \neq (u \times v) * M.$$

13.10. Let $v_1, v_2$ be unit vectors, and let $u = v_1 \times v_2$.

a. Show that the matrix

$$Rot(u) = (v_1 \cdot v_2)I + \frac{u^T * u}{1 + v_1 \cdot v_2} + u \times_- ,$$

rotates the vector $v_1$ into the vector $v_2$.

b. Interpret this result geometrically.

13.11. Show that $Mir(Q, n) = Scale(Q, n, -1)$. Interpret this result geometrically.

13.12. Show that $Mir(n) = -Rot(n, \pi)$. Interpret this result geometrically.

13.13. Let $I_{4 \times 4}$ be the $4 \times 4$ identity matrix. Show that on points $Ortho(Q, n) = I_{4 \times 4} + Mir(Q, n)$.

13.14. Let $A$ be a nonsingular affine transformation. Show that:

a. $A$ maps straight lines to straight lines.

b. $A$ preserves ratios of distances along straight lines—that is, if $P, Q, R$ are collinear, then

$$\frac{|A(R) - A(P)|}{|A(Q) - A(P)|} = \frac{|R - P|}{|Q - P|}.$$

13.15. Let $T$ be a nonsingular projective transformation. Show that:

a. $T$ maps straight lines to straight lines.

b. $T$ does not necessarily preserve ratios of distances along straight lines. That is, give an example to show that if $P, Q, R$ are collinear, then

$$\frac{|T(R) - T(P)|}{|T(Q) - T(P)|} \neq \frac{|R - P|}{|Q - P|}.$$

13.16. The *cross ratio* of four collinear points $P, Q, R, S$ is defined by

$$cr(P, Q, R, S) = \frac{|R - P||S - Q|}{|S - P||R - Q|}.$$

Show that if $T$ is a projective transformation, then $T$ preserves cross ratios—that is, show that

$$cr(T(P), T(Q), T(R), T(S)) = cr(P, Q, R, S).$$

13.17. Let $P$ be a point, and let $v_1, v_2, v_3$ be three linearly independent vectors in three-space.

a. Show that in three dimensions there is a unique affine transformation $A$ that maps $P, v_1, v_2, v_3$ to $Q, w_1, w_2, w_3$.

b. Show that if $u$ is an arbitrary vector, then

$$A(u) = \frac{Det(u \ v_2 \ v_3)w_1 + Det(v_1 \ u \ v_3)w_2 + Det(v_1 \ v_2 \ u)w_3}{Det(v_1 \ v_2 \ v_3)}.$$

c. Verify that the matrix representing this affine transformation $A$ is given by

$$A = \begin{pmatrix} v_1 & 0 \\ v_2 & 0 \\ v_3 & 0 \\ P & 1 \end{pmatrix}^{-1} * \begin{pmatrix} w_1 & 0 \\ w_2 & 0 \\ w_3 & 0 \\ Q & 1 \end{pmatrix}.$$

d. Show that Equation 13.1 is a special case of the matrix in part c.

13.18. Let $P_1, P_2, P_3, P_4$ be four noncollinear points in three-space.

a. Show that in three dimensions there is a unique affine transformation $A$ that maps four noncollinear points $P_1, P_2, P_3, P_4$ to four arbitrary points $Q_1, Q_2, Q_3, Q_4$.

b. Show that if $R$ is an arbitrary point, then

$$A(R) = \frac{Det\begin{pmatrix} R & 1 \\ P_2 & 1 \\ P_3 & 1 \\ P_4 & 1 \end{pmatrix}Q_1 + Det\begin{pmatrix} P_1 & 1 \\ R & 1 \\ P_3 & 1 \\ P_4 & 1 \end{pmatrix}Q_2 + Det\begin{pmatrix} P_1 & 1 \\ P_2 & 1 \\ R & 1 \\ P_4 & 1 \end{pmatrix}Q_3 + Det\begin{pmatrix} P_1 & 1 \\ P_2 & 1 \\ P_3 & 1 \\ R & 1 \end{pmatrix}Q_4}{Det\begin{pmatrix} P_1 & 1 \\ P_2 & 1 \\ P_3 & 1 \\ P_4 & 1 \end{pmatrix}}.$$

c. Verify that the matrix representing this affine transformation $A$ is given by

$$A = \begin{pmatrix} P_1 & 1 \\ P_2 & 1 \\ P_3 & 1 \\ P_4 & 1 \end{pmatrix}^{-1} * \begin{pmatrix} Q_1 & 1 \\ Q_2 & 1 \\ Q_3 & 1 \\ Q_4 & 1 \end{pmatrix}.$$

## Programming Projects

### 13.1. *The Turtle in Three Dimensions*

In two dimensions the turtle's state consists of a point $P$ specifying her location and a vector $u$ specifying both her heading and her step size. To navigate in three dimensions, the turtle carries around not just one vector $u$, but three mutually orthogonal vectors $u, v, w$ specifying her heading $u$, her left hand $v$, and her up direction $w$. Thus in three dimensions the turtle carries around her own three-dimensional local coordinate system. The FORWARD and MOVE commands work much as before, but in place of the TURN command, there are three new turtle commands: YAW, PITCH, and ROLL. YAW represent a rotation in the *uv-plane*, the plane perpendicular to the up direction $w$; PITCH represents a rotation in the *wu-plane*, the plane perpendicular to the left-hand vector $v$; and ROLL represents a rotation in the *vw-plane*, the plane perpendicular to the forward direction $u$. Also the RESIZE command is redefined to apply simultaneously to all three turtle vectors instead of just one turtle vector. Thus in three dimensions, LOGO has the following commands:

- *FORWARD D and MOVE D—translation*

$$P_{new} = P + D\,u.$$

- *YAW A—rotation in uv-plane*

$$u_{new} = u\cos(A) + v\sin(A) \qquad \Leftrightarrow \qquad \begin{pmatrix} u_{new} \\ v_{new} \end{pmatrix} = \begin{pmatrix} \cos(A) & \sin(A) \\ -\sin(A) & \cos(A) \end{pmatrix} * \begin{pmatrix} u \\ v \end{pmatrix}.$$
$$v_{new} = -u\sin(A) + v\cos(A)$$

- *PITCH A—rotation in wu-plane*

$$w_{new} = w\cos(A) + u\sin(A) \qquad \Leftrightarrow \qquad \begin{pmatrix} w_{new} \\ u_{new} \end{pmatrix} = \begin{pmatrix} \cos(A) & \sin(A) \\ -\sin(A) & \cos(A) \end{pmatrix} * \begin{pmatrix} w \\ u \end{pmatrix}.$$
$$u_{new} = -w\sin(A) + u\cos(A)$$

- *ROLL A—rotation in vw-plane*

$$v_{new} = v\cos(A) + w\sin(A) \qquad \Leftrightarrow \qquad \begin{pmatrix} v_{new} \\ w_{new} \end{pmatrix} = \begin{pmatrix} \cos(A) & \sin(A) \\ -\sin(A) & \cos(A) \end{pmatrix} * \begin{pmatrix} v \\ w \end{pmatrix}.$$
$$w_{new} = -v\sin(A) + w\cos(A)$$

- *RESIZE s—uniform scaling*

$$u_{new} = s\,u$$
$$v_{new} = s\,v \qquad \Leftrightarrow \qquad \begin{pmatrix} u_{new} \\ v_{new} \\ w_{new} \end{pmatrix} = \begin{pmatrix} s & 0 & 0 \\ 0 & s & 0 \\ 0 & 0 & s \end{pmatrix} * \begin{pmatrix} u \\ v \\ w \end{pmatrix}.$$
$$w_{new} = s\,w$$

a. Implement the three-dimensional turtle in your favorite programming language using your favorite API.

b. Write LOGO programs to generate the edges of simple polyhedra in three dimensions.

c. Write recursive LOGO programs to create your own novel fractals in three dimensions.

13.2. *The Turtle on the Surface of a Sphere*

The classical turtle walks on the surface of an infinite plane. But we could imagine that the turtle, like us, is actually walking on the surface of a large sphere. In this environment, when the turtle executes the command FORWARD $D$, the turtle draws the arc of a great circle on the sphere. The FORWARD and TURN commands correspond to different rotations of the sphere beneath the turtle or equivalently different rotations of the turtle in three-space. Since the turtle never leaves the surface of the sphere, her position can be stored as a vector $u$ of a fixed length from the center of the sphere. Hence the turtle's state consist of two orthogonal vectors: a position vector $u$ of a fixed length and a direction vector $v$, perpendicular to the position vector $u$, tangent to the surface of the sphere. Thus for the turtle on the surface of a sphere, LOGO consists of the following commands:

- *FORWARD D and MOVE D—translation along the sphere = rotation in the uv-plane*

$$u = |u|\left(\cos(D|v|)\frac{u}{|u|} + \sin(D|v|)\frac{v}{|v|}\right)$$
$$v = |v|\left(-\sin(D|v|)\frac{u}{|u|} + \cos(D|v|)\frac{v}{|v|}\right)$$
$$\Leftrightarrow \quad \begin{pmatrix} u_{new} \\ v_{new} \end{pmatrix} = \begin{pmatrix} u \\ v \end{pmatrix} * rot\left(\frac{u \times v}{|u \times v|}, D|v|\right).$$

- *TURN A—rotation around the vector u*

$$v = |v|\left(\cos(A)\frac{v}{|v|} + \sin(A)\frac{u \times v}{|u \times v|}\right) \quad \Leftrightarrow \quad v_{new} = v * rot\left(\frac{u}{|u|}, A\right).$$

- *RESIZE S*

$$v_{new} = Sv$$

a. Implement LOGO on the surface of a sphere in your favorite programming language using your favorite API.

b. Write turtle programs to generate simple curves like polygons, stars, and spirals on the surface of a sphere.

c. Write recursive turtle programs to generate fractals on the surface of a sphere.

13.3. *CODO in Three Dimensions*

a. Implement CODO in three dimensions in your favorite programming language using your favorite API, incorporating the three-dimensional affine transformations discussed in the text and the exercises of this chapter.

b. Write CODO programs to generate the edges of simple polyhedra in three dimensions.

c. Develop iterated function systems to create fractals in three dimensions with and without condensation sets. In particular, develop iterated function systems to generate:

   i. Three-dimensional gaskets.

   ii. Three-dimensional snowflakes.

   iii. Three-dimensional trees.

   iv. Three-dimensional fractals of your own novel design.

13.4. *Converting Three-Dimensional Recursive Turtle Programs into Three-Dimensional Iterated Function Systems*

a. Let $S = \begin{pmatrix} u & v & w & P \\ 0 & 0 & 0 & 1 \end{pmatrix}^{\mathrm{T}}$ denote the state of the three-dimensional turtle (see Programming Project 13.1). Show that the LOGO commands for the three-dimensional turtle can be implemented in the following fashion:

| Command | Exocentric | Turtle–Centric |
|---|---|---|
| FORWARD D | $S_{new} = S * Trans\,(Du)$ | $S_{new} = Trans\,(Di) * S$ |
| MOVE D | $S_{new} = S * Trans(Du)$ | $S_{new} = Trans(Di) * S$ |
| YAW A | $S_{new} = S * Rot(P, w/|w|, A)$ | $S_{new} = Rot(Origin, k, A) * S$ |

| | | |
|---|---|---|
| PITCH $A$ | $S_{new} = S * Rot(P, v/|v|, A)$ | $S_{new} = Rot(Origin, j, A) * S$ |
| ROLL $A$ | $S_{new} = S * Rot(P, u/|u|, A)$ | $S_{new} = Rot(Origin, i, A) * S$ |
| RESIZE $s$ | $S_{new} = S * Scale(P, s)$ | $S_{new} = Scale(Origin, s) * S$ |

b. Write a parser to convert recursive programs for the three-dimensional turtle into iterated function systems that generate the same fractal.

c. Give examples to illustrate the results of your parser.

13.5. *The Three-Dimensional Turtle on a Bounded Domain*

The three-dimensional turtle lives in an infinite space. Suppose, however, that the three-dimensional turtle is restricted to a finite domain bounded by walls. When the turtle hits a wall, she bounces off the wall so that her angle of incidence with the normal vector to the wall is equal to her angle of reflection, and then she continues on her way (see Chapter 2, Programming Project 2.3: *The Classical Turtle on a Bounded Domain*). To implement the three-dimensional turtle on a bounded domain, the FORWARD command must be replaced by

NEWFORWARD $D$
$\quad D_1 =$ Distance from turtle to wall in direction of turtle heading
$\quad$ IF $D < D_1$, FORWARD $D$
$\quad$ OTHERWISE
$\qquad$ FORWARD $D_1$

$$\text{ROTATE } A = \begin{pmatrix} u & v & w \\ 0 & 0 & 0 \end{pmatrix}^{T} * Rot\left(\frac{n \times u}{|n \times u|}, A\right) \{A = \text{angle of reflection}\}$$

$\qquad$ NEWFORWARD $D - D_1$
Here $n$ is the normal to the wall at the point of impact, and $\cos(A) = \dfrac{u \cdot n}{|u||n|}$.

a. Implement LOGO on a bounded domain, where the walls form:

$\quad$ i. A rectangular box

$\quad$ ii. A sphere

$\quad$ iii. An ellipsoid

b. Consider the turtle program consisting of the single command

$\qquad$ NEWFORWARD $D$.
Investigate the curves generated by this program for different shape walls, different values of $D$, and different initial positions and headings for the turtle.

c. Investigate how curves drawn by the walled in turtle differ from curves drawn by the turtle in infinite space using the same turtle programs, where FORWARD is replaced by NEWFORWARD.

13.6. *The Three-Dimensional Turtle in a Three-Dimensional Manifold.*

The three-dimensional turtle lives in a flat three-dimensional space. Here we shall investigate turtles that live on curved three-dimensional manifolds. These manifolds can be modeled by rectangular boxes, where pairs of opposite sides—front and back, left and right, top and bottom—are glued together, possibly with a twist (see Chapter 2, Programming Project 2.5: *The Turtle on Two-Dimensional Manifolds*, for the analogous two-manifolds modeled by rectangles). When the turtle encounters a wall, she does not bounce off the wall like the turtle on a bounded domain (Programming Project 13.5); rather she emerges on the opposite wall of the box heading in the same direction relative to the new wall in which she hit the opposing wall.

a. Implement LOGO on one or more different three-manifolds.

   *Note*: For each of these three-manifolds, you will need to define a command MFORWARD to replace of the command FORWARD in standard LOGO (see Programming Project 13.5).

b. Consider the turtle program consisting of the single command

   MFORWARD *D*.

   Investigate the curves generated by this program for different three manifolds, different values of *D*, and different initial positions and headings for the turtle.

c. Investigate how curves drawn by turtles on these three-dimensional manifolds differ from curves drawn by the turtle in an infinite space using the same turtle programs, where FORWARD is replaced by MFORWARD.

13.7. *Linear Algebra Package for Computer Graphics*

Implement a basic linear algebra package for Computer Graphics. Your package should include the following:

a. Algebra of Points and Vectors

   Coordinate implementations for addition, subtraction, scalar multiplication, dot product, cross product, and determinant.

b. Affine and Projective Transformations

   Matrix implementations for translation, rotation, mirror image, uniform and nonuniform scaling, shear, orthogonal projection, perspective projection, and composite transformations.

c. Distance Formulas

   Vector algebra implementations for the distance between two points, a point and a line, a point and a plane, and two parallel or two skew lines.

d. Intersection Computations

   Vector algebra implementations for the intersections of two lines, a line and a plane, two planes, or three planes.

13.8. *Animation*

a. Build a collection of three-dimensional polyhedral models.

b. Animate the models using the transformations developed in this chapter.

# Chapter 14

## Projective Space versus the Universal Space of Mass-Points

*Yet there shall be a space*

– Joshua 3:4

## 14.1 Algebra and Geometry

Geometry guides our intuitions; algebra conducts our computations. So far, in order to facilitate calculations on a computer, our coordinate representations for geometry have been driven largely by linear algebra: matrix representations for affine transformations pointed us to affine coordinates; matrix representations for projective transformations steered us to homogeneous coordinates.

Algebra expedites computations. But in the wake of all this matrix algebra many geometric questions have been left unanswered or even suppressed. The goal of this chapter is to confront these geometric issues head on and to provide some straightforward answers to help guide our intuition.

Here are a few basic questions that we have avoided till now, but should have been bothering you all along:

1. For vectors $v$, we know the geometric meaning of $cv$ for any constant $c$. But for points $P$, we have a geometric interpretation of $cP$ only for $c = 0, 1$. *What is the geometric meaning of $cP$ when $c \neq 0, 1$?*

2. In affine coordinates, each vector $v$ and each point $P$ has a unique representation: $(x, y, z, 0)$ represents a unique vector $v$, and $(x, y, z, 1)$ represents a unique point $P$. But in homogeneous coordinates, the 4-tuple $(wx, wy, wz, w)$ represents the point $P = (x, y, z)$ for any $w \neq 0$. Thus, there are infinitely many $w$'s for the same point $P$. *What is the geometric interpretation of all the homogeneous 4-tuples $(wx, wy, wz, w)$ that represent the same point $P = (x, y, z)$?*

3. We have been using two distinct, complementary models of space: points and vectors. *Is there a single consistent model of space that contains both points and vectors?*

4. Perspective projection does not map vectors to vectors. Nevertheless, we can still multiply a vector $v = (v_1, v_2, v_3, 0)$ by the $4 \times 4$ matrix $M$ that represents perspective projection. *How should we interpret the result of perspective projection applied to a vector?*

There are two sets of answers to each of these questions: one invokes projective space, the other the universal space of mass-points. In this chapter we shall investigate these two approaches to geometry, along with their advantages and disadvantages for Computer Graphics.

## 14.2   Projective Space: The Standard Model

Projective space consists of two types of points: affine points and points at infinity. An *affine point* is the same as an ordinary point in Euclidean space; *a point at infinity* is a new kind of point, defined by a direction in Euclidean space (see Figure 14.1). Points at infinity are represented by equivalence classes of vectors: two vectors that point in the same direction or in diametrically opposite directions represent the same point at infinity. Thus $v$ and $cv$ represent the same point at infinity for any constant $c \neq 0$. We write the equivalence class of the vector $v$ using homogeneous coordinates as $[v, 0]$. Hence, in projective space, $[v, 0] \equiv [cv, 0] \equiv c[v, 0]$ for any $c \neq 0$. Similarly, in projective space, we represent affine points $P$ using equivalence classes of homogeneous coordinates, so $[P, 1] \equiv [cP, c] \equiv c[P, 1]$ for any $c \neq 0$. Notice that the zero vector does not correspond to a point in projective space, since the zero vector does not define a direction in Euclidean space.

Projective space is the standard model for geometry adopted in most textbooks on Computer Graphics. Here is how projective space provides answers to the four questions listed in the introduction to this chapter:

1. *What is the geometric meaning of cP when $c \neq 0, 1$?*

   If $P = [x, y, z, 1]$, then $cP = [cx, cy, cz, c]$ is just another member of the equivalence class representing the same affine point $P$.

2. *What is the geometric interpretation of all the homogeneous 4-tuples (wx, wy, wz, w) that represent the same affine point $P = (x, y, z)$?*

   The 4-tuples $(wx, wy, wz, w)$ represent the points on the line through the origin in the direction $(x, y, z, 1)$. Thus each line through the origin in four-dimensions represent a point in projective three-space.

3. *Is there a single consistent model of space that contains both points and vectors?*

   Yes. Projective space includes all the affine points and incorporates the vectors as points at infinity.

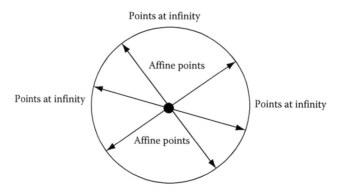

**FIGURE 14.1:**  Projective space. The affine points are the same as the standard points in Euclidean space. Points at infinity correspond to directions in Euclidean space and are glued onto the affine points at infinity. In this figure, we picture the projective plane (two-dimensional projective space), but the construction is valid in any dimension.

4. *How should we interpret the result of perspective projection applied to a vector?*

Matrix multiplication commutes with scalar multiplication. Hence, if $M$ is a $4 \times 4$ matrix and $v$ is a vector, then for any $c \neq 0$

$$(c(v,0)) * M = c((v,0) * M).$$

Thus, matrix multiplication is well defined in projective space, since matrix multiplication maps equivalence classes to equivalence classes—that is, matrix multiplication maps constant multiples to constant multiples. Notice, however, that points at infinity are not necessarily mapped to points at infinity by matrix multiplication, since the last component of $(v,0) * M$ need no longer be zero when the last column of $M$ is not $(0,0,0,1)^T$. This result accounts for our observation in Chapter 12, Section 12.5.2 that vectors are not mapped to vectors by perspective projection. In fact, under perspective projection, the point at infinity corresponding to the vector $v$ is mapped to the intersection with the projection plane $S$ of the line determined by the eye point $E$ and the point at infinity corresponding to the vector $v$ (see Figure 14.2).

We can prove our assertion that under perspective projection, the point at infinity corresponding to the vector $v$ is mapped to the intersection with the projection plane $S$ of the line through the eye point $E$ and the point at infinity corresponding to the direction $v$ in the following manner. Let $S$ be the plane defined by the point $Q$ on $S$ and the unit vector $n$ normal to $S$. In Chapter 13, Section 13.6.2 we showed that

$$Persp(E, Q, n) = \begin{pmatrix} ((E - Q) \cdot n)I - n^T * E & -n^T \\ (Q \cdot n)E & E \cdot n \end{pmatrix}.$$

Thus,

$$(v, 0) * Persp(E, Q, n) = (((E - Q) \cdot n) v - (v \cdot n)E, -v \cdot n), \qquad (14.1)$$

so dividing by $-v \cdot n$, we find that

$$v^{new} = E - \frac{(E - Q) \cdot n}{v \cdot n} v. \qquad (14.2)$$

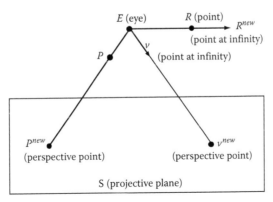

**FIGURE 14.2:** Perspective projection maps a point at infinity $v$ to the intersection of the line through the eye point $E$ and the point at infinity $v$ with the projection plane $S$. Notice too that a point $R$ on a line through $E$ parallel to $S$ is mapped to the point at infinity in the direction $R-E$.

Hence clearly, $v^{new}$ lies on the line through $E$ in the direction $v$. In projective space, this line contains the eye point $E$ and the point at infinity corresponding to the direction $v$. Moreover, $v^{new} - Q$ is perpendicular to $n$, since

$$(v^{new} - Q) \cdot n = (E - Q) \cdot n - (E - Q) \cdot n = 0. \tag{14.3}$$

Hence, $v^{new}$ also lies on the plane $S$.

The main advantage of projective space is that *projective space completes the geometry of affine space by including the points at infinity.* This completion leads to cleaner theorems by avoiding the special cases that often arise in affine geometry due to the existence of parallel lines. For example:

- *Each pair of lines in the projective plane intersect at a point.* In the projective plane, there are no parallel lines, since lines that are parallel to a vector $v$ in Euclidean space meet at the point at infinity $[v, 0]$ in projective space. Thus, every pair of lines in the same plane intersect in projective space.

- *The principle of duality holds in projective space.* For any theorem in projective space, interchanging the words

    *point* $\leftrightarrow$ *line* in two dimensions or *point* $\leftrightarrow$ *plane* in three dimensions
    *collinear* $\leftrightarrow$ *concurrent*

    results in another valid theorem in projective space. For example, Brianchon's theorem and Pascal's theorem are dual theorems in the projective plane (see Figure 14.3). Notice again that in projective space we do not have to worry about special cases arising when certain lines are parallel because lines that are parallel in Euclidean space meet at a point at infinity in projective space.

- *Bezout's theorem is valid in (complex) projective space.* Bezout's theorem counts the number of intersections between polynomial curves. Bezout's theorem asserts that if $f(x,y)$ is a polynomial of degree $m$ and $g(x,y)$ is a polynomial of degree $n$, and $f(x,y)$ and $g(x,y)$ have no common factors, then the number of intersection points (counting multiplicities) of the curves $f(x,y) = 0$ and $g(x,y) = 0$ is exactly $mn$ in *complex projective space.*

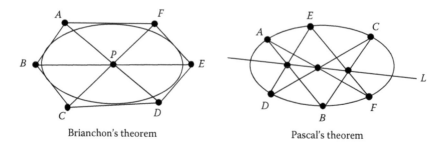

Brianchon's theorem          Pascal's theorem

**FIGURE 14.3:** Brianchon's theorem (left): *If a hexagon is circumscribed about a conic, its diagonals are concurrent.* Pascal's theorem (right): *If a hexagon is inscribed in a conic, then opposite sides intersect in collinear points.* Brianchon's theorem and Pascal's theorem are dual in projective space. Notice that in some cases lines may be parallel, so the intersections may occur at points at infinity in the projective plane.

Notice that to get the correct count, we must include intersection points at infinity as well as intersections in the complex plane. Thus, Bezout's theorem is valid in complex projective space, but not in real affine space. Bezout's theorem also extends in a natural way to higher dimensional complex projective spaces and to $n$ polynomials in $n$ variables.

- *Every* $4 \times 4$ *matrix represents a well-defined transformation in projective space.* Let $M$ be a $4 \times 4$ matrix. Then in projective space for all $c \neq 0$

$$[cP, c] * M \equiv c([P, 1] * M) \equiv [P, 1] * M,$$

$$[cv, 0] * M \equiv c([v, 0] * M) \equiv [v, 0] * M.$$

Thus, $M$ maps any two 4-tuples that represent the same point in projective space to equivalent points in projective space. In particular, perspective projection is defined for all points in projective space except for the eye point, even, as we have seen previously, for points at infinity. Transformations of projective space represented by $4 \times 4$ matrices are called *projective transformations*. Notice that projective transformations need not map affine points to affine points or points at infinity to points at infinity; affine points may be mapped to points at infinity and points at infinity may be mapped to affine points. The projective transformations that map affine points to affine points and points at infinity to points at infinity are precisely the affine transformations.

Despite these advantages over standard Euclidean spaces, there are nevertheless still problems with projective spaces that make these spaces unwieldy as the geometric foundation for Computer Graphics.

1. *Unnatural geometry.* A point on a projective line does not split the line into two disjoint segments because opposite directions along the line are connected by a point at infinity. Hence, if you were to extend your gaze along a straight line in projective space, you would see the back of your head! Thus, the geometry of projective space is not the same as the geometry of the visual world that we are trying to capture in Computer Graphics.

2. *No vectors.* Points at infinity are not the same as vectors. The points at infinity have no orientation or length. Thus, in projective space, there is no way to specify orientations or lengths.

3. *Limited algebra.* There are no affine or linear combinations in projective space because these operations are not well defined on equivalence classes. For example, in projective space,

$$[1, 2, 3, 0] = [2, 4, 6, 0] \quad \text{and} \quad [2, 1, 0, 0] = [6, 3, 0, 0],$$

but

$$[1, 2, 3, 0] + [2, 1, 0, 0] = [3, 3, 3, 0] \neq [8, 7, 6, 0] = [2, 4, 6, 0] + [6, 3, 0, 0].$$

Moreover, vector algebra—addition, subtraction, scalar multiplication, dot product, cross product, and determinant—does not apply to points at infinity because, once again, these operations are not well defined on equivalence classes. Thus, the algebraic foundation for many of our computations is no longer present in projective space. Matrix multiplication still works in projective space, since matrix multiplication commutes with scalar multiplication. For this reason, we can use matrices to represent projective transformations. But the rest of the algebra that we developed for vector spaces and affine spaces no longer applies in projective spaces.

## 14.3   Mass-Points: The Universal Model

Fortunately, there is an alternative model for geometry that has many of the advantages, but none of the drawbacks of projective spaces. This model is *the universal space of mass-points and vectors*, which we shall shorten simply to *the space of mass-points*.

A *mass-point* is a point $P$ in affine space together with a nonzero mass $m$. Rather than writing the pair $(P, m)$, we shall see shortly that it is more convenient to denote a mass-point by the pair $(mP, m)$. Here $m$ is the mass and $P = mP/m$ is the point; the expression $mP$ by itself has no meaning. Notice that the mass $m$ may be negative; nevertheless, we still use the term *mass-point*, even for points with negative mass. Vectors reside in this space as objects with zero mass, so for vectors $v$ we write $(v, 0)$.

The universal space of mass-points and vectors forms a vector space. Addition and scalar multiplication are defined by the following formulas:

$$(m_1 P_1, m_1) + (m_2 P_2, m_2) = (m_1 P_1 + m_2 P_2, m_1 + m_2), \tag{14.4}$$

$$(-mP_1, -m) + (mP_2, m) = (m(P_2 - P_1), 0), \tag{14.5}$$

$$c(mP, m) = (cmP, cm), \tag{14.6}$$

$$(v, 0) + (w, 0) = (v + w, 0), \tag{14.7}$$

$$c(v, 0) = (cv, 0), \tag{14.8}$$

$$(mP, m) + (v, 0) = (mP + v, m). \tag{14.9}$$

Thus, to add, we simply add the components of each pair; similarly, to multiply by a scalar, we multiply the components of each pair by the scalar.

Equations 14.4 through 14.9 have physical interpretations. Equation 14.4 says that the sum of two mass-points is the mass-point whose mass is the sum of the two masses located at their center of mass. To verify the last claim, we shall apply *Archimedes' Law of the Lever*, which says that two masses balance at their center of mass. Let

$$d_1 = dist\left(\frac{m_1 P_1 + m_2 P_2}{m_1 + m_2}, P_1\right) = \frac{m_2 \, |P_2 - P_1|}{m_1 + m_2},$$

$$d_2 = dist\left(\frac{m_1 P_1 + m_2 P_2}{m_1 + m_2}, P_2\right) = \frac{m_1 \, |P_1 - P_2|}{m_1 + m_2}.$$

Then it is easy to see that

$$m_1 d_1 = m_2 d_2.$$

Thus, by *Archimedes' Law of the Lever*, $(m_1 P_1, m_1) + (m_2 P_2, m_2)$ is located at the center of mass of $(m_1 P_1, m_1)$ and $(m_2 P_2, m_2)$. If $m_1 = -m_2$, then the sum of the masses is zero. In this case, Equation 14.5 says that the sum of two mass-points $(-mP_1, -m)$ and $(mP_2, m)$ with equal and opposite masses is the vector from $P_1$ to $P_2$ scaled by the mass $m$.

To multiply a mass-point $(mP, m)$ by a scalar $c$, Equation 14.6 says that we multiply the mass by $c$ and leave the position of the point unchanged.

Addition and scalar multiplication have already been defined for vectors in affine space. Equations 14.7 and 14.8 say that we just carry over these definitions in the obvious manner to the space of mass-points. Thus, the vectors from affine space form a subspace of the space of mass-points.

To complete our algebra of mass-points and vectors, we need to define how to add a vector $v$ to a mass-point $(mP, m)$. Think of the vectors as carrying momentum and think about what happens when momentum is transferred to a mass-point. Since momentum is conserved, the larger the mass, the smaller the velocity imparted to the mass-point. In fact, since momentum is conserved, the velocity varies inversely with the mass, so the velocity imparted to the mass-point $(mP, m)$ by the momentum vector $v$ is simply $v/m$. Thus, in unit time, the mass-point at $P$ is relocated to the new position $P + v/m$. Therefore, since the mass-point $(mP + v, m)$ is located at the affine point $P + v/m$, we define

$$(mP, m) + (v, 0) = (mP + v, m).$$

Thus, once again, to compute the sum, we simply add the components. Notice that if a unit mass is located at $P$, then the momentum vector $v$ moves the mass-point $(P, 1)$ to the location $P + v$, which is the location of the standard sum of a point and a vector in affine space.

It is easy to check that with these definitions, scalar multiplication distributes through addition, so the space of mass-points is indeed a vector space. The space of mass-points is four-dimensional: three of the dimensions are spatial; the fourth dimension is due to the mass.

The space of mass-points can be adopted to model the geometry of Computer Graphics. Here is how the space of mass-points provides answers to the four questions listed in Section 14.1:

1. *What is the geometric meaning of cP when $c \neq 0, 1$?*

   To multiply a mass-point by a scalar $c$, we multiply the mass by $c$ and leave the position of the point unchanged.

2. *What is the geometric interpretation of all the homogeneous 4-tuples $(wx, wy, wz, w)$ that represent the same point $P = (x, y, z)$?*

   The fourth component $w$ represents the mass of the mass-point located at the position $(x, y, z)$. Different values of $w$ represent different masses for a point at a fixed location.

3. *Is there a single consistent model of space that contains both points and vectors?*

   Yes. The space of mass-points includes the affine points as points with unit mass and incorporates the vectors as objects with zero mass.

4. *How should we interpret the result of perspective projection applied to a vector?*

   The space of mass-points is a four-dimensional vector space. Any $4 \times 4$ matrix $M$ represents a linear transformation on this four-dimensional vector space. However, a vector $v$—that is, an object with zero mass—is not necessarily mapped to a vector by matrix multiplication, since the last component of $(v, 0) * M$ need no longer be zero when the last column of $M$ is not $(0, 0, 0, 1)^T$. This result accounts for our observation in Chapter 12, Section 12.5.2, that vectors are not mapped to vectors by perspective projection. In general, under perspective projection, a vector $v$ is mapped to a mass-point. By Equations 14.2 and 14.3, this mass-point is located at the intersection of the line determined by the eye point $E$ and the vector $v$ with the projection plane $S$, and by Equation 14.1, the mass is $-v \cdot n$, where $n$ is the unit normal to the plane $S$. (see Figure 14.4). Thus, a vector is mapped to a vector only if $v \cdot n = 0$—that is, only if $v$ is perpendicular to $n$.

The main advantage of the space of mass-points is that *the space of mass-points completes the algebra of affine space by introducing mass for the points*. This completion leads to a neater algebra

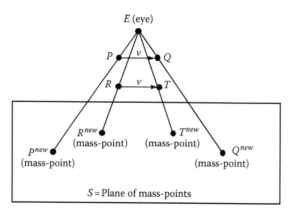

**FIGURE 14.4:** Perspective projection from an eye point $E$ to the plane of mass-points $S$. Perspective projection maps vectors to mass-points, not vectors. This result explains why perspective projection works in the space of mass-points, but not in affine space: if $T - R = Q - P$ as vectors, then $T^{new} - R^{new} = Q^{new} - P^{new}$ as mass-points, not as vectors (see Exercise 14.5). In fact, the plane $S$ is not really a plane at all: a line lies over each affine point in $S$; each point on this line represents a different mass at the same affine point in the plane. Thus, $S$ extends into three dimensions, but the third dimension is mass-like not spatial, which is why we do not see this dimension in this figure.

by permitting arbitrary linear combinations instead of just affine combinations for points. The space of mass-points also has the following advantages over projective space.

1. *Natural geometry.* The geometry of the space of mass-points coincides closely with the geometry of the visual world. There are no strange visual artifacts in this space due to points at infinity.

2. *Points and vectors.* The space of mass-points incorporates both points and vectors. Thus, we can specify orientations and directions as well as positions.

3. *Complete algebra.* In the space of mass-points, we are able to take arbitrary linear combinations of both points and vectors without any annoying restrictions on the coefficients. Moreover, vector products—dot product, cross product, and determinant—still apply to vectors because the vectors are incorporated as a subspace of the space of mass-points.

4. *Projective transformations.* Projective transformations such as perspective projection are linear transformations on the four-dimensional vector space of mass-points. Hence, projective transformations are well defined for all points and all vectors, though in general, mass-points may be mapped to vectors and vectors may be mapped to mass-points under arbitrary projective transformations.

The one disadvantage of the space of mass-points is that this space is four-dimensional, whereas the visual world we are trying to model is three-dimensional. This fourth dimension, however, is mass-like, not spatial, so it is easy to keep track of this extra dimension. One extra dimension turns out to be a small price to pay for a coherent algebra and a geometry that can represent points and vectors as well as affine and projective transformations. In the next section, we shall show how to take advantage of mass-points to derive yet another important transformation in three-dimensional Computer Graphics.

## 14.4  Perspective and Pseudoperspective

We are now going to revisit perspective projection from the point of view of mass-points and *Archimedes' Law of the Lever*. Projections destroy depth, but information on depth is necessary in order to determine which surfaces are hidden behind other surfaces. We shall therefore extend perspective to a new map called *pseudoperspective* that incorporates relative depth into the calculation of perspective projection.

### 14.4.1  Perspective and the Law of the Lever

We can get some additional insight into the formula for perspective projection from *Archimedes' Law of the Lever*. Consider an eye point $E$ and a projection plane $S$ determined by a point $Q$ and a unit normal vector $n$ (see Figure 14.5). Perspective projection of the point $P$ from the eye point $E$ onto the plane $S$ lies at the intersection point $\Delta$ of the line $EP$ with the plane $S$. To find the intersection of the line $EP$ with the plane $S$, we are now going to introduce masses $m_E$ and $m_P$ at the points $E$ and $P$ so that this intersection point $\Delta$ lies at their center of mass.

By the law of the lever, the intersection point $\Delta$ lies at the center of mass of the mass-points $(m_E E, m_E)$ and $(m_P P, m_P)$ if and only if

$$m_P d_P = m_E d_E, \tag{14.10}$$

where $d_P = dist(P, \Delta)$ and $d_E = dist(E, \Delta)$. Let $a_P = dist(P, S)$ and $a_E = dist(E, S)$. Then,

$$\frac{a_P}{d_P} = \sin\theta = \frac{a_E}{d_E},$$

so

$$\frac{m_P a_P}{m_P d_P} = \frac{m_E a_E}{m_E d_E}.$$

Therefore, the denominators are equal if and only if the numerators are equal—that is,

$$m_P d_P = m_E d_E \Leftrightarrow m_P a_P = m_E a_E. \tag{14.11}$$

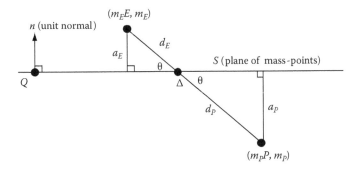

**FIGURE 14.5:**  A schematic view of the perspective projection of the point $P$ from the eye point $E$ into the plane $S$ determined by the point $Q$ and the unit normal $n$.

To find the intersection $\Delta$ of the plane $S$ with the line $EP$, we shall choose the masses $m_E$ and $m_P$ so that $m_P a_P = m_E a_E$. In particular, let

$$m_P = a_E = (E - Q) \cdot n,$$
$$m_E = a_P = (Q - P) \cdot n.$$

With this choice of the masses,

$$m_P a_P = a_E a_P = m_E a_E.$$

Therefore, by Equation 14.11,

$$m_p d_p = m_E d_E,$$

so by the law of the lever, the intersection point $\Delta$ of the line $EP$ and the plane $S$ lies at the center of mass of the mass-points $(m_E E, m_E)$ and $(m_P P, m_P)$. Hence,

$$\Delta = \frac{m_P P + m_E E}{m_P + m_E} = \frac{((E - Q) \cdot n)P + ((Q - P) \cdot n)E}{(E - P) \cdot n},$$

which is the formula for perspective projection that we derived in Chapter 12, Section 12.5.2. Notice the connection here between distance and mass.

### 14.4.2   Pseudoperspective and Pseudodepth

Perspective projection destroys depth. After applying perspective, it is no longer possible to determine which points are hidden from the eye by other points closer to the eye, since any two points along the same line of sight are mapped to the same point in the projection plane. Pseudoperspective restores relative depth to perspective projection, allowing us to determine which points are obscured from the eye by other points closer to the eye.

A scene is typically contained in a viewing frustum. The viewing frustum is the truncated cone defined by the eye and six planes: four planes—left and right, top and bottom—through the eye define the peripheral vision of the viewer; near and far clipping planes determine the nearest and furthest objects in the scene visible to the eye (see Figure 14.6). Objects outside the viewing frustum must be clipped from the model before rendering the scene.

Pseudoperspective has three goals:

1. To assist in clipping algorithms by mapping the viewing frustum to a rectangular box (see Figure 14.6) because it is much easier to determine if an object lies inside a box than to determine if an object lies inside a frustum.

**FIGURE 14.6:**   A schematic view of pseudoperspective, mapping the viewing frustum to a rectangular box.

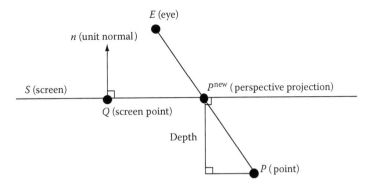

**FIGURE 14.7:** A schematic view of perspective projection, mapping a point onto the viewing plane.

2. To simplify projections by replacing perspective projection by orthogonal projection.

3. To facilitate hidden surface algorithms by preserving relative depth.

To define pseudoperspective, we shall add the depth vector back into our formula for perspective projection (see Figure 14.7). Thus, by definition,

$$Pseudoperspective = Perspective\ projection + Depth\ vector. \tag{14.12}$$

Since the depth vector is orthogonal to the plane of projection, it is easy to verify that this formula satisfies all three of our goals: mapping the viewing frustum to a rectangular box, replacing perspective projection by orthogonal projection, and preserving relative depth for hidden surface algorithms. Unfortunately, there is a problem with how to perform the addition.

Recall that the formula for perspective projection from the eye point $E$ to the plane $S$ determined by a point $Q$ and a unit normal $n$ is given by

$$P^{persp} = \frac{((E - Q) \cdot n)P + ((Q - P) \cdot n)E}{(E - P) \cdot n}.$$

Moreover, since $n$ is a unit vector, the distance from the point $P$ to the plane $S$ is

$$dist(P, S) = (Q - P) \cdot n.$$

Thus, the depth vector is simply

$$depth(P, S) = ((Q - P) \cdot n)n.$$

But if we add the depth vector to our formula for perspective projection, we find that

$$
\begin{aligned}
Perspective + Depth\ vector &= \frac{((E - Q) \cdot n)P + ((Q - P) \cdot n)E}{(E - P) \cdot n} + ((Q - P) \cdot n)n \\
&= \frac{((E - Q) \cdot n)P + ((Q - P) \cdot n)E + ((Q - P) \cdot n)\,((E - P) \cdot n)n}{(E - P) \cdot n}.
\end{aligned}
$$

The problem with this formula is that the numerator on the right-hand side is not linear in the coordinates of the point $P$; in fact, the term $((Q - P) \cdot n)((E - P) \cdot n)$ is quadratic in the coordinates of $P$. For example, if the eye point $E = (0, 0, 1)$ and $S$ is the $xy$-plane, then we can choose $Q = (0, 0, 0)$ and $n = (0, 0, 1)$. Thus, for any point $P = (x, y, z)$,

$$((Q - P) \cdot n)((E - P) \cdot n) = -z(1 - z),$$

which is quadratic in $z$. We would like to represent pseudoperspective just like any other affine or projective transformation by a $4 \times 4$ matrix, but we cannot represent a quadratic expression by matrix multiplication. We need an alternative approach.

The trick is that the addition in Equation 14.12 is not the addition of points and vectors in affine space, but rather the addition of mass-points and vectors in the space of mass-points. Recall that in the space of mass-points, denominators represent mass and vectors have zero mass, so

$$P^{persp} = (((E - Q) \cdot n)P + ((Q - P) \cdot n)E, (E - P) \cdot n),$$

and

$$depth(P, S) = (((Q - P) \cdot n)n, 0).$$

Thus,

$$P^{pseudo} = P^{persp} + depth(P, S)$$
$$= (((E - Q) \cdot n)P + ((Q - P) \cdot n)E, (E - P) \cdot n) + (((Q - P) \cdot n)n, 0).$$

Now adding in the space of mass-points yields the point

$$P^{pseudo} = \frac{((E - Q) \cdot n)P + ((Q - P) \cdot n)E}{(E - P) \cdot n} + \frac{(Q - P) \cdot n}{(E - P) \cdot n}n. \tag{14.13}$$

Define

$$pseudodepth(P, E, S) = \frac{(Q - P) \cdot n}{(E - P) \cdot n}. \tag{14.14}$$

By definition, *pseudodepth* is the signed distance from the point $P$ to the plane $S$ divided by the signed distance from the point $P$ to the plane through the eye point $E$ parallel to the plane $S$. Moreover, by Equation 14.13,

$$P^{pseudo} = P^{persp} + Pseudodepth(P, E, S)n. \tag{14.15}$$

The expression for pseudoperspective in Equation 14.13 is linear in both the numerator and the denominator, so we can represent pseudoperspective by a $4 \times 4$ matrix, just like any other affine or projective transformation. In fact, by Equation 14.15 to find the matrix representing pseudoperspective, we need only combine the matrices representing perspective projection and the term containing the pseudodepth vector. We know from Chapter 13, Section 13.6.2 that the matrix representing perspective is

$$Persp(E, Q, n) = \begin{pmatrix} ((E - Q) \cdot n)I - n^T * E & -n^T \\ (Q \cdot n)E & E \cdot n \end{pmatrix}.$$

Moreover, it is easy to verify that the matrix representing the pseudodepth vector is

$$Pseudodepth(E, Q, n) = \begin{pmatrix} -n^T * n & -n^T \\ (Q \cdot n)n & E \cdot n \end{pmatrix},$$

since

$$(P, 1) * \begin{pmatrix} -n^T * n & -n^T \\ (Q \cdot n)n & E \cdot n \end{pmatrix} = (((Q - P) \cdot n)n, (E - P) \cdot n).$$

Therefore, since the fourth columns of $Persp(E, Q, n)$ and $Pseudodepth(E, Q, n)$ representing denominators are identical, the matrix representing pseudoperspective is

$$Pseudo(E, Q, n) = \begin{pmatrix} ((E - Q) \cdot n)I - n^T * (E + n) & -n^T \\ (Q \cdot n)(E + n) & E \cdot n \end{pmatrix}.$$

Pseudodepth is not the same as depth. Nevertheless, pseudodepth preserves relative depth—that is,

$$Depth(P_2) \geq Depth(P_1) \Rightarrow Pseudodepth(P_2) \geq Pseudodepth(P_1). \tag{14.16}$$

To verify Equation 14.16, suppose that

$$Depth(P_2) \geq Depth(P_1).$$

Then

$$(Q - P_2) \cdot n \geq (Q - P_1) \cdot n,$$

so

$$(P_1 - P_2) \cdot n \geq 0.$$

Since adding the same positive constant to the numerator and the denominator increases the value of a fraction,

$$\frac{(Q - P_2) \cdot n}{(E - P_2) \cdot n} = \frac{(Q - P_1) \cdot n + (P_1 - P_2) \cdot n}{(E - P_1) \cdot n + (P_1 - P_2) \cdot n} \geq \frac{(Q - P_1) \cdot n}{(E - P_1) \cdot n};$$

hence

$$Pseudodepth(P_2) \geq Pseudodepth(P_1).$$

Thus, it is enough to know the pseudodepth in order to determine which of two points lying on the same line of sight is closer to the eye.

## 14.5 Summary

There are two potential models for the geometry of three-dimensional Computer Graphics: projective space and the space of mass-points. Projective space is based on insights from classical art; the space of mass-points is rooted in insights from classical mechanics.

Projective space consists of two kinds of points: affine points and points at infinity. Both types of points are represented by homogeneous coordinates: if $w \neq 0$, then $(wx, wy, wz, w)$ represents the affine point with rectangular coordinates $(x, y, z)$; if $c \neq 0$, then $(cx, cy, cz, 0)$ represents the point at infinity in the direction $(x, y, z)$. By introducing points at infinity, *projective space completes the geometry of affine space*, simplifying the statements and proofs of many classical theorems in geometry. But *points at infinity are not vectors*, so while the geometry of projective space is appealing, the algebra associated with projective space is severely limited. *Projective space is not a vector space.* There are no affine or linear combinations for points in projective space. Dot products, cross products, and determinants have no meaning in projective space. Thus, projective space is not a viable algebraic model for representing calculations on a computer.

The alternative to projective space is the space of mass-points. In the space of mass-points, points have mass as well as location. Mass-points are represented by pairs $(mP, m)$, where $m \neq 0$ is the mass and $P = mP/m$ is the location of the mass-point. Affine points $P$ are simply mass-points $(P, 1)$ with unit mass. Vectors $v$ are incorporated into this space as objects $(v, 0)$ with zero mass. Mass-points are added by *Archimedes' Law of the Lever*: the sum of two mass-points is the mass-point whose mass is the sum of the two masses, located at their center of mass. Vectors are added using the usual triangle rule. By introducing mass, *the space of mass-points completes the algebra of affine space*: there are no annoying restrictions on addition, subtraction, or scalar multiplication for mass-points. Thus, the space of mass-points is a particularly effective algebraic model for representing the calculations of Computer Graphics. The space of mass-points is a four-dimensional vector space: three of the dimensions are spatial; the fourth dimension is due to the mass.

We applied the algebra of mass-points to derive an explicit expression for pseudoperspective. Perspective projection destroys depth. Pseudoperspective restores relative depth to perspective projection, allowing us to determine which points are obscured from the eye by other points closer to the eye. By definition,

$$Pseudoperspective = Perspective\ projection + Depth\ vector,$$

but in order for pseudoperspective to be a projective transformation, the addition must occur in the space of mass-points, not in affine space. The formula for pseudoperspective from the eye point $E$ to the plane $S$ determined by a point $Q$ and a unit normal $n$ is given by

$$P^{pseudo} = \frac{((E-Q) \cdot n)P + ((Q-P) \cdot n)E}{(E-P) \cdot n} + \frac{(Q-P) \cdot n}{(E-P) \cdot n}n,$$

and the term

$$Pseudodepth(P, E, S) = \frac{(Q-P) \cdot n}{(E-P) \cdot n},$$

represents the relative depth of the point $P$. The $4 \times 4$ matrix representing pseudoperspective is

$$Pseudo(E, Q, n) = \begin{pmatrix} ((E-Q) \cdot n)I - n^{\mathrm{T}} * (E+n) & -n^{\mathrm{T}} \\ (Q \cdot n)(E+n) & E \cdot n \end{pmatrix}.$$

## Exercises

14.1. Show that the points of projective space correspond to equivalence classes of the objects in the space of mass-points.

14.2. Consider perspective projection from an eye point $E$ to a projection plane $S$.

    a. What is the image of the eye point $E$ in projective space?

    b. What is the image of the eye point $E$ in the space of mass-points?

14.3. Prove that in projective space, points $R$ on a line through the eye point $E$ parallel to the projection plane $S$ are mapped by perspective to points at infinity in the direction $R - E$.

14.4. Prove that in the space of mass-points:

    a. Scalar multiplication distributes through addition.

    b. Matrix multiplication commutes with scalar multiplication.

14.5. Prove that under perspective projection if $T - R = Q - P$ as vectors, then $T^{new} - R^{new} = Q^{new} - P^{new}$ as mass-points.

14.6. Calculate the entries of the matrix for pseudoperspective when the eye is located along the $z$-axis at $E = (0, 0, 1)$ and the plane of projection is the $xy$-plane.

14.7. Define the map $\pi$:*space of mass-points* $\rightarrow$ *projective space* by setting

$$\pi(mP, m) = [mP, m],$$

$$\pi(v, 0) = [v, 0].$$

Let $T$ be a projective transformation on projective space. Show that there exists a unique linear transformation $T^*$ on the space of mass-points such that the following diagram commutes:

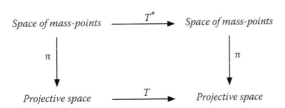

14.8. Define the map $i$:*affine space* $\rightarrow$ *space of mass-points* by setting

$i(P) = (P, 1),$
$i(v) = (v, 0).$

Let $A$ be a affine transformation on affine space. Show that there exists a unique linear transformation $A*$ on the space of mass-points such that the following diagram commutes:

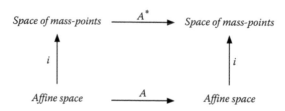

14.9.  Show that for any eye point $E$, any plane $S$, and any points $P$, $Q$:

a. $depth\left(\dfrac{P+Q}{2}, S\right) = \dfrac{depth(P, S) + depth(Q, S)}{2}$.

b. $pseudodepth\left(\dfrac{P+Q}{2}, S, E\right) = pseudodepth(P, S, E) + pseudodepth(Q, S, E),$

where the addition on the right-hand side is the addition in the space of mass-points.

14.10. In baseball, batting averages are computed by the formula:

$$\text{Batting average} = \frac{\text{Total hits}}{\text{Total at bats}}.$$

a. What are the rules for adding a player's batting averages for two different months?

b. Do batting averages resemble points in projective space or mass-points?

14.11. A fraction $a/b$ is the ratio of two integers $a$, $b$. A rational number $\{a/b\}$ is an equivalence class of fractions, where $b \neq 0$ and $a/b \sim c/d$ if and only if $ad = bc$. We can add fractions by the rule

$$\frac{a}{b} + \frac{c}{d} = \frac{a+c}{b+d},$$

and we can multiply a fraction $a/b$ by an integer $n$ by the rule

$$n\frac{a}{b} = \frac{na}{nb}.$$

Similarly, we can add rational numbers by the rule

$$\left\{\frac{a}{b}\right\} + \left\{\frac{c}{d}\right\} = \left\{\frac{ad+bc}{bd}\right\},$$

and we can multiply a rational number $\{a/b\}$ by an integer $n$ by the rule

$$n\left\{\frac{a}{b}\right\} = \left\{\frac{na}{b}\right\}.$$

a. Show that $a/b \sim c/d$ is an equivalence relation.

b. Show that addition and scalar multiplication are well defined for rational numbers.

c. Show that addition for rational numbers is associative and commutative and that multiplication by integers distributes through addition.

d. Show that addition for fractions is associative and commutative and that multiplication by integers distributes through addition.

e. What is the identity for addition:

   i. For rational numbers

   ii. For fractions

f. Which are more like projective space: fractions or rational numbers?

   i. What are the similarities? What are the differences?

g. Which are more like the space of mass-points: fractions or rational numbers?

   i. What are the similarities? What are the differences?

# Chapter 15

## Quaternions: Multiplication in the Space of Mass-Points

*Be fruitful and multiply*

– Genesis 1:28

### 15.1 Vector Spaces and Division Algebras

The space of mass-points is a four-dimensional vector space: addition, subtraction, and scalar multiplication are all well-defined operations on mass-points. But what about a more universal form of multiplication? Can we multiply together two points or more generally two mass-points?

In lower-dimensional vector spaces, we can indeed multiply any two elements in the space. Every one-dimensional vector space is isomorphic (equivalent) to the real number line. We can multiply any two points along the number line by the standard rules for multiplication of real numbers. Every two-dimensional vector space is isomorphic to the plane. If we identify each pair of rectangular coordinates $(a, b)$ in the plane with the complex number $a + bi$, then we can multiply two vectors in the plane using the standard rules for multiplication of complex numbers.

Vectors in three dimensions are endowed with two distinct products: dot product and cross product. The dot product of two vectors is a scalar; the cross product of two vectors is a vector. When we add or subtract two vectors, we get another vector. When we say that we want to multiply two vectors, we typically mean that we also want the result to be another vector. The dot product of two vectors is not a vector, so dot product is not the kind of multiplication that we seek. The cross product of two vectors is a vector; nevertheless cross product also has some undesirable properties. Multiplication is typically associative, commutative, and distributes through addition. Moreover, in order to solve simple equations, we would like to have an identity for multiplication as well as multiplicative inverses (division). Although cross product distributes through addition, the cross product is neither associative nor commutative. Moreover, there is no multiplicative identity for cross product—that is, there is no vector $u$ such that for all vectors $v$, we have $u \times v = v \times u = v$— since by definition $u \times v \perp v$. Hence, there can be no multiplicative inverses for cross product. In fact, in three dimensions, there is no entirely satisfactory notion of vector multiplication.

A *division algebra* (an algebra where we can perform division) is a set $S$ where we can add, subtract, multiply, and divide any two elements in $S$. (One exception: we cannot divide by zero, the identity for addition.) Addition and multiplication must satisfy the usual rules: addition is associative and commutative; multiplication is associative and distributes through addition. Notice that we shall not insist that multiplication is commutative, since the multiplication that we have in mind for mass-points is not commutative. There is an identity 0 for addition and an identity 1 for multiplication so that for any element $s$ in $S$:

$$s + 0 = 0 + s = s,$$
$$s \cdot 1 = 1 \cdot s = s.$$

Finally, for each element $s$ in $S$, there is an inverse for addition $(-s)$ and an inverse for multiplication $(s^{-1})$ so that:

$$s + (-s) = (-s) + s = 0,$$
$$s \cdot s^{-1} = s^{-1} \cdot s = 1 \quad (s \neq 0).$$

Thus, in a division algebra, we can subtract and divide, as well as add and multiply.

There are only three finite dimensional vector spaces over the real numbers that are also division algebras—that is, there are only three dimensions in which we can define addition, subtraction, multiplication, and division—dimensions 1, 2, and 4. These vector spaces correspond to the line (real numbers), the plane (complex numbers), and the space of mass-points (quaternions). (There is also a nonassociative multiplication in dimension 8, the Cayley numbers.) We are really unbelievably lucky that Computer Graphics is grounded in the four-dimensional vector space of mass-points.

In this chapter, we shall define quaternion multiplication in the space of mass-points and investigate applications of quaternion multiplication to Computer Graphics. We will begin, however, with a simpler and better known example: complex multiplication for vectors in the plane.

---

## 15.2　Complex Numbers

Addition, subtraction, and scalar multiplication are coordinate-free operations on vectors in the plane. Rectangular coordinates may be introduced in practice to perform these computations, but in principle there is no need for preferred directions—for coordinate axes—to define addition, subtraction, or scalar multiplication for vectors in the plane.

Complex multiplication is different. For complex multiplication, we must choose a preferred direction: the direction of the identity vector for multiplication. We denote this identity vector by 1 and we associate this vector with the $x$-axis. The unit vector perpendicular to the identity vector we denote by $i$ and we associate this vector with the $y$-axis. Relative to this coordinate system for every vector $v$ in the plane there are constants $a, b$ such that

$$v = a1 + bi.$$

Typically, we drop the symbol 1, and we write simply

$$v = a + bi. \tag{15.1}$$

In the complex number system, $i$ denotes $\sqrt{-1}$, so

$$i^2 = -1. \tag{15.2}$$

Since 1 is the identity for multiplication, we have the rules

$$1 \cdot 1 = 1, \quad i \cdot 1 = 1 \cdot i = i, \quad i \cdot i = -1. \tag{15.3}$$

Therefore, if we want complex multiplication to distribute through addition, then we must define

$$(a + bi)(c + di) = a(c + di) + bi(c + di) = (ac - bd) + (ad + bc)i. \tag{15.4}$$

Using Equation 15.4, it is easy to verify that complex multiplication is associative, commutative, and distributes through addition (see Exercise 15.1).

Every nonzero complex number has a multiplicative inverse. Let $v = a + bi$. The *complex conjugate* of $v$, denoted by $v^*$, is defined by

$$v^* = a - bi.$$

By Equation 15.4,

$$v \cdot v^* = a^2 + b^2 = |v|^2.$$

Therefore,

$$v^{-1} = \frac{v^*}{|v|^2}.$$

Equation 15.4 encapsulates the algebra of complex multiplication. But what is the geometry underlying complex multiplication?

Consider first multiplication with $i$. If $v = a + bi$, then,

$$i \cdot v = i(a + bi) = -b + ai.$$

Thus, if $v = (a, b)$, then $i \cdot v = (-b, a)$; hence,

$$i \cdot v = v_\perp.$$

Therefore, multiplication by $i$ is equivalent to rotation by $90°$. Also,

$$(-1) \cdot v = -(a + bi) = -a - bi = -v,$$

so multiplication by $-1$ is equivalent to rotation by $180°$. There seems then to be a link between rotation and complex multiplication. Let us explore this connection further.

Recall from Chapter 4, Section 4.2.2 that the matrix which rotates vectors in the plane by the angle $\phi$ is given by

$$Rot(\phi) = \begin{pmatrix} \cos(\phi) & \sin(\phi) \\ -\sin(\phi) & \cos(\phi) \end{pmatrix}.$$

Let $z = (x, y) = x + iy$ be an arbitrary vector in the plane. Then after rotation by the angle $\phi$

$$z_{new} = z * Rot(\phi) = (x \ y) * \begin{pmatrix} \cos(\phi) & \sin(\phi) \\ -\sin(\phi) & \cos(\phi) \end{pmatrix},$$

or equivalently

$$x_{new} = x\cos(\phi) - y\sin(\phi),$$

$$y_{new} = x\sin(\phi) + y\cos(\phi).$$

Now let $w(\phi)$ be the unit vector that makes an angle $\phi$ with the $x$-axis (see Figure 15.1). Then

$$w(\phi) = \cos(\phi) + i\sin(\phi),$$

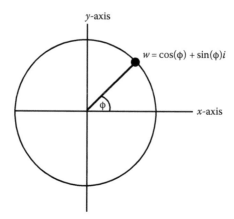

**FIGURE 15.1:**   Multiplying a complex number $z$ by a unit vector $w = \cos(\phi) + \sin(\phi)i$ is equivalent to rotating the vector $z$ by the angle $\phi$.

and

$$w(\phi)z = (\cos{(\phi)} + i\sin{(\phi)})(x + iy) = \underbrace{\left( x\cos{(\phi)} - y\sin{(\phi)} \right)}_{x_{new}} + \underbrace{\left( x\sin{(\phi)} + y\cos{(\phi)} \right)}_{y_{new}} i.$$

Thus, rather remarkably, multiplying the complex number $z$ by $w(\phi)$ is equivalent to rotating the vector $z$ by the angle $\phi$. Moreover, since complex multiplication is associative, we can compose two rotations simply by multiplying the associated complex numbers—that is, for all $z$,

$$w(\phi + \theta)z = w(\phi)w(\theta)z,$$

so

$$w(\phi + \theta) = w(\phi)w(\theta).$$

Representing rotations in the plane by complex numbers is more efficient than matrix representations for rotations in the plane. To store one complex number, we need to store only two real numbers, whereas to store one $2 \times 2$ matrix, we need to store four real numbers. To compose two rotations by complex multiplication requires only four real multiplications, whereas to compose two rotations by matrix multiplication requires eight real multiplications.

Why is rotation related so closely to complex multiplication? One explanation is that rotation in the plane closely resembles exponentiation because composing two rotations is equivalent to adding the corresponding angles. But complex multiplication is also closely related to exponentiation. Let $z$ be an arbitrary complex number. Define

$$e^z = 1 + z + \frac{z^2}{2!} + \frac{z^3}{3!} + \cdots.$$

Then

$$e^{i\phi} = 1 + i\phi + \frac{(i\phi)^2}{2!} + \frac{(i\phi)^3}{3!} + \frac{(i\phi)^4}{4!} + \frac{(i\phi)^5}{5!} + \cdots.$$

But $i^2 = -1$, so substituting and collecting terms, we arrive at the following fundamental identity:

*Euler's Formula*

$$e^{i\phi} = \left(1 - \frac{\phi^2}{2!} + \frac{\phi^4}{4!} - \cdots\right) + \left(\phi - \frac{\phi^3}{3!} + \frac{\phi^5}{5!} - \cdots\right)i = \cos(\phi) + i\sin(\phi). \tag{15.5}$$

Since

$$w(\phi) = \cos(\phi) + i\sin(\phi) = e^{i\phi},$$

we conclude that

$$w(\phi)w(\theta) = e^{i\phi}e^{i\theta} = e^{i(\phi+\theta)} = w(\phi + \theta),$$

so multiplying two complex numbers of unit length is equivalent to adding the angles that these vectors form with the $x$-axis.

Multiplying a complex number $z = x + iy$ by a complex number of unit length $w = w_1 + iw_2$ is equivalent to rotating the vector $z$ by the angle

$$\phi = \arctan(w_2/w_1),$$

since if $w$ lies along the unit circle there is an angle $\phi$ for which $w = \cos(\phi) + i\sin(\phi)$ (see Figure 15.1). But what happens if we multiply $z$ by an arbitrary complex number $w$? If $w = s$, where $s$ is a real number, then $wz = sz$, so the effect is to scale the vector $z$ by the constant $s$. Thus both rotation and scaling can be represented by complex multiplication. Finally, if $w = w_1 + iw_2$ is an arbitrary complex number, then we can write $w$ in polar form as

$$w = re^{i\phi} = rw(\phi),$$

where $r = |w|$ and $\phi = \arctan(w_2/w_1)$. Hence

$$wz = rw(\phi)z,$$

so the effect of multiplying $z$ by $w$ is to rotate $z$ by the angle $\phi$ and to scale the result by the constant $r$. Thus any conformal transformation of vectors in the plane can be represented by multiplication with a single complex number.

---

## 15.3 Quaternions

*Quaternions* are just another name for the entities in the four-dimensional space of mass-points, just as *complex numbers* are just another name for the entities in the two-dimensional space of vectors in the plane. In fact, quaternions are an extension of complex numbers to four dimensions, so we can multiply two quaternions in a manner similar to the way that we can multiply two complex numbers.

### 15.3.1 Quaternion Multiplication

To multiply two complex numbers, we need to choose a preferred direction: the direction of the identity vector for complex multiplication. Similarly, to multiply two quaternions, we must choose a preferred mass-point in the space of mass-points: the identity for quaternion multiplication. We shall denote this special point by $O$, and we will think of this point as the origin, not in the four-dimensional vector space of mass-points, but rather in the three-dimensional affine space of affine points. Thus, $O$ is a special mass-point with mass $m=1$. The choice of the point $O$ is arbitrary, just as the choice of the origin is arbitrary, but to perform quaternion multiplication, we must fix $O$ once and for all. The three-dimensional vectors lie in the quaternion subspace orthogonal to $O$. Thus, if we were to introduce rectangular coordinates into the four-dimensional space of mass-points, then we would have $O = (0, 0, 0, 1)$, and the three-dimensional vectors $v$ would be written as $v = (v_1, v_2, v_3, 0)$ or equivalently $v = v_1 i + v_2 j + v_3 k$, where $i, j, k$ are the unit vectors along the $x, y, z$ coordinate axes.

Every mass-point can be written as a scalar mass $m$ times the affine point $O$ plus a vector $v$. Suppose that $(mP, m)$ is a mass-point with mass $m$. Then,

$$(mP, m) - mO = (mP, m) - (mOrigin, m) = (m(P - Origin), 0).$$

Setting $v = (m(P - Origin), 0)$, we have

$$(mP, m) = mO + v. \tag{15.6}$$

Interpreted in terms of coordinates, Equation 15.6 means that every quaternion $q$ can be written as $q = (q_1, q_2, q_3, q_4)$ or equivalently

$$q = q_4 O + q_1 i + q_2 j + q_3 k. \tag{15.7}$$

In many texts, the letter $O$ is dropped, and quaternions are written as

$$q = q_4 + q_1 i + q_2 j + q_3 k. \tag{15.8}$$

In this notation, a quaternion $q$ seems to be a weird kind of hermaphrodite, a scalar $q_4$ added to a vector $q_1 i + q_2 j + q_3 k$. In fact, there are no mysterious hermaphrodites; the quaternion $q$ in Equation 15.8 is simply a mass-point with mass $q_4$.

Quaternion multiplication is defined by setting

$$(aO + u)(bO + v) = (ab - u \cdot v)O + (bu + av + u \times v), \tag{15.9}$$

Notice how the vector products—dot product, cross product, and scalar product—all appear in the formula for quaternion multiplication. Except for the choice of the special point $O$, quaternion multiplication is coordinate-free, since dot product, cross product, and scalar product are all coordinate-free. Using Equation 15.9, it is straightforward, though somewhat tedious, to check that quaternion multiplication is associative and distributes through addition (see Exercise 15.8). Notice, however, that quaternion multiplication is not commutative, since cross product is anti-commutative.

The point $O$ is the identity for quaternion multiplication because by Equation 15.9,

$$O(bO + v) = bO + v.$$

In particular,

$$O^2 = O.$$

Similarly, it is easy to verify from Equation 15.9 that the basis vectors $i, j, k$ multiply according the following rules:

$$i^2 = j^2 = k^2 = -O \quad \text{and} \quad ij = k = -ji, \quad jk = i = -kj, \quad ki = j = -ik. \tag{15.10}$$

More generally, if $u$, $v$ are three-dimensional vectors, then by Equation 15.9

$$uv = -(u \cdot v)O + u \times v.$$

Thus, the product of two vectors $u$, $v$ in three dimensions is a four-dimensional mass-point with mass $m = -u \cdot v$. In particular,

$$u^2 = -(u \cdot u)O = -|u|^2 O,$$

so

$$u^{-1} = \frac{-u}{|u|^2}.$$

Every nonzero quaternion has a multiplicative inverse. Let $q = m_q O + v_q$. The *conjugate* of $q$, denoted by $q^*$, is defined by

$$q^* = m_q O - v_q.$$

By Equation 15.9,

$$qq^* = \left( m_q^2 + v_q \cdot v_q \right) O.$$

But recall that in quaternion space $v_q \perp O$, so by the Pythagorean theorem

$$m_q^2 + v_q \cdot v_q = |q|^2.$$

Hence,

$$qq^* = |q|^2 O. \tag{15.11}$$

Therefore,

$$q^{-1} = \frac{q^*}{|q|^2}.$$

In terms of coordinates, if $q = (q_1, q_2, q_3, q_4)$, then

$$q = q_4 O + q_1 i + q_2 j + q_3 k,$$
$$q^* = q_4 O - q_1 i - q_2 j - q_3 k,$$
$$|q|^2 = q_1^2 + q_2^2 + q_3^2 + q_4^2.$$

Later on we shall also need the following result.

**Lemma 15.1:** $(pq)^* = q^*p^*$.

**Proof:** Let $p = m_p O + v_p$ and $q = m_q O + v_q$. Then

$$pq = (m_p O + v_p)(m_q O + v_q) = (m_p m_q - v_p \cdot v_q)O + (m_q v_p + m_p v_q + v_p \times v_q),$$
$$q^*p^* = (m_q O - v_q)(m_p O - v_p) = (m_q m_p - v_q \cdot v_p)O + (-m_q v_p - m_p v_q + v_q \times v_p).$$

Comparing the right-hand sides of these two equations and recalling that the cross product is anticommutative, we conclude that

$$(pq)^* = q^*p^*.$$

With this brief introduction to quaternion algebra, we are now ready to turn our attention to the geometric applications of quaternion multiplication.

### 15.3.2   Quaternion Representations for Conformal Transformations

Quaternion multiplication can be used to model rotations of vectors in three dimensions much like complex multiplication can be used to model rotations of vectors in the plane. In fact, quaternion multiplication can be used to model any conformal transformation—rotation, mirror image, and uniform scaling—of vectors in three dimensions.

The key to modeling conformal transformations with quaternions is *sandwiching*. Let $q$ be an arbitrary quaternion, and let $v$ be an arbitrary vector. Define

$$S_q(v) = qvq^*.  \tag{15.12}$$

Equation 15.12 is called *sandwiching* because the vector $v$ is sandwiched in between the quaternions $q$ and $q^*$. Notice that

$$(S_p \circ S_q)(v) = S_{pq}(v),  \tag{15.13}$$

because by Lemma 15.1

$$(S_p \circ S_q)(v) = S_p(S_q(v)) = S_p(qvq^*) = pqvq^*p^* = (pq)v(pq)^* = S_{pq}(v).$$

**Proposition 15.1:** *Let $q(N, \phi) = \cos(\phi/2)O + \sin(\phi/2)N$. Then a vector $v$ can be rotated around the axis $N$ through the angle $\phi$ by sandwiching $v$ between $q(N, \phi)$ and $q(N, \phi)^*$.*

**Proof:** By the Formula of Rodrigues (see Chapter 12, Section 12.3.2), if we rotate a vector $v$ around the axis $N$ through the angle $\phi$, then

$$v_{new} = \cos(\phi)v + (1 - \cos(\phi))(v \cdot N)N + \sin(\phi)N \times v.$$

Thus, we need to show that sandwiching $v$ between $q(N, \phi)$ and $q(N, \phi)^*$ yields the same result. To proceed, let $v_{\parallel}$ and $v_{\perp}$ be the components of $v$ parallel and perpendicular to $N$. Then

$$v_{\parallel} \times N = 0 \Rightarrow Nv_{\parallel} = -(v_{\parallel} \cdot N)O = v_{\parallel}N,$$
$$v_{\perp} \cdot N = 0 \Rightarrow Nv_{\perp} = N \times v_{\perp} = -v_{\perp}N.$$

To simplify our notation, let $c = \cos(\phi/2)$, $s = \sin(\phi/2)$, and $q = q(N, \phi) = cO + sN$. Now

$$qvq^* = (cO + sN)(v_\| + v_\perp)(cO - sN)$$
$$= c^2v_\| - s^2Nv_\|N + cs(\underbrace{Nv_\| - v_\|N}_{0}) + cs(\underbrace{Nv_\perp - v_\perp N}_{2Nv_\perp = 2N \times v_\perp}) + c^2v_\perp - s^2Nv_\perp N$$
$$= c^2v_\| - s^2NNv_\| + 2csN \times v_\perp + c^2v_\perp + s^2NNv_\perp.$$

But since $N$ is a unit vector,

$$N^2 = -O.$$

Hence,

$$qvq^* = (c^2 + s^2)v_\| + 2csN \times v_\perp + (c^2 - s^2)v_\perp.$$

By the half angle formulas for sine and cosine (see Exercise 15.3),

$$2cs = \sin(\phi),$$
$$c^2 - s^2 = \cos(\phi).$$

Hence,

$$qvq^* = v_\| + \cos(\phi)v_\perp + \sin(\phi)N \times v_\perp = (1 - \cos(\phi))v_\| + \cos(\phi)(v_\perp + v_\|) + \sin(\phi)(N \times v_\perp).$$

Finally,

$$v = v_\perp + v_\|,$$
$$N \times v = N \times (v_\| + v_\perp) = N \times v_\perp,$$
$$N \cdot v = N \cdot (v_\| + v_\perp) = N \cdot v_\|,$$
$$v_\| = -(NN)v_\| = -N(Nv_\|) = (N \cdot v_\|)N = (N \cdot v)N,$$

so

$$qvq^* = \cos(\phi)v + (1 - \cos(\phi))(N \cdot v)N + \sin(\phi)(N \times v).$$

Notice that the quaternion $q(N, \phi)$ is a unit quaternion because by Equation 15.11,

$$|q(N, \phi)|^2 O = q(N, \phi)q(N, \phi)^* = \left(\cos^2(\phi/2) + \sin^2(\phi/2)\right)O = O.$$

Thus, every rotation is represented by a unit quaternion. Conversely, every unit quaternion represents a rotation because if $q = m_q O + v_q$ is a unit quaternion, then $m_q^2 + |v_q|^2 = 1$. Hence, there is an angle $\theta$ and a unit vector $N$ such that $m_q = \cos(\theta)$ and $v_q = \sin(\theta)N$, so $q = \cos(\theta)O + \sin(\theta)N = q(N, 2\theta)$.

To model mirror image with sandwiching, observe that to find the mirror image of a vector $v$ in the plane perpendicular to the unit normal vector $N$, we can rotate $v$ by $180°$ around $N$ and then negate the result. But

$$q(N, \pi) = \cos(\pi/2)O + \sin(\pi/2)N = N.$$

Thus, to rotate $v$ by $180°$ around $N$, we can sandwich $v$ between $N$ and $N^*$. Since for vectors $N^* = -N$, it follows that the mirror image of $v$ in the plane perpendicular to $N$ is given by

$$-S_N(v) = NvN.$$

To verify this result, recall from Chapter 12, Section 12.3.3 that under mirror image

$$v_{new} = v - 2(v \cdot N)N.$$

**Proposition 15.2:** $-S_N(v) = NvN = v - 2(v \cdot N)N.$

*Thus, $v$ can be mirrored in the plane perpendicular to the unit normal $N$ by sandwiching $v$ with $N$ and negating.*

**Proof:**

$$-S_N(v) = NvN = ((-N \cdot v)O + N \times v)N = -(N \cdot v)N - \underbrace{((N \times v) \cdot N)}_{0} O + (N \times v) \times N$$

$$= -(N \cdot v)N + (N \cdot N)v - (N \cdot v)N$$
$$= v - 2(N \cdot v)N.$$

**Proposition 15.3:** $S_{\sqrt{c}O}(v) = cv.$

*Thus $v$ can be scaled by the factor $c$ by sandwiching $v$ with $\sqrt{c}O$.*

**Proof:** $S_{\sqrt{c}O}(v) = \sqrt{c}Ov\sqrt{c}O = cv.$

**Theorem 15.1:** *Every conformal transformation of vectors in three dimensions can be modeled by sandwiching with a single quaternion.*

**Proof:** This result follows from Propositions 15.1 through 15.3, since by Equation 15.13, composition of sandwiching is equivalent to multiplication of quaternions.

Conformal transformations of vectors in three dimensions can be computed by sandwiching the vectors with the appropriate quaternions. But what about conformal transformations on points in three dimensions? To rotate points about arbitrary lines, or to mirror points in arbitrary planes, or to scale points uniformly about arbitrary points, we cannot simply sandwich the points we wish to transform with some appropriate quaternion (see Exercise 15.21). Recall, however, that rotation, mirror image, and uniform scaling all have a fixed point $Q$ of the transformation. Since $P = Q + (P - Q)$, we can compute conformal transformations on affine points $P$ by using quaternions to transform the vectors $P - Q$ and then adding the resulting vectors to $Q$. In this way, quaternions can be applied to compute conformal transformation on points as well as on vectors.

### 15.3.3  Quaternions versus Matrices

We now have two ways to represent conformal transformations on vectors in three dimensions: $3 \times 3$ matrices and quaternions. Here we shall compare and contrast some of the advantages and disadvantages of quaternion and matrix representations for conformal transformations.

A $3 \times 3$ matrix contains nine scalar entries, whereas a quaternion can be represented by four rectangular coordinates. Thus, if memory is at a premium, quaternions are more than twice as efficient as matrices. On the other hand, by Equation 15.9, to compute the product of two quaternions requires

**TABLE 15.1:** Trade-offs between speed and memory for quaternions and $3 \times 3$ matrices.

|  | Memory | Transformation | Composition |
|---|---|---|---|
| Quaternions | 4 scalars | 28 multiplies | 16 multiplies |
| $3 \times 3$ Matrices | 9 scalars | 9 multiplies | 27 multiplies |

16 real scalar multiplications: 6 for the cross product, 3 for the dot product, 6 for the two scalar products, and 1 for the product of the two scalar masses. Similarly, since a vector has zero mass, the product of a quaternion and a vector requires 12 scalar multiplications, since the product of the masses as well as one of the scalar products is zero. Sandwiching a vector with a quaternion computes a product of a quaternion with a vector followed by the product of a quaternion with a quaternion, so the total cost for one conformal transformation using quaternions is $12 + 16 = 28$ scalar multiplications, whereas the cost for multiplying a vector by a $3 \times 3$ matrix is only 9 scalar multiplications. Thus, matrices are more than three times as fast as quaternions for computing arbitrary conformal transformations. But there is an additional computational consideration. To compose two transformations by quaternion multiplication requires 16 scalar multiplications, whereas to compose two transformations by matrix multiplication requires 27 scalar multiplications. Therefore, quaternion multiplication is more than 1.5 times faster than matrix multiplication for composing conformal transformations. Thus, some computations favor quaternions, whereas others favor matrices. We summarize these trade-offs for both memory and speed in Table 15.1.

### 15.3.4 Avoiding Distortion

Aside from considerations of memory and speed, one of the main advantages of quaternions over matrices is avoiding distortions that arise from numerical inaccuracies that are inevitably introduced by floating point computations.

For any $3 \times 3$ matrix $A$ representing an affine transformation,

$$\begin{pmatrix} i_{new} \\ j_{new} \\ k_{new} \end{pmatrix} = \begin{pmatrix} i \\ j \\ k \end{pmatrix} * A.$$

Since the matrix with rows $i, j, k$ is the identity matrix, we conclude that

$$A = \begin{pmatrix} i_{new} \\ j_{new} \\ k_{new} \end{pmatrix}. \tag{15.14}$$

Hence the rows of $A$ represent the images of the unit vectors $i, j, k$ along the coordinate axes.

Thus if a $3 \times 3$ matrix $A$ represents a conformal transformation, then the rows of $A$ must be mutually orthogonal vectors, since conformal transformations preserve angles and the original vectors $i, j, k$ are mutually orthogonal. To compose conformal transformations, we must multiply matrices. But the entries of these matrices—especially rotation matrices that contain values of sines and cosines—are typically represented by floating point numbers. Therefore, after composing several conformal transformations, the rows of the product matrix will no longer be mutually orthogonal vectors due to numerical inaccuracies introduced by floating point computations. Thus these matrices no longer represent conformal transformations, so applying these matrices will distort angles in the image.

Quaternions avoid this problem. To compose conformal transformations, we multiply the corresponding quaternions. Since every quaternion represents a conformal transformation, computing transformations by sandwiching with quaternions will never distort angles.

Suppose, however, that we want to compose only rotations. Composing rotations by matrix multiplication, we will generate distortions in both lengths and angles due to numerical inaccuracies introduced by floating point computations. We need to renormalize the resulting matrix so that the rows are mutually orthogonal unit vectors, but it is not clear how to perform this normalization.

This normalization, however, is easy to perform with quaternions. Rotations are represented by unit quaternions. When we compose rotations by multiplying unit quaternions, the result may no longer be a unit quaternion due to floating point computations. But we can normalize the resulting quaternion $q$ simply by dividing by its length $|q|$. The resulting quaternion $q/|q|$ is certainly a unit quaternion, so this normalization avoids distorting either lengths or angles in the image.

### 15.3.5   Key Frame Animation

Quaternions have one additional advantage over matrices: quaternions can be used to interpolate in between frames for key frame animation. In key frame animation, an artist draws only a few key frames in the scene, and an animator must interpolate intermediate frames to make the animation look natural.

Consider an object is tumbling through a scene. An artist will typically draw only a few frames representing rotations of the object at certain key times. An animator must then find intermediate rotations, so that the tumbling appears smooth. Suppose that from the artist's drawing we know the rotations $R_0$ and $R_1$ at times $t = 0$ and $t = 1$. How do we find the appropriate intermediate rotations at intermediate times?

If the rotations $R_0$, $R_1$ are represented by $3 \times 3$ matrices, then we seek intermediate rotation matrices $R(t)$ for $0 < t < 1$. We might try linear interpolation and set

$$R(t) = (1 - t)R_0 + tR_1.$$

The problem with this approach is that the matrices $R(t)$ are no longer rotation matrices because the rows of these matrices are no longer mutually orthogonal unit vectors. Thus the matrices $R(t)$ will introduce undesirable distortions into the animation.

Alternatively, we could use unit quaternions $q_0$, $q_1$ to represent these two rotations. The quaternion generated by linear interpolation,

$$q(t) = (1 - t)q_0 + tq_1,$$

is no longer a unit quaternion, but we can adjust the length of $q(t)$ by dividing by $|q(t)|$. Now $q(t)/|q(t)|$ is a unit quaternion representing an intermediate rotation. Unfortunately, the tumbling motions generated by this approach will still not appear smooth because linear interpolation does not generate uniformly spaced quaternions—that is, if $\phi$ is the angle between $q_0$ and $q_1$, then $t\phi$ is not generally the angle between $q_0$ and $q(t)/|q(t)|$.

To correct this problem, we apply spherical linear interpolation (SLERP, see Chapter 11, Section 11.6)—that is, we set

$$q(t) = slerp(q_0, q_1, t) = \frac{\sin((1 - t)\phi)}{\sin(\phi)}q_0 + \frac{\sin(t\phi)}{\sin(\phi)}q_1, \tag{15.15}$$

where $\phi$ is the angle between $q_0$ and $q_1$. Recall that SLERP guarantees that if $q_0$ and $q_1$ are unit quaternions, then $q(t)$ is also a unit quaternion, and if $\phi$ is the angle between $q_0$ and $q_1$, then $t\phi$ is the angle between $q_0$ and $q(t)$. Thus spherical linear interpolation applied to quaternions generates the appropriate intermediate rotations for key frame animation.

### 15.3.6 Conversion Formulas

Since we have two ways to represent conformal transformations on vectors in three dimensions—$3 \times 3$ matrices and quaternions—and since each of these representations is superior for certain applications, it is natural to develop formulas to convert between these two representations. If a quaternion represents scaling along with rotation, the scale factor can be retrieved from the magnitude of the quaternion. Similarly, if a $3 \times 3$ matrix is a composite of rotation and uniform scaling, the scale factor can be retrieved from the determinant of the matrix. Therefore, we can easily retrieve these scale factors, and we can always normalize quaternions to have unit length and conformal matrices to have unit determinant. Unit quaternions and $3 \times 3$ conformal matrices with unit determinant represent rotations. As we have just seen in Sections 15.3.4 and 15.3.5, rotations embody the most important applications of quaternions to Computer Graphics, so we will concentrate here on converting between different representations for rotations.

Consider a rotation around the unit axis vector $N$ by the angle $\phi$. The rotation matrix (Chapter 13, Section 13.3.2) and the unit quaternion (Proposition 15.1) that represent this rotation are given by

$$Rot(N, \phi) = (\cos \phi)I + (1 - \cos \phi)(N^T * N) + (\sin \phi)(N \times \_), \qquad (15.16)$$

$$q(N, \phi) = \cos (\phi/2)O + \sin (\phi/2)N. \qquad (15.17)$$

To convert between these two representations for rotation, we shall make extensive use of the following half angle and double angle formulas for sine and cosine (see Exercise 15.3):

$$\cos^2 (\phi/2) = \frac{1 + \cos \phi}{2},$$

$$\sin^2 (\phi/2) = \frac{1 - \cos \phi}{2},$$

$$\sin \phi = 2 \sin (\phi/2) \cos (\phi/2).$$

Suppose that we are given the rotation matrix

$$Rot(N, \phi) = (\cos \phi)I + (1 - \cos \phi)(N^T * N) + (\sin \phi)(N \times \_) = (R_{i,j}),$$

and we want to find the corresponding unit quaternion

$$q(N, \phi) = \cos (\phi/2)O + \sin (\phi/2)N.$$

Then we need to find $\cos (\phi/2)$ and $\sin(\phi/2)N$. Expanding Equation 15.16 in matrix form yields

$$Rot(N, \phi) = \begin{pmatrix} \cos \phi + (1 - \cos \phi)N_1^2 & (1 - \cos \phi)N_1 N_2 + \sin \phi N_3 & (1 - \cos \phi)N_1 N_3 - \sin \phi N_2 \\ (1 - \cos \phi)N_1 N_2 - \sin \phi N_3 & \cos \phi + (1 - \cos \phi)N_2^2 & (1 - \cos \phi)N_2 N_3 + \sin \phi N_1 \\ (1 - \cos \phi)N_1 N_3 + \sin \phi N_2 & (1 - \cos \phi)N_2 N_3 - \sin \phi N_1 & \cos \phi + (1 - \cos \phi)N_3^2 \end{pmatrix}. \qquad (15.18)$$

Since $N$ is a unit vector,

$$N_1^2 + N_2^2 + N_3^2 = N \cdot N = 1.$$

Therefore, summing the diagonal entries of $Rot(N, \phi)$ gives

$$R_{1,1} + R_{2,2} + R_{3,3} = 3\cos\phi + 1 - \cos\phi = 1 + 2\cos\phi.$$

Solving for $\cos\phi$, we find that

$$\cos\phi = \frac{R_{1,1} + R_{2,2} + R_{3,3} - 1}{2}. \qquad (15.19)$$

But by the half angle formula,

$$\cos(\phi/2) = \sqrt{\frac{1 + \cos\phi}{2}}.$$

Therefore, by Equation 15.19,

$$\cos(\phi/2) = \sqrt{\frac{R_{1,1} + R_{2,2} + R_{3,3} + 1}{2}}. \qquad (15.20)$$

To find $\sin(\phi/2)N$, observe from Equation 15.18 that

$$2\sin(\phi)N = (R_{2,3} - R_{3,2}, R_{3,1} - R_{1,3}, R_{1,2} - R_{2,1}). \qquad (15.21)$$

But

$$\sin(\phi) = 2\sin(\phi/2)\cos(\phi/2).$$

Substituting this result into Equation 15.21 and solving for $\sin(\phi/2)N$ yield

$$\sin(\phi/2)N = \frac{(R_{2,3} - R_{3,2}, R_{3,1} - R_{1,3}, R_{1,2} - R_{2,1})}{4\cos(\phi/2)}. \qquad (15.22)$$

Thus, by Equations 15.20 and 15.22,

$$q(N, \phi) = \frac{\sqrt{R_{1,1} + R_{2,2} + R_{3,3} + 1}}{2}O + \frac{(R_{2,3} - R_{3,2})i + (R_{3,1} - R_{1,3})j + (R_{1,2} - R_{2,1})k}{2\sqrt{R_{1,1} + R_{2,2} + R_{3,3} + 1}}. \qquad (15.23)$$

Conversely, suppose that we are given the unit quaternion:

$$q(N, \phi) = \cos(\phi/2)O + \sin(\phi/2)N = (q_1, q_2, q_3, q_4), \qquad (15.24)$$

and we want to find the matrix:

$$Rot(N, \phi) = (\cos\phi)I + (1 - \cos\phi)(N^T * N) + (\sin\phi)(N \times \_).$$

Then we need to compute the three matrices $(\cos\phi)I$, $(1 - \cos\phi)(N^T * N)$, and $(\sin\phi)(N \times \_)$. By the half angle formula,

$$\cos\phi = 2\cos^2(\phi/2) - 1.$$

Thus, by Equation 15.24, since $q(N, \phi)$ is a unit quaternion,

$$\cos \phi = 2q_4^2 - 1 = q_4^2 - q_1^2 - q_2^2 - q_3^2,$$

so

$$(\cos \phi)I = \begin{pmatrix} q_4^2 - q_1^2 - q_2^2 - q_3^2 & 0 & 0 \\ 0 & q_4^2 - q_1^2 - q_2^2 - q_3^2 & 0 \\ 0 & 0 & q_4^2 - q_1^2 - q_2^2 - q_3^2 \end{pmatrix}. \tag{15.25}$$

Moreover, since by the half angle formula,

$$\sin^2(\phi/2) = \frac{1 - \cos(\phi)}{2},$$

it follows that

$$(1 - \cos \phi)(N^{\mathrm{T}} * N) = 2 \sin(\phi/2)N^{\mathrm{T}} * \sin(\phi/2)N.$$

Hence, by Equation 15.24,

$$(1 - \cos \phi)(N^{\mathrm{T}} * N) = 2(q_1, q_2, q_3)^{\mathrm{T}} * (q_1, q_2, q_3),$$

so

$$(1 - \cos \phi)(N^{\mathrm{T}} * N) = \begin{pmatrix} 2q_1^2 & 2q_1q_2 & 2q_1q_3 \\ 2q_1q_2 & 2q_2^2 & 2q_2q_3 \\ 2q_1q_3 & 2q_2q_3 & 2q_3^2 \end{pmatrix}. \tag{15.26}$$

Finally, since

$$\sin \phi = 2 \sin(\phi/2) \cos(\phi/2),$$

we have

$$(\sin \phi)N = 2 \sin(\phi/2) \cos(\phi/2)N.$$

Hence, again by Equation 15.24,

$$(\sin \phi)N = 2q_4(q_1, q_2, q_3),$$

so

$$(\sin \phi)N \times \_ = \begin{pmatrix} 0 & 2q_4q_3 & -2q_4q_2 \\ -2q_4q_3 & 0 & 2q_4q_1 \\ 2q_4q_2 & -2q_4q_1 & 0 \end{pmatrix}. \tag{15.27}$$

Adding together Equations 15.25 through 15.27 yields

$$Rot(N, \phi) = \begin{pmatrix} q_4^2 + q_1^2 - q_2^2 - q_3^2 & 2q_1q_2 + 2q_3q_4 & 2q_1q_3 - 2q_2q_4 \\ 2q_1q_2 - 2q_3q_4 & q_4^2 - q_1^2 + q_2^2 - q_3^2 & 2q_2q_3 + 2q_1q_4 \\ 2q_1q_3 + 2q_2q_4 & 2q_2q_3 - 2q_1q_4 & q_4^2 - q_1^2 - q_2^2 + q_3^2 \end{pmatrix}. \quad (15.28)$$

---

## 15.4  Summary

Complex numbers can be used to represent vectors in the plane. A vector $z = (x, y)$ in the $xy$-plane can be rotated through the angle $\phi$ by multiplying the complex number $z = x + iy$ with the complex number $w(\phi) = e^{i\phi} = \cos(\phi) + i\sin(\phi)$, which corresponds to the unit vector that makes an angle $\phi$ with the $x$-axis. Similarly, the vector $z = (x, y)$ can be scaled by the factor $s$ by multiplying the complex number $z = x + iy$ by the real number $s$. Thus, every conformal transformation of vectors in the plane can be represented by multiplication with a single complex number $w = w_1 + iw_2$, where $s = |w|$ is the scale factor and $\phi = \arctan(w_2/w_1)$ is the angle of rotation.

Quaternions are complex numbers on steroids. Quaternion multiplication is an extension of complex multiplication from the two-dimensional space of vectors in the plane to the four-dimensional space of mass-points. The product of two quaternions incorporates dot product, cross product, and scalar product and is given by the formula:

$$(aO + u)(bO + v) = (ab - u \cdot v)O + (bu + av + u \times v),$$

where $O$ represents the origin of the points in three-dimensional affine space.

Quaternion multiplication can be used to model conformal transformations of vectors in three dimensions by sandwiching a vector $v$ between a quaternion $q$ and its conjugate $q^*$:

$$S_q(v) = qvq^*.$$

In particular, we have the following formulas for rotation, mirror image, and uniform scaling.

*Rotation—around the unit direction vector N by the angle $\phi$*

$$q(N, \phi) = \cos(\phi/2)O + \sin(\phi/2)N,$$
$$S_{q(N,\phi)}(v) = q(N, \phi)vq^*(N, \phi).$$

*Mirror image—in the plane perpendicular to the unit normal N*

$$-S_N(v) = NvN.$$

*Uniform scaling by the scale factor c*

$$S_{\sqrt{c}O}(v) = cv.$$

Quaternions provide more compact representations for conformal transformations than $3 \times 3$ matrices, since quaternions are represented by four rectangular coordinates whereas $3 \times 3$ matrices contain nine scalar entries. Composing conformal transformations by multiplying quaternions is faster than composing conformal transformations by matrix multiplication because to compute the product of two quaternions requires only 16 real scalar multiplications whereas to compose two transformations by matrix multiplication requires 27 real scalar multiplications. Nevertheless, quaternions are not a panacea. The total cost for computing one conformal transformation on a vector using quaternions is 28 scalar multiplications, whereas the cost for multiplying a vector by a $3 \times 3$ matrix is only 9 scalar multiplications. Thus, matrices are more than three times faster than quaternions for computing conformal transformations.

Quaternions are used in Computer Graphics to avoid distortions that can arise during conformal transformations due to numerical inaccuracies introduced by floating point computations. Quaternion transformations never distort angles because every quaternion represents a conformal transformation. Moreover, unlike $3 \times 3$ matrices, quaternions are easily normalized, so unit quaternions can be employed to avoid distorting distances as well as angles during rotation.

Quaternions are also used in Computer Graphics to perform key frame animation. In key frame animation, an artist draws only a few key frames in the scene, and an animator must then interpolate intermediate frames to make the animation look natural. Linear interpolation of rotation matrices does not generate rotation matrices. Linear interpolation of unit quaternions does not generate unit quaternions and even when normalized the resulting unit quaternions are not uniformly spaced along an arc in four dimensions, so the resulting motion appears jerky rather than smooth. To interpolate smoothly between two rotations represented by the unit quaternions $q_0$, $q_1$, we apply spherical linear interpolation (SLERP).

$$slerp(q_0, q_1, t) = \frac{\sin((1-t)\phi)}{\sin(\phi)} q_0 + \frac{\sin(t\phi)}{\sin(\phi)} q_1,$$

where $\phi$ is the angle between $q_0$ and $q_1$. Spherical linear interpolation avoids undesirable artifacts that arise from interpolation methods that either fail to generate rotations or build rotations that do not transition smoothly between the artist's original rotations.

Unit quaternions $q(N, \phi) = (q_1, q_2, q_3, q_4)$ and $3 \times 3$ matrices $Rot(N, \phi) = (R_{i,j})$ can both be used to represent rotations around the unit axis vector $N$ by the angle $\phi$. The following formulas can be applied to convert between these two representations:

$$q(N, \phi) = \frac{\sqrt{R_{1,1} + R_{2,2} + R_{3,3} + 1}}{2} O + \frac{(R_{2,3} - R_{3,2})i + (R_{3,1} - R_{1,3})j + (R_{1,2} - R_{2,1})k}{2\sqrt{R_{1,1} + R_{2,2} + R_{3,3} + 1}},$$

$$Rot(N, \phi) = \begin{pmatrix} q_4^2 + q_1^2 - q_2^2 - q_3^2 & 2q_1 q_2 + 2q_3 q_4 & 2q_1 q_3 - 2q_2 q_4 \\ 2q_1 q_2 - 2q_3 q_4 & q_4^2 - q_1^2 + q_2^2 - q_3^2 & 2q_2 q_3 + 2q_1 q_4 \\ 2q_1 q_3 + 2q_2 q_4 & 2q_2 q_3 - 2q_1 q_4 & q_4^2 - q_1^2 - q_2^2 + q_3^2 \end{pmatrix}.$$

## Exercises

15.1. Using Equation 15.4, verify that complex multiplication is associative, commutative, and distributes through addition.

15.2. Let $(r, \theta)$ denote the polar coordinates of a point in the plane. Show that in polar coordinates complex multiplication is equivalent to

$$(r_1, \theta_1) \cdot (r_2, \theta_2) = (r_1 r_2, \theta_1 + \theta_2).$$

15.3. Here we shall derive the half angle formulas used in the text.

    a. Using Euler's Formula (Equation 15.5), show that:

$$\cos(2\phi) + i\sin(2\phi) = (\cos(\phi) + i\sin(\phi))^2.$$

    b. Conclude from part a that:

      i. $\cos(2\phi) = \cos^2(\phi) - \sin^2(\phi) = 2\cos^2(\phi) - 1.$

      ii. $\sin(2\phi) = 2\sin(\phi)\cos(\phi).$

    c. Conclude from part b that:

      i. $\cos(\phi/2) = \sqrt{\dfrac{1 + \cos(\phi)}{2}}.$

      ii. $\sin(\phi/2) = \sqrt{\dfrac{1 - \cos(\phi)}{2}}.$

15.4. Let $z = (x, y) = x + iy$ be an arbitrary vector in the plane, and let $w(\phi) = \cos(\phi/2) + i\sin(\phi/2)$.

    a. Show that

$$z * Rot(\phi) = w(\phi)zw(\phi).$$

    b. Conclude that a vector $z$ in the plane can be rotated through the angle $\phi$ by sandwiching $z$ between $w(\phi)$ and $w(\phi)$.

15.5. Use complex multiplication to derive the following formula for *slerp* given in Chapter 11, Seciton 11.6:

$$slerp(v_0, v_1, t) = \frac{\sin((1 - t)\phi)}{\sin(\phi)} v_0 + \frac{\sin(t\phi)}{\sin(\phi)} v_1, \text{ where } \phi \text{ is the angle between } v_0 \text{ and } v_1.$$

    (Hint: Introduce complex coordinates into the plane of the vectors $v_0$, $v_1$, and set $v_0 = z_0 = e^{i\theta_0}$, $v_1 = z_1 = e^{i\theta_1}$. Now use the fact that

$$slerp(v_0, v_1, t) = slerp(z_0, z_1, t) = e^{i(\theta_0 + t(\theta_1 - \theta_0))} = e^{i\theta_0} e^{it(\theta_1 - \theta_0)}.)$$

15.6. Represent points $(x, y)$ in the plane by complex numbers $z = x + yi$.

    a. Show that for each conformal transformation $T$ on the plane, there are two complex numbers $w_{T0}$, $w_{T1}$ such that for all points $z$ in the complex plane $T(z) = w_{T0}\, z + w_{T1}$.

    b. Using part a, explain how to use complex numbers to represent conformal iterated functions systems on the plane.

    c. Represent the iterated function systems for the following fractals (see Chapter 6) in terms of complex numbers:

      i. Sierpinski gasket

      ii. Koch curve

      iii. Hangman fractal

15.7. Let $w_1$, $w_2$, $w$ be complex numbers and let $p(z) = a_n z^n + \cdots + a_0$ be a polynomial in the complex variable $z$ with real coefficients $a_0, \ldots, a_n$.

a. Show that:

    i. $(w_1 + w_2)^* = w_1^* + w_2^*$.

    ii. $(w_1 w_2)^* = w_1^* w_2^*$.

    iii. $(w^n)^* = (w^*)^n$.

b. Conclude from part a that if $p(w) = 0$, then $p(w^*) = 0$.

15.8. Using Equation 15.9 and the properties of dot product and cross product, verify that quaternion multiplication is associative and distributes through addition.

15.9. Let $a, b, x, y$ be real numbers, and let $z_1 = a + bi$, $z_2 = x + yi$ be complex numbers.

a. Verify Brahmagupta's identity:

$$(a^2 + b^2)(x^2 + y^2) = (ax - by)^2 + (ay + bx)^2.$$

b. Using Brahmagupta's identity, prove that:

$$|z_1 z_2| = |z_1||z_2|.$$

15.10. Let $a, b, c, d, w, x, y, z$ be real numbers, and let $q_1 = aO + bi + cj + dk$, $q_2 = wO + xi + yj + zk$ be quaternions.

a. Verify the following generalization of Brahmagupta's identity:

$$(a^2 + b^2 + c^2 + d^2)(w^2 + x^2 + y^2 + z^2) = (aw - bx - cy - dz)^2 + (ax + bw + cz - dy)^2$$
$$+ (ay - bz + cw + dx)^2 + (az + by - cx + dw)^2.$$

b. Using the result in part a, prove that:

$$|q_1 q_2| = |q_1||q_2|.$$

15.11. Consider a quaternion $q = q_0 O + q_1 i + q_2 j + q_3 k$.

a. Show that

$$q = (q_0 O + q_1 i) + (q_2 O + q_3 i)j.$$

b. Conclude from part a that every quaternion can be represented by a pair of complex numbers.

c. Let $a, b, c, d$ be complex numbers. Show that

$$(a + bj)(c + dj) = (ac - bd^*) + (ad + bc^*)j.$$

15.12. Let $q$ be an arbitrary quaternion. Define

$$e^q = O + q + \frac{q^2}{2!} + \frac{q^3}{3!} + \cdots$$

a. Show that if $N$ is a unit vector, then

$$e^{\lambda N} = \cos(\lambda)O + \sin(\lambda)N.$$

b. Conclude from part a and Proposition 15.1 that $v$ can be rotated around a unit vector $N$ through the angle $\phi$ by sandwiching $v$ between $q = e^{\phi N/2}$ and $q^* = e^{-\phi N/2}$.

15.13. Consider two unit vectors $v_0, v_1$. Let $\phi$ be the angle between $v_0$ and $v_1$ and let $w = \dfrac{v_0 \times v_1}{|v_0 \times v_1|}$.

a. Using the results of Exercise 15.12, show that:

i. $slerp(v_0, v_1, t) = e^{t\phi w/2} v_0 e^{-t\phi w/2}$.

ii. $v_1^* v_0 = e^{\phi w}$.

b. Conclude from part a that:

$$slerp(v_0, v_1, t) = (v_1^* v_0)^{t/2} v_0 (v_0^* v_1)^{t/2}.$$

15.14. Let $p = p_4 O + p_1 i + p_2 j + p_3 k$ and $q = q_4 O + q_1 i + q_2 j + q_3 k$. Show that:

$$pq = (p_4 q_4 - p_1 q_1 - p_2 q_2 - p_3 q_3)O + (p_4 q_1 + p_1 q_4)i + (p_4 q_2 + p_2 q_4)j + (p_4 q_3 + p_3 q_4)k$$
$$+ (p_2 q_3 - p_3 q_2)i + (p_3 q_1 - p_1 q_3)j + (p_1 q_2 - p_2 q_1)k.$$

15.15. Let $p = (p_1, p_2, p_3, p_4)$ and $q = (q_1, q_2, q_3, q_4)$ be quaternions, and let

$$L(q) = \begin{pmatrix} q_4 & q_3 & -q_2 & -q_1 \\ -q_3 & q_4 & q_1 & -q_2 \\ q_2 & -q_1 & q_4 & -q_3 \\ q_1 & q_2 & q_3 & q_4 \end{pmatrix} \quad \text{and} \quad R(q) = \begin{pmatrix} q_4 & -q_3 & q_2 & -q_1 \\ q_3 & q_4 & -q_1 & -q_2 \\ -q_2 & q_1 & q_4 & -q_3 \\ q_1 & q_2 & q_3 & q_4 \end{pmatrix}.$$

a. Using Exercise 15.14, verify that:

i. $qp = (p_1, p_2, p_3, p_4) * L(q)$.

ii. $pq = (p_1, p_2, p_3, p_4) * R(q)$.

b. Conclude from part a that:

$$S_q(p) = (p_1, p_2, p_3, p_4) * (L(p) * R(q^*)).$$

c. By direct computation, verify that:

$$L(p) * R(q^*) = \begin{pmatrix} q_4^2 + q_1^2 - q_2^2 - q_3^2 & 2q_1 q_2 + 2q_3 q_4 & 2q_1 q_3 - 2q_2 q_4 & 0 \\ 2q_1 q_2 - 2q_3 q_4 & q_4^2 - q_1^2 + q_2^2 - q_3^2 & 2q_2 q_3 + 2q_1 q_4 & 0 \\ 2q_1 q_3 + 2q_2 q_4 & 2q_2 q_3 - 2q_1 q_4 & q_4^2 - q_1^2 - q_2^2 + q_3^2 & 0 \\ 0 & 0 & 0 & q_1^2 + q_2^2 + q_3^2 + q_4^2 \end{pmatrix}.$$

d. Conclude from parts b and c that if $q = q(N, \phi)$, then:

$$Rot(N, \phi) = \begin{pmatrix} q_4^2 + q_1^2 - q_2^2 - q_3^2 & 2q_1 q_2 + 2q_3 q_4 & 2q_1 q_3 - 2q_2 q_4 \\ 2q_1 q_2 - 2q_3 q_4 & q_4^2 - q_1^2 + q_2^2 - q_3^2 & 2q_2 q_3 + 2q_1 q_4 \\ 2q_1 q_3 + 2q_2 q_4 & 2q_2 q_3 - 2q_1 q_4 & q_4^2 - q_1^2 - q_2^2 + q_3^2 \end{pmatrix}.$$

15.16. Let $q = q_4 O + q_1 i + q_2 j + q_3 k$.

a. By direct computation, show that:

i. $qiq^* = (q_4^2 + q_1^2 - q_2^2 - q_3^2, 2q_1q_2 + 2q_3q_4, 2q_1q_3 - 2q_2q_4)$.

ii. $qjq^* = (2q_1q_2 - 2q_3q_4, q_4^2 - q_1^2 + q_2^2 - q_3^2, 2q_2q_3 + 2q_1q_4)$.

iii. $qkq^* = (2q_1q_3 + 2q_2q_4, 2q_2q_3 - 2q_1q_4, q_4^2 - q_1^2 - q_2^2 + q_3^2)$.

b. Conclude from part a that if $q = q(N, \phi)$, then:

$$Rot(N, \phi) = \begin{pmatrix} q_4^2 + q_1^2 - q_2^2 - q_3^2 & 2q_1q_2 + 2q_3q_4 & 2q_1q_3 - 2q_2q_4 \\ 2q_1q_2 - 2q_3q_4 & q_4^2 - q_1^2 + q_2^2 - q_3^2 & 2q_2q_3 + 2q_1q_4 \\ 2q_1q_3 + 2q_2q_4 & 2q_2q_3 - 2q_1q_4 & q_4^2 - q_1^2 - q_2^2 + q_3^2 \end{pmatrix}.$$

15.17. Show that:

a. $q(N, \phi)^{-1} = q(N, \phi)^* = q(N, -\phi) = q(-N, \phi)$.

b. $Rot(N, \phi)^{-1} = Rot(N, \phi)^{\mathrm{T}} = Rot(N, -\phi) = Rot(-N, \phi)$.

15.18. Let $v$ be a vector and $q$ be a unit quaternion. Show that:

a. $S_{-q}(v) = S_q(v)$.

b. If $S_q(v)$ rotates the vector $v$ around the unit vector $N$ by the angle $\theta$, then $S_{q^*}(v)$ rotates the vector $v$ around the unit vector $N$ by the angle $-\theta$.

15.19. Let $v$, $w$ be unit vectors.

a. Show that:

i. $vw$ is a unit quaternion.

ii. $S_{vw}(w)$ lies in the plane of the vectors $v$, $w$.

iii. $v$ bisects the angle between $w$ and $S_{vw}(w)$.

b. Prove that mirroring a vector $u$ in the plane perpendicular to $w$ and then mirroring the resulting vector in the plane perpendicular to $v$ is equivalent to rotating the vector $u$ around the vector $v \times w$ by twice the angle between $w$ and $v$ (see also Exercise 15.23).

15.20. In this exercise, we will study the composite of two rotations.

a. Using Lemma 15.1, show that if $p$ and $q$ are quaternions, then $|pq| = |p|\,|q|$.

b. Conclude from part a that if $p$ and $q$ are unit quaternions, then $pq$ is a unit quaternion.

c. Using part b, show that:

$$q(N_1, \phi_1)q(N_2, \phi_2) = q(N, \phi),$$

where:

i. $\cos(\phi/2) = \cos(\phi_1/2)\cos(\phi_2/2) - \sin(\phi_1/2)\sin(\phi_2/2)(N_1 \cdot N_2)$.

ii. $\sin(\phi/2)N = \cos(\phi_2/2)\sin(\phi_1/2)N_1 + \cos(\phi_1/2)\sin(\phi_2/2)N_2$
$\qquad\qquad + \sin(\phi_1/2)\sin(\phi_2/2)N_1 \times N_2$.

d. Conclude from part c that

$$q(N, \phi_1)q(N, \phi_2) = q(N, \phi_1 + \phi_2).$$

e. Using parts c and d, show that:

    i. $Rot(N_2, \phi_2) * Rot(N_1, \phi_1) = Rot(N, \phi)$.

    ii. $Rot(N, \phi_1) * Rot(N, \phi_2) = Rot(N, \phi_1 + \phi_2)$.

15.21. In this exercise, we explore the effect of sandwiching on points in affine space.

a. Show that for any quaternions $q, a, b$:

$$S_q(a + b) = S_q(a) + S_q(b).$$

b. Show that:

    i. $S_{q(N, \theta)}(O) = O$.

    ii. $-S_N(O) = -O$.

    iii. $S_{\sqrt{c}O}(O) = cO$.

c. Conclude from parts a and b that if $P$ is a point in affine space, then:

    i. $S_{q(N, \phi)}(P)$ rotates the point $P$ by the angle $\phi$ around the line $L$ through the origin parallel to the vector $N$.

    ii. $-S_N(P)$ is not the mirror image of the point $P$ in the plane through the origin perpendicular to the unit normal vector $N$. Give a geometric interpretation of $-S_N(P)$.

    iii. $S_{\sqrt{c}O}(P)$ scales the mass of $P$ by the factor $c$, but leaves the location of $P$ unchanged.

15.22. In this exercise, we will develop formulas for some non-conformal affine transformations (see Chapter 12) using quaternions.

a. Show that:

    i. $(N \cdot v)O = \dfrac{-(vN + Nv)}{2}$.

    ii. $N \times v = \dfrac{Nv - vN}{2}$.

b. Using the results of part a, verify that:

    i. *Orthogonal projection—onto the plane perpendicular to the unit normal N*

$$v_{new} = \frac{v - S_N(v)}{2}.$$

    ii. *Parallel projection—onto the plane perpendicular to the unit normal N in the direction parallel to w*

$$v_{new} = v - \frac{vNw + Nvw}{wN + Nw}.$$

   iii. *Nonuniform scaling—along the direction w by the scale factor c*

$$v_{new} = \frac{v - S_w(v)}{2} + c\frac{v + S_w(v)}{2}.$$

15.23. In this exercise, we will show that every rotation is the composite of two reflections. Let $N$ be a unit vector and let $u, v$ be two unit vectors in the plane perpendicular to $N$ with $\text{sgn}(u, v, N) > 0$ and $\angle(u, v) = \phi/2$. Show that:

a. $q(N, \phi) = -vu$.

b. $S_{q(N,\phi)}(w) = (vu)w(uv)$.

c. $S_{q(N,\phi)} = S_v \circ S_u$.

d. Conclude from part c that every rotation is the composite of two reflections.

---

## Programming Projects

15.1. *Conformal Iterated Function Systems in the Complex Plane*

   a. In your favorite programming language using your favorite API, implement conformal iterated functions systems for generating fractals in the plane, using only addition and multiplication of complex numbers. (Hint: See Exercise 15.6.)

   b. Using your implementation in part a, generate the following fractals:

     i. Sierpinski gasket

     ii. Koch curve

     iii. Hangman fractal

   c. Generate your own novel fractals in the plane, using only addition and multiplication of complex numbers.

15.2. *The Mandelbrot Set*

   Given a complex number $z_0$, define a sequence of complex numbers $\{z_k(z_0)\}$ by setting:

$$z_0(z_0) = z_0,$$

$$z_{n+1}(z_0) = z_n^2(z_0) + z_0.$$

For some complex numbers $z_0$, the sequence $\{z_k(z_0)\}$ will diverge; for other complex numbers $z_0$, the sequence $\{z_k(z_0)\}$ will converge. The *Mandelbrot set* is the set of all complex numbers $z_0$ for which the sequence $\{z_k(z_0)\}$ does not diverge to infinity.

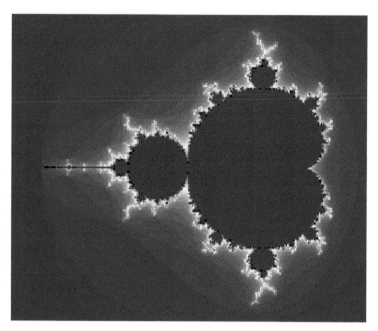

**FIGURE 15.2: (See color insert following page 354.)** The Mandelbrot set, generating by iterating the expression $z_{n+1}(z_0) = z_n^2(z_0) + z_0$.

    a. In your favorite programming language using your favorite API, draw the Mandelbrot set by painting the points that converge in one color and the points that diverge to infinity in another color. (Hint: The points $z_0$ for which the sequence $\{z_k(z_0)\}$ does not diverge to infinity are contained in the rectangle $[-2, 25, 1.0] \times [-1.25, 1.25]$. For each complex number in this rectangle, compute $z_n(z_0)$ for some large value of $n$, say $n = 100$. If $|z_n(z_0)|$ becomes large, then the sequence is diverging to infinity.)

    b. Develop code that allows you to zoom in on any part of the boundary of the Mandelbrot set and display the associated points in the Mandelbrot set.

15.3. *Key Frame Animation*

    a. In your favorite programming language using your favorite API, develop a key frame animation system based on quaternions.

    b. Build a collection of polyhedral models and animate your models using your system for key frame animation.

# Part III

# Three-Dimensional Computer Graphics: Realistic Rendering

# Chapter 16

## Color and Intensity

*and he made him a coat of many colours*

– Genesis 37:3

## 16.1 Motivation

To display a picture using Computer Graphics, we need to compute the color and intensity of the light at each point in the picture. The purpose of this chapter is to explain how to represent color and intensity numerically and then to develop illumination models that allow us to compute the color and intensity at each point once we know the color, location, and intensity of the light sources and the optical characteristics of the objects in the scene.

## 16.2 The RGB Color Model

Color forms a three-dimensional vector space. The three primary colors—red, green, and blue—are a basis for this vector space: every color can be represented as a linear combination of red, green, and blue.

The unit cube is often used to represent color space (see Figure 16.1). Black is located at one corner of the cube usually associated with the origin of the coordinate system, and the three primary colors are placed along the three orthogonal axes. Intensity varies along the edges of the cube, with the full intensity of each primary color corresponding to a unit distance along the associated edge. Every combination of color and intensity is then represented by some linear combination of red, green, and blue, where the coefficients of each primary color lie between 0 and 1. Thus, for each color $c$, there is a unique set of coordinates $(r, g, b)$ inside the unit cube that represents the color $c$. The numerical values $(r, g, b)$ represent intensities: the higher the value of $r$ or $g$ or $b$, the greater the contribution of red or green or blue to the color $c$. This model is known as the *RGB color model* and is one of the most common color models in Computer Graphics.

Pairs of primary colors combine to form colors complementary to the missing primary. Thus, *red + green = yellow*, which is the color complementary to blue. Similarly, *blue + green = cyan*, which is the color complementary to red, and *blue + red = magenta*, which is the color complementary to green. Combining all three primaries at full intensity yields *white = red + green + blue*, which lies at the corner of the cube diagonally opposite to black. Shades of gray are represented along the diagonal of the cube joining black to white.

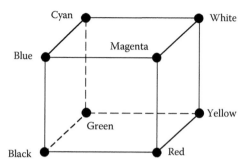

**FIGURE 16.1:** The RGB color cube. Every color and intensity is represented by a linear combination of red, green, and blue, where the coefficients of each primary color lie between 0 and 1.

In the remainder of this chapter, we shall develop illumination models. These illumination models allow us to render a scene by computing the $(r, g, b)$ color intensities for each surface point once we know the light sources and the optical characteristics of the objects in the scene. We shall consider three illumination models: ambient light, diffuse reflections, and specular highlights.

## 16.3   Ambient Light

*Ambient light* is the light that is reflected into a scene off outside surfaces. For example, sunlight that enters a room through a window by bouncing off nearby buildings is ambient light. Ambient light softens harsh shadows generated by point light sources. Thus ambient light helps to make scenes rendered by Computer Graphics appear more natural.

Typically we assume that the intensity $I_a$ of the ambient light in a scene is a constant. What we need to compute is the intensity $I$ of the ambient light reflected to a viewer from each surface point in the scene. The formula for this intensity $I$ is simply

$$I = k_a I_a, \tag{16.1}$$

where $0 \leq k_a \leq 1$ is the *ambient reflection coefficient*. The ambient reflection coefficient $k_a$ is a property of the color and material of the surface, which can be determined experimentally or simply set by the programmer. Notice that the intensity $I$ is independent of the position of the viewer.

Equation 16.1 is really three equations, one for each primary color red, green, and blue. If we set $I_a = (I_a^r, I_a^g, I_a^b)$ and $k_a = (k_a^r, k_a^g, k_a^b)$, where $(I_a^r, I_a^g, I_a^b)$ are the ambient intensities for red, green, and blue, and $(k_a^r, k_a^g, k_a^b)$ are the ambient reflection coefficients for red, green, and blue, then Equation 16.1 becomes

$$I = (r, g, b) = \left(k_a^r I_a^r, k_a^g I_a^g, k_a^b I_a^b\right). \tag{16.2}$$

Notice that if the ambient light is white, then $I_a = (1, 1, 1)$, so the color of the light reflected from a surface is the color of the surface. However, if the color of the ambient light is blue and the color of the surface is red, then $I_a = (0, 0, 1)$ and $k_a = (k_a^r, 0, 0)$, so $I = (r, g, b) = (0, 0, 0)$ and the color perceived by the viewer is black.

## 16.4 Diffuse Reflection

*Diffuse light* is the light reflected off dull surfaces like cloth. Light dispersed from dull surfaces is reflected from each point by the same amount in all directions. Thus the intensity of the reflected light is independent of the position of the viewer. Here we shall compute the diffuse light reflected off a dull surface from a point light source.

Let $I_p$ be the intensity of the point light source. We need to compute the intensity $I$ of the diffuse light reflected to a viewer from each point on a dull surface. Let $L$ be the unit vector from the point on the surface to the point light source, and let $N$ be the outward pointing unit normal at the point on the surface (see Figure 16.2). Then the formula for the diffuse intensity $I$ is simply

$$I = k_d(L \cdot N)I_p, \tag{16.3}$$

where $0 \le k_d \le 1$ is the *diffuse reflection coefficient*. The diffuse reflection coefficient $k_d$, like the ambient reflection coefficient $k_a$, is a property of the color and material of the surface, which can be determined experimentally or simply set by the programmer.

Equation 16.3 is a consequence of Lambert's Law. Consider a point light source far away from a small surface facet (see Figure 16.3). Let $I_{facet}$ denote the intensity of light on the facet, and let $I_{source}$ denote the intensity of the light source. Then

$$I_{facet} = \frac{\text{Light}}{\text{Unit area}} = \frac{\text{Beam cross section}}{\text{Facet area}} \times I_{source}.$$

But we can see from Figure 16.3 that

$$\frac{\text{Beam cross section}}{\text{Facet area}} = \cos(\theta),$$

where $\theta$ is the angle between the normal to the facet and the vector to the light source. Thus we arrive at the following result:

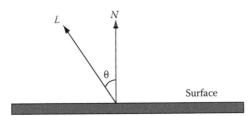

**FIGURE 16.2:** The unit normal vector $N$ to a point on the surface, and a unit vector $L$ pointing in the direction of a point light source.

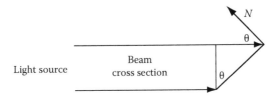

**FIGURE 16.3:** Lambert's Law: $I_{facet} = \cos(\theta)\, I_{source}$.

*Lambert's Law*

$$I_{facet} = \cos(\theta) I_{source}. \tag{16.4}$$

In Equation 16.3, $L$ and $N$ are unit vectors. Hence, $L \cdot N = \cos(\theta)$, where $\theta$ is the angle between the normal to the surface ($N$) and the direction to the light source ($L$). Thus Equation 16.3 is simply Equation 16.4 attenuated by the factor $k_d$.

Equation 16.3 is really three equations, one for each primary color red, green, and blue. If we set $I_p = (I_p^r, I_p^g, I_p^b)$ and $k_d = (k_d^r, k_d^g, k_d^b)$, where $(I_p^r, I_p^g, I_p^b)$ are the intensities of the point light source for red, green, and blue, and $(k_d^r, k_d^g, k_d^b)$ are the diffuse reflection coefficients for red, green, and blue, then Equation 16.3 becomes

$$I = (r, g, b) = (L \cdot N)\left(k_d^r I_p^r, k_d^g I_p^g, k_d^b I_p^b\right). \tag{16.5}$$

When there are many light sources, we simply add the contributions from each light source.

As with ambient light, the color perceived by the viewer depends both on the color of the light source and the color of the surface. If the light source is white, then $I_p = (1, 1, 1)$, so the color of the reflected light is the color of the surface. However, if the color of the light source is blue and the color of the surface is red, then $I_p = (0, 0, 1)$ and $k_d = (k_d^r, 0, 0)$, so $I = (r, g, b) = (0, 0, 0)$ and once again the color perceived by the viewer is black.

---

## 16.5 Specular Reflection

*Specular reflections* are highlights reflected off shiny surfaces. Unlike ambient and diffuse reflections, specular highlights are sensitive to the position of the viewer. Here we shall compute the specular highlights reflected off a shiny surface from a point light source.

Let $I_p$ be the intensity of the point light source. We need to compute the intensity $I$ of the specular highlight reflected to a viewer from each point on a shiny surface. Let $L$ be the unit vector from the point on the surface to the point light source, and let $N$ be the outward pointing unit normal at the point on the surface. Let $R$ be the image of the vector $L$ as the light from the point light source in the direction $L$ bounces off the surface—that is, $R$ is the vector defined by setting the angle of incidence of $L$ equal to the angle of reflection of $R$. Finally, let $V$ be a unit vector from the point on the surface to the viewer, and let $\alpha$ be the angle between $V$ and $R$ (see Figure 16.4). Then the formula for the intensity $I$ of the specular highlight seen by a viewer in the direction $V$ is given by

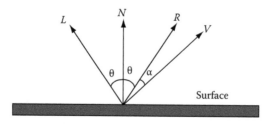

**FIGURE 16.4:** The unit normal vector $N$ to a point on the surface, and a unit vector $L$ pointing in the direction of a point light source. The unit vector $V$ points to the viewer, and the unit vector $R$ is the image of the light vector $L$ reflected off the surface. The angle $\theta$ is the angle between the vectors $L$ and $N$, and the angle $\alpha$ is the angle between the vectors $R$ and $V$.

$$I = k_s \cos^n(\alpha)I_p = k_s(R \cdot V)^n I_p, \tag{16.6}$$

where
  $0 \leq k_s \leq 1$ is the *specular reflection coefficient*
  $n \geq 0$ is the *specular exponent*, a constant that controls the concentration of the specular highlight.

The specular reflection coefficient $k_s$, like the ambient reflection coefficient $k_a$ and the diffuse reflection coefficient $k_d$, is a property of the color and material of the surface, which can be determined experimentally or simply set by the programmer.

Notice that unlike ambient and diffuse reflections, specular highlights depend on the position of the viewer. The intensity of the specular highlight falls off rapidly as the angle $\alpha$ between the reflected vector $R$ and the vector to the viewer $V$ increases, since $R \cdot V = \cos(\alpha) \to 0$ as $\alpha \to 90°$. Moreover, increasing the value of the specular exponent $n$ further concentrates the specular highlight around the direction $R$, since high powers of $\cos(\alpha)$ approach zero more rapidly than low powers of $\cos(\alpha)$. As $n \to \infty$, the surface approaches a mirror; only when $\alpha = 0$ is any effect visible to the viewer. Conversely, when $n \to 0$, neither the position of the viewer nor the position of the light source matters, and the effect is much the same as for ambient light.

In Equation 16.6, the vectors $L, V, N$ are typically known, since the position of the light source, the location of the viewer, and the shape of the surface are controlled by the programmer. The parameters $I_p$ and $k_s$—the intensity of the light source and the specular reflection coefficient—are also controlled by the programmer. The reflection vector $R$, however, must be computed. To find $R$, let $L_\parallel$ and $L_\perp$ denote the components of $L$ parallel and perpendicular to $N$ (see Figure 16.5). Then

$$L = L_\parallel + L_\perp,$$

and

$$R = L_\parallel - L_\perp.$$

Now recall that

$$L_\parallel = (L \cdot N)N,$$

$$L_\perp = L - L_\parallel = L - (L \cdot N)N.$$

Therefore

$$R = 2(L \cdot N)N - L.$$

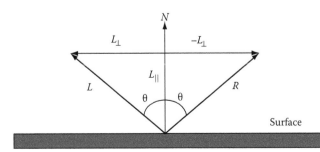

**FIGURE 16.5:** Computation of the vector $R$ from the vectors $N$ and $L$: $L = L_\parallel + L_\perp$ and $R = L_\parallel - L_\perp$.

In the calculation of specular reflection, we shall often assume that the light sources are located at infinity. Notice that if the light source is far away and if the surface is planar, then $L$, $N$, $R$ are constants, and only $R \cdot V$ needs to be recalculated for each point on the surface. Thus these assumptions help to speed up the calculation of specular reflections.

Equation 16.6, like Equations 16.1 and 16.3, is really three equations, one for each primary color red, green, and blue. If we set $I_p = (I_p^r, I_p^g, I_p^b)$ and $k_s = (k_s^r, k_s^g, k_s^b)$, where $(I_p^r, I_p^g, I_p^b)$ are the intensities of the point light source for red, green, and blue, and $(k_s^r, k_s^g, k_s^b)$ are the specular reflection coefficients for red, green, and blue, then Equation 16.6 becomes

$$I = (r, g, b) = (R \cdot V)^n \left( k_s^r I_p^r, k_s^g I_p^g, k_s^b I_p^b \right). \tag{16.7}$$

When there are many light sources, we simply add the contributions from each light source.

Specular highlights come in two types, depending on whether the surface material is homogeneous or inhomogeneous. When the surface material is homogeneous, like a pure metal, then the specular highlight takes on the color of the surface much like diffuse reflections. Thus, if the light source is white, then $I_p = (1, 1, 1)$, so the color of the specular highlight is the color of the surface. However, if the color of the light source is blue and the color of the surface is red, then $I_p = (0, 0, 1)$ and $k_d = (k_s^r, 0, 0)$, so $I = (r, g, b) = (0, 0, 0)$ and no specular highlight is visible to the viewer.

When the surface material is inhomogeneous, like a plastic, then unlike ambient light and diffuse reflections, the color of the specular highlights perceived by the viewer depends only on the color of the light source and not on the color of the surface. For example, if the color of the light source is blue and the color of the surface is red, then the color of the specular highlight is blue not red or black.

---

## 16.6   Total Intensity

To find the total intensity at each point, we must add the ambient, diffuse, and specular components. Thus,

$$Total\ Intensity = Ambient\ Intensity + Diffuse\ Intensity + Specular\ Intensity.$$

For each scene, there is only one ambient intensity, but for the diffuse and specular intensities, we need to add the contributions from each light source. Therefore,

$$Total\ Intensity = k_a I_a + \Sigma_p k_d (L \cdot N) I_p + \Sigma_p k_s (R \cdot V)^n I_p, \tag{16.8}$$

where the sums are taken over all point light sources. As usual, Equation 16.8 is really three equations, one for each primary color. Since in the RGB color model every color and intensity is represented by a point inside the unit cube, when there are many light sources it may be necessary to normalize the total intensities to lie between 0 and 1 by dividing the intensity at each point by the maximum intensity in the scene.

## 16.7 Summary

In the RGB color model, each color and intensity is represented by three coordinates $(r, g, b)$ in the unit cube, each coordinate representing the contribution of the associated primary color red, green, or blue, to the given color.

There are three illumination models for computing the color and intensity at each surface point: ambient intensity, diffuse reflection, and specular highlights. The formulas for these illumination models are collected here for easy reference.

*Illumination Models*

$$\text{Ambient Intensity} = k_a I_a,$$
$$\text{Diffuse Reflection} = k_d (L \cdot N) I_p,$$
$$\text{Specular Reflection} = k_s (R \cdot V)^n I_p,$$
$$\text{Total Intensity} = k_a I_a + \Sigma_p k_d (L \cdot N) I_p + \Sigma_p k_s (R \cdot V)^n I_p,$$

where
  $L$ is the unit vector from the point on the surface to the point light source
  $N$ is the outward pointing unit normal at the point on the surface
  $R = 2(L \cdot N)N - L$ is the image of the vector $L$ as the light from the point light source in the direction $L$ bounces off the surface (see Figure 16.5)
  $V$ is a unit vector from the point on the surface to the viewer
  $I_a$ is the ambient intensity
  $I_p$ is the intensity of the light source located at the point $p$
  $k_a$ is the ambient reflection coefficient
  $k_d$ is the diffuse reflection coefficient
  $k_s$ is the specular reflection coefficient
  $n$ is the specular reflection exponent

and the sums are taken over all the point light sources. The distinguishing properties of these three illumination models are summarized in Table 16.1.

**TABLE 16.1:** How the color and intensity at a point depend on the position of the viewer as well as the color of the surface and the color of the light source for ambient, diffuse, and specular reflections.

| Illumination Model | Color | Viewer Location |
|---|---|---|
| Ambient intensity | Surface | Independent |
| Diffuse reflection | Surface | Independent |
| Specular highlights (homogeneous) | Surface | Dependent |
| Specular highlights (inhomogeneous) | Light source | Dependent |

## Exercises

16.1. Suppose that the color of the light source is yellow and the color of the surface is cyan. What is the color perceived by the viewer for:

    a. Diffuse reflection.

    b. Specular reflection (homogenous material).

    c. Specular reflection (inhomogeneous material).

16.2. Show that the reflection vector $R$ in Figure 16.5 can be computed in each of the following ways:

    a. By finding the mirror image of the light vector $L$ in the plane perpendicular to $L_\perp$.

    b. By rotating the light vector $L$ by $180°$ around the axis vector $N$.

    c. By finding the mirror image of the light vector $L$ in the plane perpendicular to $N$ and then negating the result.

# Chapter 17

## Recursive Ray Tracing

*Where is the way where light dwelleth?*

$-$ Job 38:19

### 17.1 Raster Graphics

Typical graphics terminals today are raster displays. A raster display renders a picture scan line by scan line. A *scan line* is a horizontal line consisting of many small, tightly packed dots called *pixels*. To display a scene using a raster display, we need to compute the color and intensity of the light at each pixel. The purpose of this chapter is to explain how to compute the color and intensity of light for each pixel given a three-dimensional model of the objects in the scene and the location, color, and intensity of the light sources.

### 17.2 Recursive Ray Tracing

For realistic rendering, we need the following setup: an eye point $E$, a viewing screen $S$, and one or more point light sources with known location, color, and intensity (see Figure 17.1). We also need to specify the locations and the shapes of all the objects in the scene, together with their reflectivity, transparency, and indices of refraction. Our goal is to display the two-dimensional projection onto the screen of the three-dimensional scene as viewed from the eye point, much like an artist painting the scene or a still photographer capturing a vista. We shall accomplish this task by tracing light rays from the scene back to the eye. Here is the basic ray tracing algorithm.

*Ray Tracing Algorithm*

For each pixel:

1. Find all the intersections of the ray from the eye through the pixel with every object in the scene.

2. Keep the intersection closest to the eye.

3. Compute the color and intensity of the light at this intersection point.

Render the scene by displaying the color and intensity at each pixel.

Step 1 of the ray tracing algorithm requires us to compute the intersections of a straight line with lots of different surfaces. The difficulty of this intersection problem depends on the complexity of

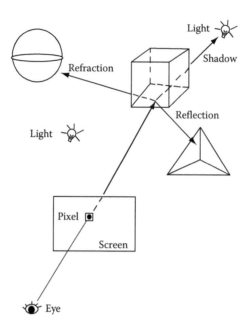

**FIGURE 17.1:**   To determine the color and intensity at a pixel, a ray is cast from the eye point through the pixel. The color and intensity of the point where the ray first strikes a surface is computed by summing the contributions from the direct illumination of the unobstructed light sources and the contributions of reflected and refracted rays.

the surfaces in the scene. For planar surfaces, these intersections are easy to compute: we simply solve one linear equation in one unknown. But for more complicated surfaces, we shall need more sophisticated tools. We will study line–surface intersection algorithms for a variety of different surfaces in Chapters 18 and 19.

Step 2 of the intersection algorithm is straightforward; in this chapter, we shall concentrate on Step 3 of the algorithm.

To find the color and intensity of the light at a surface point, we need to consider not only the light shining directly on the surface from the light sources, but also the light reflected and refracted from these sources off of other surfaces. Thus, the total intensity $I$ at a surface point is given by

$$I = I_{direct} + k_s I_{reflected} + k_t I_{refracted}, \tag{17.1}$$

where, as we observed in Chapter 16,

$$I_{direct} = I_{ambient} + I_{diffuse} + I_{specular}, \tag{17.2}$$

and $0 \le k_s, k_t \le 1$ are the reflection and refraction (transparency) coefficients. These coefficients depend on the optical characteristics of the surface, which can be determined experimentally or simply set by the programmer. Again the light reflected and refracted from other surfaces depends on the light shining directly on these surfaces as well as on the light reflected and refracted onto these surfaces from other surfaces. These observations lead to the binary light tree depicted in Figure 17.2.

Notice that for each object in the scene, we must decide whether it is reflecting, transparent, semitransparent, or opaque—that is, we need to assign values for reflectivity ($k_s$) and transparency ($k_t$). Moreover, for transparent or semitransparent objects, we need to assign an index of refraction ($c_i$). For opaque objects, $c_i = 0$; for air, $c_i = 1$. Note too that for transparent objects $I_{direct} = 0$.

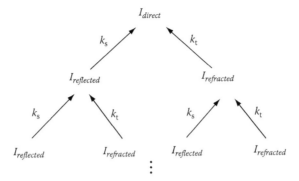

**FIGURE 17.2:** Binary light tree. The intensity of light at a surface point (apex) is the sum (depicted here by arrows) of the intensities of the light from the point light sources shining directly on the surface and the intensity of the light reflected and refracted off of other surfaces. The intensity of the light reflected and refracted from other surfaces also depends on the light shining directly on these surfaces and the light reflected and refracted to these surfaces from other surfaces. This recursive computation is summarized graphically in the binary light tree depicted in this figure. This tree must be truncated at a prespecified depth in order to avoid infinite recursion.

## 17.3   Shadows

Just as we may not see every object in a scene because some objects are hidden from view by other objects closer to the eye, so too not every object in the scene is necessarily illuminated by every light source in the scene because some objects may lie in the shadow of other objects. That is, if an opaque object lies between a light source and a surface, then the surface will not be illuminated by the light source.

To determine whether or not a surface point is illuminated by a point light source, we treat the surface point as a virtual eye and we cast a virtual ray from the surface point to the point light source. As in the ray casting algorithm, we compute all the intersections of this virtual ray with every object in the scene. If the virtual ray intersects an opaque object before the virtual ray hits the light source, then the surface point lies in shadow; otherwise the surface point is illuminated by the light source. Next we summarize this basic shadow casting algorithm.

*Shadow Casting Algorithm*

For each surface point visible to the eye, cast a virtual ray to each light source.

1. If the virtual ray hits an opaque object before the virtual ray hits a light source, then omit the contribution of this light source. That is, for this light source, $I_{diffuse} = I_{specular} = 0$.

2. If the virtual ray hits a transparent or semitransparent object before the virtual ray hits the light source, then scale the contribution of this light source and continue to look for further intersections.

Note that in Step 1, an object may be self-shadowed—that is, a point on an object (e.g., a sphere) may lie in the object's own shadow—so we must compute intersections of the virtual ray even with the object containing the point of interest.

Since we need to compute the intersections of the virtual ray with every object in the scene, shadow computations can be quite time consuming. To speed up shadow computations, we can take advantage of *shadow coherence*. The idea behind shadow coherence is that if a point $P$ lies in the shadow of an object $O$, then a nearby point $Q$ is also likely to lie in the shadow of the same object $O$. Therefore, to speed up the shadow calculations, we store the current shadowing object and we test first for intersections with this object when we cast the virtual ray from the next point in the scene.

Shadow rays are not refracted; if shadow rays were refracted, these rays would not hit the light source! We could solve this problem by casting rays starting from the light sources instead of from the surface points taking refraction into account, but this approach is very expensive and not generally done. Typically this problem is ignored, since the resulting scenes still appear realistic.

---

## 17.4   Reflection

To find the color and intensity of the light at a surface point, we need to calculate the light reflected off this point from other surfaces in the scene. To compute the color and intensity of this reflected light, we consider the surface point as a virtual eye and we cast a secondary ray from the surface point treating the surface as a mirror (see Figure 17.3). As in the standard ray casting algorithm, we find all the intersections of this secondary ray with every object in the scene, keep the closest intersection point, and calculate the color and intensity of the light at this intersection point recursively. We then add a scaled value of this color and intensity to the color and intensity at the original point. Next we summarize this reflection algorithm.

*Reflection Algorithm*

For each visible point on a reflecting surface:

1. Use the law of mirrors—angle of incidence = angle of reflection—to calculate a reflected *secondary ray* (see Figure 17.3).

2. Find all the intersections of this reflected secondary ray with every object in the scene.

3. Keep the intersection closest to the surface point.

4. Compute the color and intensity of the light at this intersection point recursively and add a scaled value of this contribution, $k_s I_{reflected}$, to the color and intensity at the original point.

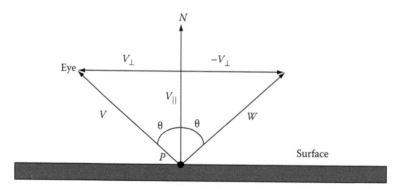

**FIGURE 17.3:**   Law of the mirror: angle of incidence = angle of reflection. Thus, if $V$ is the unit vector from the point $P$ to the eye, then the reflected vector $W$ is given by $W = V_{||} - V_{\perp}$, where $V_{||}$ and $V_{\perp}$ denote the components of $V$ parallel and perpendicular to the surface normal $N$ at the point $P$.

To find the reflected secondary ray $R(t)$, let

$$V = \frac{E - P}{|E - P|},$$

denote the unit vector from the point $P$ to the eye point $E$, and let $V_{\parallel}$ and $V_{\perp}$ denote the components of $V$ parallel and perpendicular to the surface normal $N$ at the point $P$ (see Figure 17.3). Then

$$V = V_{\parallel} + V_{\perp},$$

so the reflected vector $W$ is given by

$$W = V_{\parallel} - V_{\perp}.$$

But

$$V_{\parallel} = (V \cdot N)N,$$
$$V_{\perp} = V - V_{\parallel} = V - (V \cdot N)N.$$

Hence

$$W = 2(V \cdot N)N - V,$$

and

$$R(t) = P + tW.$$

Note that the calculation here of the reflected vector $W$ is the same as the calculation of the reflected vector $R$ for specular reflection (see Chapter 16, Section 16.5).

---

## 17.5   Refraction

To find the color and intensity of the light at a surface point, we need to calculate not only the light reflected off this point from other surfaces in the scene, but also for transparent or semitransparent surfaces the light refracted through the surface at this point. Refraction is caused by light traveling at different speeds in different mediums: the denser the medium, the slower the speed of light. This change in speed gives the illusion that objects immersed in two different mediums bend because the light bends due to the change in speed when passing between the two mediums.

To compute the color and intensity of refracted light, we again consider the surface point as a virtual eye and we cast a secondary ray from the surface point treating the surface as a refracting surface (see Figure 17.4). As in the standard ray casting algorithm, we again find all the intersections of this secondary refracted ray with every object in the scene, keep the closest intersection point, and calculate the color and intensity of the light at this intersection point recursively. We then add a scaled value of this color and intensity to the color and intensity at the original point. Next we summarize this refraction algorithm.

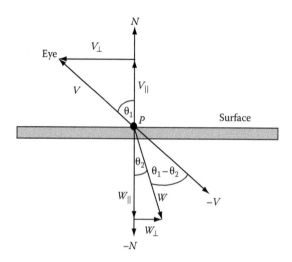

**FIGURE 17.4:**   Refraction. The unit vector $V$ from the point $P$ to the eye, the surface normal $N$ at the point $P$, and the refracted vector $W$. Here $\theta_1$ is the angle between the surface normal $N$ and the vector $V$, and $\theta_2$ is the angle between the refracted vector $W$ and the vector $-N$. By Snell's Law, if the indices of refraction on either side of the surface are $c_1$ and $c_2$, then $\frac{c_2}{c_1} = \frac{\sin(\theta_2)}{\sin(\theta_1)}$.

*Refraction Algorithm*

For each visible point on a transparent object:

1. Use Snell's Law to calculate the refracted secondary ray (see Figure 17.4).

2. Find all the intersections of this refracted ray with every object in the scene.

3. Keep the intersection closest to the surface point.

4. Compute the color and intensity of the light at this intersection point recursively and add a scaled value of this contribution, $k_t I_{refracted}$, to the color and intensity at the original point.

To find the refracted secondary ray $R(t)$, again let

$$V = \frac{E - P}{|E - P|},$$

denote the unit vector from the point $P$ to the eye point $E$, and let $W$ denote the direction of the refracted ray (see Figure 17.4). To compute the refracted ray $R(t)$, we need to calculate the refracted vector $W$ in terms of the known vectors $N$ and $V$. To perform this computation, we shall apply the following rule from optics:

*Snell's Law*

Let $\theta_1$ be the angle between the unit vector $V$ and the surface normal $N$ at the point $P$, and let $\theta_2$ be the angle between the refracted vector $W$ and the unit normal $-N$ at the point $P$ pointing away from the eye. If the indices of refraction on either side of the surface are $c_1$ and $c_2$, then

$$\frac{c_2}{c_1} = \frac{\sin(\theta_2)}{\sin(\theta_1)}. \tag{17.3}$$

There are several ways to find the refracted vector $W$ in terms of the known vectors $N$ and $V$. For example, we could rotate $-N$ about the vector $N \times V$ through the angle $\theta_2$ and then apply Snell's Law to eliminate $\theta_2$ (see Exercise 17.4). Alternatively, we could compute $W = slerp(-N, -V, \theta_2/\theta_1)$, and again apply Snell's Law to eliminate $\theta_2$ (see Exercise 17.3). Here we present a simpler method due to S. Schaefer that avoids both rotation and SLERP.

Let $V_{\|}$ and $V_{\perp}$ denote the parallel and perpendicular components of $V$ relative to $N$, and let $W_{\|}$ and $W_{\perp}$ denote the parallel and perpendicular components of $W$ relative to $-N$ (see Figure 17.4). Then, by construction,

$$
\begin{aligned}
W_{\|} \| &- V_{\|}, \\
W_{\perp} \| &- V_{\perp}.
\end{aligned}
\tag{17.4}
$$

Moreover, since $V$ and $W$ are unit vectors,

$$
\begin{aligned}
|V_{\|}| &= \cos(\theta_1), & |V_{\perp}| &= \sin(\theta_1), \\
|W_{\|}| &= \cos(\theta_2), & |W_{\perp}| &= \sin(\theta_2).
\end{aligned}
\tag{17.5}
$$

Therefore, by Equations 17.4 and 17.5 and Snell's Law (Equation 17.3),

$$
W_{\|} = \frac{\cos(\theta_2)}{\cos(\theta_1)}(-V_{\|}) = -\frac{\cos(\theta_2)}{(V \cdot N)}V_{\|},
\tag{17.6}
$$

$$
W_{\perp} = \frac{\sin(\theta_2)}{\sin(\theta_1)}(-V_{\perp}) = -\frac{c_2}{c_1}V_{\perp}.
\tag{17.7}
$$

But recall that

$$
\begin{aligned}
V_{\|} &= (V \cdot N)N, \\
V_{\perp} &= V - (V \cdot N)N.
\end{aligned}
$$

Substituting these formulas for $V_{\|}$ and $V_{\perp}$ into Equations 17.6 and 17.7 yields

$$
W_{\|} = -\frac{\cos(\theta_2)}{(V \cdot N)}(V \cdot N)N = -\cos(\theta_2)N,
$$

$$
W_{\perp} = -\frac{c_2}{c_1}(V - (V \cdot N)N),
$$

so

$$
W = W_{\|} + W_{\perp} = \left(\frac{c_2}{c_1}(V \cdot N) - \cos(\theta_2)\right)N - \frac{c_2}{c_1}V.
\tag{17.8}
$$

It remains only to solve for $\cos(\theta_2)$ in terms of $N$ and $V$. We proceed by applying the trigonometric identity $\cos^2(\theta) + \sin^2(\theta) = 1$ together with Snell's Law:

$$
\cos(\theta_2) = \sqrt{1 - \sin^2(\theta_2)} = \sqrt{1 - \frac{c_2^2}{c_1^2}\sin^2(\theta_1)} = \sqrt{1 - \frac{c_2^2}{c_1^2}(1 - \cos^2\theta_1)}
$$

$$
= \sqrt{1 - \left(\frac{c_2}{c_1}\right)^2(1 - (N \cdot V)^2)}.
$$

Thus, we conclude that

$$W = \left( \frac{c_2}{c_1}(N \cdot V) - \cos(\theta_2) \right) N - \frac{c_2}{c_1} V, \tag{17.9}$$

where

$$\cos(\theta_2) = \sqrt{1 - \left( \frac{c_2}{c_1} \right)^2 (1 - (N \cdot V)^2)}. \tag{17.10}$$

The equation of the refracted ray is simply

$$R(t) = P + tW.$$

Notice that for each color, we use the same index of refraction and therefore the same refracted ray. Technically, this procedure is incorrect, but this tactic saves tremendously on computation time and the results still look realistic, so most ray tracing programs adopt this approach.

There are some angles $\theta_1$ between the vector $V$ to the eye and the normal vector $N$ to the surface at which refraction cannot occur. This phenomenon is called *total reflection*. Total reflection may occur when $c_2/c_1 > 1$. In this case, for some angles $\theta_1$,

$$\frac{c_2}{c_1} \sin(\theta_1) > 1,$$

so by Snell's law,

$$\sin(\theta_2) > 1,$$

which is impossible. To check for total reflection, test

$$\frac{c_2^2}{c_1^2}(1 - (N \cdot V)^2) > 1 \Leftrightarrow 1 - (N \cdot V)^2 > \left( \frac{c_1}{c_2} \right)^2 \quad \Leftrightarrow \quad 1 > (N \cdot V)^2 + \left( \frac{c_1}{c_2} \right)^2.$$

In this case, $\cos(\theta_2)$ in Equation 17.9 is imaginary and the refracted vector $W$ does not exist. Be careful to test for total reflection, otherwise your ray tracing program may crash.

---

## 17.6  Summary

Recursive ray tracing is one of the most powerful tools in Computer Graphics for generating realistic images of virtual scenes. Shadows, reflections, and refractions are all modeled effectively by recursively tracing the paths of light rays through the scene. In recursive ray tracing, the total intensity at each surface point is given by the following recursive computation:

*Total Intensity*

$$I = I_{direct} + k_s I_{reflected} + k_t I_{refracted} \quad 0 \le k_s, k_t \le 1$$
$$I_{direct} = I_{ambient} + I_{diffuse} + I_{specular}.$$

To find the reflected and refracted rays at each surface point, let $N$ represent the unit normal vector at the surface point and let $V$ represent the unit vector from the surface point to the eye. In addition, let $c_1$ and $c_2$ denote the indices of refraction on opposite sides of the surface. Then we have the following formulas:

*Reflected Vector*

$$W = 2(V \cdot N)N - V.$$

*Refracted Vector*

$$W = \left( \frac{c_2}{c_1}(N \cdot V) - \cos(\theta_2) \right) N - \frac{c_2}{c_1}V,$$

$$\cos(\theta_2) = \sqrt{1 - \left( \frac{c_2}{c_1} \right)^2 (1 - (N \cdot V)^2)}.$$

*Total Reflection*

$$(N \cdot V)^2 + \left( \frac{c_1}{c_2} \right)^2 < 1.$$

To implement recursive ray tracing, we need to be able to calculate two things: surface normals and line–surface intersections. We shall take up the calculation of these items for a variety of different surfaces in the next two chapters.

---

## Exercises

17.1. Show that the reflection vector $W$ in Figure 17.3 can be computed in each of the following ways:

   a. By finding the mirror image of the vector $V$ in the plane perpendicular to $V_\perp$.

   b. By rotating the vector $V$ by $180°$ around the normal vector $N$.

   c. By finding the mirror image of the vector $V$ in the plane perpendicular to the normal vector $N$ and then negating the result.

17.2. Using the result of Exercise 17.1c, show that light following the law of the mirror—angle of incidence = angle of reflection—takes the shortest path from the light source through the mirror to the eye.

17.3. Consider the refraction vector $W$ in Figure 17.4.

   a. Show that $W = slerp(-N, -V, t)$, for $t = \theta_2/\theta_1$ (see Chapter 11, Section 11.6).

   b. Use the result of part a together with Snell's Law to derive Equation 17.9 for the refraction vector $W$.

17.4. Consider the refraction vector $W$ in Figure 17.4.

 a. Show that $W$ is the result of rotating the vector $-N$ thorough the angle $\theta_2$ about the axis vector $\frac{N \times V}{|N \times V|}$.

 b. Using part a, show that

$$W = \frac{\sin(\theta_2 - \theta_1)N - \sin(\theta_2)V}{\sin(\theta_1)}.$$

 c. Use the result of part b together with Snell's Law to derive Equation 17.9 for the refraction vector $W$.

17.5. Show that light following Snell's Law: $\frac{c_2}{c_1} = \frac{\sin(\theta_2)}{\sin(\theta_1)}$, where $c_1$ and $c_2$ are the speeds of light in the two mediums, takes the path of shortest time from the initial point to the eye.

 (Hint: Express the total time in terms of the constants $a, b, d, c_1, c_2$ and either the variable $x$ or the variables $\theta_1, \theta_2$ (see Figure 17.5) and then differentiate to show that the path of minimum time satisfies Snell's Law.)

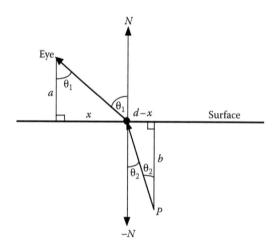

**FIGURE 17.5:** Refraction of light from a point $P$ to the eye. Here $a$ is the distance from the eye to the surface, $b$ is the distance from the point $P$ to the surface, and $d$ is the horizontal distance between the point $P$ and the eye. The value of the distance $x$ and the angles $\theta_1, \theta_2$ are the unknowns.

# Chapter 18

## Surfaces I: The General Theory

*Thou hast set all the borders of the earth*

– Psalm 74:17

## 18.1 Surface Representations

Recursive ray tracing is a technique for displaying realistic images of objects bordered by surfaces. To render surfaces via ray tracing, we require two essential procedures: we must be able to compute surface normals and we need to calculate ray–surface intersections.

There are four standard ways to represent surfaces in Computer Graphics: implicit equations, parametric equations, deformations of known surfaces, and specialized procedures. In this chapter, we shall review each of these general surface types and in each case explain how to compute surface normals and how to calculate ray–surface intersections. To facilitate further analysis, at the end of this chapter we shall also provide explicit formulas for mean and Gaussian curvature.

### 18.1.1 Implicit Surfaces

Planes, spheres, cylinders, and tori are examples of implicit surfaces. An *implicit surface* is the collection of all points $P$ satisfying an implicit equation of the form $F(P) = 0$. Typically, the function $F$ is a polynomial in the coordinates $x$, $y$, $z$ of the points $P$. For example, a unit sphere can be represented by the implicit equation:

$$F(x, y, z) \equiv x^2 + y^2 + z^2 - 1 = 0.$$

Generating lots of points along an implicit surface may be difficult because for complicated expressions $F$ it may be hard to find points $P$ for which $F(P) = 0$. On the other hand, determining if a point $P$ lies on an implicit surface is easy, since we need only check if $F(P) = 0$. Moreover, points on different sides of an implicit surface are distinguished by the sign of $F$; for closed surfaces, $F(P) < 0$ may indicate points on the inside, whereas $F(P) > 0$ may indicate points on the outside. This ability to distinguish inside from outside is important in solid modeling applications.

### 18.1.2 Parametric Surfaces

Planes, spheres, cylinders, and tori are also examples of parametric surfaces. A *parametric surface* is a surface represented by parametric equations—that is, to each point $P$ on the surface, we assign a pair of parameter values $s$, $t$ so that there is a formula $P = P(s, t)$ for computing points along the surface. In terms of rectangular coordinates, the equation $P = P(s, t)$ is equivalent to three parametric equations for the coordinates: $x = p_1(s, t)$, $y = p_2(s, t)$, $z = p_3(s, t)$. In Computer Graphics, the functions $p_1(s, t)$, $p_2(s, t)$, $p_3(s, t)$ are typically either polynomials or rational functions—ratios of polynomials—in the parameters $s, t$. For example, the unit sphere can be represented by the parametric equations:

$$x(s, t) = \frac{2s}{1 + s^2 + t^2}, \quad y(s, t) = \frac{2t}{1 + s^2 + t^2}, \quad z(s, t) = \frac{1 - s^2 - t^2}{1 + s^2 + t^2},$$

since it is easy to verify that

$$x^2(s, t) + y^2(s, t) + z^2(s, t) - 1 = 0.$$

Generating lots of points along a parametric surface is straightforward: simply substitute lots of different parameter values $s$, $t$ into the expression $P = P(s, t)$. Thus parametric surfaces are easy to display, so parametric surfaces are a natural choice for Computer Graphics. However, determining if a point $P$ lies on a parametric surface is not so simple, since it may be difficult to determine whether or not there are parameters $s$, $t$ for which $P = P(s, t)$.

### 18.1.3  Deformed Surfaces

An ellipsoid is a deformed sphere; an elliptical cone is a deformed circular cone. Any surface generated by deforming another surface is called a *deformed surface*. The advantage of deformed surfaces is that deformations often permit us to represent complicated surfaces in terms of simpler surfaces. This device can lead to easier analysis algorithms for complicated shapes. For example, we shall see in Chapter 19, Section 19.4.4 that ray tracing an ellipsoid can be reduced to ray tracing a sphere.

If the original surface has an implicit representation $F_{old}(P) = 0$ or a parametric representation $P = S_{old}(s, t)$, then the deformed surface also has an implicit representation $F_{new}(P) = 0$ or a parametric representation $P = S_{new}(s, t)$; moreover, we can compute $F_{new}(P)$ and $S_{new}(s, t)$ from $F_{old}(P)$ and $S_{old}(s, t)$. In fact, if $M$ is a nonsingular transformation matrix that maps the original surface into the deformed surface, then:

$$F_{new}(P) = F_{old}\left(P * M^{-1}\right), \tag{18.1}$$

$$S_{new}(s, t) = S_{old}(s, t) * M. \tag{18.2}$$

Equation 18.2 is easy to understand because $M$ maps points on $S_{old}$ to points on $S_{new}$. Equation 18.1 follows because $P$ is a point on the deformed surface if and only if $P * M^{-1}$ is a point on the original surface. Thus,

$$F_{new}(P) = 0 \Leftrightarrow F_{old}(P * M^{-1}) = 0.$$

Equations 18.1 and 18.2 are valid for any affine transformation $M$ of the original surface. Thus the map $M$ can be a rigid motion—a translation or a rotation—so we can use Equations 18.1 and 18.2 to reposition as well as to deform a surface.

### 18.1.4  Procedural Surfaces

Fractal surfaces are examples of procedural surfaces. Typically, fractals are not represented by explicit formulas; instead, fractals are represented either by recursive procedures or by iterated function systems. Any surface represented by a procedure instead of a formula is called a *procedural surface*. Often surfaces that blend between other surfaces are represented by procedures for generating the blend rather than by explicit formulas, so blends are another important class of procedural surfaces. Though procedural surfaces are important in Computer Graphics, ray tracing procedural surfaces requires specialized algorithms that depend on the particular procedure used to generate the surface. Since we are interested here in general ray tracing algorithms, we shall not have much to say about ray tracing for procedural surfaces.

## 18.2  Surface Normals

For implicit, parametric, and deformed surfaces, there are straightforward, explicit formulas for calculating normal vectors. Next we shall present these general formulas.

### 18.2.1  Implicit Surfaces

For an implicit surface $F(x, y, z) = 0$, the normal vector $N$ is given by the gradient—that is,

$$N = \nabla F = \left( \frac{\partial F}{\partial x}, \frac{\partial F}{\partial y}, \frac{\partial F}{\partial z} \right). \tag{18.3}$$

The proof that the gradient is normal to the surface is just the chain rule. Suppose that $P(t) = (x(t), y(t), z(t))$ is a parametric curve on the surface $F(x, y, z) = 0$. Then

$$F(x(t), y(t), z(t)) = 0.$$

Hence, by the chain rule,

$$\frac{\partial F}{\partial x} \frac{dx}{dt} + \frac{\partial F}{\partial y} \frac{dy}{dt} + \frac{\partial F}{\partial z} \frac{dz}{dt} = \frac{dF}{dt} = 0.$$

Therefore,

$$\underbrace{\nabla F}_{\text{Gradient}} \cdot \underbrace{P'(t)}_{\text{Derivative}} = \underbrace{\left( \frac{\partial F}{\partial x}, \frac{\partial F}{\partial y}, \frac{\partial F}{\partial z} \right)}_{\text{Normal vector}} \cdot \underbrace{\left( \frac{dx}{dt}, \frac{dy}{dt}, \frac{dz}{dt} \right)}_{\text{Tangent vector}} = 0,$$

so

$$N = \nabla F = \left( \frac{\partial F}{\partial x}, \frac{\partial F}{\partial y}, \frac{\partial F}{\partial z} \right).$$

### 18.2.2  Parametric Surfaces

For a parametric surface $P(s, t) = (x(s, t), y(s, t), z(s, t))$, the normal vector $N$ is given by the cross product of the partial derivatives of $P(s, t)$—that is,

$$N = \frac{\partial P}{\partial s} \times \frac{\partial P}{\partial t} = \left( \frac{\partial x}{\partial s}, \frac{\partial y}{\partial s}, \frac{\partial z}{\partial s} \right) \times \left( \frac{\partial x}{\partial t}, \frac{\partial y}{\partial t}, \frac{\partial z}{\partial t} \right). \tag{18.4}$$

To derive Equation 18.4, observe that if we fix $t = t_0$, then

$$P(s) = P(s, t_0) = (x(s, t_0), y(s, t_0), z(s, t_0))$$

is a curve on the surface $P(s,t)$, and the tangent to this curve is given by

$$\frac{\partial P}{\partial s} = \left( \frac{\partial x}{\partial s}, \frac{\partial y}{\partial s}, \frac{\partial z}{\partial s} \right).$$

Similarly, if we fix $s = s_0$, then

$$P(t) = P(s_0, t) = (x(s_0, t), y(s_0, t), z(s_0, t))$$

is another curve on the surface $P(s,t)$, and the tangent to this curve is given by

$$\frac{\partial P}{\partial t} = \left( \frac{\partial x}{\partial t}, \frac{\partial y}{\partial t}, \frac{\partial z}{\partial t} \right).$$

Since the normal to the surface is orthogonal to the surface, the normal vector must be perpendicular to the tangent vector for every curve on the surface. Therefore, since the cross product of two vectors is orthogonal to the two vectors,

$$N = \frac{\partial P}{\partial s} \times \frac{\partial P}{\partial t} = \left( \frac{\partial x}{\partial s}, \frac{\partial y}{\partial s}, \frac{\partial z}{\partial s} \right) \times \left( \frac{\partial x}{\partial t}, \frac{\partial y}{\partial t}, \frac{\partial z}{\partial t} \right).$$

### 18.2.3   Deformed Surfaces

Consider a surface $S_{new}$ that is the image of another surface $S_{old}$ under a nonsingular $4 \times 4$ affine transformation matrix $M$. Recall that if

$$M = \begin{pmatrix} M_u & 0 \\ w & 1 \end{pmatrix},$$

then only the upper $3 \times 3$ linear transformation matrix $M_u$ affects tangent vectors to the surface, since vectors are unaffected by translation. Thus, if $v_{old}$ is tangent to $S_{old}$, then the corresponding tangent $v_{new}$ to the surface $S_{new}$ is given by

$$v_{new} = v_{old} * M_u.$$

For rotations, normal vectors transform in the same way as tangent vectors, since rigid motions preserve orthogonality. Similarly, uniform scaling preserves orthogonality, since uniform scaling preserves angles even though scaling does not preserve length. Thus, for all conformal transformations—that is, for all composites of rotation, translation, and uniform scaling—the normal vector transforms in the same way as the tangent vector. Hence, for conformal maps,

$$v_{new} = v_{old} * M_u$$

$$N_{new} = N_{old} * M_u.$$

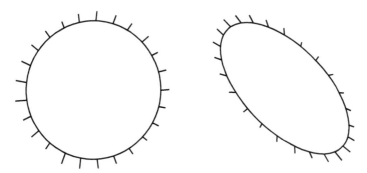

**FIGURE 18.1:** A circle with its normal vectors, and the circle transformed into an ellipse by scaling nonuniformly by the factor 1/2 along the minor axis. Notice that, in general, nonuniform scaling does not map normal vectors to the circle to normal vectors to the ellipse.

Nevertheless, perhaps surprisingly, the general transformation rule for normal vectors is different from the general transformation rule for tangent vectors, since nonconformal maps such as nonuniform scaling do not preserve orthogonality (see Figure 18.1). For arbitrary transformations $M$, if we know the normal vector $N_{old}$ to the surface $S_{old}$, then we can compute the corresponding normal vector $N_{new}$ to the surface $S_{new}$ by the formula:

$$N_{new} = N_{old} * M_u^{-T}.$$

For rotations, the transpose is equivalent to the inverse, so for rotations $M_u^{-T} = M_u$. Similarly, if $M_u$ is uniform scaling by the scale factor $s$, then $M_u^{-T}$ is uniform scaling by the scale factor $1/s$. However, for arbitrary transformations, $M_u^{-T}$ can be very different from $M_u$.

The general transformation formula for the normal vector is a consequence of the following observations. Since $N_{old}$ is orthogonal to $S_{old}$, we know that the normal vector $N_{old}$ is perpendicular to the tangent vector $v_{old}$. Therefore, by the properties of the dot product,

$$v_{old} * N_{old}^T = v_{old} \cdot N_{old} = 0.$$

Now we seek a vector $N_{new}$ such that

$$v_{new} * N_{new}^T = v_{new} \cdot N_{new} = 0.$$

But we can easily verify that if $N_{new} = N_{old} * M_u^{-T}$, then $v_{new} * N_{new}^T = 0$ because

$$v_{new} * N_{new}^T = (v_{old} * M_u) * (N_{old} * M_u^{-T})^T = (v_{old} * M_u) * (M_u^{-1} * N_{old}^T) = v_{old} * N_{old}^T = 0.$$

To summarize: using affine coordinates, points and tangent vectors transform by the same rule:

$$(P_{new}, 1) = (P_{old}, 1) * M,$$

$$(v_{new}, 0) = (v_{old}, 0) * M,$$

but tangent vectors and normal vectors transform by different rules:

$$v_{new} = v_{old} * M_u,$$

$$N_{new} = N_{old} * M_u^{-T}.$$

Moreover, inserting a zero affine coordinate for normal vectors will not work because the fourth coordinate of $(N_{old}, 0) * M^{-T}$ is not zero. Thus, normal vectors are qualitatively different from tangent vectors. The single term *vector* is not rich enough to capture the full diversity of different kinds of vectors. Both in differential geometry and in modern physics, this distinction between tangent vectors and normal vectors is emphasized by calling tangent vectors *covariant vectors* and normal vectors *contravariant vectors*.

## 18.3 Ray–Surface Intersections

Ray tracing a surface requires computing the intersection points of an arbitrary line with the surface as well as calculating the corresponding parameter values along the line of these intersection points. For implicit, parametric, and deformed surfaces, there are straightforward algorithms for finding these intersection points along with their corresponding parameter values.

### 18.3.1 Implicit Surfaces

Many of the surfaces encountered in recursive ray tracing have implicit representations $F(P) = 0$. Let $L(t) = Q + tv$ be the parametric equation of a line. Then we can compute the parameters along the line of the intersection points of the ray and the implicit surface by solving the equation $F(L(t)) = 0$. Substituting the roots of this equation into the parametric equation $L(t) = Q + tv$ for the line yields the actual intersection points. Thus we have the following intersection algorithm:

*Ray–Surface Intersection Algorithm—Implicit Surfaces*

1. Solve $F(L(t)) = 0$.

   The roots $t = t_1, \ldots, t_n$ are the parameter values along the line of the actual intersection points.

2. Compute $R_i = L(t_i) = Q + t_i v$, $i = 1, \ldots, n$.

   The values $R_1, \ldots, R_n$ are the actual intersection points of the line and the surface.

The bottleneck in this algorithm is clearly Step 1, since the equation $F(L(t)) = 0$ may be difficult to solve. When the function $F$ is a polynomial in the coordinates $x$, $y$, $z$, then $F(L(t)) = 0$ is a univariate polynomial equation. For low degree polynomials (degree $< 5$), there are explicit formulas for the roots; for high degree polynomials, a numerical method can be used to solve this equation. Thus implicit polynomial surfaces are a natural choice for recursive ray tracing.

### 18.3.2 Parametric Surfaces

Ray tracing parametric surfaces is more difficult than ray tracing implicit surfaces. Let $S(u, v) = (x(u, v), y(u, v), z(u, v))$ be a parametric surface, and let $L(t) = P + tv$ be the parametric equation of a line. The line and the surface intersect whenever there are parameter values $t$, $u$, $v$ for which $S(u, v) = L(t)$. This equation really represents three equations, one for each coordinate, in three unknowns $t$, $u$, $v$. Thus to find the intersection points of the line and the surface, we must solve three equations in three unknowns. Substituting the solutions in $t$ into the parametric equation $L(t) = P + tv$ for the line or the solutions in $u,v$ into the parametric equations

$S(u, v) = (x(u, v), y(u, v), z(u, v))$ of the surface yields the actual intersection points. Thus we have the following intersection algorithm:

*Ray–Surface Intersection Algorithm—Parametric Surfaces*

1. Solve $S(u, v) = L(t)$ for the parameters $t, u, v$. That is, solve the three simultaneous equations:

$$x(u, v) = p_1 + tv_1, \quad y(u, v) = p_2 + tv_2, \quad z(u, v) = p_3 + tv_3.$$

The roots $t = t_1, \ldots, t_n$ are the parameter values along the line of the actual intersection points.

2. Compute $R_i = L(t_i) = P + t_i v$, $i = 1, \ldots, n$, or equivalently $R_i = S(u_i, v_i)$, $i = 1, \ldots, n$.

The values $R_1, \ldots, R_n$ are the actual intersection points of the line and the surface.

The bottleneck in this algorithm is again clearly Step 1, since the simultaneous equations $S(u, v) = L(t)$ may be difficult to solve. We can simplify from three equations in three unknown to two equations in two unknowns in the following fashion. Let $N_1, N_2$ be two linearly independent vectors perpendicular to $v$. Then $v \cdot N_1 = v \cdot N_2 = 0$. Therefore, dotting the equation $S(u, v) = L(t)$ with $N_1, N_2$ eliminates $t$, the coefficient of $v$. We are left with two equations in two unknowns:

$$(S(u, v) - P) \cdot N_1 = 0,$$

$$(S(u, v) - P) \cdot N_2 = 0,$$

which we now must solve for $u, v$. Even with this simplification, however, we must still solve two nonlinear equations in two unknowns. Thus, typically ray tracing for parametric surfaces, even for parametric polynomial surfaces, requires more sophisticated numerical root finding techniques than ray tracing for implicit surfaces.

### 18.3.3 Deformed Surfaces

Ray tracing deformed surfaces is often easy because the deformed surface is usually the image of a simple surface, which we already know how to ray trace. Suppose that $S_{new}$ is the image under a nonsingular transformation matrix $M$ of the surface $S_{old}$—that is, $S_{new} = S_{old} * M$. For ray tracing, the key observation is that intersecting a line $L$ with the surface $S_{new}$ is equivalent to intersecting the line $L * M^{-1}$ with the surface $S_{new} * M^{-1} = S_{old}$. More precisely, a line $L$ and the surface $S_{new}$ intersect at a point $P$ if and only if the line $L * M^{-1}$ and the surface $S_{new} * M^{-1} = S_{old}$ intersect at the point $P * M^{-1}$ Thus, for deformed surfaces, we have the following intersection algorithm:

*Ray–Surface Intersection Algorithm—Deformed Surfaces*

1. Transform the line $L$ by the matrix $M^{-1}$.

If $L(t) = P + tv$, then $L(t) * M^{-1} = P * M^{-1} + t(v * M^{-1})$.

2. Find the intersection points $Q_1, \ldots, Q_n$ and the corresponding parameter values $t = t_1, \ldots, t_n$ of the intersection of the line $L * M^{-1}$ and the surface $S_{old}$.

3. Compute $R_i = L(t_i) = P + t_i v$, $i = 1, \ldots, n$ or equivalently $R_i = Q_i * M$, $i = 1, \ldots, n$.

The values $R_1, \ldots, R_n$ are the actual intersection points of the line $L$ and the surface $S_{new}$, and the parameters $t = t_1, \ldots, t_n$ are the corresponding parameter values along the line $L$.

The main difficulty in this algorithm is Step 2, since we must know how to find the intersections of an arbitrary line with the surface $S_{old}$. This intersection algorithm works well whenever the surface $S_{old}$ is easy to ray trace. Notice that the parameter values of the intersection points, but not the intersection points themselves, are the same for the surface $S_{new}$ and the line $L$, and the surface $S_{old}$ and the line $L * M^{-1}$.

---

## 18.4   Mean and Gaussian Curvature

To fully analyze surfaces, in addition to expressions for the normal vectors curvature formulas are often required. For easy reference, we present here without proof formulas for the mean and Gaussian curvatures of implicit and parametric surfaces. Rigorous definitions and derivations are beyond the scope of this chapter. Formal mathematical foundations for surface curvature can be found in most classical books on differential geometry.

### 18.4.1   Implicit Surfaces

For an implicit surface $F(x, y, z) = 0$, let $\nabla F$ denote the gradient and $hess(F)$ denote the Hessian of $F$—that is, let

$$\nabla F = \left( \frac{\partial F}{\partial x} \quad \frac{\partial F}{\partial y} \quad \frac{\partial F}{\partial z} \right)$$

$$hess(F) = \begin{pmatrix} \dfrac{\partial^2 F}{\partial x^2} & \dfrac{\partial^2 F}{\partial x \partial y} & \dfrac{\partial^2 F}{\partial x \partial z} \\[2ex] \dfrac{\partial^2 F}{\partial x \partial y} & \dfrac{\partial^2 F}{\partial y^2} & \dfrac{\partial^2 F}{\partial y \partial z} \\[2ex] \dfrac{\partial^2 F}{\partial x \partial z} & \dfrac{\partial^2 F}{\partial y \partial z} & \dfrac{\partial^2 F}{\partial z^2} \end{pmatrix}.$$

In addition, let $hess^*(F)$ denote the adjoint of $hess(F)$—the matrix generated by replacing each entry of $hess(F)$ by its cofactor. Then the Gaussian curvature is given by

$$K = \frac{\nabla F * hess^*(F) * \nabla F^{\mathrm{T}}}{|\nabla F|^4} = -\frac{\mathrm{Det}\begin{pmatrix} hess(F) & \nabla F^{\mathrm{T}} \\ \nabla F & 0 \end{pmatrix}}{|\nabla F|^4}, \tag{18.5}$$

and the mean curvature is given by

$$H = \frac{|\nabla F|^2 \, Trace(hess(F)) - \nabla F * hess(F) * \nabla F^{\mathrm{T}}}{2|\nabla F|^3} = \nabla \cdot \left( \frac{\nabla F}{|\nabla F|} \right), \tag{18.6}$$

where $\nabla \cdot G = \dfrac{\partial G}{\partial x} + \dfrac{\partial G}{\partial y} + \dfrac{\partial G}{\partial z}$ denotes the divergence of $G$.

For planar implicit curves $F(x, y) = 0$, all of these formulas for the mean and Gaussian curvature reduce to the standard curvature formula for a planar curve.

## 18.4.2  Parametric Surfaces

For a parametric surface $P(s, t) = (x(s, t), y(s, t), z(s, t))$, define the first and second fundamental forms by

$$I = \begin{pmatrix} \dfrac{\partial P}{\partial s} \cdot \dfrac{\partial P}{\partial s} & \dfrac{\partial P}{\partial s} \cdot \dfrac{\partial P}{\partial t} \\[2ex] \dfrac{\partial P}{\partial s} \cdot \dfrac{\partial P}{\partial t} & \dfrac{\partial P}{\partial t} \cdot \dfrac{\partial P}{\partial t} \end{pmatrix}$$

$$II = \begin{pmatrix} \dfrac{\partial^2 P}{\partial s^2} \cdot N & \dfrac{\partial^2 P}{\partial s \partial t} \cdot N \\[2ex] \dfrac{\partial^2 P}{\partial s \partial t} \cdot N & \dfrac{\partial^2 P}{\partial t^2} \cdot N \end{pmatrix},$$

where

$$N(s, t) = \frac{\dfrac{\partial P}{\partial s} \times \dfrac{\partial P}{\partial t}}{\left| \dfrac{\partial P}{\partial s} \times \dfrac{\partial P}{\partial t} \right|}$$

is a unit normal to the surface. Then the Gaussian curvature

$$K = \frac{\mathrm{Det}(II)}{\mathrm{Det}(I)} \tag{18.7}$$

and the mean curvature

$$H = \frac{Trace(I * II^*)}{2\mathrm{Det}(I)}, \tag{18.8}$$

where $II^*$ is the adjoint of $II$.

## 18.4.3  Deformed Surfaces

If $M$ is a nonsingular transformation matrix that maps some original surface in implicit or parametric form into a deformed surface, then by Equations 18.1 and 18.2:

$$F_{new}(P) = F_{old}(P * M^{-1}),$$

$$S_{new}(u, v) = S_{old}(u, v) * M.$$

Since we already have formulas for the mean and Gaussian curvatures both for implicit and for parametric surfaces, we can calculate the mean and Gaussian curvature of deformed surfaces from their implicit or parametric equations—that is, from $F_{new}$ or $S_{new}$.

## 18.5  Summary

The three most common types of surfaces in Computer Graphics are implicit, parametric, and deformed surfaces. For each of these surface representations, we have explicit formulas for the normal vectors and curvatures as well as straightforward algorithms for calculating ray–surface intersections. We summarize these formulas and algorithms here for easy reference.

### 18.5.1 Implicit Surfaces

*Surface Representation*

$$F(x, y, z) = 0.$$

*Normal Vector*

$$N = \nabla F = \left( \frac{\partial F}{\partial x}, \frac{\partial F}{\partial y}, \frac{\partial F}{\partial z} \right).$$

*Hessian*

$$hess(F) = \begin{pmatrix} \dfrac{\partial^2 F}{\partial x^2} & \dfrac{\partial^2 F}{\partial x \partial y} & \dfrac{\partial^2 F}{\partial x \partial z} \\[2mm] \dfrac{\partial^2 F}{\partial x \partial y} & \dfrac{\partial^2 F}{\partial y^2} & \dfrac{\partial^2 F}{\partial y \partial z} \\[2mm] \dfrac{\partial^2 F}{\partial x \partial z} & \dfrac{\partial^2 F}{\partial y \partial z} & \dfrac{\partial^2 F}{\partial z^2} \end{pmatrix}.$$

*Gaussian Curvature*

$$K = \frac{\nabla F * hess^*(F) * \nabla F^{\mathrm{T}}}{|\nabla F|^4} = -\frac{\mathrm{Det}\begin{pmatrix} hess(F) & \nabla F^{\mathrm{T}} \\ \nabla F & 0 \end{pmatrix}}{|\nabla F|^4}.$$

*Mean Curvature*

$$H = \frac{|\nabla F|^2 Trace(hess(F)) - \nabla F * hess(F) * \nabla F^{\mathrm{T}}}{2|\nabla F|^3} = \nabla \cdot \left( \frac{\nabla F}{|\nabla F|} \right).$$

*Ray–Surface Intersection Algorithm*

1. Solve $F(L(t)) = 0$.

   The roots $t = t_1, \ldots, t_n$ are the parameter values along the line of the actual intersection points.

2. Compute $R_i = L(t_i) = P + t_i v$, $i = 1, \ldots, n$.

   The values $R_1, \ldots, R_n$ are the actual intersection points of the line and the surface.

### 18.5.2 Parametric Surfaces

*Surface Representation*

$$P(s, t) = (x(s, t), y(s, t), z(s, t)).$$

*Normal Vector*

$$N = \frac{\partial P}{\partial s} \times \frac{\partial P}{\partial s} = \left( \frac{\partial x}{\partial s}, \frac{\partial y}{\partial s}, \frac{\partial z}{\partial s} \right) \times \left( \frac{\partial x}{\partial t}, \frac{\partial y}{\partial t}, \frac{\partial z}{\partial t} \right).$$

*First Fundamental Form*

$$I = \begin{pmatrix} \dfrac{\partial P}{\partial s} \cdot \dfrac{\partial P}{\partial s} & \dfrac{\partial P}{\partial s} \cdot \dfrac{\partial P}{\partial t} \\ \dfrac{\partial P}{\partial s} \cdot \dfrac{\partial P}{\partial t} & \dfrac{\partial P}{\partial t} \cdot \dfrac{\partial P}{\partial t} \end{pmatrix}.$$

*Second Fundamental Form*

$$II = \begin{pmatrix} \dfrac{\partial^2 P}{\partial s^2} \cdot N & \dfrac{\partial^2 P}{\partial s \partial t} \cdot N \\ \dfrac{\partial^2 P}{\partial s \partial t} \cdot N & \dfrac{\partial^2 P}{\partial t^2} \cdot N \end{pmatrix}.$$

*Gaussian Curvature*

$$K = \frac{\text{Det}(II)}{\text{Det}(I)}.$$

*Mean Curvature*

$$H = \frac{\text{Trace}(I * II^*)}{2 \, \text{Det}(I)}.$$

*Ray–Surface Intersection Algorithm*

1. Solve $S(u,v) = L(t)$ for the parameters $t$, $u$, and $v$. That is, solve the three simultaneous equations:

$$x(u, v) = p_1 + tv_1, \quad y(u, v) = p_2 + tv_2, \quad z(u, v) = p_3 + tv_3,$$

or equivalently the two simultaneous equations

$$(S(u, v) - P) \cdot N_1 = 0,$$
$$(S(u, v) - P) \cdot N_2 = 0,$$

where $N_1$ and $N_2$ are two linearly independent vectors perpendicular to $v$, and $P = (p_1, p_2, p_3)$.

The roots $t = t_1, \ldots, t_n$ are the parameter values along the line of the actual intersection points.

2. Compute $R_i = L(t_i) = P + t_i v$, $i = 1, \ldots, n$, or equivalently $R_i = S(u_i, v_i)$, $i = 1, \ldots, n$.

The values $R_1, \ldots, R_n$ are the actual intersection points of the line and the surface.

### 18.5.3   Deformed Surfaces

*Surface Representation*

An implicit surface $F_{old}(P) = 0$ or a parametric surface $S_{old}(u, v) = (x(u, v), y(u, v), z(u, v))$ together with a deformation matrix $M = \begin{pmatrix} M_u & 0 \\ v & 1 \end{pmatrix}$.

*Implicit Representation*

$$F_{new}(P) = F_{old}\left(P * M^{-1}\right).$$

*Parametric Representation*

$$S_{new}(u, v) = S_{old}(u, v) * M.$$

*Normal Vector*

$$N_{new} = N_{old} * M_u^{-T}.$$

*Mean and Gaussian Curvatures*

From implicit or parametric representations.

*Ray–Surface Intersection Algorithm*

1. Transform the line $L$ by the matrix $M^{-1}$.

   If $L(t) = P + tv$, then $L(t) * M^{-1} = P * M^{-1} + t(v * M^{-1})$.

2. Find the intersection points $Q_1, \ldots, Q_n$ and the corresponding parameter values $t = t_1, \ldots, t_n$ of the intersection of the line $L(t) * M^{-1}$ and the surface $F_{old}$ or $S_{old}$.

3. Compute $R_i = L(t_i) = P + t_i v$, $i = 1, \ldots, n$, or equivalently $R_i = Q_i * M$, $i = 1, \ldots, n$. The values $R_1, \ldots, R_n$ are the actual intersection points of the line $L$ and the surface $F_{new}$ or $S_{new}$, and the parameters $t = t_1, \ldots, t_n$ are the corresponding parameter values along the line $L$.

---

## Exercises

18.1. Let $M$ be a nonsingular $3 \times 3$ matrix. Show that for all $u, v, M$

$$(u \times v) * M^{-T} = \frac{(u * M) \times (v * M)}{\text{Det}(M)}.$$

(Hint: Compute the dot product of both sides with $u * M$, $v * M$, $(u \times v) * M$.)

Explain why this result is plausible geometrically.

18.2. Let $M$ be a nonsingular $3 \times 3$ matrix, and let $M^*$ denote the adjoint of $M$.

  a. Give an example to show that:

$$(u \times v) * M \neq (u * M) \times (v * M).$$

  b. Show that:

  i. $M^* = \text{Det}(M)M^{-\text{T}}$.

  ii. $M^{**} = \text{Det}(M)M$.

  c. Conclude from Exercise 18.1 and part b that:

  i. $(u \times v) * M^* = (u * M) \times (v * M)$.

  ii. $(u \times v) * M = \dfrac{(u * M^*) \times (v * M^*)}{\text{Det}(M)}$.

18.3. Let $D(s) = (x(s), y(s), z(s))$ be a parametrized curve lying in a plane in three-space, and let $v = (v_1, v_2, v_3)$ be a fixed vector in three-space not in the plane of $D(s)$. The parametric surface:

$$C(s, t) = D(s) + tv,$$

is called the *generalized cylinder* over the curve $D(s)$. Compute the normal $N(s,t)$ at an arbitrary point on the cylinder $C(s,t)$.

18.4. Let $D(s) = (x(s), y(s), z(s))$ be a parametrized curve lying in a plane in three-space, and let $Q = (q_1, q_2, q_3)$ be a fixed point in three-space not in the plane of $D(s)$. The parametric surface:

$$C(s, t) = (1 - t)D(s) + tQ,$$

is called the *generalized cone* over the curve $D(s)$. Compute the normal $N(s, t)$ at an arbitrary point on the cone $C(s,t)$.

18.5. Let $U_1(s)$, $U_2(s)$ be two curves in three-space. The parametric surface:

$$R(s, t) = (1 - t)U_1(s) + tU_2(s),$$

is called the *ruled surface* generated by the curves $U_1(s)$, $U_2(s)$. Compute the normal $N(s,t)$ at an arbitrary point on the ruled surface $R(s,t)$.

18.6. Consider the surface defined by the implicit equation:

$$xyz - 1 = 0.$$

  a. Find the normal vector to this surface at the point $P = (2, 1, 0.5)$.

  b. Find the implicit equation of this surface after rotating the surface by $\pi/6$ radians around the $z$-axis and then translating the surface by the vector $v = (1, 3, 2)$.

18.7. Let $S$ be a sphere of radius $R$. The mean and Gaussian curvature of $S$ are given by

$$H = \frac{1}{R} \quad \text{and} \quad K = \frac{1}{R^2}.$$

Derive these curvature formulas in two ways:

a. From the implicit equation:

$$x^2 + y^2 + z^2 - R^2 = 0.$$

b. From the parametric equations:

$$x(s,t) = \frac{2Rs}{1+s^2+t^2}, \quad y(s,t) = \frac{2Rt}{1+s^2+t^2}, \quad z(s,t) = \frac{R(1-s^2-t^2)}{1+s^2+t^2}.$$

# Chapter 19

## Surfaces II: Simple Surfaces

*He hath put down the mighty from their seats, and exalted them of low degree.*

– Luke 1:52

### 19.1 Simple Surfaces

Some surfaces are easy to describe without equations. A *sphere* is the locus of points at a fixed distance from a given point; a *cylinder* is the locus of points at a fixed distance from a given line. Thus, a sphere can be represented simply by a center point and a radius; a cylinder by an axis line and a radius. For these simple surfaces, we shall see that we can compute surface normals and ray–surface intersections directly from their geometry, without resorting to implicit or parametric equations. Thus ray tracing these surfaces is particularly easy.

In this chapter, we will investigate ray tracing strategies for surfaces defined geometrically rather than algebraically. These surfaces include the plane and the natural quadrics—the sphere, the right circular cylinder, and the right circular cone. We shall also explore ray tracing for general quadric (second degree) surfaces as well as for the torus, which is a surface of degree 4. General surfaces of revolution will also be investigated here. These surfaces are among the simplest and most common surfaces in Computer Graphics, so they deserve special attention.

### 19.2 Intersection Strategies

Special strategies exist for ray tracing simple surfaces. One of the most common tactics is to apply a transformation to reduce the ray–surface intersection problem to a simpler problem either by projecting to a lower dimension, or by repositioning to a canonical location, or by rescaling to a more symmetric shape.

Consider intersecting a curve $C$ and a surface $S$. If a point $P$ lies on the intersection of the curve $C$ and the surface $S$, then for any affine or projective transformation $M$, the point $P * M$ lies on the intersection of the curve $C * M$ and the surface $S * M$. If the transformation $M$ is cleverly chosen, it is easier, in many cases, to intersect the curve $C * M$ with the surface $S * M$ rather than the original curve $C$ with the original surface $S$. Moreover, if the transformation $M$ is invertible, then

$$P \in C \cap S \Leftrightarrow P * M \in C * M \cap S * M.$$

Therefore, when $M$ is invertible, we can find the intersection points of the original curve $C$ and the original surface $S$ by computing the intersection points of the transformed curve $C * M$ with the transformed surface $S * M$ and then mapping back by $M^{-1}$. This method is essentially the deformation technique discussed in Chapter 18, Section 18.3.3, though $M$ may also be a rigid motion as well as a shear or a scale.

But what if the transformation $M$ is not invertible? For example, what if $M$ is an orthogonal or perspective projection? Remarkably, for parametric curves and surfaces, the same strategy can still work.

Consider intersecting a parametric curve $C = C(t)$ and a parametric surface $S = S(u,v)$. The curve $C$ and the surface $S$ intersect whenever there are parameter values $t^*, u^*, v^*$ for which

$$C(t^*) = S(u^*, v^*).$$

Applying an affine or projective transformation $M$ yields

$$C(t^*) * M = S(u^*, v^*) * M.$$

Thus if the original curve and surface intersect at the parameters $t^*, u^*, v^*$, so will the transformed curve and surface. The intersection points may change, but the parameter values are unaffected. This observation is the key to our intersection strategy.

If no new intersection points are introduced by applying the transformation $M$, then we can find the intersection points of the original curve and surface by solving for the parameters $t^*$ or $(u^*, v^*)$ of the intersection points of the transformed curve and surface and then substituting these parameter values back into the original parametric equations $C = C(t^*)$ or $S = S(u^*, v^*)$ to find the intersection points of the original curve and surface.

Moreover, even if we do introduce some new spurious intersection points by applying a singular transformation, we never lose any of the old intersection points. So, at worst, we would need to check whether the parameters representing intersection points on the transformed curve and surface also represent intersection points on the original curve and surface. But we can always verify potential intersections by substituting the intersection parameters back into the parametric equations of the original curve and surface and checking whether or not we get the same point on the curve and on the surface—that is, by verifying that $C(t^*) = S(u^*, v^*)$. We shall see many examples of this general transformation strategy for computing intersections throughout this chapter.

---

## 19.3   Planes and Polygons

A plane can be described by a single point $Q$ on the plane and a unit vector $N$ normal to the plane. A point $P$ lies on the plane if and only if the vector $P - Q$ is perpendicular to the normal vector $N$ or equivalently if and only if $N \cdot (P - Q) = 0$.

A line $L(t) = P + tv$ intersects the plane determined by the point $Q$ and the normal vector $N$ at the parameter value $t^*$, where

$$N \cdot (L(t^*) - Q) = N \cdot (P + t^*v - Q) = 0.$$

If $N \cdot v = 0$, then the line is parallel to the plane, so the line and plane do not intersect. Otherwise, solving for the parameter $t^*$ yields

$$t^* = \frac{N \cdot (Q - P)}{N \cdot v}. \tag{19.1}$$

Substituting the parameter $t^*$ back into the equation of the line, we find the intersection point

$$R = L(t^*) = P + t^*v. \tag{19.2}$$

Thus we have a simple algorithm to intersect a line with an infinite plane.

Often, however, we want to find the intersection of a line with a finite polygon, rather than with an infinite plane. For example, when we ray trace a cube, the faces of the cube are squares, not infinite planes. Even if a line intersects the plane containing the polygon, the line might not intersect the plane at a point inside the polygon.

For convex polygons like triangles and rectangles, there is a straightforward test for determining if a point $R$ on a plane of the polygon lies inside the polygon. Consider a closed convex polygon with vertices $P_i$, $i = 1, \ldots, m+1$, where $P_{m+1} = P_1$. Let $N_{i,i+1}$ be the vector normal to the edge $P_i P_{i+1}$ and pointing into the polygon (see Figure 19.1).

A point $R$ in the plane of the polygon lies inside the polygon if and only if

$$N_{i,i+1} \cdot (R - P_i) \geq 0, \quad i = 1, \ldots, m. \tag{19.3}$$

To find the normal vectors $N_{i,i+1}$, observe that if the vertices of the polygon are oriented counterclockwise with respect to the normal $N$ to the plane of the polygon, then

$$N_{i,i+1} = N \times (P_{i+1} - P_i), \quad i = 1, \ldots, m,$$

and Equation 19.3 reduces to

$$\text{Det}(N, P_{i+1} - P_i, R - P_i) \geq 0, \quad i = 1, \ldots, m; \tag{19.4}$$

if the vertices are oriented clockwise with respect to the normal $N$, then

$$N_{i,i+1} = (P_{i+1} - P_i) \times N, \quad i = 1, \ldots, m,$$

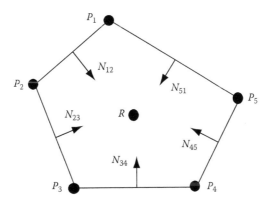

**FIGURE 19.1:** A point $R$ in the plane of a convex polygon lies inside the polygon if and only if $R$ lies on the positive side of the inward pointing normal vector to each boundary line.

and Equation 19.3 reduces to

$$\text{Det}(R - P_i, P_{i+1} - P_i, N) \geq 0, \quad i = 1, \ldots, m. \tag{19.5}$$

The tests in Equations 19.3 through 19.5 work only for convex polygons. For arbitrary, possibly non-convex, planar polygons, there are more general inside–outside tests based either on ray casting or on winding numbers. The details of these tests are provided in Chapter 11, Section 11.7. In most cases, however, we shall be interested only in convex polygons, so we shall not generally require these more sophisticated tests.

---

## 19.4   Natural Quadrics

The *natural quadrics* consist of the sphere, the right circular cylinder, and the right circular cone. These three second-degree surfaces are among the most common surfaces in Computer Graphics, so we shall develop special ray tracing algorithms for each of these surfaces. Moreover, for these three surfaces, we shall reduce the ray–surface intersection computation to a single algorithm for computing the intersection of a line and a circle.

### 19.4.1   Spheres

Next to planes, spheres are the simplest and most common surfaces in Computer Graphics. Therefore we would like to have a simple way to represent spheres as well as a fast, robust ray–sphere intersection algorithm.

A sphere $S$ can be represented simply by a center point $C$ and a scalar radius $R$. The vector $N$ from the center point $C$ to a point $P$ on the surface of the sphere $S$ is normal to the surface of the sphere at the point $P$. Therefore, the outward pointing normal vector $N$ at the point $P$ is given by

$$N_{sphere} = P - C.$$

To simplify the ray–sphere intersection calculation, we shall reduce the three-dimensional ray–sphere intersection problem to a two-dimensional ray–circle intersection problem. Suppose we want to intersect the line $L$ with the sphere $S$. Consider the plane containing the center point $C$ and the line $L$ (see Figure 19.2). This plane intersects the sphere $S$ in a circle with center $C$ and radius $R$, and the line $L$ intersects this circle in the same points that the line $L$ intersects the sphere.

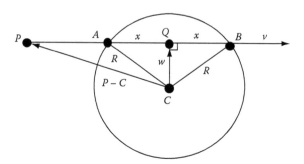

**FIGURE 19.2:**   A line determined by a point $P$ and a vector $v$ intersecting a circle with center $C$ and radius $R$.

### 19.4.1.1 Intersecting a Line and a Circle

To find the intersection points of a line and a circle, we can take a simple geometric approach. Let the line $L$ be determined by a point $P$ and a direction vector $v$. First we check the distance from the center point $C$ to the line $L$. If this distance is greater than the radius $R$, then there is no intersection between the line and the circle. Thus we have a fast test for rejection. Otherwise, we proceed by first finding the point $Q$ on the line $L$ closest to the center point $C$. We then use the Pythagorean Theorem to calculate the distance from $Q$ to the intersection points $A$, $B$ of the line and the circle. Here is pseudocode for this line–circle intersection algorithm.

*Line–Circle Intersection Algorithm*
 *Input*
  $C =$ circle center
  $R =$ circle radius
  $P =$ point on the line
  $v =$ vector in the direction of the line

 *Algorithm*
  If $Dist^2(C, L) > R^2$, there are no intersection points (fast reject).
  Otherwise
   Normalize the direction vector $v$ to a unit vector:

$$v \rightarrow \frac{v}{|v|}.$$

   Find the orthogonal projection $w$ of the vector $P - C$ on the line vector $v$:

$$w = (P - C)_\perp = (P - C) - ((P - C) \cdot v)v.$$

   Find the point $Q$ on the line $L$ closest to the center point $C$:

$$Q = C + w.$$

   Using the Pythagorean Theorem, find the distance $x$ from the point $Q$ to the intersection points $A$, $B$:

$$x^2 = R^2 - |w|^2.$$

   Compute the intersection points $A$, $B$:

$$A = Q - xv,$$
$$B = Q + xv.$$

### 19.4.1.2 Inversion Formulas for the Line

The preceding algorithm finds the intersection points of a line and a circle or equivalently a line and a sphere. But in ray tracing we also need to know the parameter values along the line corresponding to these intersection points. A formula for finding the parameter value corresponding to a point along a line is called an *inversion formula* for the line. The exact form of the inversion formula depends on the particular form of the parametric equation of the line.

Consider, for example, a line $L$ determined by a point $P$ and a vector $v$; then $L(t) = P + tv$. If $Q$ is a point on the line $L(t)$, then for some value $t$

$$Q = P + tv.$$

Subtracting $P$ from both sides, dotting with $v$, and solving for $t$ yields:

*Inversion Formula (Linear Parametrization)*

$$t = \frac{(Q - P) \cdot v}{v \cdot v}. \tag{19.6}$$

When $v$ is a unit vector, then $v \cdot v = 1$ and this inversion formula simplifies to:

*Inversion Formula (Linear Parametrization with Unit Vector $v$)*

$$t = (Q - P) \cdot v. \tag{19.7}$$

If the line $L$ is determined by two points $P_0$, $P_1$ rather than a point $P$ and a vector $v$, then

$$L(t) = (1 - t)P_0 + tP_1 = P_0 + t(P_1 - P_0).$$

Therefore, to find the parameter value corresponding to a point $Q$ on the line $L(t)$, we can use the first inversion formula (Equation 19.6) with $v = P_1 - P_0$.

Suppose, however, that $L$ is the image under perspective projection of a line in three-space. Perspective projection maps points and vectors into mass-points. Therefore if the original line is represented either by a point and a vector or by two points, then the image $L$ is represented by two mass-points $(X_0, m_0)$ and $(X_1, m_1)$. Here if $m_k \neq 0$, then $X_k = m_k P_k$, where $P_k$ is a point in affine space, whereas if $m_k = 0$, then $X_k = v_k$, where $v_k$ is a vector. In either case, since $L$ is determined by mass-points, to find the affine points on the projected line $P + tv \rightarrow (X_0, m_0) + t(X_1, m_1)$, we must divide by the mass. Thus $L$ has a rational linear parametrization:

$$L(t) = \frac{X_0 + tX_1}{m_0 + tm_1}.$$

To find an inversion formula for a line $L(t)$ with a rational linear parametrization, let $Q$ be a point on the line $L(t)$. Then for some value of $t$,

$$Q = \frac{X_0 + tX_1}{m_0 + tm_1}.$$

Multiplying both sides by $m_0 + tm_1$ yields

$$(m_0 + tm_1)Q = X_0 + tX_1$$

or equivalently

$$(m_0 Q - X_0) = t(X_1 - m_1 Q).$$

Dotting both sides with $X_1 - m_1Q$, and solving for $t$, we arrive at:

*Inversion Formula (Rational Linear Parametrization)*

$$t = \frac{(m_0Q - X_0) \cdot (X_1 - m_1Q)}{(X_1 - m_1Q) \cdot (X_1 - m_1Q)}. \tag{19.8}$$

If $m_0, m_1 \neq 0$, then $(X_k, m_k) = (m_k P_K, m_k)$ and this formula reduces to:

*Inversion Formula (Rational Linear Parametrization with $m_0, m_1 \neq 0$)*

$$t = \frac{m_0(Q - P_0) \cdot (P_1 - Q)}{m_1(P_1 - Q) \cdot (P_1 - Q)}. \tag{19.9}$$

We shall have occasion to apply these inversion formulas for rational linear parametrizations of the line in Section 19.4.3 when we ray trace the cone.

### 19.4.2 Cylinders

A sphere is the locus of points equidistant from a given point; a cylinder is the locus of points equidistant from a given line. Thus, a cylinder can be represented simply by an axis line $L$ and a scalar radius $R$. Typically the axis line $L$ is itself represented by a point $Q$ on the line and a unit direction vector $A$ parallel to the line (see Figure 19.3).

The normal $N$ to the cylinder at a point $P$ on the surface is just the orthogonal component of the vector $P - Q$ with respect to the axis vector $A$ (see Figure 19.3). Therefore,

$$N_{cylinder} = (P - Q)_\perp = (P - Q) - ((P - Q) \cdot A)A.$$

#### 19.4.2.1 Intersecting a Line and an Infinite Cylinder

As with the sphere, we can reduce the ray–cylinder intersection problem to a ray–circle intersection problem. The strategy here is to project both the cylinder and the line orthogonally onto the plane determined by the point $Q$ and the axis vector $A$. The cylinder projects to the circle with center $Q$ and radius $R$; the line projects to another line. Now we can use the algorithm in Section 19.4.1.1 to intersect the projected line with the circle. However, orthogonal projection is not

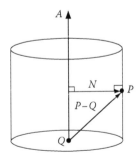

**FIGURE 19.3: (See color insert following page 354.)** A right circular cylinder with an axis line $L$ determined by a point $Q$ and a unit direction vector $A$. The normal vector $N$ to the surface of the cylinder at the point $P$ is the orthogonal component of the vector $P - Q$ with respect to the axis vector $A$.

an invertible map, so we cannot use the inverse map to recover the actual intersection points of the line and the cylinder.

What saves us here is our observation in Section 19.2 that the parameters of the intersection points of the projected line and the projected cylinder (the circle) are the same as the parameters of the intersection points of the unprojected line and the unprojected cylinder.

Thus to intersect a line and a cylinder, we can proceed in the following manner. Let the line $L$ be determined by a point $P$ and a unit direction vector $v$. If the distance from the cylinder axis to the line $L$ is greater than the cylinder radius $R$ or if the direction of the line vector $v$ is parallel to the direction of the cylinder axis vector $A$, then there is no intersection between the line and the cylinder. Thus, once again, we have fast tests for rejection. If the line and the cylinder do intersect, then we project the line $L$ orthogonally into the plane determined by the point $Q$ and the axis vector $A$, and we intersect the projected line with the projection of the cylinder, the circle with center $Q$ and radius $R$. We then apply the inversion formula for the projected line to find the corresponding parameter values. Finally, we substitute these parameter values into the equation of the original line $L$ to find the intersection points of the line and the cylinder. Here is the pseudocode for this line–cylinder intersection algorithm. Notice that this algorithm computes both the intersection points and the parameter values along the line corresponding to these intersection points.

*Line–Cylinder Intersection Algorithm (Infinite Cylinder)*
    *Input*
        $Q$ = point on the cylinder axis
        $A$ = unit direction vector parallel to the cylinder axis
        $R$ = cylinder radius
        $P$ = point on the line $L$
        $v$ = unit vector in the direction of the line $L$

    *Algorithm*
        If *Dist(Axis Line, L)* > $R$ or $v \| A$ ($|v \cdot A| \approx 1$), there are no intersection points (fast reject).
        Otherwise
            Project the line $L$ into the plane determined by the point $Q$ and the axis vector $A$, and find the equation of the projected line $L_\perp$:

$$L_\perp(t) = P_\perp + tv_\perp.$$

- $v_\perp = v - (v \cdot A)A$

- $P_\perp = P - ((P - Q) \cdot A)A$

Intersect the projected line $L_\perp$ with the circle with center $Q$ and radius $R$.
            Apply the line–circle intersection algorithm in Section 19.4.1.1 to find the intersection points $D_1, D_2$.
            Use the inversion formula in Section 19.4.1.2 for the line $L_\perp(t) = P_\perp + tv_\perp$ (or alternatively apply Exercise 19.8) to find the corresponding parameter values $t = t_1, t_2$:

$$t_i = \frac{(D_i - P_\perp) \cdot v_\perp}{v_\perp \cdot v_\perp} = \frac{(D_i - P) \cdot v_\perp}{v \cdot v_\perp}, \quad i = 1, 2.$$

Substitute the parameter values $t = t_1, t_2$ into the equation of the line $L$ to find the intersection points $R_1, R_2$ of the line and the cylinder:

$$R_i = L(t_i) = P + t_i v, \quad i = 1, 2.$$

### 19.4.2.2 Intersecting a Line and a Bounded Cylinder

The preceding algorithm intersects a line with an infinite cylinder, but most of the cylinders that appear in Computer Graphics are closed and bounded. Thus additional computations are needed to find the intersections of the ray with the disks at the top and the bottom of the cylinder. Fortunately these computations are easy to perform.

To intersect a line with a bounded cylinder, first intersect the line with the planes of the two bounding disks; then test to see whether or not these intersections points lie within a distance of the cylinder radius $R$ to the centers of these disks. If both intersection points lie inside the bounding disks, then the line cannot intersect the surface of the bounded cylinder; otherwise use the algorithm in the preceding section to find the intersection points of the line with the infinite cylinder and discard those intersection points that are not within the bounds of the finite cylinder. Here is the pseudocode for the line–cylinder intersection algorithm for bounded cylinders.

*Line–Cylinder Intersection Algorithm (Bounded Cylinder)*

  *Input*

    $Q$ = point on the cylinder axis at the base of the cylinder
    $A$ = unit direction vector parallel to the cylinder axis
    $R$ = cylinder radius
    $H$ = cylinder height
    $P$ = point on the line $L$
    $v$ = unit vector in the direction of the line $L$

  *Algorithm*

    If the line $L$ is parallel to the plane determined by the point $Q$ and the unit normal vector $A$—that is, if $v \cdot A \approx 0$—then:

      If the distance from the line $L$ to the plane determined by the point $Q + (H/2)A$ and the normal vector $A$ is greater than $H/2$, then there are no intersections between the line $L$ and the bounded cylinder.

        If $|(P - (Q + (H/2)A)) \cdot A| > H/2$, then the line $L$ does not intersect the cylinder.

      Otherwise use the algorithm in Section 19.4.2.1 to intersect the line $L$ with the infinite cylinder and store the results.

    Otherwise intersect the line $L$ with the planes of the two disks bounding the cylinder.

      Use the results in Section 19.3 to find the intersection points $E_1$, $E_2$ of the line $L$ and the two planes determined by the points $Q$ and $Q + HA$ and the unit normal vector $A$.

    If the distance of these intersection points $E_1$, $E_2$ from the centers $Q, Q + HA$ of the corresponding disks are less than the cylinder radius $R$, then save these intersection points and stop.

        Test: $Dist^2(Q, E_1) < R^2$ and $Dist^2(Q + HA, E_2) < R^2$.

    Otherwise intersect the line $E(t) = E_1 + t(E_2 - E_1)$ with the infinite cylinder.

      Use the algorithm in Section 19.4.2.1 for intersecting a line and an infinite cylinder to find the intersection points $D_1$, $D_2$ and the corresponding parameter values $t_1$, $t_2$. Discard any intersection point $D_i$ with parameter value $t_i < 0$ or $t_i > 1$, $i = 1, 2$, since any intersections of the line $L$ with the bounded cylinder must lie between $E_1$ and $E_2$.

      Now there are two cases:

      Case 1: *Both points $E_1$, $E_2$ lie outside the corresponding disks.*

        Either both $D_1$ and $D_2$ lie on the bounded cylinder or the line $L$ does not intersect the bounded cylinder.

      Case 2: *One of the points $E_1$ or $E_2$ lies inside the corresponding disk and the other point lies outside the corresponding disk.*

        Exactly one of the points $D_1$ or $D_2$ lies on the bounded cylinder and one of the points $E_1$ or $E_2$ lies on one of the bounding disks.

Notice that in both cases we need to test only one of the parameters $t_1, t_2$ to see if the parameter lies in the interval $[0, 1]$. Whether we will need to keep or discard the other parameter is completely determined by the case; no further testing is required.

Warning: $E(t)$ and $L(t)$ represent the same line, but not the same parametrization of the line. Therefore, the parameter values of the intersection points of the cylinder and the line $E(t)$ are not the same as the parameters of the intersection points of the cylinder and the line $L(t)$. In ray tracing, to determine which surface a ray strike first, you need the parameters of the intersection points along the original line $L(t)$. Thus after you find the coordinates of the intersection points of the cylinder and the line $E(t)$, you need to compute the parameters of these intersection points along the line $L(t)$ by using the inversion formula for the line $L(t)$.

### 19.4.3    Cones

A cone can be represented simply by a vertex $V$, a unit axis vector $A$, and a cone angle $\alpha$ (see Figure 19.4). Notice that a cylinder is a special case of a cone, where the vertex $V$ is located at infinity.

The normal $N$ to the cone at a point $P$ on the surface is the sum of a component along the cone axis $A$ and a component along the vector $P - V$ from the cone vertex $V$ to the point $P$ (see Figure 19.4). By straightforward trigonometry, the length of the component of $N$ along $A$ is $\frac{|P - V|}{\cos(\alpha)}$. Therefore,

$$N_{cone} = (P - V) - \left\{ \frac{|P - V|}{\cos(\alpha)} \right\} A.$$

As with the cylinder, we can reduce the ray–cone intersection to a ray–circle intersection by projecting both the cone and the line onto the plane determined by the point $Q = V + A$ and the axis vector $A$. Here, however, instead of using orthogonal projection, we use perspective projection from the vertex of the cone. Under perspective projection, the cone projects to the circle with center $Q$ and, since $Dist(V, Q) = 1$, with radius $R = \tan(\alpha)$. The line projects to another line, but since perspective projection introduces mass, the projected line has a rational linear parametrization (see Section 19.4.1.2). Nevertheless, we can now proceed much as we did in the ray–cylinder intersection algorithm. We intersect the projected line with the circle to find the parameter values corresponding to the intersection points, and we substitute these parameter values back into the equation of the original line to find the intersection points of the line with the cone. Here is the pseudocode for

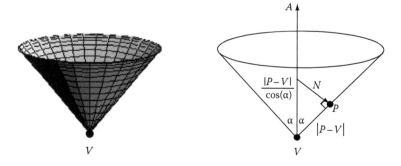

**FIGURE 19.4: (See color insert following page 354.)**   A right circular cone with a vertex $V$, a unit axis vector $A$, and a cone angle $\alpha$. The normal vector $N$ to the surface of the cone at the point $P$ can be decomposed into the sum of two components: one along the axis of the cone and the other along the vector from the cone vertex $V$ to the point $P$. By straightforward trigonometry, the length of the component of $N$ along $A$ is $|P - V| / \cos(\alpha)$.

this line–cone intersection algorithm. As usual, this algorithm computes both the intersection points and the parameter values along the line corresponding to these intersection points.

*Line–Cone Intersection Algorithm (Double Infinite Cone)*

    *Input*

        $V =$ cone vertex

        $A =$ unit direction vector parallel to the cone axis

        $\alpha =$ cone angle

        $P =$ point on the line $L$

        $u =$ unit vector in the direction of the line $L$

    *Algorithm*

    Project the line $L$ into the plane determined by the point $Q = V + A$ and the normal vector $A$, and find the rational parametrization of the projected line $L^*(t)$:

$$L^*(t) = \frac{X_0 + tX_1}{m_0 + tm_1},$$

$$X_0 = ((V - Q) \cdot A)P + ((Q - P) \cdot A)V = -P + ((Q - P) \cdot A)V$$

$$m_0 = (V - P) \cdot A = (Q - P) \cdot A - 1$$

$$X_1 = ((V - Q) \cdot A)u - (u \cdot A)V = -(u + (u \cdot A)V)$$

$$m_1 = -u \cdot A$$

Intersect the projected line $L^*$ with the circle with center $Q = V + A$ and radius $R = \tan(\alpha)$. Apply the line–circle intersection algorithm from Section 19.4.1.1 to find the intersection points $D_1, D_2$. There are four cases to consider, depending on the masses $m_0, m_1$:

    Case 1: $m_0 = m_1 = 0$. The line $L$ projects to the line at infinity, so there are no intersection points between the circle and the line $L^*(t)$.

    Case 2: $m_0, m_1 \neq 0$. The line $L^*(t)$ is determined by the two affine points $\frac{X_0}{m_0}, \frac{X_1}{m_1}$ or equivalently by the point $P = \frac{X_0}{m_0}$ and the vector $v = \frac{X_0}{m_0} - \frac{X_1}{m_1}$.

    Case 3: $m_0 \neq 0, m_1 = 0$. The line $L^*(t)$ is determined by the affine point $P = \frac{X_0}{m_0}$ and the vector $v = X_1$.

    Case 4: $m_0 = 0, m_1 \neq 0$. The line $L^*(t)$ is determined by the affine point $P = \frac{X_1}{m_1}$ and the vector $v = X_0$.

Use the inversion formula for the rational linear parametrization of the line $L^*(t)$ (or alternatively apply Exercise 19.8) to find the corresponding parameter values $t = t_1, t_2$:

$$t_i = \frac{(m_0 D_i - X_0) \cdot (X_1 - m_1 D_i)}{(X_1 - m_1 D_i) \cdot (X_1 - m_1 D_i)}, \quad i = 1, 2.$$

Substitute the parameter values $t = t_1, t_2$ into the equation of the line $L$ to find the intersection points $R_1, R_2$ of the line and the cone:

$$R_i = L(t_i) = P + t_i v, \quad i = 1, 2.$$

This algorithm intersects a line with a double infinite cone. To intersect a line with a finite cone bounded by two disks (or one disk and the cone vertex), proceed as in the algorithm in Section 19.4.2.2 for intersecting a line with a bounded cylinder. First intersect the line with the planes of the two bounding disks; then test to see whether or not these intersections points lie inside these disks. If both intersection points lie inside the bounding disks, then the line cannot intersect the surface of the finite cone (here we assume that the two bounding disks are either both above a both below the vertex of the cone); otherwise use the preceding algorithm to find the intersection points of the line

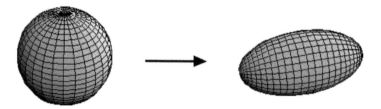

**FIGURE 19.5: (See color insert following page 354.)** The ellipsoid is the image of the sphere under nonuniform scaling.

with the infinite cone and discard those intersections that are not within the bounds of the finite cone. Details are left as an exercise for the reader (see Exercise 19.3).

### 19.4.4 Ellipsoids, Elliptical Cylinders, and Elliptical Cones

The ellipsoid is a scaled sphere; the elliptical cylinder and the elliptical cone are scaled variants of the right circular cylinder and right circular cone. One simple strategy for ray tracing these elliptical surfaces is to apply nonuniform scaling to transform these surfaces to the corresponding natural quadrics, perform the surface normal and ray–surface intersection calculations on these natural quadrics, and then map the normal vectors and the ray–surface intersection points back to the elliptical quadrics.

For example, the ellipsoid can be represented by a center point $C$, three orthogonal unit axis vectors $u$, $v$, $w$, and three scalar lengths $a$, $b$, $c$. Scaling nonuniformly about the center point $C$ by $a$ in the $u$ direction, $b$ in the $v$ direction, and $c$ in the $w$ direction maps the sphere with center $C$ and unit radius to the ellipsoid (see Figure 19.5). The inverse transformation that maps the ellipsoid to the unit sphere scales nonuniformly about the center point $C$ by $1/a$ in the $u$ direction, $1/b$ in the $v$ direction, and $1/c$ in the $w$ direction.

To find the normal vector to the ellipsoid at an arbitrary point, let $M$ be the nonuniform scaling transformation that maps the sphere to the ellipsoid, and let $N_E$, $N_S$ denote the normal vectors to the ellipsoid and the sphere at corresponding points. We know for any point $P$ on the sphere how to compute $N_S = P - C$. Now we can easily compute $N_E$, since by the results in Chapter 18, Section 18.2.3,

$$N_E = N_S * M_u^{-T},$$

where $M_u$ is the upper $3 \times 3$ submatrix of $M$. Note that in this case $M_u^{-T} = M_u^{-1}$.

Intersection points between a ray $L$ and an ellipsoid $E$ are similarly easy to compute. Simply intersect the ray $L * M^{-1}$ with the unit sphere centered at $C$ using the algorithm in Section 19.4.1. If $R_1$, $R_2$ are the intersection points of $L * M^{-1}$ with the unit sphere, then $R_1 * M$, $R_2 * M$ are the intersection points of $L$ with the ellipsoid. Note that the parameter values of the intersection points are the same for the sphere and the ellipsoid.

Analogous strategies can be applied to find the normal vectors and the ray–surface intersection points for the elliptical cylinder and the elliptical cone by applying nonuniform scaling transformations to the right circular cylinder and right circular cone. We leave the details as simple exercises for the reader (see Exercises 19.2 and 19.4).

### 19.5 General Quadric Surfaces

A general quadric surface is the collection of all points satisfying a second-degree equation in $x$, $y$, $z$—that is, an implicit equation of the form:

$$Q(x, y, z) \equiv Ax^2 + By^2 + Cz^2 + 2Dxy + 2Eyz + 2Fxz + 2Gx + 2Hy + 2Iz + J = 0.$$

We can rewrite this equation in matrix notation. Let

$$P = (x, y, z, 1),$$

$$Q = \begin{pmatrix} A & D & F & G \\ D & B & E & H \\ F & E & C & I \\ G & H & I & J \end{pmatrix}.$$

Then it is easy to verify that

$$P * Q * P^{\mathrm{T}} = Ax^2 + By^2 + Cz^2 + 2Dxy + 2Eyz + 2Fxz + 2Gx + 2Hy + 2Iz + J.$$

Thus, in matrix notation, the equation of a quadric surface is

$$P * Q * P^{\mathrm{T}} = 0.$$

For this reason, we will typically represent quadric surfaces by symmetric $4 \times 4$ matrices $Q$.
   The normal vector to an implicit surface is parallel to the gradient. For a quadric surface

$$Q(x, y, z) = Ax^2 + By^2 + Cz^2 + 2Dxy + 2Eyz + 2Fxz + 2Gx + 2Hy + 2Iz + J,$$

the gradient

$$\nabla Q(x, y, z) = 2(Ax + Dy + Fz + G, Dx + By + Ez + H, Fx + Ey + Cz + I).$$

Therefore, the normal vector

$$N(x, y, z) \parallel (Ax + Dy + Fz + G, Dx + By + Ez + H, Fx + Ey + Cz + I).$$

Equivalently, in matrix notation,

$$N(x, y, z) \parallel P * Q_{4 \times 3}, \tag{19.10}$$

where $Q_{4 \times 3}$ consists of the first three columns of $Q$.
   To intersect a line $L(t) = P + tv$ with a quadric surface $Q$, we can use the quadratic formula to solve the second-degree equation:

$$L(t) * Q * L(t)^{\mathrm{T}} = 0.$$

The real roots of this equation represent the parameter values along the line $L(t)$ of the intersection points. The actual intersection points can be computed by substituting these parameter values into the parametric equation $L(t) = P + tv$ for the line.
   Affine and projective transformations map quadric surfaces to quadric surfaces. To find the image of a quadric surface $Q$ under a nonsingular transformation matrix $M$, recall from Chapter 18, Section 18.1.3 that for any implicit surface $F(P) = 0$, the equation of the transformed surface is

$$F_{new}(P) = F(P * M^{-1}). \tag{19.11}$$

For a quadric surface represented by a symmetric $4 \times 4$ matrix $Q$, Equation 19.11 becomes

$$P * Q_{new} * P^{\mathrm{T}} = \left(P * M^{-1}\right) * Q * \left(P * M^{-1}\right)^{\mathrm{T}} = P * \left(M^{-1} * Q * M^{-\mathrm{T}}\right) * P.$$

Therefore,

$$Q_{new} = M^{-1} * Q * M^{-\mathrm{T}} \qquad (19.12)$$

is the matrix representation of the transformed surface.

There are many different types of quadric surfaces. In addition to the natural quadrics—the sphere, the right circular cylinder, and the right circular cone—and their elliptical variants, there are also parabolic and hyperbolic cylinders, elliptical and hyperbolic paraboloids, and hyperboloids of one or two sheets. In Table 19.1, we list the equations of the different types of quadric surfaces in canonical position—that is, with center at the origin and axes along the coordinate axes.

Special ray tracing algorithms for the natural quadrics and their elliptical variants are presented in Section 19.4. The reason that these special ray tracing techniques are effective is that it is easy to represent the natural quadrics in arbitrary positions geometrically without resorting to equations. But it is not so natural to represent general quadric surfaces such as paraboloids and hyperboloids in arbitrary positions geometrically without resorting to equations, and even if such geometric representations were feasible, straightforward intersection algorithms are not readily available. Rather than start with a quadric surface in general position, typically one starts with a quadric surface in canonical position and then repositions the surface by applying a rigid motion $M$.

Now there are two ways we can proceed to ray trace a quadric surface in general position. We can compute the equation of the surface and the surface normal in general position from the equation of the surface in canonical position using Equations 19.10 and 19.12. We can then solve a general

**TABLE 19.1:**   Equations of quadric surfaces in canonical position.

| Types of Quadric Surfaces | Canonical Equation |
|---|---|
| Sphere | $x^2 + y^2 + z^2 - R^2 = 0$ |
| Right circular cylinder | $x^2 + y^2 - R^2 = 0$ |
| Right circular cone | $x^2 + y^2 - c^2 z^2 = 0$ |
| Ellipsoid | $\dfrac{x^2}{a^2} + \dfrac{y^2}{b^2} + \dfrac{z^2}{c^2} - 1 = 0$ |
| Elliptical cylinder | $\dfrac{x^2}{a^2} + \dfrac{y^2}{b^2} - 1 = 0$ |
| Elliptical cone | $\dfrac{x^2}{a^2} + \dfrac{y^2}{b^2} - z^2 = 0$ |
| Parabolic cylinder | $x^2 - 4py = 0$ |
| Hyperbolic cylinder | $\dfrac{x^2}{a^2} - \dfrac{y^2}{b^2} - 1 = 0$ |
| Elliptical paraboloid | $\dfrac{x^2}{a^2} + \dfrac{y^2}{b^2} - z = 0$ |
| Hyperbolic paraboloid | $\dfrac{x^2}{a^2} - \dfrac{y^2}{b^2} - z = 0$ |
| Hyperboloid of two sheets | $-\dfrac{x^2}{a^2} - \dfrac{y^2}{b^2} + \dfrac{z^2}{c^2} - 1 = 0$ |
| Hyperboloid of one sheet | $\dfrac{x^2}{a^2} + \dfrac{y^2}{b^2} - \dfrac{z^2}{c^2} - 1 = 0$ |

second-degree equation to find the intersection of the ray and the quadric surface. Alternatively, if we store the transformation matrix $M$ along with the equation of the surface in canonical position, then we can compute the normal to the surface in arbitrary position from the normal to the surface in canonical position from the formula:

$$N_{new} = N_{old} * M_u^{-T}.$$

The normal $N_{old}$ to the surface in canonical position is parallel to the gradient, and this gradient is easy to compute because the canonical equation is simple. Similarly, to intersect a line $L$ with a quadric surface in general position, we can intersect the line $L * M^{-1}$ with the quadric surface in canonical position and then map these intersection points back to the desired intersection points by the transformation $M$. Again since the equation of a quadric in canonical position is simple, the quadratic equation for the intersection of a line and a quadric in canonical position is also simple.

## 19.6   Tori

The torus is the next most common surface in Computer Graphics, after the plane and the natural quadrics. A torus can be represented geometrically by a center point $C$, a unit axis vector $A$, and two scalar radii $d, a$ (see Figure 19.6). Cutting the torus by the plane through $C$ orthogonal to $A$ generates two concentric circles with center $C$ on the surface of the torus. The scalar $d$ represents the radius of a third concentric circle midway between these two circles—that is, a circle inside the torus. We call this circle the *generator* of the torus. The surface of the torus consist of all points at a fixed distance $a$ from this generator. Thus, the scalar $a$ is the radius of the toroidal tube.

To find the implicit equation of the torus, project an arbitrary point $P$ on the torus orthogonally into the plane of the generator. Then the point $P$, the image point $Q$, and the point $R$ lying on the generator between $Q$ and $C$ form a right triangle whose hypotenuse has length $a$ (see Figure 19.6). Thus, by the Pythagorean Theorem,

$$|P - Q|^2 + |Q - R|^2 = a^2. \tag{19.13}$$

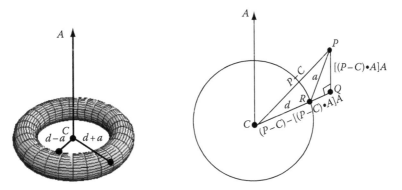

**FIGURE 19.6: (See color insert following page 354.)**   The torus. The circle at a distance $d$ from the center of the torus in the plane perpendicular to the axis $A$ is the generator for the torus. The radius of the inner tube is $a$.

But $|P - Q|$ is just the distance from the point $P$ to the plane determined by the center point $C$ and the unit axis vector $A$, which in turn is just the length of the parallel projection of the vector $P - C$ on $A$. Therefore,

$$|P - Q| = |(P - C) \cdot A|. \tag{19.14}$$

Moreover,

$$|Q - R| = |Q - C| - d, \tag{19.15}$$

and $|Q - C|$ is just the length of the perpendicular component of the vector $P - C$ with respect to $A$. Therefore,

$$|Q - C| = |(P - C)_\perp| = \sqrt{|P - C|^2 - [(P - C) \cdot A]^2}. \tag{19.16}$$

Substituting Equations 19.14 through 19.16 into Equation 19.13, we arrive at the following implicit equation for the torus in general position:

$$\left( \underbrace{(P - C) \cdot A}_{\text{Distance to plane}} \right)^2 + \left( \underbrace{\sqrt{|P - C|^2 - ((P - C) \cdot A)^2}}_{\text{Distance of projection to center}} - \underbrace{d}_{\text{Radius of generator}} \right)^2 = a^2. \tag{19.17}$$

Because of the square root, the degree of this equation is actually 4 in the coordinates $(x, y, z)$ of the point $P$, since to clear the square root we must square the equation.

We can also find the normal $N$ to the surface of the torus at the point $P$ by a geometric argument. Cut the torus by the plane determined by the point $P$ and the axis of the torus. The intersection of this plane and the torus consists of two circles; one of these circles passes through the point $P$. The center of this circle is a point $R$ on the generator of the torus, and the vector $P - R$ from $R$ to $P$ is orthogonal to the surface of the torus. Thus we need to find the point $R$.

To locate the point $R$, observe from Figure 19.6 that $R$ is on the line from $Q$ to $C$, and

$$Q - C = (P - C) - ((P - C) \cdot A)A.$$

Let $q$ be the distance from $Q$ to $C$. Then, by the Pythagorean Theorem,

$$q = |Q - C| = \sqrt{|P - C|^2 - ((P - C) \cdot A)^2}.$$

Thus, since $R$ lies on the generator of the torus,

$$R = C + (d/q)(Q - C).$$

Therefore the normal to the torus is given by

$$N \parallel (P - R) = (P - C) - (d/q)(Q - C). \tag{19.18}$$

To intersect a line $L(t) = P + tv$ with a torus, we simply substitute the parametric equations of the line into the implicit equation for the torus and solve the fourth-degree equation:

$$((L(t) - C) \cdot A)^2 + \left( \sqrt{|L(t) - C|^2 - ((L(t) - C) \cdot A)^2} - d \right)^2 = a^2. \tag{19.19}$$

The real roots of this equation represent the parameter values along the line of the intersection points. As usual, the actual intersection points can be computed by substituting these parameter values into the parametric equation $L(t) = P + tv$ for the line.

The implicit equation for the torus in general position is somewhat cumbersome. Often, as with general quadric surfaces, it is easier to begin with a torus in canonical position—with center at the origin and axis along the z-axis—and then apply a rigid motion to relocate the torus to an arbitrary position and orientation. This approach also works for an elliptical torus, since we can include nonuniform scaling along with the rigid motion.

To find the implicit equation of the torus in canonical position, substitute $C = (0, 0, 0)$ and $A = (0, 0, 1)$ into Equation 19.17, the implicit equation of the torus in general position. Squaring to eliminate the square root and simplifying the algebra, we arrive at the fourth-degree implicit equation for the torus in canonical position:

$$T(x, y, z) \equiv \left(x^2 + y^2 + z^2 - d^2 - a^2\right)^2 + 4d^2z^2 - 4a^2d^2 = 0. \tag{19.20}$$

The normal vector $N(x, y, z)$ to the surface $T(x, y, z)$ is given by the gradient. Therefore,

$$N(x, y, z) = \left(\frac{\partial T}{\partial x}, \frac{\partial T}{\partial y}, \frac{\partial T}{\partial z}\right) = \left(4xG(x, y, z), 4yG(x, y, z), 4z\left(G(x, y, z) + 2d^2\right)\right),$$

$$G(x, y, z) = x^2 + y^2 + z^2 - d^2 - a^2. \tag{19.21}$$

To reposition the torus so that the center is at $C$ and the axis is oriented along the direction $A$, rotate the z-axis to $A$ and then translate the origin to $C$. These transformations are given explicitly by the matrices (see Chapter 13, Sections 13.3.1, and 13.3.2):

$$Trans(C) = \begin{pmatrix} I & 0 \\ C & 1 \end{pmatrix},$$

$$Rot(u, \theta) = \cos(\theta)I + (1 - \cos(\theta))\left(u^T * u\right) + \sin(\theta)(u \times_-).$$

where $u = \dfrac{k \times A}{|k \times A|}$ and $\cos(\theta) = k \cdot A$.

Now we can find normal vectors on the surface of the torus in general position simply by transforming the corresponding normal vectors from canonical position. Thus,

$$N_{general\ position} = N_{cononical\ position} * Rot(u, \theta).$$

(Recall that $Rot(u, \theta)^{-T} = Rot(u, \theta)$, so we do not have to take the inverse transpose.) Similarly, we can intersect lines with the torus in general position by transforming the lines to canonical position via the inverse transformation:

$$M^{-1} = Trans(-C) * Rot(u, -\theta),$$

intersecting the transformed lines with the torus in canonical position, and then transforming the intersection points back to general position using the forward transformation matrix:

$$M = \begin{pmatrix} Rot(u, \theta) & 0 \\ Trans(C) & 1 \end{pmatrix}.$$

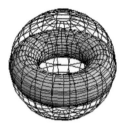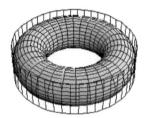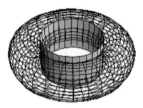

**FIGURE 19.7: (See color insert following page 354.)** A sphere (left) and a cylinder (center) bounding a torus on the outside, and a cylinder (right) surrounding the hole to the torus on the inside. Notice that the cylinder is a tighter fit than the sphere to the torus on the outside.

### 19.6.1 Bounding the Torus

Ray–torus intersections are expensive to compute because to find the intersection points of a line and a torus, we must solve a fourth-degree equation. However, the torus is a bounded surface, so many lines fail to intersect the torus. Thus to speed up our computations, we would like to have fast rejection tests to determine when a line does not intersect the torus. We can develop such fast tests by bounding the torus with simpler surfaces.

Consider a torus represented by a center point $C$, a unit axis vector $A$, and two scalar radii $d, a$ as in Figure 19.6. We can bound this torus on the outside by a sphere with center $C$ and radius $d + a$ (see Figure 19.7, left). Any line that fails to intersect this sphere will surely fail to intersect the torus. Since we have fast tests for determining whether or not a line intersects a sphere, this approach provides a fast rejection test when intersecting a line and a torus.

Nevertheless, lines that do intersect this sphere may still fail to intersect the torus by passing above or below the planes bounding the torus or by going through the hole at the center of the torus. To eliminate many of the lines passing through the hole in the torus, consider the cylinder of radius $d - a$, whose axis line is the axis of the torus and whose height above and below the center point $C$ is $a$ (see Figure 19.7, right). Any line that passes through the interior of this cylinder will fail to intersect the torus. To detect this case, consider the disks at the top and the bottom of this bounded cylinder. Any line that intersects both of these disks cannot intersect the surface of this bounded cylinder and hence cannot intersect the torus. Since we have fast tests for determining if a line intersects these disks (see Section 19.4.2.2), this approach provides a fast test for rejecting many of the lines that pass through the hole in the torus.

We can also use a cylinder to bound the torus on the outside. Consider the cylinder of radius $d + a$, whose axis line is the axis of the torus and whose height above and below the center point $C$ is $a$. Any line that fails to intersect this bounded cylinder will surely fail to intersect the torus (see Figure 19.7, center). To test whether or not a line intersects this bounded cylinder, we first compute the intersection points of the line with the surface of the infinite cylinder. If both intersection points lie above the plane bounding the cylinder from above or below the plane bounding the cylinder from below, then the line cannot intersect the bounded cylinder. Since there are fast tests for determining if a point lies above or below a plane, this approach provides another fast test for rejecting lines that fail to intersect the torus. This cylinder is often a tighter bound to the torus than the sphere, so this approach may speed up the ray tracing computation by eliminating more lines. Here is the pseudocode for the fast rejection tests for the line–torus intersection algorithm.

*Fast Rejection Tests for Line–Torus Intersection*
  *Input*
    $C$ = center point of the torus
    $A$ = unit direction vector parallel to the axis of the torus
    $d$ = radius of the generator of the torus

$a =$ tube radius
$P =$ point on the line $L$
$v =$ unit vector in the direction of the line $L$

*Algorithm*

If there are no intersections between the line $L$ and the sphere with center $C$ and radius $d + a$ bounding the torus, then there are no intersections between the line $L$ and the torus.

If $Dist^2(C, L) > (d + a)^2$, then the line $L$ does not intersect the torus.

Otherwise if the line $L$ is parallel to the plane determined by the point $C$ and the unit normal vector $A$—that is, if $v \cdot A \approx 0$—then:

If the distance from the line $L$ to the plane determined by the point $C$ and the normal vector $A$ is greater than $a$, there are no intersections between the line $L$ and the torus.

If $v \cdot A \approx 0$ and $|(P - C) \cdot A| > a$, then the line $L$ does not intersect the torus.

Otherwise, find the intersections $E_1, E_2$ of the line $L$ with the planes determined by the points $C - aA$ and $C + aA$ and the normal vector $A$ that bound the torus above and below.

If the points $E_1, E_2$ both lie inside the disks bounding the inner cylinder above and below, then there are no intersections between the line $L$ and the torus.

If $Dist^2(E_1, C - aA) < (d - a)^2$ and $Dist^2(E_2, C + aA) < (d - a)^2$, then the line $L$ does not intersect the torus.

Otherwise, if the points $E_1, E_2$ both lie outside the disks bounding the outer cylinder above and below:

If $Dist^2(E_1, C - aA) > (d + a)^2$ and $Dist^2(E_2, C + aA) > (d + a)^2$, then:

Find one intersection point $D$ and the corresponding parameter $t$ of the line $E(t) = E_1 + t(E_2 - E_1)$ with the cylindrical surface bounding the torus on the outside—that is, the infinite cylinder with radius $d + a$ and the same axis line as the torus.

If $t < 0$ or $t > 1$, then the line $L$ does not intersect the torus.

If these fast rejection tests all fail, then you will need to intersect the line with the torus. To intersect a line and a torus, you must solve Equation 19.19, an equation of degree 4 in the parameter $t$. There are two ways to proceed: either solve analytically or perform a binary search. Solving analytically requires using the formula for the roots of a fourth-degree equation; this formula can be found in standard algebra texts. To perform a binary search, first find the parameter values where the line intersects the bounding sphere or the bounding cylinder. Then using substitution, search for intermediate parameter values where the implicit equation (Equation 19.19) changes sign. An intersection point must occur between any two parameter values where Equation 19.19 takes on opposite signs. You will need to pick a tolerance at which to terminate the search. Note that for each line you may find anywhere between 0 and 4 intersection points.

## 19.7   Surfaces of Revolution

A *surface of revolution* is a surface generated by rotating a planar curve called the *directrix* about a straight line called the *axis of revolution*. Thus the cross sections of a surface of revolution by planes orthogonal to the axis of revolution are circles. Spheres, right circular cylinders, right circular cones, and tori are all examples of surfaces of revolution (see Table 19.2).

So far we have studied special surfaces using either a geometric approach or implicit equations. Surfaces of revolution, however, are modeled most easily using parametric equations.

**TABLE 19.2:** Directrix curves that are lines and circles together with the corresponding surfaces of revolution. The axis of revolution is the $z$-axis.

| Directrix | Surface of Revolution |
|---|---|
| Line: $D(u) = (1, 0, u)$ | Cylinder: $R(u, v) = (\cos(v), \sin(v), u)$ |
| Line: $D(u) = (u, 0, u)$ | Cone: $R(u, v) = (u \cos(v), u \sin(v), u)$ |
| Circle: $D(u) = (\cos(u), 0, \sin(u))$ | Sphere: $R(u, v) = (\cos(u) \cos(v), \cos(u) \sin(v), \sin(u))$ |
| Circle: $D(u) = (d + a \cos(u), 0, a \sin(u))$ | Torus: $R(u, v) = \begin{pmatrix} (d + a \cos(u)) \cos(v), \\ (d + a \cos(u)) \sin(v), a \sin(u) \end{pmatrix}$ |

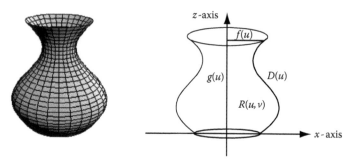

**FIGURE 19.8: (See color insert following page 354.)**   A surface of revolution $R(u, v)$ with a directrix $D(u)$ in the $xz$-plane. The axis of revolution is the $z$-axis.

Consider a parametric curve

$$D(u) = (f(u), 0, g(u)),$$

in the $xz$-plane. Rotating this curve about the $z$-axis generates the parametric surface

$$R(u, v) = (f(u) \cos(v), \ f(u) \sin(v), g(u)). \tag{19.22}$$

(see Figure 19.8). If we fix a parameter $u = u^*$, then the plane $z = g(u^*)$ intersects the surface $R(u, v)$ in the circle

$$C(v) = (f(u^*) \cos v, f(u^*) \sin v, g(u^*)),$$

with center $(0, 0, g(u^*))$ and radius $f(u^*)$, since $x^2 + y^2 = f^2(u^*)$. Thus cross sections of the surface $R(u, v)$ by planes perpendicular to the $z$-axis are indeed circles, so the parametric surface $R(u, v)$ truly is a surface of revolution. Table 19.2 provides some examples of simple directrix curves and the corresponding surfaces of revolution. Using this table together with Equation 19.22 allows us to generate simple parametric equations for cylinders, cones, spheres, and tori.

We are not restricted to surfaces of revolution whose axis of revolution is the $z$-axis. To extend our parametric equations to general position, we first rewrite the parametric equations of the directrix and the surface of revolution in vector notation:

$$D(u) = Origin + f(u)i + g(u)k,$$

$$R(u,v) = Origin + f(u) \cos(v)i + f(u) \sin(v)j + g(u)k.$$

Rotating the unit coordinate axis vectors $i, j, k$ to three arbitrary orthogonal unit vectors $C, B, A$ and then translating the origin to an arbitrary point $E$, we arrive at the parametric equations of the directrix and the surface of revolution in general position:

$$D(u) = E + f(u)C + g(u)A,$$

$$R(u, v) = E + f(u)\cos(v)C + f(u)\sin(v)B + g(u)A.$$

Notice that here the axis of revolution is parallel to the vector $A$, and the directrix lies in the plane of the vectors $A$ and $C$. The vectors $B$ and $C$ are orthogonal unit vectors perpendicular to the axis $A$.

There is not much difference between working in canonical position and working in general position for surfaces of revolution, since we can always transform from canonical position to general position by the rigid motion:

$$M = \begin{pmatrix} C & 0 \\ B & 0 \\ A & 0 \\ E & 1 \end{pmatrix}.$$

Therefore, for the remainder of this discussion, we will consider surfaces of revolution in canonical position, where the axis of revolution is the $z$-axis.

The normal vector to a parametric surface is parallel to the cross product of the partial derivatives (see Chapter 18, Section 18.2.2). For the surface of revolution in Equation 19.22,

$$R(u,v) = (f(u)\cos(v), f(u)\sin(v), g(u)),$$

so

$$\frac{\partial R}{\partial u} = (f'(u)\cos(v), f'(u)\sin(v), g'(u)),$$

$$\frac{\partial R}{\partial v} = (-f(u)\sin(v), f(u)\cos(v), 0).$$

Therefore,

$$N(u,v) \parallel \frac{\partial R}{\partial u} \times \frac{\partial R}{\partial v} = (-f(u)g'(u)\cos(v), -f(u)g'(u)\sin(v), f(u)f'(u)).$$

To ray trace a surface of revolution, we need to intersect a surface of revolution with straight lines. A surface of revolution:

$$R(u,v) = (f(u)\cos(v), f(u)\sin(v), g(u))$$

intersects a line:

$$L(t) = At + B = \left( A_x t + B_x, A_y t + B_y, A_z t + B_z \right),$$

at parameters where

$$R(u, v) = L(t),$$

or equivalently where

$$(f(u)\cos(v), f(u)\sin(v), g(u)) = (A_x t + B_x,\ A_y t + B_y,\ A_z t + B_z).$$

Thus to find the intersection of a line and a surface of revolution, we need to solve three simultaneous nonlinear equations with three unknowns $t$, $u$, $v$. The simplest of these three equations is the $z$-equation, since only two parameters $u$, $t$ appear in this equation. Solving the $z$-equation:

$$A_z t + B_z = g(u),$$

for $t$ yields

$$t = \frac{g(u) - B_z}{A_z},\qquad\qquad (19.23)$$

so we can eliminate one of the variables. (If $L(t)$ is horizontal, then $A_z = 0$ and we can solve $g(u) = B_z$ directly for $u$.) Next, summing the squares of the $x$ and $y$ equations and invoking the trigonometric identity:

$$\cos^2(v) + \sin^2(v) = 1,$$

yields

$$f^2(u) = (A_x t + B_x)^2 + (A_y t + B_y)^2.\qquad\qquad (19.24)$$

Since we already know $t$ as a function of $u$ from Equation 19.23, we can reduce Equation 19.24 to one unknown. We then solve Equation 19.24 for $u$, usually by numerical methods. The corresponding $t$ parameters can now be computed from Equation 19.23. Finally, taking the ratio of the $x$ and $y$ equations yields

$$\tan(v) = \frac{A_y t + B_y}{A_x t + B_x},$$

so

$$v = \tan^{-1}\left(\frac{A_y t + B_y}{A_x t + B_x}\right).\qquad\qquad (19.25)$$

The only bottleneck in this intersection algorithm is solving Equation 19.24. If $f(u)$, $g(u)$ are trigonometric functions, as is the case for the sphere and the torus in Table 19.2, then this equation contains transcendental functions and may be difficult to solve.

There is a trick, however, for replacing sines and cosines by rational functions. Let $s = \tan(u/2)$. Then,

$$\cos(u) = \cos^2(u/2) - \sin^2(u/2) = 2\cos^2(u/2) - 1 \Rightarrow \cos^2(u/2) = \frac{1 + \cos(u)}{2}$$

so

$$s^2 + 1 = \tan^2(u/2) + 1 = \sec^2(u/2) = \frac{1}{\cos^2(u/2)} = \frac{2}{1 + \cos(u)}.$$

Solving for cos (*u*), we find that

$$\cos(u) = \frac{1 - s^2}{1 + s^2} \quad \text{and} \quad \sin(u) = \frac{2s}{1 + s^2}.$$

Replacing sine and cosine with rational functions often leads to faster computations. Moreover, it is generally easier to solve polynomial equations than to solve trigonometric equations.

If the directrix is an unbounded curve, then the method outlined previously intersects a line with an unbounded surface of revolution. But just like cylinders and cones, most surfaces of revolution that appear in Computer Graphics are closed and bounded. To intersect a line with a surface of revolution bounded above and below by a pair of disks, we must discard any intersections that are above or below the planes that contain these disks. We must also intersect the line with the planes of the two bounding disks and then test to see whether or not these intersection points lie inside these disks. The details are similar to the ray tracing algorithms for the bounded cylinder and the bounded cone and are therefore left as an exercise for the reader (see Exercise 19.17).

## 19.8   Summary

In this chapter, we have concentrated on the simplest mathematical surfaces, those surfaces most common in Computer Graphics:

- Planes and convex polygons

- Natural quadrics and their elliptical variants

   – Spheres and ellipsoids

   – Right circular cylinders and elliptical cylinders

   – Right circular cones and elliptical cones

- General quadric surfaces both parabolic and hyperbolic

   – Parabolic cylinders and hyperbolic cylinders

   – Paraboloids and hyperboloids

- Tori and surfaces of revolution

For each of these surfaces, we have shown how to compute their surface normals as well as how to calculate their intersections with straight lines. With these results in hand, ray tracing these surfaces is a straightforward task.

In Chapter 20, we shall extend our discussion from surface modeling to solid modeling. Solids are bounded by surfaces, so many of the surfaces that we have discussed here will reappear in our discussion of solid modeling. Ray tracing, which is a powerful technique for rendering surfaces, will also reappear as a potent technique for analyzing solid models.

## Exercises

19.1. Show that a point $R$ in the plane of $\Delta P_1 P_2 P_3$ lies inside $\Delta P_1 P_2 P_3$ if and only if all three of the barycentric coordinates of $R$ with respect to $\Delta P_1 P_2 P_3$ are positive. (See Chapter 4, Exercise 4.24 for a definition of barycentric coordinates.)

19.2. Explain in detail how to ray trace an elliptical cylinder. In particular, explain how you would:

a. Model the elliptical cylinder.

b. Compute the surface normals to the elliptical cylinder.

c. Find ray–surface intersections for the elliptical cylinder.

Treat both the bounded and unbounded elliptical cylinder.

19.3. Develop an algorithm to intersect a straight line with a right circular cone bounded by:

a. Two disks.

b. One disk and the cone vertex.

19.4. Explain in detail how to ray trace an elliptical cone. In particular, explain how you would:

a. Model the elliptical cone.

b. Compute the surface normals to the elliptical cone.

c. Find ray–surface intersections for the elliptical cone.

Treat both the bounded and the unbounded elliptical cone.

19.5. Let $D(s) = (x(s), y(s), z(s))$ be a parametrized curve lying in a plane in three-space, and let $v = (v_1, v_2, v_3)$ be a fixed vector in three-space not in the plane of $D(s)$. The parametric surface

$$C(s, t) = D(s) + tv$$

is called the *generalized cylinder* over the curve $D(s)$ (see too Chapter 18, Exercise 18.3). Suppose that you already have an algorithm to compute the intersection points of any line in the plane of $D(s)$ with the curve $D(s)$. Based on this algorithm, develop a procedure to find the intersection points of an arbitrary line in three-space with the surface $C(s,t)$.

19.6. Let $D(s) = (x(s), y(s), z(s))$ be a parametrized curve lying in a plane in three-space, and let $Q = (q_1, q_2, q_3)$ be a fixed point in three-space not in the plane of $D(s)$. The parametric surface

$$C(s, t) = (1 - t)D(s) + tQ$$

is called the *generalized cone* over the curve $D(s)$ (see too Chapter 18, Exercise 18.4). Suppose that you already have an algorithm to compute the intersection points of any line in the plane of $D(s)$ with the curve $D(s)$. Based on this algorithm, develop a procedure to find the intersection points of an arbitrary line in three-space with the surface $C(s,t)$.

19.7. Let $U_1(s)$, $U_2(s)$ be two curves in three-space. The parametric surface

$$R(s, t) = (1 - t)U_1(s) + tU_2(s)$$

is called the *ruled surface* generated by the curves $U_1(s)$, $U_2(s)$ (see too Chapter 18, Exercise 18.5). Explain how you would find the intersection points of a ruled surface $R(s,t)$ with an arbitrary straight line.

19.8. Let $R(s) = R + su$ and $P(t) = P + tv$ be two intersecting lines in three-space. Let $v_\perp$ be the component of $v$ perpendicular to $u$.

a. Show that the lines $R(s)$ and $P(t)$ intersect at the parameter value

$$t = \frac{(R - P) \cdot v_\perp}{v \cdot v_\perp}.$$

b. Use the result of part a to find the parameter value where the line $L(t)$ and the cylinder intersect without resorting to the inversion formula for the line $L_\perp(t)$.

c. Use the result of part a to find the parameter value where the line $L(t)$ and the cone intersect without resorting to the inversion formula for the rational linear parametrization of the line $L^*(t)$.

19.9. Let $R = (x_0, y_0, z_0, 1)$ be a point on a quadric surface represented by a symmetric $4 \times 4$ matrix $Q$, and let $P = (x, y, z, 1)$. Show that the equation of the plane tangent to the quadric $Q$ at the point $R$ is given by $R * Q * P^T = 0$.

19.10. Find the symmetric $4 \times 4$ matrices representing each of the quadric surfaces in Table 19.1.

19.11. Compute the mean and Gaussian curvature (see Chapter 18, Section 18.4.1) for each of the quadric surfaces in Table 19.1.

19.12. Compute the mean and Gaussian curvature (see Chapter 18, Sections 18.4.1 and 18.4.2) for a torus in canonical position in two ways:

a. From the implicit equation.

b. From the parametric equation.

19.13. A *cyclide* is a surface whose lines of constant curvature are all either circles or straight lines. (The torus is a special case of a cyclide.) The equation of a cyclide in canonical position— centered at the origin with axes aligned along the coordinate axes—is

$$F(x, y, z) = (x^2 + y^2 + z^2 - d^2 + b^2)^2 - 4(ax - cd)^2 - 4b^2 y^2 = 0,$$

where $a$, $b$, $c$, $d$ are constants such that $c, d \geq 0$, $b > 0$, and $a^2 = b^2 + c^2$.

a. For the cyclide in canonical position, explain how you would:

i. Compute the normal vector at the point $(x, y, z)$.

ii. Find the intersection points with an arbitrary straight line.

b. For a cyclide in arbitrary position—a cyclide with center at the point $C$ and axes aligned along three mutually orthogonal unit vectors $u$, $v$, $w$—explain how you would:

   i. Model the cyclide.

   ii. Compute the normal to the cyclide at the point $(x, y, z)$.

   iii. Find the intersection points of the cyclide with an arbitrary straight line.

19.14. Consider a surface whose cross sections by planes perpendicular to a fixed line $L$ are ellipses, whose centers lie on the line $L$, and whose major axes are all parallel to each other. Explain how you would:

a. Model this surface.

b. Compute the normal to this surface at an arbitrary point on the surface.

c. Find the intersection of this surface with an arbitrary straight line.

19.15. Consider a surface of revolution

$$R(u, v) = (f(u) \cos (v), f(u) \sin (v), g(u)).$$

Suppose that there is a function $F(u)$ such that

$$F(g(u)) = f^2(u).$$

a. Show that the implicit equation of $R(u,v)$ is given by

$$x^2 + y^2 = F(z).$$

b. Find the function $F(u)$ for each of the surfaces of revolution in Table 19.2.

c. Using the results of part b, find the implicit equation for each of the surfaces of revolution in Table 19.2.

19.16. Consider the parametrizations in Table 19.2 of the cylinder, the cone, the sphere, and the torus.

a. Using the substitution $t = \tan (v/2)$, find parametrizations for the cylinder and the cone that do not involve trigonometric functions.

b. Using in addition the substitution $s = \tan (u/2)$, find parametrizations for the sphere and the torus that do not involve trigonometric functions.

19.17. Develop an algorithm to intersect a straight line with a surface of revolution bounded by two disks.

---

## Programming Projects

19.1. *Recursive Ray Tracing*

Implement a recursive ray tracer in your favorite programming language using your favorite API.

a. Include at least the following surfaces:

   i. Tetrahedra and cubes

   ii. Natural quadrics and their elliptical variants

   iii. Tori

b. Use several light sources along with the following lighting models:

   i. Ambient light

   ii. Diffuse reflection

   iii. Specular reflection

c. Incorporate both reflection and refraction.

d. Incorporate a spotlight (see the following description).

A *spotlight* is a focused point light source at a finite distance from the scene. The light cone of the spotlight is specified by the following parameters (see Figure 19.9):

   i. $C$ = cone vertex = location of light source

   ii. $A$ = cone unit axis vector

   iii. $\gamma$ = cone angle

The lighting formulas for a spotlight are the same as the lighting formulas for a point light source, except that the diffuse and specular components are attenuated by a factor $c_{spot}^e$. Thus for a spotlight:

   i. $I_{diffuse} = c_{spot}^e I_p k_d (L \cdot N)$

   ii. $I_{specular} = c_{spot}^e I_p k_s (R \cdot V)^n$

where
    $L$ is the direction to light source
    $N$ is the unit vector normal to the surface
    $V$ is the direction to eye

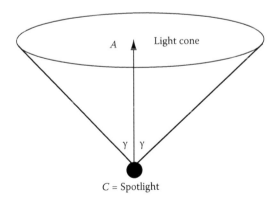

**FIGURE 19.9:** The light cone for a spotlight located at the point $C$ with cone angle $\gamma$ and cone axis vector $A$.

$R$ is the direction of reflection vector $= 2(L \cdot N)N - L$

$A$ is the spotlight cone axis vector

$\gamma$ is the spotlight cone angle

$$
\begin{aligned}
c_{spot} &= -L \cdot A & -L \cdot A \geq \cos\gamma \\
&= 0 & \text{otherwise}
\end{aligned}
$$

$e$ is the exponent controlling the falloff in intensity from the center of spotlight (user defined).

## 19.2. *Camera Obscura*

A *camera obscura* is the technical term for a pinhole camera. In a camera obscura, the film is behind the aperture. Placing the film behind the aperture is equivalent to placing the screen behind the eye, rather than in front of the eye.

a. Use recursive ray tracing to simulate a camera obscura.

b. How does the image differ from the image generated by standard ray tracing?

# Chapter 20

## Solid Modeling

*...a cubit on the one side, and a cubit on the other side*

– Exodus 26:13

## 20.1   Solids

A *solid* is a three-dimensional shape with two well-defined sides: an inside and an outside. Thus, for any point in space it is possible, at least in principle, to determine whether the point lies on the inside, or on the outside, or on the boundary of a solid.

Solid modeling allows us to represent more complicated shapes than surface modeling. With solid models, we can compute mass properties such as volume and moments of inertia; we can check for interference and detect collisions; we can also apply finite element analysis to calculate stress and strain for solid models.

Three types of solid modeling techniques are common in Computer Graphics: constructive solid geometry (CSG), boundary representations (B-Reps), and octrees. Constructive solid geometry relies heavily on ray tracing both for rendering and for analysis, so constructive solid geometry is closest to methods with which we are already familiar from surface modeling. Boundary representations often rely on parametric patches such as Bezier patches or B-spline surfaces, freeform surfaces which we shall study later in this book (see Chapters 24 through 28). In contrast, octrees are an efficient spatial enumeration technique, a method for approximating smooth three-dimensional shapes with a simple, compact collection of rectangular boxes. In this chapter we shall give a brief survey of each of these solid modeling methods, and we shall compare and contrast the advantages and disadvantages of each approach.

## 20.2   Constructive Solid Geometry (CSG)

In constructive solid geometry, we begin with a small collection of simple primitive solids—boxes, spheres, cylinders, cones, and tori (see Figure 20.1)—and we build up more complicated solids by applying boolean operations—union, intersection, and difference (see Figures 20.2 and 20.3)—to these primitive solids. For example, if we want to drill some holes in a block, then we can model this solid by subtracting a few cylinders from a box (see Figure 20.3).

Solids are stored in binary trees called *CSG trees*. The leaves of a CSG tree store primitive solids; the interior nodes store either boolean operations or nonsingular transformations. Nodes storing boolean operations have two children; nodes storing transformations have only one child. Boolean

**FIGURE 20.1: (See color insert following page 354.)** Primitives for a solid modeling system: a box, a sphere, a cylinder, a cone, and a torus.

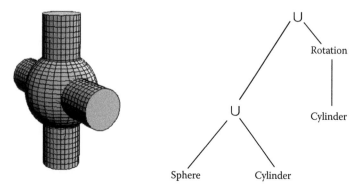

**FIGURE 20.2: (See color insert following page 354.)** A spherical tank with two cylindrical pipes (left), modeled by a CSG tree consisting of the union of a sphere and two cylinders (right).

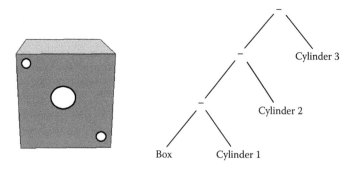

**FIGURE 20.3: (See color insert following page 354.)** A solid block with three cylindrical holes (left), modeled by a CSG tree that subtracts three cylinders from a box (right).

operations allow us to build up more complicated solids from primitive solids. Transformations allow us to reposition the solid from a canonical location to an arbitrary location or to rescale the solid from a symmetric shape to a more general shape.

   Each node of a CSG tree defines a solid. The solid defined by a leaf node is the primitive solid stored in the leaf. The solids at interior nodes are defined recursively. If the interior node stores a transformation, then the solid corresponding to the node is the solid generated by applying the transformation in the node to the solid stored by the node's child. If the interior node stores a boolean operation, then the solid corresponding to the node is the solid generated by applying the boolean operation in the node to the solids stored in the node's two children. The solid corresponding to the root of the tree is the solid represented by the CSG tree (see Figures 20.2 and 20.3).

Ray tracing can be applied to render solids represented by CSG trees. However, unlike ray tracing for surface models where only isolated intersection points are computed, ray tracing for solid models proceeds by finding the parameter intervals for which each line lies inside the solid. We need these parameter intervals for two reasons. First, even if a ray intersects a primitive solid in one of the leaf nodes, there is no guarantee that this intersection point lies on the boundary of the solid represented by the CSG tree, since we may in fact be subtracting this primitive from the solid. Keeping track of parameter intervals inside the solid instead of merely the intersection points on the surfaces bounding primitive solids allows us to keep track of which intersection points actually lie on the boundary of the solid. Second, later on we plan to use these parameter intervals to help compute mass properties such as the volume of a solid.

Thus to ray trace a solid, for each line from the eye through a pixel, we first compute the parameter intervals along the line for which the ray lies inside the solid. We then display the point corresponding to the parameter closest to the eye. Finding these parameter intervals for the primitive solids is usually straightforward, since these solids are typically chosen so that they are easy to ray trace. To find these parameter intervals for interior nodes, we must consider two cases. If the node stores a transformation $T$, then the parameter intervals for the node along the line $L$ are the same as the parameter intervals for the line $T^{-1}(L)$ and node's child; if the node stores a boolean operation, then the parameter intervals can be found by applying the boolean operation at the node to the parameter intervals for the node's two children. Pseudocode for ray tracing a CSG tree is presented in the following algorithm.

*Ray Tracing Algorithm for CSG Trees*
For each pixel:
    Construct a line $L$ from the eye through the pixel.
    If the solid is a primitive, compute all the intersections of the line with the primitive.
    Otherwise if the solid is a CSG tree:
        If the root stores a transformation $T$,
            Recursively find all the intervals where the line $T^{-1}(L)$ lies inside the root's child.
        Otherwise if the root stores a boolean operation:
            Recursively find all the intervals in which the line intersects the left and right subtrees.
            Combine these intervals using the boolean operation at the root.
    Display the closest intersection point.

Ray casting can also be applied to compute the volume of solids represented by CSG trees. Instead of firing a ray from the eye through each pixel, fire closely spaced parallel rays at the solid. Each ray represents a solid beam with a small cross-sectional area $\Delta A$. If we multiply the length of each interval $I_k$ inside the solid by $\Delta A$ and add the results, then we get a good approximation to the volume of the solid. Thus,

$$Volume \approx \sum_k Length(I_k)\Delta A.$$

Alternatively, we can apply Monte Carlo methods to compute the volume of a solid represented by a CSG tree. Monte Carlo methods are stochastic methods based on probabilistic techniques. Enclose the solid inside a rectangular box. If we select points at random, uniformly distributed inside the box, then some of the points will fall inside the solid and some of the points will fall outside the solid. The probability of a point falling inside the solid is equal to the ratio of the volume of the solid to the volume of the box. Thus,

$$\frac{\#\,Points\ inside\ Solid}{\#\,Points\ Selected} \approx \frac{Volume(Solid)}{Volume(Box)},$$

so

$$Volume(Solid) \approx \frac{\#\,Points\ inside\ Solid}{\#\,Points\ Selected} \times Volume(Box).$$

Therefore to find the volume of a solid, all we need is a test for determining if a point lies inside or outside the solid.

There are two ways to perform this inside–outside test: by ray casting or by analyzing the CSG tree recursively. To apply ray casting, observe that a point lies inside a solid if and only if an arbitrary ray emanating from the point intersects the solid an odd number of times; otherwise, if the ray intersects the solid an even number of times, then the point lies outside the solid. Now the same interval analysis we applied in the ray casting algorithm for rendering solids represented by CSG trees can be applied to determine the number of times a ray intersects a solid and hence too whether a point lies inside or outside a solid.

However, for CSG trees built from simple primitives, it is easier to determine whether a point lies inside or outside a solid by analyzing the CSG tree recursively. To begin, one can easily determine whether a point lies inside or outside simple primitives like boxes, spheres, cylinders, and cones by using the appropriate distance formulas (see Exercises 20.1 through 20.3). For example, a point $P$ lies inside a sphere if and only if the distance from the point $P$ to the center $C$ of the sphere is less than the radius $R$ of the sphere—that is, if and only if $|P - C|^2 < R^2$. For a solid such as a torus defined by an implicit equation $F(x, y, z) = 0$, if the gradient $\nabla F = (\partial F/\partial x, \partial F/\partial y, \partial F/\partial z)$ represents the outward pointing normal, then a point $P$ lies inside the solid if and only if $F(P) < 0$; if the gradient $\nabla F$ represents the inward pointing normal, then a point $P$ lies inside the solid if and only if $F(P) > 0$ (see Exercise 20.4).

Once we have an inside–outside test for each of the primitives, we can generate an inside–outside test for each CSG tree recursively in the following fashion. If the root stores a transformation $T$, then a point $P$ lies inside the corresponding solid if and only if the point $T^{-1}(P)$ lies inside the solid represented by the root's child. If the root stores a boolean operation, then to determine if a point $P$ lies inside the corresponding solid, first perform tests to determine if the point $P$ lies inside the solids represented by each of the root's two children and then apply the boolean operation at the root to the outcome of these two tests. Pseudocode for this inside–outside test on a CSG tree is presented in the following algorithm.

*Algorithm for Determining if a Point P Lies Inside or Outside a Solid Represented by a CSG Tree*
> If the solid is a primitive:
>> Use the appropriate distance formula or the implicit equation associated with the primitive to determine whether the point $P$ lies on the inside or the outside of the primitive.
>
> Otherwise if the solid is a CSG tree:
>> If the root stores a transformation $T$,
>>> Recursively determine if the point $T^{-1}(P)$ lies inside or outside the root's child.
>>
>> Otherwise if the root stores a boolean operation:
>>> Recursively determine if the point lies inside or outside the root's two children.
>>> There are three cases to consider:
>>> *Union*
>>> If $P$ lies inside the child at either node, then $P$ lies inside the solid.
>>> *Intersection*
>>> If $P$ lies inside the children at both nodes, then $P$ lies inside the solid.
>>> *Difference* $(A - B)$
>>> If $P$ lies inside the child at the left node $(A)$ and outside the child at the right node $(B)$, then $P$ lies inside the solid.
>>> Otherwise, in all three cases, $P$ lies outside the solid.

Constructive solid geometry based on CSG trees has many advantages. The representation is compact, the data structure is robust, and the user interface based on boolean operations is simple and natural. Transformation nodes in the CSG tree permit parameterized objects based on canonical positions or canonical shapes. The analysis of CSG trees is also straightforward: ray casting can be used for rendering, calculating volume, and testing whether points lie inside or outside a solid.

Nevertheless, constructive solid geometry also has certain disadvantages. There is no adjacency information in a CSG tree. That is, there is no information about which surfaces are next to a given surface. But adjacency information is crucial in manufacturing, especially for numerically controlled (NC) machining. Also, in a CSG model there is no direct access to the vertices and edges of the solid. Thus, it is difficult for a designer to select specific parts of an object as referents, nor can a designer tweak the model by adjusting the location of vertices and edges. Worse yet, it is hard to extract salient features of the solid such as holes and slots, which are important for manufacturing. Finally, the CSG representation of a solid is not unique. After subtracting a primitive, there is nothing to prevent us from unioning the primitive back into the solid. Thus it is difficult to determine if two CSG trees represent the same solid. Therefore even though CSG representations are simple, natural, and robust, other types of representations for solids have been developed to overcome some of the shortcomings of constructive solid geometry.

## 20.3 Boundary Representations (B-Rep)

A boundary representation for a solid stores the topology as well as the geometry of the solid. Geometry models position, size, and orientation; topology models adjacency and connectivity. Constructive solid geometry models the geometry of a solid, but the topology of a solid is not present in a CSG tree. Therefore we shall concentrate our discussion here on topology, which is the novel component of the boundary file representation.

Topology consists of vertices, edges, and faces along with pointers storing connectivity information describing which of these entities are either on or adjacent to other entities. An edge joins two vertices; a face is surrounded by one or more closed loops of edges. Topological information is binary data: a vertex $V$ is either on the edge $E$ or not on the edge $E$, a face $F_1$ is either adjacent to another face $F_2$ or not adjacent to the face $F_2$. Topology allows us to answer basic topological queries such as: Find all the edges adjacent to the edge $E$, or list all the faces surrounding the face $F$. This topological information is what is missing from CSG representations. Topology does not store any information about the position, size, and orientation of the vertices, edges, or faces; this numerical data is stored in the geometry.

The geometry of a solid consists of points, curves, and surfaces along with numerical data describing the position, size, and orientation of these entities: for each point, coordinates; for each curve or surface, data for generating implicit or parametric equations. Geometry is not concerned with adjacency or connectivity; this information is stored in the topology. Topology and geometry are tied together by pointers: vertices point to points, edges to curves, faces to surfaces. The data structure that encompasses the geometry, the topology, and the pointers from the topology to the geometry is called a *boundary file representation (B-Rep)* of a solid.

The topology of a solid can be quite elaborate. There are nine potential sets of pointers connecting vertices ($V$), edges ($E$), and faces ($F$) (see Figure 20.4). Maintaining this complete data structure would be complicated and time consuming. Typically fewer pointers are actually stored; trade-offs are made between the speed with which each topological query can be answered and the amount of space required to store the topology.

The *winged-edge* data structure is a standard representation for topology, which makes a reasonable compromise between time and space. Most of the pointers in a winged-edge data

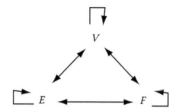

**FIGURE 20.4:** The nine potential sets of pointers for a topological data structure with vertices $V$, edges $E$, and faces $F$.

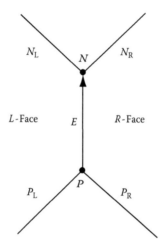

**FIGURE 20.5:** The winged-edge data structure. Every edge $E$ points to two vertices ($P$ and $N$), four edges ($P_L, P_R, N_L, N_R$), and two faces ($L$-Face and $R$-Face).

structure are stored with the edges. For solids bounded by two-dimensional manifolds, each edge lies on two faces, and each edge typically connects two distinct vertices. In the winged-edge data structure, edges are oriented and each edge points to:

2 vertices—previous ($P$) and next ($N$) (orientation),

2 faces—left ($L$) and right ($R$),

4 edges—$P_R$, $N_R$ and $P_L$, $N_L$.

In addition, each vertex points to one edge that contains the vertex, and each face points to one edge that lies on the face (see Figure 20.5).

With the winged-edge topology, we can answer all questions of the form: Is $A$ on or adjacent to $B$. Typical queries are:

Find all the edges surrounding a face.

Find all the faces adjacent to a face.

Find all the vertices lying on a face.

Find all the edges passing through a vertex.

We leave it as a straightforward exercise to the readers to convince themselves that these queries can all be answered quickly and easily from the data stored in the winged-edge topology (see Exercise 20.9). The main advantages of the winged-edge topology are that in addition to supporting fast

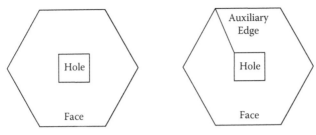

**FIGURE 20.6:**   A face with a hole. The face is surrounded by two loops (left). An auxiliary edge connects the two loops (right). Notice that the auxiliary edge has the same face on both sides.

retrieval of topological information, the winged-edge data structure is of moderate size and is relatively easy to maintain during boolean operations.

One problem with the winged-edge topology is that this data structure does not support faces with holes—that is, faces defined by one exterior loop and one or more interior loops. To solve this problem auxiliary edges are typically added to the model in order to connect the inner loops to the outer loop. These auxiliary edges have the same face on both sides (see Figure 20.6).

Boundary file representations are more complicated than CSG trees, so it is easy to make a mistake in the topology when updating a model during a boolean operation. One way to check the consistency of the topology in the boundary file is to verify Euler's formula.

Euler's formula for solids bounded by two-dimensional manifolds asserts that

$$V - E + F - H = 2(C - G), \tag{20.1}$$

where
   $V=$ the number of vertices
   $E=$ the number of edges
   $F=$ the number of faces
   $H=$ the number of holes in faces
   $C=$ the number of connected components
   $G=$ the number of holes in the solid (genus)

For example, for a cube, $V = 8$, $E = 12$, $F = 6$, $H = 0$, $C = 1$, and $G = 0$. Therefore,

$$V - E + F - H = 2 = 2(C - G).$$

If Euler's formula is not satisfied, then there is a error in the model. Moreover, it is unlikely (though possible) that an invalid model will actually satisfy Euler's formula. Therefore Euler's formula provides a robust check for the consistency of the topological model.

Topology plays no essential role in rendering. Therefore, ray tracing for solids represented by boundary files is essentially the same as ray tracing for surface models: simply intersect each ray from the eye with each surface on the boundary of the solid and display the closest intersection point.

Volume for solids with boundary file representations can be computed in three different ways: by ray casting, by Monte Carlo methods, and by the divergence theorem. The ray casting method for boundary file representations is similar to the ray casting method for constructive solid geometry. Fire closely spaced parallel rays at the solid, but instead of using the CSG tree, use the boundary file representation to determine the intervals $I_k$ along each line that lie inside the solid by intersecting each line with the surfaces that bound the solid. Each ray represents a solid beam with a small cross-sectional area $\Delta A$, so if we multiply the length of each interval $I_k$ inside the solid by $\Delta A$ and add the results, then we get a good approximation to the volume of the solid.

Alternatively, we can apply Monte Carlo methods to find the volume of a solid by enclosing the solid in a rectangular box. If we select points at random, uniformly distributed inside the box, then the probability of a point falling inside the solid is equal to the ratio of the volume of the solid to the volume of the box. We can use ray casting at a boundary file representation to determine if a point lies inside a solid: a ray emanating from a point lies inside a solid if and only if the ray intersects the surfaces bounding the solid an odd number of times.

The divergence theorem provides a novel technique for computing the volume of a solid with a boundary file representation, which is not available for solids represented by CSG trees. If the solid $V$ is bounded by the surfaces $S = \{S_k\}$ with outward pointing normals $N = \{N_k\}$, then by the divergence theorem:

$$Volume(V) = \frac{1}{3} \oiint_S (P \cdot N)dS = \frac{1}{3} \sum_k \iint_{S_k} (P \cdot N_k)dS_k, \qquad (20.2)$$

where $P = (x, y, z)$ represents an arbitrary point on the bounding surfaces. If the surfaces bounding the solid are sufficiently simple, then we can compute these integrals analytically (see Exercise 20.10); otherwise we can use quadrature methods to approximate these integrals.

Boolean operations, unlike rendering or volume computations, are much more difficult for boundary file representations than for solids represented by CSG trees because for boundary file representations not only the geometry but also the topology must be updated. Unlike constructive solid geometry where boolean operations are performed simply by updating a CSG tree, in a boundary file representation the curves and surfaces of the first solid must be intersected explicitly with the curves and surfaces of the second solid to form the new vertices, edges, and faces for the topology of the combined solid. These intersection computations are much more difficult than ray–surface intersections, so boolean operations on boundary file representations are much more difficult than boolean operations on CSG representations. Here is the pseudocode for performing boolean operations on boundary file representations.

*Boolean Operations on Boundary File Representations*
*Input*: Boundary file representations of two solids $A, B$.

*Algorithm*
Intersect each surface (face) of $A$ with each surface (face) of $B$ to form new curves (edges) on $A$ and $B$. {Difficult Computations}
Intersect each new curve with existing edges on the old faces to form new vertices and edges on $A$ and $B$.
Insert the new faces, edges, and vertices into the topology of $A$ and $B$.
{Update Topological Data Structures}
Combine the boundary topologies based on the particular boolean operation:

$$\partial(A \cup B) = (\partial A - \partial A_{inB}) \cup (\partial B - \partial B_{inA}),$$
$$\partial(A \cap B) = \partial A_{inB} \cup \partial B_{inA},$$
$$\partial(A - B) = (\partial A - \partial A_{inB}) \cup \partial B_{inA},$$
$$\partial(B - A) = (\partial A_{inB}) \cup (\partial B - \partial B_{inA}).$$

While this pseudocode may seem straightforward, the actual computations can be quite difficult. Computing accurate surface–surface intersections is hard; maintaining correct boundary files is time consuming as well as susceptible to numerical errors because to speed up the computations the intersections are typically calculated using floating point arithmetic. Many man years are generally required to code robust boolean operations on boundary file representations.

The main advantage of boundary file representations is that adjacency information is readily available. This adjacency information is important for manufacturing operations such as numerically controlled (NC) machining. Adjacency information also permits the designer to modify the model by tweaking vertices and edges, and to extract important features for manufacturing such as holes and slots. Moreover, unlike CSG representations, the boundary file representation is unique. Thus it is possible to decide whether two solids are identical by comparing their boundary file representations. Finally, we can verify the correctness of the topological model by checking Euler's formula.

The main disadvantage of the boundary file representation is that maintaining the topology requires maintaining a large and complicated data structure. Therefore boolean operations on boundary file representations are slow and cumbersome to compute. Moreover, floating point inaccuracies can cause disagreements between geometry and topology; tangencies are particularly hard to handle. Hence, it is difficult to maintain robust boundary file representations for solids bounded by curved surfaces.

Thus even though boundary file representations facilitate the solution of several important problems in design and manufacturing, boundary files also introduce additional problems of their own. Therefore, next we shall investigate yet another, much simpler type of representation for solids that has some advantages over boundary file representations.

## 20.4 Octrees

*Octrees* are an efficient spatial enumeration technique. In naive spatial enumeration, space is subdivided into a large collection of small disjoint cubes all of the same size and orientation—usually with faces parallel to the coordinate planes—and each solid is defined by a list of those cubes lying inside or on the boundary of the solid. The accuracy of spatial enumeration for solids bounded by curved surfaces depends on the size of the cubes used in the spatial enumeration, so there are trade-offs between the accuracy of the model and the space necessary to store the model.

To save on storage, octrees vary the size of the cubes so that lots of small cubes get coalesced into larger cubes. Large cubes are used to model the interior of the solid, while small cubes are used to model the boundary of the solid. The cubes in an octree are called *voxels*. In an octree representation of a solid the number of voxels is typically proportional to the surface area; most of the subdivision of large cubes occurs to capture the boundary of the solid. Octrees to model solids are generated by the following divide and conquer algorithm. Figure 20.7 illustrates a two-dimensional analogue of an octree, a quadtree for the area enclosed by a planar curve.

*Octree Algorithm*
Start with a large cube containing the solid. The faces of the large cube should be oriented parallel to the coordinate planes.
Divide this large cube into 8 octants.
    Label the cube in each octant as either $E$ = empty, $F$ = full, or $P$ = partially full, depending on whether the octant is outside, inside, or overlaps the boundary of the solid.
    Recursively subdivide the $P$ nodes until their descendents are labeled either $E$, or $F$, or a lowest level of resolution (cutoff depth) is reached.

Since an octree is just a collection of different size cubes, many algorithms for analyzing solids represented by octrees reduce to simple algorithms for analyzing cubes. For example, to render an octree by ray tracing, simply intersect each ray from the eye with all the $F$ voxels and accept the

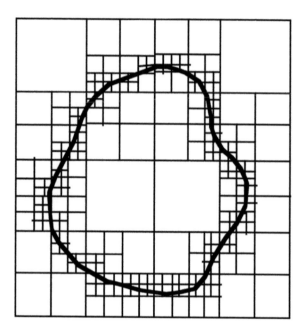

**FIGURE 20.7:** A quadtree model for the area enclosed by a planar curve. Notice that most of the subdivision of the large squares into smaller squares occurs along the bounding curve.

nearest hit. To find the volume of an octree, simply sum the volume of all the $F$ voxels. Simple pseudocode for boolean operations on octrees is presented in the following algorithm.

*Boolean Operations for Octree Representations*
*Input*
    Octree representations of two solids $A, B$.
*Algorithm*
    Apply the boolean operation to corresponding nodes in the octree representations for $A, B$. There are three cases to consider:
    *Union*
        If either node is labeled $F$, label the result with $F$.
        Otherwise if both nodes are labeled $E$, label the result with $E$.
        Otherwise, label the new node as $P$ and recursively inspect the children.
    *Intersection*
        If either node is labeled $E$, label the result with $E$.
        Otherwise if both nodes are labeled $F$, label the result with $F$.
        Otherwise, label the node as $P$ and recursively inspect the children.
    *Difference* $(A - B)$
        If the $A$-node is labeled $E$ or the B-node is labeled $F$, label the result with $E$.
        Otherwise, if the $A$-node is labeled $F$ and the B-node is labeled $E$, label the result with $F$.
        Otherwise, label the node as $P$ and recursively inspect the children.
    If all the children are labeled $E$ ($F$), then change the label $P$ to the label $E$ ($F$).

The main advantages of octree representations are that they are easy to generate and they are much more compact than naive spatial enumeration. Moreover, octree representations can achieve any desired accuracy simply by allowing smaller and smaller voxels. Many analysis algorithms for octrees, such as ray tracing, volume computations, and boolean operations, reduce to simple

**TABLE 20.1:** Comparison of different solid modeling methods.

| | CSG | B-REP | Octree |
|---|---|---|---|
| Accuracy | Good | Good | Mediocre |
| Domain | Large | Small | Large |
| Uniqueness | No | Yes | Yes |
| Validity | Easy | Hard | Easy |
| Closure | Yes | No | Yes |
| Compactness | Good | Bad | Mediocre |

algorithms for analyzing cubes. Scaling along the coordinate axes and orthogonal rotations around coordinate axes are also easy to perform simply by scaling and rotating the voxels. Finally, octrees facilitate fast mesh generation for solids.

The main disadvantage of octree representations is that the voxels are usually aligned with the coordinate axes. Thus most octree models are coordinate dependent. Therefore, it is difficult to perform an arbitrary affine transformation on an octree representation of a solid. Also, when highly accurate models are required, octree representations will use lots of storage to model curved boundaries, so for highly accurate models, octree representations can be extremely inefficient.

## 20.5 Summary

There are three standard techniques in Computer Graphics for modeling solids: CSG trees, boundary files (B-Rep), and octrees. Each of these representations has certain advantages and disadvantages. For easy reference, in Table 20.1 we compare and contrast six of the main properties of these three representations for solids, including (1) the accuracy of the model, (2) the size of the domain of the solids that are readily represented by the model, (3) the uniqueness of the model, (4) the ease of verifying the validity of the model, (5) whether or not the model is closed under boolean operations, and finally (6) the compactness of the model.

## Exercises

20.1. Consider a box with a corner at $Q$, edges parallel to the unit vectors $a$, $b$, $c$, with length, width, and height given by the scalars $l$, $w$, $h$ (see Figure 20.8). Show that a point $P$ lies inside the box if and only if the following three conditions are satisfied:

a. $0 < (P - Q) \cdot a < l$.

b. $0 < (P - Q) \cdot b < w$.

c. $0 < (P - Q) \cdot c < h$.

Interpret these formulas geometrically.

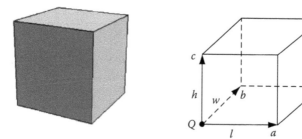

**FIGURE 20.8: (See color insert following page 354.)** A box with a corner at $Q$, edges parallel to the unit vectors $a, b, c$, with length, width, and height given by the scalars $l, w, h$.

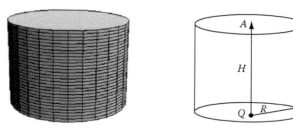

**FIGURE 20.9: (See color insert following page 354.)** A bounded cylinder of radius $R$ and height $H$ with an axis line $L$ determined by a base point $Q$ and a unit direction vector $A$.

20.2. Consider a bounded cylinder of radius $R$ and height $H$ with an axis line $L$ determined by a base point $Q$ and a unit direction vector $A$ (see Figure 20.9). Show that a point $P$ lies inside the cylinder if and only if the following two conditions are satisfied:

a. $(P - Q) \cdot (P - Q) - ((P - Q) \cdot A)^2 < R^2$.

b. $\left| (P - (Q + \frac{H}{2}A)) \cdot A \right| < \dfrac{H}{2}$.

Interpret these formulas geometrically.

20.3. Consider a bounded cone with vertex $V$, cone angle $\alpha$, height $H$, and an axis line $L$ parallel to a unit direction vector $A$ (see Figure 20.10). Show that a point $P$ lies inside the cone if and only if the following two conditions are satisfied:

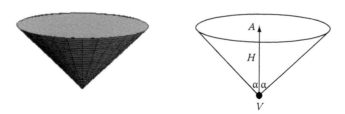

**FIGURE 20.10: (See color insert following page 354.)** A bounded cone with vertex $V$, cone angle $\alpha$, and height $H$, with an axis line parallel to the unit direction vector $A$.

a. $((P - V) \cdot A)^2 > |P - V|^2 \cos^2{(\alpha)}.$

b. $0 < (P - V) \cdot A < H.$

Interpret these formulas geometrically.

20.4. Consider the torus with center at the origin and axis parallel to the $z$-axis, with a generator of radius $d$ and a tube radius $a$. Recall from Chapter 19, Section 19.6 that this torus is described by the implicit equation:

$$F(x, y, z) \equiv \left(x^2 + y^2 + z^2 - d^2 - a^2\right)^2 + 4d^2 z^2 - 4a^2 d^2 = 0.$$

Show that:

a. $\nabla F = \left(\frac{\partial F}{\partial x}, \frac{\partial F}{\partial y}, \frac{\partial F}{\partial z}\right) = (4xG(x, y, z), 4yG(x, y, z), 4z(G(x, y, z) + 2d^2))$ is the outward pointing normal, where $G(x, y, z) = x^2 + y^2 + z^2 - d^2 - a^2.$

b. A point $P = (x, y, z)$ lies inside this torus if and only if $F(P) < 0.$

20.5. Consider a torus with center $C$ and axis parallel to the unit vector $A$, with a generator of radius $d$ and a tube radius $a$ (see Figure 20.11). Show that a point $P$ lies inside this torus if and only if

$$((P - C) \cdot A)^2 + \left(\sqrt{|P - C|^2 - ((P - C) \cdot A)^2} - d\right)^2 < a^2.$$

Interpret this result geometrically.

20.6. Verify Euler's formula for the cube, the tetrahedron, and the octahedron.

20.7. Verify Euler's formula for a cube with a rectangular hole passing through the top face and ending at the center of the cube.

20.8. Verify Euler's formula for a cube with:

a. A rectangular hole from top to bottom.

b. Two rectangular holes: one from top to bottom and one from front to back.

c. Three rectangular holes: top to bottom, front to back, and side to side.

 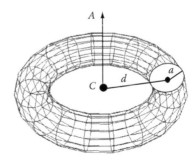

**FIGURE 20.11: (See color insert following page 354.)** A torus with center $C$ and axis parallel to the unit vector $A$, with a generator of radius $d$ and a tube radius $a$.

20.9. Develop algorithms based on the winged-edge topology to answer the following queries:

a. Find all the edges surrounding a face.

b. Find all the faces adjacent to a face.

c. Find all the vertices lying on a face.

d. Find all the edges passing through a vertex.

20.10. Let $T$ be a polyhedron bounded by planar polygonal faces $\{S_k\}$. Let $Q_k$ be a fixed point on $S_k$ and let $N_k$ be the outward pointing unit vector normal to $S_k$. Using the divergence theorem (Equation 20.2), show that

$$Vol(T) = \frac{1}{3}\Sigma_k (Q_k \cdot N_k) Area(S_k).$$

20.11. Develop an efficient algorithm to determine if a given point lies inside or outside of a solid represented by an octree.

---

## Programming Projects

20.1. *Constructive Solid Geometry*

Implement a solid modeler based on constructive solid geometry in your favorite programming language using your favorite API.

a. Include the following primitive solids:

i. Boxes and wedges

ii. Spheres, cylinders, and cones

iii. Tori

b. Incorporate the following boolean operations:

i. Union

ii. Intersection

iii. Difference

c. Develop algorithms to render these solids.

d. Develop algorithms to compute the volume of these solids.

20.2. *Octrees*

Implement a solid modeler based on octrees in your favorite programming language using your favorite API.

a. Incorporate the following boolean operations:

   i. Union

   ii. Intersection

   iii. Difference

b. Develop algorithms to render these solids.

c. Develop algorithms to compute the volume of these solids.

# *Chapter 21*

## *Shading*

*put your trust in my shadow.*

– Judges 9:15

### 21.1  Polygonal Models

Polygonal models are one of the most common representations for geometry in Computer Graphics. Polygonal models are popular because the underlying surface of each polygon is a plane and planes are the easiest surfaces to analyze: the normal vector along a plane is constant, and the points on a plane satisfy a linear equation. Thus rendering algorithms are particularly simple for polygonal models.

But models consisting solely of polygons also have severe limitations because planar polygons cannot model smooth curved surfaces. To overcome this deficiency, many small polygons are used to simulate curved shapes. But even with lots of small polygons, the human eye is adept at seeing the edges between adjacent polygons. Thus surfaces modeled by polygons do not look smooth. *The purpose of shading algorithms is to smooth the appearance of polygonal models by reducing the impact of sharp edges in order to give the eye the impression of a smooth curved surface.*

We shall study three different shading algorithms: uniform shading, Gouraud shading, and Phong shading. As in ray tracing, we will consider three illumination models: ambient reflection, diffuse reflection, and specular reflection. Since a computer model of a complicated curved shape may contain millions of polygons, the emphasis in shading algorithms is on speed. A good shading algorithm must be fast and make the model look smooth; shading algorithms are not necessarily intended to be physically accurate models of the behavior of light.

Uniform shading is fast, but makes no attempt to smooth the appearance of the model. Thus we introduce uniform shading only to compare the results with Gouraud and Phong shading. Gouraud shading smooths the appearance of the model by linearly interpolating intensities along scan lines; Phong shading simulates curved surfaces by linearly interpolating normal vectors along scan lines.

To speed up our computations, we shall make several simplifying assumptions. In addition to assuming that curved surfaces are approximated by planar polygons, we shall assume that all light sources are point light sources and that both the light sources and the eye are located far from the model (essentially at infinity). These assumptions mean that the vectors to each of the light sources as well as the vector to the eye are constant along each polygon. Together with the fact that the normal vector is constant along a polygon, these assumptions help to simplify the computation of diffuse and specular reflections.

### 21.1.1 Newell's Formula for the Normal to a Polygon

Intensity calculations require normal vectors. Polygons are typically represented by their vertices, so to calculate the intensity of light on a polygon, we need to compute the normal vector of a polygon from the vertices of the polygon. Newell's formula is a numerically robust way to compute the normal vector of a polygon from the vertices of the polygon.

Naively, we could compute the normal vector $N$ to a polygon from its vertices $P_0, \ldots, P_n$ by selecting two adjacent edge vectors $P_{i+1} - P_i, P_{i+2} - P_{i+1}$ and setting

$$N = (P_{i+1} - P_i) \times (P_{i+2} - P_{i+1}),$$

since this cross product is perpendicular to the plane of the polygon. The problem with this approach is that two of the vertices might be very close together so one of the edge vectors $P_{i+1} - P_i$, $P_{i+2} - P_{i+1}$ might be very short; worse yet the three points $P_i$, $P_{i+1}, P_{i+2}$ might be collinear or almost collinear. In each of these cases, the computation of the unit normal $N/|N|$ is numerically unstable.

Newell's formula overcomes this difficulty by using all the vertices of the polygon to calculate the normal to the polygon. (For a derivation of Newell's formula, see Chapter 11, Section 11.4.2.2.)

*Newell's Formula*

$$N = \sum_{k=0}^{n} P_k \times P_{k+1} \quad \{P_{n+1} = P_0\}. \tag{21.1}$$

Recall from Chapter 11, Section 11.4.2.2 that the magnitude of the normal vector in Newell's formula is related to the area of the associated polygon. In fact,

$$|N| = 2 \times Area(Polygon).$$

Thus Newell's formula avoids degeneracies due to collinear vertices.

---

## 21.2 Uniform Shading

In *uniform shading or flat shading* no attempt is made to smooth the appearance of a polygonal model. Each individual polygon is displayed using the intensity formula (see Chapter 16):

$$I_{uniform} = \underbrace{I_a k_a}_{\text{ambient}} + \underbrace{I_p k_d (L \cdot N)}_{\text{diffuse}} + \underbrace{I_p k_s (R \cdot V)^n}_{\text{specular}}, \tag{21.2}$$

where
 $N$ is the unit vector normal to the polygon
 $L$ is the unit vector pointing from the polygon to the light source
 $V$ is the unit vector pointing from the polygon to the eye
 $R = 2(L \cdot N)N - L$ is the direction of the light vector $L$ after it bounces off the polygon

By assumption, the vectors $N, L, R, V$ are constant along each polygon, so the intensity $I_{uniform}$ is also constant along each polygon (see Figure 21.1). For multiple light sources, we add the contributions of the individual light sources. Thus uniform shading is fast, since we need to compute only one intensity for each polygon.

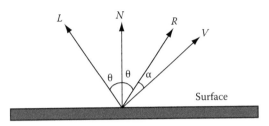

**FIGURE 21.1:** The vectors $N$, $L$, $R$, $V$ are constant along each polygonal surface.

Unfortunately, with uniform shading individual polygons are highly visible because the vectors $N$, $L$, $R$, $V$ and hence the intensity $I_{uniform}$ is different for each polygon. Therefore, discontinuities in intensity appear along the edges of the polygons, and these discontinuities are heightened by the physiology of the eye. Discontinuities in intensity along polygon edges are called *Mach bands*. The purpose of the shading algorithms that we shall study next is to eliminate these Mach bands.

## 21.3 Gouraud Shading

The purpose of Gouraud shading is to eliminate Mach bands; the method employed by Gouraud shading is linear interpolation of intensities. The underlying strategy here for each polygon is to compute the intensities for the pixels in three steps:

1. First, calculate the intensities at the individual vertices of the polygon.

2. Next, interpolate these vertex intensities along the edges of the polygon.

3. Finally, interpolate these edge intensities along scan lines to the pixels in the interior of the polygon (see Figure 21.2).

We shall now elaborate on each of these steps in turn.

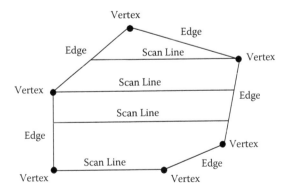

**FIGURE 21.2:** Gouraud shading. First compute the intensity at each vertex. Then interpolate these vertex intensities along each edge. Finally interpolate these edge intensities to the pixels in the interior of the polygon by interpolating edge intensities along scan lines.

To compute the intensity at a vertex, we need to know the unit normal vector at the vertex. Since each vertex may belong to many polygons, we first use Newell's formula (Equation 21.1) to calculate the unit normal for each polygon containing the vertex. We then calculate the unit normal vector at the vertex by averaging the unit normals of all the polygons containing the vertex:

$$N_{vertex} = \frac{\Sigma_{vertex \,\in\, polygon} N_{polygon}}{\left|\Sigma_{vertex \,\in\, polygon} N_{polygon}\right|}. \tag{21.3}$$

Finally, we apply Equation 21.2 to compute the intensity at the vertex.

Once we have the intensities at the vertices, we can apply linear interpolation to compute intensities first along edges and then along scan lines in the following fashion. Suppose that we know the intensities $I_1, I_2$ at two points $P_1, P_2$. The parametric equation of the line through $P_1, P_2$ is

$$L(t) = (1 - t)P_1 + tP_2 = P_1 + t(P_2 - P_1).$$

*Linear interpolation* means that we compute the intensity $I$ at a parameter $t$ by the analogous formula:

$$I(t) = (1 - t)I_1 + tI_2 = I_1 + t(I_2 - I_1).$$

To speed up the computation of intensity, we shall reduce this computation to one addition per pixel by computing $I(t)$ incrementally. For the line

$$L(t + \Delta t) = P_1 + (t + \Delta t)(P_2 - P_1) = P_1 + t(P_2 - P_1) + \Delta t(P_2 - P_1),$$
$$L(t) = P_1 + t(P_2 - P_1),$$

so

$$\Delta L = \Delta t(P_2 - P_1),$$
$$L(t + \Delta t) = L(t) + \Delta L. \tag{21.4}$$

Thus

$$L_{next} = L_{current} + \Delta L. \tag{21.5}$$

In terms of coordinates, Equation 21.4 for $\Delta L$ is really three equations, one for each coordinate:

$$\Delta x = \Delta t(x_2 - x_1), \quad \Delta y = \Delta t(y_2 - y_1), \quad \Delta z = \Delta t(z_2 - z_1). \tag{21.6}$$

Similarly, for intensity

$$I(t + \Delta t) = I_1 + (t + \Delta t)(I_2 - I_1) = I_1 + t(I_2 - I_1) + \Delta t(I_2 - I_1),$$
$$I(t) = I_1 + t(I_2 - I_1),$$

so

$$\Delta I = \Delta t(I_2 - I_1),$$
$$I(t + \Delta t) = I(t) + \Delta I. \tag{21.7}$$

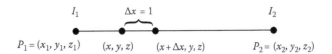

**FIGURE 21.3:**   Moving along a scan line $y = constant$ to the next pixel: $\Delta x = 1$.

Thus

$$I_{next} = I_{current} + \Delta I. \tag{21.8}$$

Since we know $I_1$ and $I_2$, if we know $\Delta t$, then we can compute $\Delta I$ and we can update $I_{next}$ by a single addition per pixel.

Now consider what happens in two cases: along a scan line and along an edge. Along a scan line, when we move to the next pixel $\Delta x = 1$ (see Figure 21.3). Since by Equation 21.6 $\Delta x = \Delta t(x_2 - x_1)$, it follows that along a scan line

$$\Delta t = 1/(x_2 - x_1),$$
$$\Delta I = (I_2 - I_1)\Delta t = (I_2 - I_1)/(x_2 - x_1).$$

To get to the next scan line, we move along an edge. When we move to the next scan line $\Delta y = 1$ (see Figure 21.4). Since by Equation 21.6 $\Delta y = \Delta t(y_2 - y_1)$, it follows that along an edge

$$\Delta t = 1/(y_2 - y_1),$$
$$\Delta I = (I_2 - I_1)\Delta t = (I_2 - I_1)/(y_2 - y_1).$$

Thus in each case, we can reduce the computation of intensity to one addition per pixel.

Notice that for adjacent polygons, intensities necessarily agree along common edges because the intensities agree at common vertices. Moreover, on both sides of an edge, the value of the intensity near the edge is close to the value of the intensity along the edge. Thus linear interpolation of intensities along scan lines blurs the difference in intensities between adjacent polygons. This blurring eliminates Mach bands and provides the appearance of smooth curved surfaces.

The Gouraud shading algorithm is an example of a scan line algorithm. *A scan line algorithm* is an algorithm that evaluates values at pixels scan line by scan line, taking advantage of the natural coordinate system induced by the orientation of the scan lines. One problem that arises with a scan line approach to shading is that the method is coordinate dependent—that is, the intensities depend not only on the lighting, but also on the orientation of the polygons relative to the coordinate system. Consider, for example, the square in Figure 21.5. Oriented one way, half the square is white,

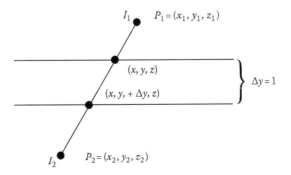

**FIGURE 21.4:**   Moving along a polygonal edge to the next scan line: $\Delta y = 1$.

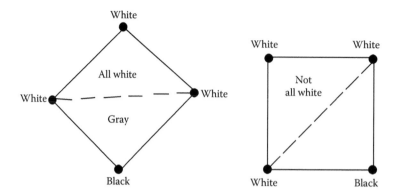

**FIGURE 21.5:** The same square oriented differently relative to the coordinate system has different shading.

whereas if the coordinate systems is rotated by 90° relative to the polygon, the same half of the square is gray.

One way to fix this problem is to subdivide all the polygons into triangles. Scan line algorithms for triangles are independent of the orientation of the triangles relative to the coordinate system. We can establish this independence in the following manner. Consider $\Delta P_1 P_2 P_3$ in Figure 21.6. Since $P_1, P_2, P_3$ are not collinear, the vectors $P_3 - P_1, P_3 - P_2$ span the plane. Therefore any point $P$ inside the triangle can be written uniquely as

$$P = P_1 + \beta_2(P_2 - P_1) + \beta_3(P_3 - P_1) = \beta_1 P_1 + \beta_2 P_2 + \beta_3 P_3,$$

where $\beta_1 + \beta_2 + \beta_3 = 1$. The values $\beta_1, \beta_2, \beta_3$ are called the *barycentric coordinates* (see Chapter 4, Exercise 4.24) of the point $P$ relative to $\Delta P_1 P_2 P_3$. We can compute the barycentric coordinates $\beta_1, \beta_2, \beta_3$ by performing linear interpolation twice: first along the edges of the triangle and then along a horizontal line (see Figure 21.6). Using this approach, we find that

$$P = \underbrace{((1 - u)(1 - t) + u(1 - s))}_{\beta_1} P_1 + \underbrace{((1 - u)t)}_{\beta_2} P_2 + \underbrace{(us)}_{\beta_3} P_3.$$

But linear interpolations of intensities is performed in exactly the same way. Thus

$$I = \underbrace{((1 - u)(1 - t) + u(1 - s))}_{\beta_1} I_1 + \underbrace{((1 - u)t)}_{\beta_2} I_2 + \underbrace{(us)}_{\beta_3} I_3.$$

Since the barycentric coordinates $\beta_1, \beta_2, \beta_3$ are unique, this result is independent of the orientation of the triangle relative to the coordinate system.

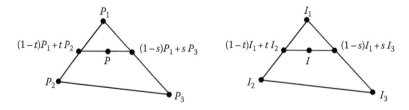

**FIGURE 21.6:** Linear interpolation for triangles: for points along edges (left) and for intensities along edges (right).

Nevertheless, even though for triangles the pixel intensities are independent of the orientation of the triangles relative to the coordinate system, the pixel intensities for polygons still depend on exactly how we subdivide the polygons into triangles; different subdivisions will generate different intensities for the same pixels.

---

## 21.4 Phong Shading

Gouraud shading reduces the visibility of polygonal edges by blurring the intensity across the boundaries of adjacent polygons. But blurring these intensities also blurs specular reflections. To reproduce sharp specular reflections, we need a better way to simulate curved surfaces.

Phong shading simulates curved surfaces by interpolating normal vectors instead of intensities. Normal vectors vary along a curved surface, so Phong shading provides a better approximation to the intensity for curved surfaces modeled by planar polygons. We begin by presenting a naive version of Phong shading; more sophisticated enhancements will be presented in subsequent subsections.

### 21.4.1 Naive Phong Shading

The underlying strategy of Phong shading is similar to the underlying strategy for Gouraud shading, except that intensities are replaced by normal vectors. The normal vectors for each polygon are computed in three steps:

1. First, calculate the normals at the individual vertices of the polygon.

2. Next, interpolate these normals along the edges of the polygon.

3. Finally, interpolate these edge normals along scan lines to the pixels in the interior of the polygon.

Once the normal vectors are computed, the intensities are calculated in the usual manner using Equation 21.2.

As in Gouraud shading, we calculate the unit normal vector at a vertex by Equation 21.3, averaging the unit normals of all the polygons to which the vertex belongs. The unit normal for each polygon is computed by Newell's formula (Equation 21.1).

Once we have the unit normals at the vertices, we can apply linear interpolation to compute normals first along edges and then along scan lines just as we calculated intensities for Gouraud shading. If $N_1, N_2$ are the unit normals at the end points $P_1, P_2$ of the line $P(t) = P_1 + t(P_2 - P_1)$, then using linear interpolation the normal $N(t)$ at the point $P(t)$ is calculated by

$$N(t) = (1 - t)N_1 + tN_2 = N_1 + t(N_2 - N_1).$$

To speed up the computation of normal vectors, we reduce this computation to three additions (one for each coordinate) per pixel by computing normals incrementally, using the same approach we used to compute intensities incrementally in Gouraud shading. Replacing intensities by normals in Equation 21.7 leads to

$$\Delta N = \Delta t(N_2 - N_1),$$
$$N(t + \Delta t) = N(t) + \Delta N. \tag{21.9}$$

Thus

$$N_{next} = N_{current} + \Delta N. \tag{21.10}$$

Since we know $N_1$ and $N_2$, if we know $\Delta t$, then we can compute $\Delta N$, and we can update $N_{next}$ by three additions (one for each coordinate) per pixel. However, to calculate intensities with Equation 21.2, we need not just a normal, but a unit normal. Thus after we calculate $N_{next} = N_{current} + \Delta N$, we need to renormalize—that is, we must calculate

$$N_{next}^{unit} = \frac{N_{current} + \Delta N}{|N_{current} + \Delta N|}.$$

We compute the normals for Phong shading, just like we computed the intensities for Gouraud shading, in two stages: along scan lines and along polygonal edges. Recall that when we move to the next pixel along a scan line, $\Delta x = 1$ (see Figure 21.3). Since by Equation 21.6 $\Delta x = \Delta t(x_2 - x_1)$, it follows that along scan lines

$$\Delta t = 1/(x_2 - x_1),$$
$$\Delta N = (N_2 - N_1)\Delta t = (N_2 - N_1)/(x_2 - x_1).$$

To get to the next scan line, we move along an edge. Now recall that when we move to the next scan line, $\Delta y = 1$ (see Figure 21.4). Since by Equation 21.6 $\Delta y = \Delta t(y_2 - y_1)$, it follows that along edges

$$\Delta t = 1/(y_2 - y_1),$$
$$\Delta N = (N_2 - N_1)\Delta t = (N_2 - N_1)/(y_2 - y_1).$$

Thus naive Phong shading is very similar to Gouraud shading: we simply replace intensities by normal vectors in our calculations along edges and along scan lines. There are, however, two differences: we must renormalize the normal vectors after the incremental computation, and then we must use these renormalized normals in Equation 21.2 to recalculate the intensity at each pixel. These additional computations make naive Phong shading considerably slower than Gouraud shading. In the next three subsections, we shall develop more sophisticated techniques to increase the speed of Phong shading.

Phong shading, like Gouraud shading, is coordinate dependent—that is, the normals depend not only on the polygons, but also on the orientations of the polygons relative to the coordinate system. Once again we can fix this problem by subdividing the polygons into triangles, since scan line algorithms for triangle are independent of the orientation of the triangles relative to the coordinate system. Nevertheless, as with Gouraud shading, the normal vectors and hence the pixel intensities for polygons will still depend on exactly how we subdivide the polygons into triangles; different subdivisions will still generate different intensities for the same pixels.

### 21.4.2   Fast Phong Shading and Diffuse Reflection

Linear interpolation for scalars is faster and cheaper than linear interpolation for vectors. For diffuse reflection, we can reduce at least part of the computation for Phong shading from a vector calculation to a scalar calculation.

Recall that the formula for diffuse reflection is (see Chapter 16, Section 16.4)

$$I_{diffuse} = I_p k_d (L \cdot N).$$

For Phong shading, the normal vector $N$ is calculated by linear interpolation—that is,

$$N(t) = (1 - t)N_1 + tN_2.$$

Thus the unit normal is given by

$$N^{unit}(t) = \frac{(1 - t)N_1 + tN_2}{|(1 - t)N_1 + tN_2|}.$$

Substituting this formula for the unit normal into the formula for diffuse reflection yields

$$I_{diffuse} = I_p k_d \frac{(1 - t)(L \cdot N_1) + t(L \cdot N_2)}{|(1 - t)N_1 + tN_2|}.$$

Now we can consider the numerator and the denominator independently.

We begin with the numerator. Let $J(t) = L \cdot N(t)$. Then the numerator is

$$J(t) = (1 - t)(L \cdot N_1) + t(L \cdot N_2) = (1 - t)J_1 + tJ_2, \qquad (21.11)$$

where $J_1 = L \cdot N_1$ and $J_2 = L \cdot N_2$. Equation 21.11 for $J(t)$ is just linear interpolation of the scalar valued function $J(t) = L \cdot N(t)$. We can calculate this scalar valued function at each pixel in exactly the same way that we calculate the intensities for Gouraud shading: first calculate $J(t) = L \cdot N(t)$ at the vertices, then interpolate along edges, and finally interpolate along scan lines. This computation is line for line the exact same computation as Gouraud shading: simply apply the computation to the linear function $J(t)$ instead of the intensity $I(t)$.

Now consider the denominator. Let

$$N(t) = (1 - t)N_1 + tN_2.$$

Then the denominator is given by

$$d(t) = |N(t)| = \sqrt{N(t) \cdot N(t)}.$$

From naive Phong shading (Equation 21.9), we know that

$$\Delta N = \Delta t(N_2 - N_1).$$

Therefore by Equation 21.10

$$N_{next} = N_{current} + \Delta N;$$

hence

$$d_{next} = \sqrt{N_{next} \cdot N_{next}}. \qquad (21.12)$$

We can calculate $\Delta N$ and $N_{next}$ quickly using the scan line approach to linear interpolation presented in Section 21.4.1; what slows us down here is the square root in Equation 21.12.

To find $d_{next}$ quickly, we can apply Newton's method (see Chapter 7, Section 7.3.1) to calculate this square root. Let

$$F(x) = x^2 - d_{next}^2 = x^2 - N_{next} \cdot N_{next}.$$

Newton's method starts with an initial guess $x_0$ for a root of $F(x)$—the value of $d_{next}$—and computes subsequent guesses $x_{n+1}$ from previous guesses $x_n$ by the formula:

$$x_{n+1} = x_n - \frac{F(x_n)}{F'(x_n)} = x_n - \frac{x_n^2 - d_{next}^2}{2x_n}.$$

Since $N$ changes only by a small amount from pixel to pixel, $\Delta N$ is typically small, so we can start the iteration with the initial guess $x_0 = d_{current}$. Because our initial guess is already close to $d_{next}$, Newton's method will typically converge to a good approximation to the required square root after only a few iterations. Notice that in this implementation, we need to compute both $N$ and $d$ for each pixel. The scan line algorithm is used to update $N$ and $d^2 = N \cdot N$; Newton's method is used only to compute $d = \sqrt{d^2}$.

### 21.4.3  Fast Phong Shading and Specular Reflection

Recall that the formula for specular reflection is (see Chapter 16, Section 16.5)

$$I_{specular} = I_p k_s (R \cdot V)^n,$$

where

$$R = 2(L \cdot N)N - L.$$

Substituting this expression for $R$ into the formula for specular reflection yields

$$I_{specular} = I_p k_s (2(L \cdot N)(N \cdot V) - (L \cdot V))^n. \tag{21.13}$$

Now recall again that for Phong shading the normal vector $N$ is calculated by linear interpolation, so

$$N(t) = (1 - t)N_1 + tN_2,$$

and

$$N^{unit}(t) = \frac{(1 - t)N_1 + tN_2}{|(1 - t)N_1 + tN_2|}.$$

Substituting this formula for the unit normal into Equation 21.13 for specular reflection yields

$$I_{specular} = I_p k_s \left( \frac{2\{(1 - t)(L \cdot N_1) + t(L \cdot N_2)\}\{(1 - t)(V \cdot N_1) + t(V \cdot N_2)\}}{|(1 - t)N_1 + tN_2|^2} - (L \cdot V) \right)^n. \tag{21.14}$$

To analyze this formula, let $J(t) = L \cdot N(t)$ and $K(t) = V \cdot N(t)$. Then the first term in the numerator on the right-hand side of Equation 21.14 is the product of the expressions:

$$J(t) = (1 - t)(L \cdot N_1) + t(L \cdot N_2) = (1 - t)J_1 + tJ_2,$$
$$K(t) = (1 - t)(V \cdot N_1) + t(V \cdot N_2) = (1 - t)K_1 + tK_2,$$

where $J_1 = L \cdot N_1, J_2 = L \cdot N_2, K_1 = V \cdot N_1$, and $K_2 = V \cdot N_2$. These expressions are just linear interpolation for the scalar valued functions $J(t) = L \cdot N(t)$ and $K(t) = V \cdot N(t)$. Once again we can calculate these scalar valued functions at each pixel in exactly the same way that we calculate the intensities for Gouraud shading: first calculate $J(t)$ and $K(t)$ at the vertices, then interpolate along edges, and finally interpolate along scan lines. This computation is line for line the exact same computation as Gouraud shading, with the functions $J(t)$ and $K(t)$ in place of the intensity $I(t)$. Moreover, the term $L \cdot V$ is a constant, so we need to compute this value only once. Thus, we have

$$I_{specular} = I_p k_s \left( \frac{2 \underbrace{\{(1-t)(L \cdot N_1) + t(L \cdot N_2)\}}_{\text{Similar to Gouraud}} \underbrace{\{(1-t)(V \cdot N_1) + t(V \cdot N_2)\}}_{\text{Similar to Gouraud}}}{|(1-t)N_1 + tN_2|^2} - \underbrace{(L \cdot V)}_{\text{Constant}} \right)^n.$$

We are left to compute only the denominator of the first term on the right-hand side. This denominator is given by

$$D(t) = |N(t)|^2 = N(t) \cdot N(t).$$

Notice, in particular, that unlike the calculation of Phong shading for diffuse reflection, there is no square root in this denominator. Now from naive Phong shading (Equation 21.9), we know that

$$\Delta N = \Delta t (N_2 - N_1).$$

Therefore by Equation 21.10

$$N_{next} = N_{current} + \Delta N,$$

and

$$D_{next} = N_{next} \cdot N_{next}.$$

We can calculate $\Delta N$ and $N_{next}$ quickly using the scan line approach to Phong shading presented in Section 21.4.1, and there is no square root here to slow us down as in the calculation of diffuse reflection: to compute $D_{next}$ from $N_{next}$ only one additional dot product is required.

### 21.4.4 Phong Shading and Spherical Linear Interpolation

In Phong shading we perform linear interpolation on unit normals. But linear interpolation does not preserve length. Therefore for every pixel, we must renormalize the length of these normal vectors. This renormalization takes time and involves a square root, substantially slowing down the shading computation.

Linear interpolation is unnatural for unit vectors because linear interpolation does not preserve length. But there is a natural way to preserve length while transitioning between two unit vectors: we can simply rotate one vector into the other in the plane of the two vectors. Rotation preserves length, so no renormalization is required. If we rotate at a uniform speed, this approach is exactly *spherical linear interpolation* or *slerp*. (For a derivation of the formula for spherical linear interpolation, see Chapter 11, Section 11.6.)

*Spherical Linear Interpolation*

$$slerp(N_1, N_2, t) = \frac{\sin((1-t)\phi)}{\sin(\phi)} N_1 + \frac{\sin(t\phi)}{\sin(\phi)} N_2, \tag{21.15}$$

where $\phi = \cos^{-1}(N_1 \cdot N_2)$.

Spherical linear interpolation eliminates the need to compute square roots (renormalization), but at the cost of computing the trigonometric functions $\sin((1-t)\phi)$ and $\sin(t\phi)$. This trade-off hardly seems worthwhile. Fortunately, we can compute these trigonometric functions incrementally, and thus avoid computing sines for every pixel.

To see how to compute the unit normal incrementally, suppose we have already computed

$$N(t) = \frac{\sin((1-t)\phi)}{\sin(\phi)} N_1 + \frac{\sin(t\phi)}{\sin(\phi)} N_2, \tag{21.16}$$

and we want to compute

$$N(t + \Delta t) = \frac{\sin((1 - t - \Delta t)\phi)}{\sin(\phi)} N_1 + \frac{\sin((t + \Delta t)\phi)}{\sin(\phi)} N_2.$$

First notice that the denominator, $\sin(\phi)$, is a constant. Moreover, since $N_1, N_2$ are unit vectors,

$$\cos(\phi) = N_1 \cdot N_2,$$

so

$$\sin(\phi) = \sqrt{1 - \cos^2(\phi)} = \sqrt{1 - (N_1 \cdot N_2)^2}.$$

The square root here is not a problem, since this square root is computed only once. We do not need to repeat this computation for each pixel; rather we compute $\sin(\phi)$ once and store the result.

Next observe that by the trigonometric identities for the sine of the sum or difference of two angles:

$$\sin((t + \Delta t)\phi) = \sin(t\phi)\cos(\Delta t\phi) + \cos(t\phi)\sin(\Delta t\phi), \tag{21.17}$$

$$\sin((1 - t - \Delta t)\phi) = \sin((1 - t)\phi)\cos(\Delta t\phi) - \cos((1 - t)\phi)\sin(\Delta t\phi). \tag{21.18}$$

Since $\Delta t, \phi$ are constants, the values $\cos(\Delta t\phi), \sin(\Delta t\phi)$ are also constants, so we can also compute these values once and store them for subsequent use. Furthermore, we can assume that we have already calculated $\sin(t\phi)$, $\sin((1-t)\phi)$ for the current pixel, so we do not need to recalculate these values for the next pixel. But we still need $\cos(t\phi)$, $\cos((1-t)\phi)$. Fortunately, these values can also be computed incrementally from the trigonometric identities for the cosine of the sum or difference of two angles:

$$\cos((t + \Delta t)\phi) = \cos(t\phi)\cos(\Delta t\phi) - \sin(t\phi)\sin(\Delta t\phi) \tag{21.19}$$

$$\cos((1 - t - \Delta t)\phi) = \cos((1 - t)\phi)\cos(\Delta t\phi) + \sin((1 - t)\phi)\sin(\Delta t\phi). \tag{21.20}$$

Using Equations 21.17 through 21.20, after computing the constants $\cos(\Delta t\phi)$, $\sin(\Delta t\phi)$, we can compute all the necessary sines and cosines incrementally with just four additions/subtractions and eight multiplications. No trigonometric evaluations are necessary, nor are any square roots required.

Now we can proceed in the usual fashion for Phong shading:

i. First, calculate the unit normals at the individual vertices of each polygon.

ii. Next, apply spherical linear interpolation, using Equations 21.17 through 21.20 with $\Delta t = 1/(y_2 - y_1)$ to calculate the unit normals incrementally along the edges of each polygon.

iii. Next, apply spherical linear interpolation to the unit normals along the edges, using Equations 21.17 through 21.20 with $\Delta t = 1/(x_2 - x_1)$ to calculate the unit normals incrementally along scan lines for the pixels in the interior of each polygon.

iv. Finally, use these unit normals to calculate the intensity at each pixel using Equation 21.2.

The preceding algorithm corresponds to naive Phong shading with spherical linear interpolation replacing standard linear interpolation. This procedure eliminates square roots, but if we were to follow this approach, in step (iv) we would need to compute two dot products—$L \cdot N$ and $V \cdot N$— for every pixel. To eliminate these dot products, we can proceed as in fast Phong shading by setting $J(t) = L \cdot N(t)$ and $K(t) = V \cdot N(t)$. Substituting the right-hand side of Equation 21.16 for $N(t)$ into these expression for $J(t)$ and $K(t)$ yields the scalar valued functions:

$$J(t) = \frac{\sin((1-t)\phi)}{\sin(\phi)} L \cdot N_1 + \frac{\sin(t\phi)}{\sin(\phi)} L \cdot N_2,$$

$$K(t) = \frac{\sin((1-t)\phi)}{\sin(\phi)} V \cdot N_1 + \frac{\sin(t\phi)}{\sin(\phi)} V \cdot N_2.$$

Now we can proceed as in the previous algorithm, but instead of updating the unit normal vector $N(t)$ incrementally along scan lines using Equations 21.17 through 21.20, we use these equations to incrementally update the scalar valued functions $J(t)$ and $K(t)$ along scan lines. This approach eliminates the need to compute two dot products at every pixel, but notice that we still need to update the normal vector $N(t)$ incrementally along each edge, since we need the angle $\phi$ between normal vectors at opposite edges along each scan line in order to calculate the functions $J(t)$ and $K(t)$ along the scan lines. Finally, notice that we can eliminate the division by $\sin(\phi)$ at each pixel by absorbing this division into the constants $I_p k_d$ and $I_p k_s$ along each scan line.

---

## 21.5 Summary

The purpose of shading algorithms is to smooth the appearance of polygonal models by reducing the impact of sharp edges in order to give the eye the impression of a smooth curved surface. Gouraud shading blurs these sharp edges by applying linear interpolation to the intensity for each polygon along scan lines. But Gouraud shading also blurs specular reflections. Phong shading provides a better model for curved surfaces by performing linear interpolation of normal vectors along scan lines. Phong shading is slower than Gouraud shading, but Phong shading is considerably better for specular reflections.

We investigated several ways to speed up Phong shading, including calculating diffuse and specular reflections independently by linear interpolation and using Newton's method to compute the square root needed to find the magnitude of the normal vector for diffuse reflection. We also examined spherical linear interpolation of unit normals in order to avoid the square root altogether.

To speed up these algorithms, the interpolation steps in both Gouraud and Phong shading are performed incrementally. Here we summarize these incremental formulas for both Gouraud and Phong shading.

*Gouraud Shading*

$$I_{next} = I_{current} + \Delta I \qquad \{\text{intensity}\}$$
$$\Delta I = (I_2 - I_1)/(x_2 - x_1) \quad \{\text{along scan lines}\}$$
$$\Delta I = (I_2 - I_1)/(y_2 - y_1) \quad \{\text{along edges}\}$$

*Phong Shading*

   a. *Naive Phong Shading*

$$N_{next} = \frac{N_{current} + \Delta N}{|N_{current} + \Delta N|} \qquad \text{\{unit normal\}}$$

$$\Delta N = (N_2 - N_1)/(x_2 - x_1) \qquad \text{\{along scan lines\}}$$

$$\Delta N = (N_2 - N_1)/(y_2 - y_1) \qquad \text{\{along edges\}}$$

   b. *Fast Phong Shading*

   i. *Diffuse Reflection*

$$J(t) = L \cdot N(t) \qquad \text{\{numerator\}}$$

$$J_{next} = J_{current} + \Delta J$$

$$\Delta J = (J_2 - J_1)/(x_2 - x_1) \qquad \text{\{along scan lines\}}$$

$$\Delta J = (J_2 - J_1)/(y_2 - y_1) \qquad \text{\{along edges\}}$$

$$d_{next} = \sqrt{N_{next} \cdot N_{next}} \qquad \text{\{denominator\}}$$

$$N_{next} = N_{current} + \Delta N$$

$$\Delta N = (N_2 - N_1)/(x_2 - x_1) \qquad \text{\{along scan lines\}}$$

$$\Delta N = (N_2 - N_1)/(y_2 - y_1) \qquad \text{\{along edges\}}$$

   ii. *Specular Reflection*

$$J(t) = L \cdot N(t)$$
$$K(t) = V \cdot N(t) \qquad \text{\{numerator\}}$$

$$J_{next} = J_{current} + \Delta J$$

$$\Delta J = (J_2 - J_1)/(x_2 - x_1) \qquad \text{\{along scan lines\}}$$

$$\Delta J = (J_2 - J_1)/(y_2 - y_1) \qquad \text{\{along edges\}}$$

$$K_{next} = K_{current} + \Delta K$$

$$\Delta K = (K_2 - K_1)/(x_2 - x_1) \qquad \text{\{along scan lines\}}$$

$$\Delta K = (K_2 - K_1)/(y_2 - y_1) \qquad \text{\{along edges\}}$$

$$D_{next} = N_{next} \cdot N_{next} \qquad \text{\{denominator\}}$$

$$N_{next} = N_{current} + \Delta N$$

$$\Delta N = (N_2 - N_1)/(x_2 - x_1) \qquad \text{\{along scan lines\}}$$

$$\Delta N = (N_2 - N_1)/(y_2 - y_1) \qquad \text{\{along edges\}}$$

c. *Spherical Linear Interpolation*

$$N_{next} = N(t + \Delta t) = \frac{\sin((1 - t - \Delta t)\phi)}{\sin(\phi)}N_1 + \frac{\sin((t + \Delta t)\phi)}{\sin(\phi)}N_2 \qquad \{\text{along edges only}\}$$

$$J_{next} = J(t + \Delta t) = \frac{\sin((1 - t - \Delta t)\phi)}{\sin(\phi)}L \cdot N_1 + \frac{\sin((t + \Delta t)\phi)}{\sin(\phi)}L \cdot N_2$$

$$K_{next} = K(t + \Delta t) = \frac{\sin((1 - t - \Delta t)\phi)}{\sin(\phi)}V \cdot N_1 + \frac{\sin((t + \Delta t)\phi)}{\sin(\phi)}V \cdot N_2$$

$$\sin((t + \Delta t)\phi) = \sin(t\phi)\cos(\Delta t\phi) + \cos(t\phi)\sin(\Delta t\phi)$$

$$\sin((1 - t - \Delta t)\phi) = \sin((1 - t)\phi)\cos(\Delta t\phi) - \cos((1 - t)\phi)\sin(\Delta t\phi)$$

$$\cos((t + \Delta t)\phi) = \cos(t\phi)\cos(\Delta t\phi) - \sin(t\phi)\sin(\Delta t\phi)$$

$$\cos((1 - t - \Delta t)\phi) = \cos((1 - t)\phi)\cos(\Delta t\phi) + \sin((1 - t)\phi)\sin(\Delta t\phi)$$

$$\Delta t = 1/(x_2 - x_1) \qquad \{\text{along scan lines}\}$$

$$\Delta t = 1/(y_2 - y_1) \qquad \{\text{along edges}\}$$

## Exercises

21.1. A table of sines states that $\sin(24°) = 0.40674$ and $\sin(25°) = 0.42262$. Use linear interpolation to estimate $\sin(24.3°)$.

21.2. Show how to compute $e^t$ incrementally using only one multiplication per pixel.

21.3. Show how to compute $t^2$ incrementally using only two additions and no multiplications per pixel. (Hint: At each pixel compute both $t^2$ and $2t\Delta t + (\Delta t)^2$, and update each of these values with only one addition per pixel.)

21.4. Show how to further simplify Phong shading for specular reflection using spherical linear interpolation by eliminating the computation of the square root in the calculation of $\sin(\phi)$.

## Programming Project

21.1. *Shading Algorithms*

Implement both Gouraud and Phong shading in your favorite programming language using your favorite API.

a. Use your implementation to shade several large polygonal models.

b. Compare the relative speeds of the following algorithms for each of your models:

    i. Gouraud shading

    ii. Naive Phong shading

    iii. Fast Phong shading

    iv. Phong shading using spherical linear interpolation

# Chapter 22

## Hidden Surface Algorithms

*thou didst hide thy face, and I was troubled.*

– Psalm 30:7

## 22.1 Hidden Surface Algorithms

Surfaces can be hidden from view by other surfaces. *The purpose of hidden surface algorithms is to determine which surfaces are obstructed by other surfaces in order to display only those surfaces visible to the eye.* In theory, hidden surface algorithms are required for all types of surfaces; in practice, we shall restrict our attention here to polygonal models.

In this chapter, we shall assume that all the surfaces are planar polygons and that all the polygons are opaque. We shall not assume that the polygons necessarily enclose a solid nor that the polygons form a manifold. Our objective is: given any collection of polygons, to display only those polygons visible to the eye.

Polygonal models of complicated shapes may contain millions of polygons, so just like shading algorithms the emphasis in hidden surface algorithms is on speed. A good hidden surface algorithm must be fast as well as accurate. Sorting, tailored data structures, and pixel coherence are all employed to speed up hidden surface algorithms.

There are two standard types of hidden surface algorithms: *image space algorithms* and *object space algorithms*. Image space algorithms work in pixel space (the frame buffer) and are applied after pseudoperspective; object space algorithms work in model space and are applied before pseudoperspective. These algorithms have the following generic structures:

| | |
|---|---|
| *Image Space Algorithm* | *Object Space Algorithm* |
| For each pixel: | For each polygon: |
|    Find the closest polygon. |    Find the unobstructed part of the polygon. |
|    Render the pixel with the color. |    Render this part of the polygon. |
|    and intensity of this polygon. | |

The speed of image space algorithms is $O(nN)$, where $N$ is the number of pixels and $n$ is number of polygons, because for each pixel we must examine every polygon. The speed of object space algorithms is $O(n^2)$, where $n$ is the number of polygons, because to determine which part of each polygon is visible, we need to compare every polygon with every other polygon.

Many hidden surface algorithms have been developed, each with their own advantages and disadvantages. We shall examine five of the most common hidden surface algorithms: z-buffer, scan line, ray casting, depth sort, and binary space partitioning tree (bsp-tree). The z-buffer and scan line algorithms are image space algorithms; the depth sort and bsp-tree algorithms are object space algorithms. Ray casting can work both in image space and in object space. We will compare and contrast the relative merits and limitations of each of these algorithms, and we shall see that there is no single solution to the hidden surface problem that is optimal in all situations.

## 22.2   The Heedless Painter

The heedless painter displays the polygons in a scene in the order in which they occur in some list. Polygons that appear early in the list can be overpainted by polygons that appear later in the list.

The heedless painter is slow because the same pixel may be painted many times, once for each polygon that lies over the pixel. Worse the scene displayed by the heedless painter is often incorrect because polygons that appear later in the list may overpaint polygons that appear earlier in the list, even though the later polygons actually lie behind the earlier polygons.

The heedless painter procedure is not a true hidden surface algorithm. Nevertheless, valid hidden surface algorithms can be generated by fixing the problems in the heedless painter procedure. We begin with two such hidden surface algorithms: $z$-buffer and scan line.

## 22.3   $z$-Buffer (Depth Buffer)

The $z$-buffer or depth buffer algorithm employs a special data structure called the *z-buffer* or *depth buffer*. The $z$-buffer is a large memory array which is the same size as the frame buffer—that is, the $z$-buffer stores an entry for each pixel. This entry consists of the current depth as well as the current color or intensity of the corresponding pixel.

Using this data structure, the $z$-buffer algorithm proceeds much as the heedless painter visiting each polygon in turn, but with one crucial difference: a pixel is overpainted—that is, a new color or intensity is stored in the $z$-buffer—only if

*depth of current polygon at pixel < current depth of pixel in z-buffer.*

The $z$-buffer begins with the color or intensity of each pixel initialized to the background color or intensity, and the depth of each pixel initialized to infinity. To make the $z$-buffer algorithm as fast as possible, the depth of the pixels in each polygon can be computed incrementally.

To compute the depth of each pixel incrementally, we proceed almost exactly as we did in Chapter 21 for Gouraud and Phong shading, where we computed the intensities and normal vectors at the pixels of a polygon incrementally.

1. First, compute the depth at the vertices of the polygon. After pseudoperspective, this depth is typically just the value of the $z$-coordinate at each vertex.

2. Next, compute the depth along the edges of the polygon.

3. Finally, compute the depth along scan lines for the pixels in the interior of the polygon.

The explicit details of this computation are given below.

Recall that for a line $L(t)$ passing through the points $P_1, P_2$

$$L(t) = (1 - t)P_1 + tP_2 = t(P_2 - P_1),$$
$$L(t + \Delta t) = P_1 + (t + \Delta t)(P_2 - P_1).$$

Subtracting the first equation from the second equation yields

$$\Delta L = \Delta t(P_2 - P_1),$$

or equivalently

$$\Delta x = \Delta t(x_2 - x_1), \quad \Delta y = \Delta t(y_2 - y_1), \quad \Delta z = \Delta t(z_2 - z_1).$$

Thus along any line we can compute depth incrementally by setting

$$z_{next} = z_{current} + \Delta z.$$

Along a scan line

$$\Delta x = 1 \quad \Rightarrow \quad \Delta t = \frac{1}{x_2 - x_1} \quad \Rightarrow \quad \Delta z = \frac{z_2 - z_1}{x_2 - x_1},$$

and when we move to the next scan line along a polygonal edge

$$\Delta y = 1 \quad \Rightarrow \quad \Delta t = \frac{1}{y_2 - y_1} \quad \Rightarrow \quad \Delta z = \frac{z_2 - z_1}{y_2 - y_1}.$$

If these computations are not completely familiar, you should review Gouraud shading in Chapter 21, Section 21.3.

The advantages of the $z$-buffer algorithm are that the $z$-buffer is simple to understand and easy to implement; the disadvantages are that the $z$-buffer algorithm is memory intensive and relatively slow, since this algorithm may repaint the same pixel many times. The next algorithm we shall study avoids both of these drawbacks by introducing more complicated data structures.

## 22.4   Scan Line

The scan line algorithm paints the pixels scan line by scan line. To decide which polygon to paint for each pixel, the scan line algorithm for hidden surfaces maintains two special data structures: an *active edge list* and an *edge table*.

An edge is said to be active if the edge intersects the current scan line. For each active edge with end points $P_1 = (x_1, y_1, z_1), P_2 = (x_2, y_2, z_2)$, the active edge list stores the following data:

$$Edge\ Data = (Y_{max}, current\ X_{int}, current\ Z_{int}, \Delta x, \Delta z, pointer\ to\ polygon),$$

where

$Y_{max} = max(y_1, y_2) = $ maximum value of $y$ along the edge

*current* $X_{int} = x$-coordinate of the point where the edge intersects the current scan line

*current* $Z_{int} = z$-coordinate (depth) of the point where the edge intersects the current scan line

$$\Delta x = \frac{x_2 - x_1}{y_2 - y_1} = \text{change in } x \text{ along the edge when moving to the next scan line}$$

$$\Delta z = \frac{z_2 - z_1}{y_2 - y_1} = \text{change in } z \text{ along the edge when moving to the next scan line}$$

pointer to polygon—points to the polygon containing the edge

The expressions for $\Delta x$ and $\Delta z$ can be derived in the same way as the formula for $\Delta z$ in the $z$-buffer algorithm.

For each scan line, we also introduce an *edge table* consisting of a list of those edges whose lower vertex lies on the scan line. The data stored for each edge in the edge table is the same as the data that is stored for each edge in the active edge list.

New edges must be introduced into the active edge list as they become active, and old edges must be removed as they become inactive. Updating the active edge list is performed by the following algorithm.

*Updating the Active Edge List*
For each edge in the active edge list:
    If $Y_{scan\ line} > Y_{max}$, delete the edge from the active edge list.
    Otherwise update the values of *current* $X_{int}$ and *current* $Z_{int}$:

$$current\ X_{int} = current\ X_{int} + \Delta x,$$
$$current\ Z_{int} = current\ Z_{int} + \Delta z.$$

Insert each edge in the edge table for the current scan line $(Y_{scan\ line})$ into the active edge list. Sort the active edge list by increasing values of current $X_{int}$.

We can now use the following scan line algorithm to paint the scene and avoid hidden surfaces.

*Scan Line Algorithm*
    For each scan line:
        Update the active edge list (see the preceding algorithm).
        For each polygon $P$ with an edge in the active edge list, compute $\Delta z_P$ along the current scan line (see the discussion following this algorithm).
        Initialize
            *Current Polygon* = polygon corresponding to the first edge in the active edge list.
            *Current Pixel* = current $X_{int}$ for the first edge in the active edge list.
        Repeat until the scan line reaches the pixel on the last edge in the active edge list:
            For each polygon $P$ corresponding to an edge in the active edge list with

$$current\ X_{int} > Current\ Pixel:$$

        Compute the $z$-value $z_P$ of the polygon $P$ at the *Current Pixel* (see the discussion following this algorithm).
        Determine the pixel $x_P$ (if any) where the polygon $P$ crosses in front of the *Current Polygon* (see the discussion following this algorithm).
        If no polygon crosses in front of the *Current Polygon*:
            Fill the pixels along the scan line from the *Current Pixel* to the end of the *Current Polygon* with the color and intensity of the *Current Polygon*.
            If the last filled in pixel lies inside at least one other polygon with an edge in the active edge list, set
                *Current Pixel* = last filled in pixel.
                *Current Polygon* = the polygon with an edge in the active edge list that has the smallest value of $z$ at the *Current Pixel*.

Otherwise set

    *Current Polygon* = the polygon corresponding to the first edge in the active edge list with *current* $X_{int}$ greater than or equal to the last filled in pixel.

    *Current Pixel* = *current* $X_{int}$ for the first edge in the active edge list with *current* $X_{int}$ greater than or equal to the last filled in pixel.

Otherwise

    Set

        $x_s$ = the smallest value of $x_P$

        $Q$ = polygon corresponding to $x_s$

        Fill the pixels along the scan line from the *Current Pixel* to $x_s$ with the color and intensity of the *Current Polygon*.

    Set

        *Current Polygon* = $Q$

        *Current Pixel* = $x_s$.

To be active, a polygon $P$ must have at least two edges in the active edge list. The value $\Delta z_P$ for the polygon $P$ along the current scan line is given by

$$\Delta z_P = \frac{z_2 - z_1}{x_2 - x_1},$$

where

$$z_j = \text{current } Z_{int} \quad \text{and} \quad x_j = \text{current } X_{int},$$

for the two edges of the polygon $P$ in the active edge list. The $z$-value $z_P$ of a polygon $P$ at the *Current Pixel* is given by

$$z_P = \text{current } Z_{int} - (\text{current } X_{int} - \text{Current Pixel})\Delta z_P,$$

where *current* $Z_{int}$ and *current* $X_{int}$ are the values in the edge table for any one of the edges in the active edge list belonging to the polygon $P$. (For an alternate approach, see Exercise 22.1.)

The only real difficulty in the scan line algorithm is to determine where to switch polygons (see Figure 22.1). Mathematically our problem is: given the depths $z_P, z_Q$ of two polygons $P$, $Q$ at the same point along a scan line, find the pixel at which the polygons intersect. We can compute depth incrementally along a scan line just as we did in the $z$-buffer algorithm by adding $\Delta z$. Let $N$ be the number of pixels from the *Current Pixel* to the pixel where the crossover occurs. At the pixel where the crossover occurs, the depths of the two polygons are equal. Therefore,

$$z_P + N\Delta z_P = z_Q + N\Delta z_Q,$$

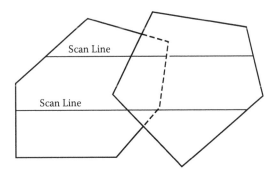

**FIGURE 22.1:** Crossover: the pixel where a new polygon from the active edge list for a scan line crosses in front of the current polygon in the active edge list.

so

$$N = \frac{z_Q - z_P}{\Delta z_P - \Delta z_Q}.$$

The number $N$ tells us how many more pixels we must visit before we switch to the next polygon. Thus in the Scan Line Algorithm:

$$x_p = Current\ Pixel\ +\ N.$$

Note that at most one crossover can occur between any pair of polygons, so once we switch polygons we never switch back. Of course, if *Current Pixel* $+ N$ is greater than the $x$-coordinate of last pixel in the *Current Polygon*, then no crossover occurs.

The main advantage of the scan line algorithm is speed. The scan line algorithm is faster than the $z$-buffer algorithm for two reasons: the scan line algorithm avoids overpainting pixels—each pixel is visited and painted only once—and the scan line algorithm takes advantage of pixel coherence—adjacent pixels typically lie on the same polygon and accordingly can be painted with the same color and intensity. Another important advantage of the scan line algorithm is that the scan line algorithm is compatible with the algorithms for Gouraud and Phong shading. If we use the scan line algorithm, then hidden surfaces and shading can both be computed in the same pass, since both algorithms proceed scan line by scan line. The main disadvantage of the scan line algorithm is that this algorithm requires more complicated data structures than the $z$-buffer algorithm and the algorithm itself is a good deal more difficult to implement.

---

## 22.5   Ray Casting

The ray casting algorithm for hidden surfaces employs no special data structures. Instead, as in recursive ray tracing, a ray is fired from the eye through each pixel on the screen in order to locate the polygon in the scene closest to the eye. The color and intensity of this polygon is displayed at the pixel.

*Ray Casting Algorithm*
Through each pixel:
    Fire a ray from the eye.
    Intersect this ray with the plane of each polygon.
    Reject intersections that lie outside the polygon.
    Accept the closest remaining intersection—that is, the intersection with the smallest value of the parameter along the line.

Ray casting is easy to implement for polygonal models because the only calculation required is the intersection of a line with a plane. Let $L$ be the line determined by the eye point $E$ and a pixel $F$, and let $S$ be a plane determined by a unit normal $N$ and a point $Q$. Then the parametric equation of the line is $L(t) = E + tv$, where $v = F - E$, and the implicit equation of the plane is $N \cdot (P - Q) = 0$. Thus the line intersects the plane when

$$N \cdot (L(t) - Q) = N \cdot (E + tv - Q) = 0.$$

Solving for $t$ yields

$$t = \frac{N \cdot (Q - E)}{N \cdot v}. \qquad (22.1)$$

The actual intersection point can be found by substituting this value of $t$ into the parametric equation of the line. For polygonal models, we still need to determine if this intersection point actually lies inside the polygon. For convex polygons such tests are provided in Chapter 19, Section 19.3; for arbitrary polygons, inside–outside tests are discussed in Chapter 11, Section 11.7.

Ray casting can be applied either before or after pseudoperspective. The only difference is that after pseudoperspective the eye is mapped to infinity, so the rays are fired perpendicular to the plane of the pixels rather than through the eye. Thus in Equation 22.1 $v$ is the normal to the pixel plane and the eye point $E$ is replaced by the pixel $F$; the remainder of the algorithm is unchanged. Employed after pseudoperspective, ray casting is an image space algorithm; invoked before pseudoperspective, ray casting is an object space algorithm.

The main advantage of the ray casting algorithm for hidden surfaces is that ray casting can be used even with nonplanar surfaces. All that is needed to implement the ray casting algorithm for hidden surfaces is a line-surface intersection algorithm for each distinct surface type in the model.

The main disadvantage of ray casting is that the method is slow. Ray casting is a brute force technique that makes no use of pixel coherence. For adjacent pixels it is quite likely that the same polygon is closest to the eye, but ray casting makes no attempt to test for this pixel coherence. Instead ray casting employs dense sampling, so the ray casting algorithm for hidden surfaces is typically slower than other methods that employ more sophisticated data structures.

## 22.6 Depth Sort

Depth sort is another variation on the heedless painter's algorithm. The data structure for depth sort is a list of polygons sorted by increasing depth. Two polygons that overlap in depth are said to *conflict*. Conflicts are resolved by a sorting algorithm which is done in object space, so depth sort is an object space algorithm. After all the conflicts are resolved, the polygons are painted in order from back to front.

*Depth Sort Algorithm*
> Sort the polygons by the furthest vertex from the screen:
>> For each polygon, find the minimum and maximum values of $z$ (depth) at the vertices.
>> Sort the polygons in order of decreasing maximum $z$-coordinates.
>> Resolve $z$-overlaps (see the procedure for *Resolving Conflicts* presented below).
> Paint the polygons in order from farthest to nearest.

The main difficulty in implementing depth sort is resolving $z$-overlaps. Notice that without this step, the depth sort algorithm would overpaint near polygons by far polygons just like the heedless painter's algorithm.

Next we present a five-step procedure for resolving conflicts ($z$-overlaps). Here the terms *z-extents* refers to the minimum and maximum values of the $z$-coordinates; *x-extents* and *y-extents* are defined in an analogous manner.

*Resolving Conflicts*
For every pair of polygons $P$ and $Q$:

> If the $z$-extents of $P$ and $Q$ overlap, then perform the following five tests:

> i. Do the $x$-extents fail to overlap?

> ii. Do the $y$-extents fail to overlap?

> iii. Does every vertex of $P$ lie on the far side of $Q$?

iv. Does every vertex of $Q$ lie on near side of $P$?

v. Do the $xy$-projections of $P$ and $Q$ fail to overlap?

If any test succeeds, $P$ does not obscure $Q$, so render $P$ before $Q$.
Otherwise we cannot resolve the conflict, so split the polygon $P$ by the plane of the polygon $Q$ (see Section 22.6.1).

The five tests are ordered by difficulty from easy to hard. The $x$-extents and $y$-extents of a polygon are easy to determine by examining the $x$ and $y$ coordinates of all the vertices of the polygon. If the $x$-extents or $y$-extents of $P$ and $Q$ fail to overlap, then it does not matter which polygon is painted first, since the projections of $P$ and $Q$ on the screen will not overlap (see Figure 22.2).

If every vertex of $P$ lies on the far side of the screen from $Q$ or if every vertex of $Q$ lies on the near side of the screen from $P$, then $P$ does not obscure $Q$ so we can safely paint $P$ before $Q$ (see Figures 22.3 and 22.4). We can determine if $P$ lies on the far side of $Q$ by computing one dot product for every vertex of $P$. Fix a vertex $V_Q$ of $Q$, and let $N_Q$ be the outward pointing normal—the normal pointing to the far side of the screen, the side of the screen away from the eye—of the plane containing the polygon $Q$. Then $P$ lies on the far side of $Q$ if and only if for every vertex $V_P$ of $P$, $(V_P - V_Q) \cdot N_Q > 0$ (see Figure 22.3). Similarly, fix a vertex $V_P$ of $P$, and let $N_P$ be the outward pointing normal of the plane containing the polygon $P$. Then $Q$ lies on the near side of $P$ if and only if for every vertex $V_Q$ of $Q$, $(V_P - V_Q) \cdot N_P > 0$ (see Figure 22.4). Notice in Figure 22.4 that $Q$ lies on the near side of $P$, even though $P$ does not lie on the far side of $Q$, so the third and fourth tests are independent. If tests *iii* and *iv* fail, we can reverse the roles of $P$ and $Q$ and retest.

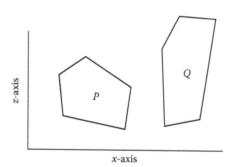

**FIGURE 22.2:** The $z$-extents of $P$ and $Q$ overlap, but the $x$-extents do not overlap. Therefore the projections of $P$ and $Q$ into the $xy$-plane will not overlap.

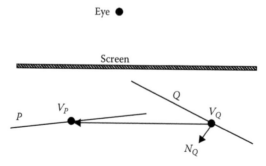

**FIGURE 22.3:** The polygon $P$ lies on the far side of the polygon $Q$: $(V_P - V_Q) \cdot N_Q > 0$.

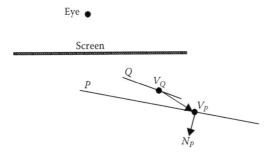

**FIGURE 22.4:** The polygon $Q$ lies on the near side of the polygon $P$: $(V_P - V_Q) \cdot N_P > 0$. Notice that the polygon $P$ does not lie on the far side of the polygon $Q$.

If the first four tests fail, then we can apply inside–outside tests to determine if the projections of $P$ and $Q$ on the screen overlap. If any vertex in the projection of $P$ lies inside the projection of $Q$ or if any vertex in the projection of $Q$ lies inside the projection of $P$, then the projections of $P$ and $Q$ on the screen overlap. For convex polygons inside–outside tests are provided in Chapter 19, Section 19.3; for arbitrary polygons, inside–outside tests are discussed in Chapter 11, Section 11.7.

If the projections of $P$ and $Q$ do not overlap, then once again it does not matter which polygon is painted first. However, if the projections of $P$ and $Q$ do overlap, then the conflict cannot be resolved, and we must split one of the polygons, say $P$, by intersecting the polygon $P$ with the plane of the polygon $Q$. Since splitting will be used again for bsp-trees, we present the details of this splitting procedure in a separate subsection (see Section 22.6.1).

This method of resolving conflicts works for conflicts between pairs of polygons, but not all conflicts can be resolved in this manner because not all conflicts are localized to pairs of polygons. There can be cyclic conflicts where $P$ obscures $Q$, $Q$ obscures $R$, and $R$ obscures $P$, even though there is no conflict between any pair of polygons (see Figure 22.5). These special situations can only be avoided by marking and splitting the troublesome polygons.

The main advantages of the depth sort algorithm for hidden surfaces is that the data structure is simple—just an ordered list of polygons—and the rendering algorithm is straightforward—just paint the polygons in the order in which they appear in the list.

The main disadvantage of depth sort is that conflicts may be difficult to resolve. In the worst case if all five tests fail, then we need to resort to a splitting algorithm. For arbitrary nonconvex polygons, splitting can be a bit tricky. Moreover, finding and resolving cyclic conflicts may also be difficult to do.

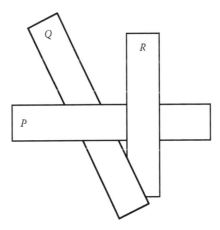

**FIGURE 22.5:** A cyclic conflict: $P$ obscures $Q$, $Q$ obscures $R$, $R$ obscures $P$. Notice, however, that there is no conflict between any pair of these three polygons.

### 22.6.1    Polygon Splitting

To split a polygon that lies on both sides of a fixed plane, we must first find the line where the given plane intersects the plane of the polygon. We then intersect this line with the edges of the polygon to split the polygon into polygonal segments that lie on either side of the splitting plane.

Consider then two nonparallel planes:

$$N_1 \cdot (P - Q_1) = 0,$$
$$N_2 \cdot (P - Q_2) = 0.$$

These planes intersect in a line $L$ determined by a point $Q$ and a direction vector $N$. By Chapter 11, Section 11.5.3, we can choose

$$N = N_1 \times N_2,$$

$$Q = \frac{(N_1 \cdot Q_1)((N_2 \cdot N_2)N_1 - (N_1 \cdot N_2)N_2) + (N_2 \cdot Q_2)((N_1 \cdot N_1)N_2 - (N_1 \cdot N_2)N_1)}{|N_1 \times N_2|^2}.$$

In fact, it is easy to see that the vector $N$ is perpendicular to both planes, since $N = N_1 \times N_2 \perp N_1, N_2$. Hence $N$ is parallel to the line $L$. Similarly, it is straightforward to verify that the point $Q$ satisfies the equations of both planes (see Exercise 22.2), and hence the point $Q$ lies on the line $L$.

Once we have a point $Q$ and a direction vector $N$ for the line $L$, we can split the polygon by intersecting the line $L$ with every edge of the polygon. Two coplanar lines:

$$L_1(s) = P + su,$$
$$L_2(t) = Q + tv,$$

intersect when $L_1(s) = L_2(t)$—that is, when

$$P + su = Q + tv,$$

or equivalently when

$$su - tv = Q - P.$$

Dotting both sides first with $u$ and then with $v$ yields two linear equations in two unknowns:

$$s(u \cdot u) - t(v \cdot u) = (Q - P) \cdot u,$$
$$s(u \cdot v) - t(v \cdot v) = (Q - P) \cdot v.$$

Solving for $s, t$ gives the parameters of the intersections points, and substituting $s$ into the expression for $L_1$ or $t$ into the expression for $L_2$ gives the coordinates of the intersection point. (For further details, see Chapter 11, Section 11.5.1.)

Since edges are bounded line segments, if the polygon is convex, the line $L$ actually intersects only two edges of the polygon. If $L_1(s)$ represents the line corresponding to an edge $P_iP_{i+1}$ of the polygon and $u = P_{i+1} - P_i$, then $L_2(t)$ intersects the edge $P_iP_{i+1}$ if and only if at the intersection point of the two lines $0 \leq s \leq 1$. Thus it is easy to detect which edges of the polygon are intersected by the line $L$ and to split the polygon accordingly (see Figure 22.6).

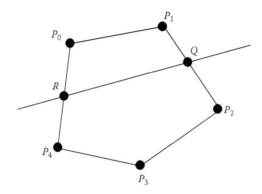

**FIGURE 22.6:** A polygon $P_0P_1P_2P_3P_4$ split by a straight line into two polygons $P_0P_1QR$ and $QP_2P_3P_4R$.

## 22.7   BSP-Tree

A *binary space partitioning tree* (*bsp-tree*) is a binary tree whose nodes contain polygons. For each node in a bsp-tree all the polygons in the left subtree lie behind the polygon at the node, while all the polygons in the right subtree lie in front of the polygon at the node. Each polygon has a fixed normal vector, and front and back are measured relative to this fixed normal. Once a bsp-tree is constructed for a scene, the polygons are rendered by an in order traversal of the bsp-tree. Recursive algorithms for generating a bsp-tree and then using the bsp-tree to render a scene are presented below.

*Algorithm for Generating a BSP-Tree*
    Select any polygon (plane) in the scene for the root.
    Partition all the other polygons in the scene to the back (left subtree) or the front (right subtree).
        Split any polygons lying on both sides of the root (see Section 22.6.1).
    Build the left and right subtrees recursively.

*BSP-Tree Rendering Algorithm* (In Order Tree Traversal)
    If the eye is in front of the root, then
        Display the left subtree (behind)
        Display the root
        Display the right subtree (front)
    If the eye is in back of the root, then
        Display the right subtree (front)
        Display the root
        Display the left subtree (back)

To generate a bsp-tree, we must be able to determine on which side of a plane a polygon lies and to split polygons that lie of both sides of a plane. Consider a plane defined by a point $Q$ and a normal vector $N$. For any point $P$ there are three possibilities:

i. $P$ lies in front of the plane $\Leftrightarrow N \cdot (P - Q) > 0$.

ii. $P$ lies on the plane $\Leftrightarrow N \cdot (P - Q) = 0$.

iii. $P$ lies behind the plane $\Leftrightarrow N \cdot (P - Q) < 0$.

**TABLE 22.1:**   Properties of hidden surface algorithms.

| Algorithm | Type | Sorting | Coherence |
|---|---|---|---|
| $z$-Buffer | Image space | None | Depth calculations |
| Scan line | Image space | Edges in the active edge list | Intensity calculations |
| Ray casting | Image/object space | None | None |
| Depth sort | Object space | Polygons | Intensity calculations |
| BSP-Tree | Object space | Polygons in bsp-tree | Intensity calculations |

If all the vertices of a polygon lie in front of a plane, then the entire polygon lies in front of the plane; if all the vertices of a polygon lie behind a plane, then the entire polygon lies behind the plane. Otherwise one part of the polygon lies in front and another part lies behind the plane. In this case, we split the polygon by the plane, using the procedure presented in Section 22.6.1.

The main advantage of the bsp-tree algorithm for hidden surfaces is that we can use the same bsp-tree for different positions of the eye. Thus when we want to move around in a scene the bsp-tree is the preferred approach to hidden surfaces. The main disadvantage of bsp-trees is the work involved in splitting polygons.

## 22.8 Summary

We have discussed five different hidden surface algorithms: $z$-buffer, scan line, ray casting, depth sort, and bsp-tree. Table 22.1 contains a summary of the main features of each of these algorithms. Two key ideas are applied to help increase the speed of these algorithms: sorting edges by depth, and pixel coherence for depth and intensity calculations. We can speed up some of these algorithms still further by storing bounding boxes for each polygonal face and avoiding interference tests when the bounding boxes do not overlap.

Algorithms for finding hidden surfaces and procedures for producing shadows use essentially the same computations: hidden surface algorithms find surfaces invisible to the eye, shadow procedures find surfaces invisible to a light source. Thus we can use the same algorithms for computing hidden surfaces and for calculating shadows. Moreover, we can often save time by reusing calculations. We can move the eye point and reuse shadow calculations; similarly, we can move the light sources and reuse hidden surface computations.

## Exercises

22.1. For each polygon $P$, let $A_P x + B_P y + C_P z + D_P = 0$ be the equation of the plane of the polygon $P$. Using these equations, find:

    a. The depth $z_P$ of the polygon $P$ at $x = Current\ Pixel, y = Y_{scan\ line}$.

    b. $\Delta z_P =$ the change in depth of the polygon $P$ as $x$ moves from pixel to pixel along $y = Y_{scan\ line}$.

    c. The $x$-coordinate where the polygon $Q$ intersects the polygon $P$ along $y = Y_{scan\ line}$.

22.2. Consider two planes

$$N_1 \cdot (P - Q_1) = 0,$$
$$N_2 \cdot (P - Q_2) = 0.$$

a. Show that if

$$Q = \frac{(N_1 \cdot Q_1)((N_2 \cdot N_2)N_1 - (N_1 \cdot N_2)N_2) + (N_2 \cdot Q_2)((N_1 \cdot N_1)N_2 - (N_1 \cdot N_2)N_1)}{|N_1 \times N_2|^2},$$

then

$$N_1 \cdot (Q - Q_1) = 0,$$
$$N_2 \cdot (Q - Q_2) = 0.$$

b. Conclude that the point $Q$ lies on the line representing the intersection of these two planes.

---

## Programming Projects

22.1. *Hidden Surface Algorithms*

Implement the five hidden surface algorithms discussed in this chapter in your favorite programming language using your favorite API.

a. Use your implementations to render several large polygonal models.

b. Compare the relative speeds of the five algorithms for each of your models.

22.2. *Scan Line Algorithms*

Integrate the scan line hidden surface algorithm with Gouraud or Phong shading to simultaneously remove hidden surfaces and eliminate Mach bands.

22.3. *BSP-Trees*

Implement the bsp-tree hidden surface algorithm and use this procedure to display the same models from several different points of view.

# *Chapter 23*

## *Radiosity*

*And he shall be as the light of the morning ...*

– 2 Samuel 23:4

## 23.1   Radiosity

Radiosity models the transfer of light between surfaces. Recursive ray tracing also models light bouncing off surfaces, but ray tracing makes several simplifying assumptions that often give scenes a harsh, unnatural look. For example, ray tracing assumes that all light sources are point sources and that the ambient light is constant throughout the scene. These assumptions lead to stark images with sharp shadows. In contrast, radiosity models all surfaces as both emitters and reflectors. This approach softens the shadows and provides a more realistic model for ambient light.

Informally, radiosity is the rate at which light energy leaves a surface. There are two contributions to radiosity: emission and reflection. Hence,

*Radiosity = Emitted Energy + Reflected Energy.*

For the purpose of display, we shall identify intensity with radiosity.

Radiosity computations typically take much longer than recursive ray tracing because the model of light is much more complex. To simplify these computations, we shall model only diffuse reflections; we shall not attempt to model specular reflections with radiosity. Since radiosity replaces ambient and diffuse intensity, radiosity is view independent. Thus, we can reuse the same computation for every viewpoint once we compute the radiosity of all the surfaces. View-dependent calculations are required only to compute hidden surfaces.

## 23.2   The Radiosity Equations

We will begin with a very general integral equation called the *Rendering Equation* based on energy conservation. We will then repeatedly simplify and discretize this equation till we get a large system of linear equations—the *Radiosity Equations*—which we can solve numerically for the radiosity of each surface. Radiosity is identified with intensity, so once we compute the radiosity for each surface we can render the scene.

### 23.2.1 The Rendering Equation

Energy conservation for light is equivalent to

$$Total\ Illumination = Emitted\ Energy + Reflected\ Energy.$$

We can rewrite these innocent looking words as an integral equation called the *Rendering Equation*.

*Rendering Equation*

$$I(x,x') = E(x,x') + \int_S \rho(x,x',x'')I(x',x'')dx'', \qquad (23.1)$$

where
  $I(x',x'')$ is the total energy passing from $x''$ to $x'$.
  $E(x,x')$ is the energy emitted directly from $x'$ to $x$.
  $\rho(x,x',x'')$ is the reflection coefficient—the percentage of the energy transferred from $x''$ to $x'$ that is passed on to $x$.

Essentially all of the computations in Computer Graphics that involve light are summarized in the Rendering Equation. Notice, in particular, that the Rendering Equation is precisely the setup for recursive ray tracing!

### 23.2.2 The Radiosity Equation: Continuous Form

The continuous form of the Radiosity Equation is just the Rendering Equation restricted to diffuse reflections. Once again, by conservation of energy,

$$Radiosity = Emitted\ Energy + Reflected\ Energy.$$

Now, however, since we are dealing only with diffuse reflections, we can be more specific about the form of the reflected energy. Restricting to diffuse reflections leads to the following integral equation for radiosity.

*Radiosity Equation—Continuous Form*

$$B(x) = E(x) + \rho_d(x)\int_S B(y)\frac{\cos\theta\cos\theta'}{\pi r^2(x,y)}V(x,y)dy, \qquad (23.2)$$

where
  $B(x)$ is the radiosity at the point $x$, which we identify with the intensity or energy reflecting off a surface in any direction—that is, the total power leaving a surface/unit area/solid angle. This energy is uniform in all directions, since we are assuming that the scene has only diffuse reflectors.
  $E(x)$ is the energy emitted directly from a point $x$ in any direction. This energy is uniform in all directions, since we are assuming that the scene has only diffuse emitters.
  $\rho_d(x)$ is the diffuse reflection coefficient—the percentage of energy reflected in all directions from the surface at a point $x$. By definition, $0 \le \rho_d(x) \ge 1$.
  $V(x,y)$ is the visibility term:
    $V(x,y) = 0$ if $x$ is not visible from $y$
    $V(x,y) = 1$ if $x$ is visible from $y$.

$\theta$ is the angle between the surface normal ($N$) at $x$ and the light ray ($L$) from $x$ to $y$.
$\theta'$ is the angle between surface normal ($N'$) at $y$ and the light ray ($L$) from $y$ to $x$.
$r(x, y)$ is the distance from $x$ to $y$.

In the Radiosity Equation, the term

$$\int_S B(y) \frac{\cos \theta \cos \theta'}{\pi r^2(x, y)} V(x, y) dy = Energy \ reaching \ the \ point \ x \ from \ all \ other \ points \ y,$$

so

$$\rho_d(x) \int_S B(y) \frac{\cos (\theta) \cos (\theta')}{\pi r^2(x, y)} V(x, y) dy = Total \ energy \ reflected \ from \ x.$$

The factor $1/r^2(x, y)$ models an inverse square law, since the intensity of light varies inversely as the square of the distance. The factors $\cos \theta, \cos \theta'$ come from Lambert's Law (see Chapter 16, Section 16.4) and represent projections of the flux onto the emitting and reflecting surfaces (see Figure 23.1 and the accompanying discussion). The appearance of the factor $\pi$ in the denominator will be explained shortly in the subsequent discussion.

To understand the cosine terms better, consider two small surface patches. Recall that the intensity (or radiosity) on any facet from any other facet is given by

$$I_{receptor} = \frac{Light \ Deposited}{Unit \ Area} = \frac{Beam \ Cross \ Section}{Receptor \ Facet \ Area} \times I_{source},$$

where

$$I_{source} = \frac{Light \ Emitted}{Unit \ Area} = \frac{Beam \ Cross \ Section}{Source \ Facet \ Area} \times I_{emitter}.$$

But from Figure 23.1,

$$\frac{Beam \ Cross \ Section}{Facet \ Area} = \cos (\theta), \cos (\theta').$$

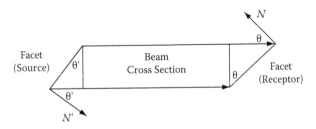

**FIGURE 23.1:** Lambert's Law. Intensity is given by the ratio of the beam cross section to the facet area, which, in turn, is equal to the cosine of the angle between the beam and the surface normal. Two cosines appear: one for the source and one for the receptor. Thus, Lambert's law accounts for the two cosines that appear in the Radiosity Equation.

Therefore

$$I_{receptor} = \cos\theta\cos\theta' I_{emitter}.$$

The factor $\pi$ in the denominator of the second term on the right-hand side of Equation 23.2 arises for the following reason. Recall that in the Rendering Equation,

$\rho(x',x,x'') =$ the percentage of the energy transferred from $x''$ to $x$ that is passed on to a single point $x'$. (Note that here we have reversed the roles of $x$ and $x'$.)

whereas in the Radiosity Equation,

$\rho_d(x) =$ the percentage of energy reflected in all directions from the surface at a point $x$.

Thus, we need to get from $\rho(x',x,x'')$ in the Rendering Equation to $\rho_d(x)$ in the Radiosity Equation. We shall now show that these two functions differ by a factor of $\pi$.

Directions can be represented by points on the unit sphere, and points on the unit sphere can be parametrized by two angles $\theta$, $\phi$. The angle $\theta$ represents the angle between the $z$-axis and the vector from the center of the sphere to the point $(\theta,\phi)$ on the sphere, and the angle $\phi$ represents the amount of rotation around the equator from the $x$-axis to the great circle of constant longitude $\phi$. Since $\rho_d(x)$ is the percentage of energy reflected in all direction whereas $\rho(x',x,x'')$ is the percentage of energy reflected in one fixed direction, it follows that $\rho_d(x)$ is the integral of $\rho(x',x,x'')$ over the unit hemisphere—that is, over all the directions that make an acute angle with the normal vector at $x$, which we identify with the $z$-axis. A factor of $\cos\theta$ must appear in this integral for the same reason (projection) that this factor appears in Lambert's Law. Thus,

$$\rho_d(x) = \int_H \rho(x',x,x'')\cos(\theta)\,dS,$$

where $H$ is a unit hemisphere centered at $x$. Since we are dealing with diffuse reflection, the function $\rho(x',x,x'')$ is the same in all directions. Thus, we can pull $\rho(x',x,x'')$ outside the integral, so

$$\rho_d(x) = \rho(x',x,x'')\int_H \cos(\theta)\,dS. \tag{23.3}$$

It remains then to compute $\int_H \cos(\theta)\,dS$.

To compute this integral, we use the parametrization $(\theta,\phi)$ of the unit sphere provided by spherical coordinates—that is, by setting

$$S(\theta,\phi) = (\sin(\theta)\cos(\phi),\sin(\theta)\sin(\phi),\cos(\theta)). \tag{23.4}$$

With this parametrization, a differential area (parallelogram) on the unit sphere is given by

$$dS = \left|\frac{\partial S(\theta,\phi)}{\partial\theta}\,d\theta \times \frac{\partial S(\theta,\phi)}{\partial\phi}\,d\phi\right| = \left|\frac{\partial S(\theta,\phi)}{\partial\theta} \times \frac{\partial S(\theta,\phi)}{\partial\phi}\right|d\theta d\phi.$$

But by Equation 23.4,

$$\frac{\partial S(\theta, \phi)}{\partial \theta} = (\cos(\theta)\cos(\phi), \cos(\theta)\sin(\phi), -\sin(\theta)),$$

$$\frac{\partial S(\theta, \phi)}{\partial \theta} = (-\sin(\theta)\sin(\phi), \sin(\theta)\cos(\phi), 0),$$

so by direct computation,

$$\left| \frac{\partial S(\theta, \phi)}{\partial \theta} \times \frac{\partial S(\theta, \phi)}{\partial \phi} \right| = \sin(\theta).$$

Therefore,

$$\int_H \cos(\theta)\,dS = \int_0^{2\pi}\int_0^{\pi/2} \cos(\theta)\sin(\theta)\,d\theta\,d\phi = \int_0^{2\pi} \left.\frac{\sin^2(\theta)}{2}\right|_0^{\pi/2} d\phi = \frac{1}{2}\int_0^{\pi/2} d\phi = \pi.$$

We conclude then from Equation 23.3 that

$$\rho_d(x) = \pi\rho(x', x, x'').$$

Thus, when we replace $\rho(x', x, x'')$ in the Rendering Equation by $\rho_d(x)$ in the Radiosity Equation, we must divide by $\pi$. This accounts for the factor $\pi$ in the denominator of the second term on the right-hand side of the Radiosity Equation.

### 23.2.3   The Radiosity Equation: Discrete Form

To find the radiosity at any point, we must solve the Radiosity Equation for $B(x)$—that is, we must calculate the integral on the right-hand side of Equation 23.2. But this integral is not easy to compute. Worse yet $B(u)$ appears on both sides of the Radiosity Equation; thus, we must know $B(u)$ to find $B(u)$!

The solution to both problems is to discretize the Radiosity Equation by breaking the surfaces in the scene into small patches. Since radiosity is approximately constant over a small patch, we shall be able to replace integrals by sums and continuous functions by discrete values. The integral equation then reduces to a large system of linear equations, which we shall be able to solve by numerical methods.

To discretize the Radiosity Equation, we begin by breaking the surfaces $S$ into small patches $P_j, j = 1, \ldots, N$. Over a small patch $P_j$, the radiosity $B(y)$ is approximately a constant $B_j$, and the integral over $S$ breaks up into a sum of integrals over the patches $P_j$. Therefore, by Equation 23.2,

$$B(x) = E(x) + \rho_d(x)\int_S B(y)\frac{\cos(\theta)\cos(\theta')}{\pi r^2(x, y)}V(x, y)dy,$$

$$\approx E(x) + \rho_d(x)\sum_{j=1}^N B_j \int_{P_j} \frac{\cos(\theta)\cos(\theta')}{\pi r^2(x, y)}V(x, y)dA_j. \tag{23.5}$$

We still need to discretize $B(x)$ on the left-hand side and $E(x)$ on the right-hand side of Equation 23.5. We can approximate the radiosity and energy of single patch $P_i$ by an area weighted average. Thus,

$$B_i \approx (1/A_i) \int_{P_i} B(x) dA_i,$$

$$E_i \approx (1/A_i) \int_{P_i} E(x) dA_i.$$

Integrating Equation 23.5 over the patch $P_i$ and dividing by the patch area $A_i$ yields

$$B_i = E_i + \rho_i \sum_{j=1}^{N} B_j (1/A_i) \int_{P_i} \int_{P_j} \frac{\cos \theta \cos \theta'}{\pi r^2(x, y)} V(x, y) dA_i \, dA_j.$$

Notice that the double integral

$$F_{ij} = (1/A_i) \int_{P_i} \int_{P_j} \frac{\cos \theta \cos \theta'}{\pi r^2(x, y)} V(x, y) dA_j \, dA_i, \tag{23.6}$$

is independent of the radiosity, and depends only on the geometry of the facets. The parameters $F_{ij}$ are called *form factors*; we shall have a lot more to say about these form factors shortly.

We have now reduced the Radiosity Equation to the following discrete form:

*Radiosity Equations—Discrete Form*

$$B_i = E_i + \rho_i \sum_{j=1}^{N} F_{ij} B_j, \quad i = 1, \ldots, N, \tag{23.7}$$

where

$B_i$ is the radiosity of patch $P_i$, which we identify with intensity.

$E_i$ is the energy emitted from patch $P_i$ (uniform in all directions, since by assumption we are dealing only with diffuse emitters).

$F_{ij}$ are the form factors, which depend only on the geometry of the scene and are independent of the lighting.

$\rho_i$ is the diffuse reflection coefficient for patch $P_i$, $0 \le \rho_i \le 1$.

Notice, in particular, that

$$\sum_{j=1}^{N} F_{ij} B_j = \textit{Energy reaching patch } P_i \textit{ from all the other patches,}$$

$$\rho_i \sum_{j=1}^{N} F_{ij} B_j = \textit{Energy reflected from the patch } P_i.$$

The discrete form of the Radiosity Equations given in Equation 23.7 are the equations that we are going to solve for the unknown radiosities $B_i$. These linear equations for $B_i$ are typically the starting point for most discussions of radiosity. The energies $E_i$ and the diffuse reflection coefficients $\rho_i$ are usually specified by the user; the form factors $F_{ij}$ depend on the geometry of the scene and must be computed. Therefore, before we can solve the Radiosity Equations for the radiosities $B_i$, we need to compute the form factors $F_{ij}$.

## 23.3 Form Factors

To gain a better understanding of the form factors, we are going to provide both a physical and a geometric interpretation for these constants. We begin with a physical interpretation.

**Theorem 23.1:** *The form factor $F_{ij}$ is the fraction of the energy leaving patch $P_i$ arriving at patch $P_j$.*

**Proof:** Let $B_{ij}$ denote the radiosity transferred from $P_i$ to $P_j$. From the Radiosity Equation,

$$B_{ij} = F_{ji}B_i.$$

Since radiosity is energy per unit area, to find the total energy transferred from $P_i$ to $P_j$, we must multiply the radiosity transferred from $P_i$ to $P_j$ by the area of $P_j$. Let $A_i$ be the area of patch $P_i$, and let $A_j$ be the area of patch $P_j$. Then the total energy transferred from $P_i$ to $P_j$ is

$$A_jB_{ij} = A_jF_{ji}B_i.$$

But by Equation 23.6,

$$A_jF_{ji} = \int\limits_{P_i} \int\limits_{P_j} \frac{\cos\theta \cos\theta'}{\pi r^2(x, y)} V(x, y)dA_j \, dA_i = A_iF_{ij}.$$

Therefore

$$A_jB_{ij} = A_iB_iF_{ij},$$

so

$$F_{ij} = \frac{A_jB_{ij}}{A_iB_i}. \tag{23.8}$$

Since the radiosity $B_i$ is the energy radiated in all directions from $P_i$ per unit area, the denominator $A_iB_i$ represents the total energy radiated from $P_i$. But we have already seen that the numerator $A_jB_{ij}$ represents the total energy transferred from $P_i$ to $P_j$. Therefore, it follows from Equation 23.8 that the form factor $F_{ij}$ is the fraction of the energy that leaves $P_i$ and arrives at $P_j$.

**Corollary 23.1:** *For each value of i, the form factors $F_{ij}$ form a partition of unity. That is,*

$$\sum_j F_{ij} = 1. \tag{23.9}$$

**Proof:** This result follows from conservation of energy. By Theorem 23.1, the form factor $F_{ij}$ is the fraction of the energy leaving patch $P_i$ arriving at patch $P_j$. Since energy is conserved, the energy leaving patch $P_i$ must arrive somewhere—that is, at one of the patches $P_j$. Therefore,

$$\sum_j F_{ij} = 1.$$

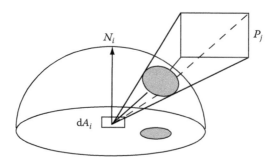

**FIGURE 23.2:** Projecting a patch $P_j$ first onto the unit hemisphere centered at the patch $dA_i$ and then onto the base of the hemisphere centered at $dA_i$.

To provide a geometric interpretation for the form factors, we must first discretize still further. Over a small patch $P_i$, we can treat the inner integral as roughly constant, so

$$F_{ij} = (1/A_i) \iint_{P_i P_j} \frac{\cos \theta \cos \theta'}{\pi r^2(x,y)} V(x,y) dA_j \ dA_i \approx \int_{P_j} \frac{\cos \theta \cos \theta'}{\pi r^2(x,y)} V(x,y) dA_j. \tag{23.10}$$

We shall now investigate the meaning of the differential $\frac{\cos \theta \cos \theta'}{\pi r^2(x,y)} dA_j$.

The product $\frac{\cos \theta'}{r^2} dA_j$ is the projection of the differential area $dA_j$ onto the unit hemisphere centered at the patch $dA_i$. Similarly, the product $(\cos \theta)(\frac{\cos \theta'}{r^2} dA_j)$ is the projection of the differential area $\frac{\cos \theta'}{r^2} dA_j$ onto the base of the hemisphere centered at $dA_i$, the plane perpendicular to the normal of $dA_i$ (see Figure 23.2).

In three dimensions, the cosine terms in these projections are not so easy to visualize, so to get a feel for these cosine terms, let us look instead in two dimensions. To project a differential length $dL$ onto a circle of radius $r$, where $r$ is the distance between the patches, we multiply $dL$ by $\cos(\theta')$ where $\theta'$ is the angle between the tangent to the circle of radius $r$ and the tangent to the curve $dL$ (see Figure 23.3a). Of course, the angle between the tangents is the same as the angle between the normals. If we think of $dL$ as one patch and if the second patch is located at the center of the circle of radius $r$, then the angle between the normals is the same as the angle $\theta'$ between the normal to $dL$ and the vector between the patches. Furthermore, to project the circle of radius $r$ onto the unit circle,

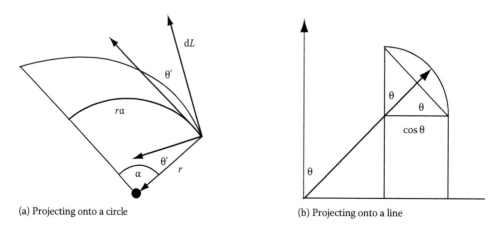

(a) Projecting onto a circle                    (b) Projecting onto a line

**FIGURE 23.3:**  (a) Projecting a differential length onto a circle is equivalent to multiplying by $\cos(\theta')$ and (b) projecting a differential length onto a line is equivalent to multiplying by $\cos(\theta)$.

we simply divide by $r$. Thus, to project a differential length $dL$ on one patch onto the unit circle centered at the other patch, we multiply $dL$ by $\cos \theta'/r$. (In three dimensions, we must scale uniformly in two directions, so the scale factor $r$ is replaced by $r^2$.)

Now to project from the unit circle to the $x$-axis, we must multiply by $\cos(\theta)$, where $\theta$ is the angle between the tangent vectors to the patch at the origin and the patch along the unit circle. Since the $x$-axis represents the plane of the first patch, the angle $\theta$ is also the angle between the normal vector to the first patch and the vector between the two patches (see Figure 23.3b). Therefore, the product $\frac{\cos \theta \cos \theta'}{r^2} dL$ is the product of the projection of $dL$, first onto the unit circle and then onto the $x$-axis. Essentially, the same analysis holds in three dimensions. Notice by the way that the factor of $\pi$ in the denominator of the integrand is simply the area of the unit circle at the base of the hemisphere. Hence, $\frac{\cos \theta \cos \theta'}{r^2} dA_j$ is the projection of $dA_j$ first onto the unit sphere centered at $dA_i$ and then onto the hemispherical plane of $dA_i$. Integrating $\frac{\cos \theta \cos \theta'}{r^2} dA_j$ over the patch $P_j$ yields the following results.

**Theorem 23.2:** *The form factor $F_{ij}$ is, up to division by $\pi$, the projection of the area of $P_j$ first onto the unit hemisphere centered at $dA_i$ and then onto the hemispherical plane of $dA_i$.*

**Corollary 23.2:** *Two surfaces $P_j, P_{j'}$ with the same projection onto the unit hemisphere centered at a small patch $dA_i$ have the same form factor—that is, $F_{ij} \approx F_{ij'}$.*

Corollary 23.2 is the main result of all this analysis: two patches with equal projections on the unit hemisphere centered around a small patch $dA_i$ will necessarily have the same form factor $F_{ij}$. We can use this insight to compute the form factors once and for all for some simple surface and then find the form factors for arbitrary surfaces by projecting onto the known surface.

### 23.3.1 Hemi-Cubes

A hemi-cube is the upper half of a cube with sides of length two, centered at a small patch $P_i$ (see Figure 23.4). Here we are going to compute explicit formulas for the form factors for small patches on the surface of the hemi-cube. To find the form factors for arbitrary patches, we will project these patches onto the hemi-cube and use Corollary 23.2, which says that patches with equal projections have equal form factors.

To compute form factors for small patches on the surface of the hemi-cube, recall from Equation 23.10 that in general the form factor for a small patch $\Delta A_i$ is given by

**FIGURE 23.4: (See color insert following page 354.)**   A hemi-cube. The lengths of the sides are twice the height of the hemi-cube.

$$F_{di,j} = \int_{P_j} \frac{\cos\theta\cos\theta'}{\pi r^2(x,y)} V(x,y) dA_j.$$

For a very small patch with area $\Delta A_j$, the integrand can be approximated by a constant, so

$$F_{di,dj} = \frac{\cos\theta\cos\theta'}{\pi r^2} \Delta A_j.$$

To find an explicit formula for the form factor $F_{di,dj}$, we need to find explicit formulas for $\cos\theta$, $\cos\theta'$, and $r$, where:

- $\theta$ is the angle between the normal $N$ to the patch $P_i$ and the vector from the center of $P_i$ to the center of $P_j$.

- $\theta'$ is the angle between the normal $N'$ to the patch $P_j$ and the vector from the center of $P_i$ to the center of $P_j$.

- $r$ is the distance from the center of $P_i$ to the center of $P_j$.

For the hemi-cube, there are two kinds of patches to consider: patches on the top of the hemi-cube and patches along the sides of the hemi-cube. To compute the form factors for these patches, choose a coordinate system with the origin at the center of the hemi-cube and the $z$-axis aligned with the normal to the patch $P_i$ at the center of the hemi-cube. For a small patch on the top face of the hemi-cube centered at the point $P = (x, y, 1)$, it follows from simple trigonometry (see Figure 23.5a) that

$$\cos\theta = \cos\theta' = 1/r.$$

Therefore, for small patches with area $\Delta A_j$ on the top face of the hemi-cube centered at the point $P = (x, y, 1)$,

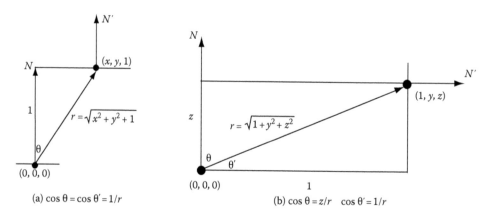

FIGURE 23.5: Schematic views of the top and side faces of a hemi-cube: (a) For the top face of the hemi-cube, the vectors $N$, $N'$ are parallel. Therefore, by simple trigonometry, $\cos\theta = \cos\theta' = 1/r$. (b) For the side faces of the hemi-cube, the vectors $N$, $N'$ are orthogonal. Therefore, by simple trigonometry, $\cos\theta = z/r$ and $\cos\theta' = 1/r$.

$$F_{di,dj} = \frac{\cos\theta\cos\theta'}{\pi r^2}\Delta A_j = \frac{\Delta A_j}{\pi r^4} = \frac{\Delta A_j}{\pi(x^2 + y^2 + 1)^2}.$$

Similarly, for a small patch on the side face of the hemi-cube parallel to the *yz*-plane centered at the point $P = (\pm 1, y, z)$, it again follows by simple trigonometry (see Figure 23.5b) that

$$\cos\theta = z/r, \quad \cos\theta' = 1/r.$$

Therefore, for small patches on side faces of the hemi-cube parallel to the *yz*-plane,

$$F_{di,dj} = \frac{\cos\theta\cos\theta'}{\pi r^2}\Delta A_j = \frac{z\Delta A_j}{\pi r^4} = \frac{z\Delta A_j}{\pi(y^2 + z^2 + 1)^2}.$$

Finally, for a small patch on the side face of the hemi-cube parallel to the *xz*-plane centered at the point $P = (x, \pm 1, z)$, it follows by an analogous argument that

$$\cos\theta = z/r, \quad \cos\theta' = 1/r,$$

$$F_{di,dj} = \frac{\cos\theta\cos\theta'}{\pi r^2}\Delta A_j = \frac{z\Delta A_j}{\pi r^4} = \frac{z\Delta A_j}{\pi(x^2 + z^2 + 1)^2}.$$

Thus, we have explicit formulas for the form factors for all the patches on the hemi-cube.

Once we have the form factors for the hemi-cube surrounding each patch, we can compute the form factors for the patches using the following algorithm.

*Form Factor Algorithm*

Compute the form factors for each cell of each hemi-cube, and store all of these form factors. To find the form factors $F_{ij}$ for the patches $P_j$ relative to the patch $P_i$,

For each face of the hemi-cube surrounding $P_i$:

1. Clip the scene to the frustum determined by the center of $P_i$ and the face of the hemi-cube. (See Chapter 14, Section 14.4.2 on how to map a frustum to a box.)

2. For each cell of the hemi-cube face:

   Find the nearest polygon in the scene. (Apply a *z*-buffer algorithm—see Chapter 22, Section 22.3.)

   Label each cell with the closest polygon.

3. For each polygon $P_j$, sum the form factors of the hemi-cube cells labeled *j*:

$$F_{di,j} = \sum_{q=j} F_q.$$

The calculation of form factors takes most of the time in the computation of radiosity. We can reduce this computation almost by half using the following identity.

*Reciprocity Relationship*

$$A_i F_{ij} = A_j F_{ji}. \tag{23.11}$$

By the reciprocity relationship, once we calculate $F_{ij}$, we can compute $F_{ji}$ with very little additional work. The reciprocity relationship holds because by Equation 23.6,

$$A_i F_{ij} = \int_{P_i} \int_{P_j} \frac{\cos\theta \cos\theta'}{\pi r^2} V(x,y) dA_j dA_i = A_j F_{ji}.$$

Note that we have already used this reciprocity relationship in the proof of Theorem 23.1. To use the reciprocity relationship to compute the form factors, we need to know the areas $A_i, A_j$. For planar polygons, these areas are easy to compute. By Newell's Formula (Chapter 11, Section 11.4.2.2) if $P_1, \ldots, P_n$ are the vertices of a planar polygon $P$, then

$$Area(P) = (1/2) \sum_{j=1}^{n} |P_j \times P_{j+1}|.$$

---

## 23.4   The Radiosity Rendering Algorithm

We introduced radiosity in order to develop more realistic looking images using Computer Graphics. Now let us put together what we now know about the Radiosity Equations, form factors, hemi-cubes, and shading to develop a rendering algorithm based on radiosity.

*Radiosity Rendering Algorithm*

1. Mesh the surfaces.

   Break each surface into small surface patches.

2. Compute the form factors for each pair of surface patches.

   Use the hemi-cube algorithm.

3. Solve the linear system (Equation 23.7) for the radiosities:

$$B_i = E_i + \rho_i \sum_{j=1}^{N} F_{ij} B_j, \quad i = 1, \ldots, N.$$

   (See Section 23.5.)

4. Compute the radiosity at the vertices of the patches. (See the discussion following this algorithm.)

5. Pick a viewpoint.

6. Determine which surfaces are visible.

   Use any hidden surface algorithm.

7. Apply Gouraud shading to the visible surfaces.

The only steps that require further elaboration are Step 3 and Step 4. In Step 3, we must solve a large system of linear equations. For large linear systems, standard techniques like Gaussian

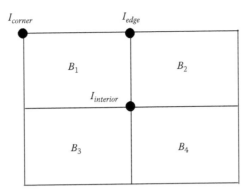

**FIGURE 23.6:** The intensities at the vertices depend on the radiosities of the adjacent patches. There are three kinds of vertices: interior vertices, edge vertices, and corner vertices. Each vertex type has a slightly different formula for intensity based on the radiosities of the adjacent patches.

elimination are slow and unstable. We shall provide instead two alternative robust numerical methods for solving these equations, but we defer this discussion till Section 23.5. In Step 4, we need to find the radiosity for each vertex so that we can perform Gouraud shading in Step 7 to eliminate discontinuities in intensity between adjacent patches.

For regular meshes consisting of rectangular patches, there are three kinds of vertices: interior, boundary, and corner (see Figure 23.6). For interior vertices, it is natural simply to set the intensity to the average of the radiosities at the four adjacent patches:

$$I_{interior} = \frac{B_1 + B_2 + B_3 + B_4}{4}.$$

For edges and corners, there are two competing strategies: either set

$$I_{edge} = \frac{B_1 + B_2}{2},$$
$$I_{corner} = B_1,$$

or set

$$\frac{I_{edge} + I_{interior}}{2} = \frac{B_1 + B_2}{2},$$
$$\frac{I_{corner} + I_{interior}}{2} = B_1,$$

so that

$$I_{edge} = B_1 + B_2 - I_{interior} = \frac{3B_1 + 3B_2 - B_3 - B_4}{4},$$
$$I_{corner} = 2B_1 - I_{interior} = \frac{7B_1 - B_2 - B_3 - B_4}{4}.$$

Both strategies yield reasonable results for Gouraud shading.

## 23.5   Solving the Radiosity Equations

The Radiosity Equations are

$$B_i = E_i + \rho_i \sum_{j=1}^{N} F_{ij}B_j, \quad i = 1, \dots, N.$$

Bringing all the radiosities to the left-hand side yields

$$\sum_{j=1}^{N} (\delta_{ij} - \rho_i F_{ij})B_j = E_i, \quad i = 1, \dots, N,$$

or equivalently

$$(1 - \rho_1 F_{11})B_1 - \rho_1 F_{12}B_2 - \cdots - \rho_1 F_{1N}B_N = E_1$$
$$-\rho_2 F_{21}B_1 + (1 - \rho_2 F_{22})B_2 - \cdots - \rho_2 F_{2N}B_N = E_2$$
$$\vdots \qquad\qquad\qquad\qquad\qquad \vdots$$
$$-\rho_N F_{N1}B_1 - \rho_N F_{N2}B_2 - \cdots + (1 - \rho_N F_{NN})B_N = E_N.$$

We can rewrite these equations in matrix form as

$$\underbrace{\begin{pmatrix} 1 - \rho_1 F_{11} & -\rho_1 F_{12} & \cdots & -\rho_1 F_{1N} \\ -\rho_2 F_{21} & 1 - \rho_2 F_{22} & \cdots & -\rho_2 F_{2N} \\ \vdots & \vdots & \ddots & \vdots \\ -\rho_N F_{N1} & -\rho_N F_{N2} & \cdots & 1 - \rho_N F_{NN} \end{pmatrix}}_{M} \underbrace{\begin{pmatrix} B_1 \\ B_2 \\ \vdots \\ B_N \end{pmatrix}}_{B} = \underbrace{\begin{pmatrix} E_1 \\ E_2 \\ \vdots \\ E_N \end{pmatrix}}_{E}.$$

Our goal is to solve for the unknown radiosities $B = (B_1, \dots, B_N)^{\mathrm{T}}$. We shall investigate two methods for solving these equations: gathering and shooting. Both methods take roughly the same total amount of time, but shooting facilitates faster viewing of intermediate solutions.

### 23.5.1   Gathering

To solve a large system of linear equations, we can employ standard relaxation techniques. Relaxation techniques are fixed point methods for linear equations. We studied these fixed point methods in Chapter 7, Section 7.3.2; we shall now briefly review these methods.

Suppose that

$$\sum_{j=1}^{N} M_{ij}B_j = E_i, \quad i = 1, \dots, N. \tag{23.12}$$

Solving for $B_i$ yields

$$B_i = \frac{E_i}{M_{ii}} - \sum_{j \neq i} \frac{M_{ij}}{M_{ii}} B_j, \quad i = 1, \dots, N.$$

In relaxation methods, we start with an initial guess

$$B^0 = \begin{pmatrix} B_1^0 \\ B_2^0 \\ \vdots \\ B_N^0 \end{pmatrix}.$$

Usually either

$$B^0 = \begin{pmatrix} E_1 \\ E_2 \\ \vdots \\ E_N \end{pmatrix} \quad \text{or} \quad B^0 = \begin{pmatrix} 0 \\ 0 \\ \vdots \\ 0 \end{pmatrix},$$

but since we are using a fixed point method, typically we can choose any value for $B^0$. We then apply a relaxation technique to compute the next guess $B^p$ from the previous guess $B^{p-1}$. Under certain simple conditions (see the following discussion), these relaxation methods are guaranteed to converge to the solution of the linear system. The two most common relaxation methods are due to Jacobi and to Gauss–Seidel.

*Jacobi Relaxation*

$$B_i^p = \frac{E_i}{M_{ii}} - \sum_{j \neq i} \frac{M_{ij}}{M_{ii}} B_j^{p-1}, \quad i = 1, \ldots, N.$$

*Gauss–Seidel Relaxation*

$$B_i^p = \frac{E_i}{M_{ii}} - \sum_{j=1}^{i-1} \frac{M_{ij}}{M_{ii}} B_j^p - \sum_{j=i+1}^{N} \frac{M_{ij}}{M_{ii}} B_j^{p-1}, \quad i = 1, \ldots, N.$$

In Jacobi relaxation, we use the values of the guess at the previous level to generate the values of the guess at the next level. In Gauss–Seidel relaxation, we use the values of the guess at the previous level along with the values of the guess already computed at the current level to compute the next value of the guess at the current level. Gauss–Seidel relaxation is a bit more complicated than Jacobi relaxation, but Gauss–Seidel relaxation typically converges faster to the solution of the linear system. (For additional details, see Chapter 7, Section 7.3.2.)

Convergence is guaranteed in both relaxation methods for any initial guess when the system is diagonally dominant. A system of equations such as Equation 23.12 is *diagonally dominant* if

$$|M_{ii}| \geq \sum_{j \neq i} |M_{ij}|.$$

It follows easily from Corollary 23.1 that the Radiosity Equations (Equations 23.7) are diagonally dominant (see Exercise 23.1), so we can use these relaxation methods to solve the Radiosity Equations.

Solving for the radiosity using relaxation methods is called *gathering* because we gather the radiosity from all the surfaces simultaneously. The disadvantage of gathering is that we must compute all the form factors to get an intermediate result. Thus we must solve for all the hemi-cubes radiosities before we can begin to render the scene. Typically, solving for all the hemi-cube radiosities and computing all the form factors takes a long time, so if there is some error in the scene or in the code, we will have to wait a long time to detect the mistake. Therefore, we shall consider another technique called *shooting*, where we can render intermediate results progressively without waiting to compute all the form factors and all the radiosities for each hemi-cube.

### 23.5.2   Shooting: Progressive Refinement

Each patch contributes to the radiosity of every other patch. To compute radiosity progressively, we fix one particular patch and compute the radiosity of every other patch due to the radiosity of the fixed patch. We can then display the scene and repeat the process by choosing another patch. In this way, we get to see intermediate results quickly without waiting to finish the entire computation.

From the Radiosity Equations,

$$B_i = E_i + \sum_{j=1}^{N} \rho_i F_{ij} B_j, \quad i = 1, \dots, N,$$

so the radiosity $B_i$ due to $B_j$ is $\rho_i F_{ij} B_j$. To solve the Radiosity Equations by gathering, we need to compute all the form factors for every patch; thus we need one hemi-cube for each patch to initiate the computation.

But in shooting, we are interested initially only in the radiosity $B_j$ due to one fixed radiosity $B_i$. The radiosity $B_j$ due to $B_i$ is $\rho_j F_{ji} B_i$. To compute the form factors $F_{ji}$ directly for each $j$, we would still need to introduce one hemi-cube computation for each patch. But recall that by the reciprocity relationships (Equation 23.11),

$$A_i F_{ij} = A_j F_{ji};$$

therefore,

$$B_j \text{ due to } B_i = \rho_j F_{ji} B_i = \rho_j F_{ij} B_i \frac{A_i}{A_j}.$$

Now to compute $F_{ij}$ for each $j$, we need the form factors $F_{ij}$ only for the single patch $P_i$ in order to update all the patches $P_j$! Thus we need to introduce only one hemi-cube at a time to update all the patch radiosities in a single pass. This reduction can save a lot of time.

In shooting, we think of radiosity as accumulating on each patch until we shoot this radiosity out to all the other patches. Let $\Delta B_j$ denote the unshot radiosity of the patch $P_j$. Then from the Radiosity Equations, it follows that after we shoot the accumulated radiosity $\Delta B_i$ from the patch $P_i$:

$$\Delta B_j = \left( \Delta B_j \right)_{old} + \rho_j F_{ij} \Delta B_i \frac{A_i}{A_j},$$

$$B_j = \left( B_j \right)_{old} + \Delta B_j = (B_j)_{old} + \rho_j F_{ij} \Delta B_i \frac{A_i}{A_j}, \tag{23.13}$$

$$\Delta B_i = 0.$$

These equations lead to the following progressive refinement procedure.

*Shooting Algorithm*

1. Initialize the radiosities.

$$B^0 = \begin{pmatrix} E_1 \\ E_2 \\ \vdots \\ E_N \end{pmatrix} = \Delta B^0.$$

2. Repeat until for all $j$, $\Delta B_j = 0$:

   a. Select the patch $P_i$ for which the total power $A_i \Delta B_i$ is maximal.

   b. Compute the form factors $F_{ij}$, using the hemi-cube computation for the patch $P_i$.

   c. Update $B_j$ and $\Delta B_j$ for all the patches $P_j$, using Equation 23.13.

   d. Display the scene using the current radiosity values $B_j$.

At the start of the shooting algorithm most of the radiosity is still unshot. Therefore, initially, the scene will appear quite dark. To get brighter pictures, we shall introduce an ambient correction term to account for the unshot radiosity. This correction term is for display only; the correction should not be added to the actual radiosities in the execution of the shooting algorithm.

To account for the ambient light due to unshot radiosity, we will first compute an average reflection coefficient and an average unshot radiosity. We will then add this average ambient reflection term to the radiosity of each patch.

We introduce an average diffuse reflection coefficient by taking an area weighted average of all the diffuse reflection coefficients:

$$\rho_{av} = \sum_{i=1}^{N} \rho_i A_i \Big/ \sum_{i=1}^{N} A_i.$$

We weight by area because larger patches will reflect more light. Next we compute the total amount of reflection $R$ by summing all the recursive reflections for the light bouncing repeatedly off all the patches:

$$R = 1 + \rho_{av} + \rho_{av}^2 + \cdots = \frac{1}{1 - \rho_{av}}.$$

The average unshot radiosity $\Delta B_{av}$ is the average of all the unshot radiosity weighted by area:

$$\Delta B_{av} = \sum_{i=1}^{N} \Delta B_i A_i \Big/ \sum_{i=1}^{N} A_i.$$

Finally, the ambient light $A$ is the reflection of all the unshot light. Thus,

$$A = R \Delta B_{av}.$$

For the purposes of display only, we add this ambient correction term to the radiosity of each patch, so that when we display the scene we set

$$B_i = B_i + \rho_i A, \quad i = 1, \ldots, N.$$

Although shooting allows us to view intermediate stages of the scene without computing all the form factors, in the end shooting is no faster than gathering. Ultimately, both gathering and shooting need to introduce the same number of hemi-cubes to find all the form factors for all the patches.

---

## 23.6  Summary

Radiosity is a classical energy transfer technique adapted to rendering surfaces in Computer Graphics. The advantages of radiosity over other standard rendering methods such as recursive ray tracing are that radiosity provides better photorealistic effects such as softer shadows and color bleeding. Radiosity computations are also view independent, since typically radiosity models only ambient light and diffuse reflection.

The main disadvantages of radiosity are that radiosity computations are expensive both in time and in space. Good accuracy demands lots of small patches to model curved surfaces, leading to lots of form factors and lots of hemi-cubes as well as a very large system of linear equations for the radiosities. In addition, radiosity does not typically model specular reflections, since specular reflections are view dependent.

Enhancements to standard radiosity include a two pass algorithm for computing specular reflections, more realistic light sources, participating mediums such as atmospheric fog, adaptive mesh generation for computing more accurate form factors, and finite element methods, or wavelets for calculating more accurate approximations to radiosity. The interested reader can find these subjects in the literature; we shall not pursue these topics here.

Below for easy reference we review the Rendering Equation along with the continuous and discrete forms of the Radiosity Equation and the form factors.

*Rendering Equation*

$$I(x, x') = E(x, x') + \int_S \rho(x, x', x'') I(x', x'') dx'', \tag{23.1}$$

where

$I(x', x'')$ is the total energy passing from $x''$ to $x'$.
$E(x, x')$ is the energy emitted directly from $x'$ to $x$.
$\rho(x, x', x'')$ is the reflection coefficient—the percentage of the energy transferred from $x''$ to $x'$ that is passed on to $x$.

*Radiosity Equation—Continuous Form*

$$B(x) = E(x) + \rho_d(x) \int_S B(y) \frac{\cos \theta \cos \theta'}{\pi r^2(x, y)} V(x, y) dy, \tag{23.2}$$

where

$B(x)$ is the radiosity at the point $x$, which we identify with the intensity or energy—that is, the total power leaving a surface/unit area/solid angle.
$E(x)$ is the energy emitted directly from a point $x$. This energy is uniform in all directions, since we are assuming that the scene has only diffuse emitters.
$\rho_d(x)$ is the diffuse reflection coefficient, $0 \le \rho_d(x) \le 1$.

$V(x, y)$ is the visibility term:

$V(x, y) = 0$ if $x$ is not visible from $y$.

$V(x, y) = 1$ if $x$ is visible from $y$.

$\theta$ is the angle between the surface normal ($N$) at $x$ and the light ray ($L$) from $x$ to $y$.

$\theta'$ is the angle between surface normal ($N'$) at $y$ and the light ray ($L$) from $y$ to $x$.

$r(x, y)$ is the distance from $x$ to $y$.

*Radiosity Equations—Discrete Form*

$$B_i = E_i + \rho_i \sum_{j=1}^{N} F_{ij}B_j, \quad i = 1, \ldots, N, \tag{23.7}$$

where

$B_i$ is the radiosity of patch $P_i$, which we identify with intensity.

$E_i$ is the energy emitted from patch $P_i$ (uniform in all directions, since by assumption we are dealing only with diffuse emitters).

$F_{ij}$ are the form factors, which depend only on the geometry of the scene and are independent of the lighting.

$\rho_i$ is the diffuse reflection coefficient for patch $P_i$, $0 \leq \rho_i \leq 1$.

*Form Factors*

$$F_{ij} = (1/A_i) \iint_{P_i P_j} \frac{\cos\theta \cos\theta'}{\pi r^2(x, y)} V(x, y) dA_j \, dA_i. \tag{23.6}$$

---

# Exercises

23.1. Prove that the Radiosity Equations are diagonally dominant.

23.2. Using Theorem 23.1, show that $F_{i,j \cup k} = F_{i,j} + F_{i,k}$.

23.3. Let $M = (M_{ij})$, $B = (B_1, \ldots, B_N)^{\mathrm{T}}$, $E = (E_1, \ldots, E_N)^{\mathrm{T}}$, and let $D$ be the diagonal matrix defined by

$$D_{ij} = \begin{cases} M_{ii}, & i = j, \\ 0, & i \neq j. \end{cases}$$

Consider the system of linear equations:

$$
\begin{aligned}
M_{11}B_1 + M_{12}B_2 + \cdots + M_{1N}B_N &= E_1 \\
M_{21}B_1 + M_{22}B_2 + \cdots + M_{2N}B_N &= E_2 \\
&\vdots \qquad\qquad \Leftrightarrow \quad M * B = E. \\
M_{N1}B_1 + M_{N2}B_2 + \cdots + M_{NN}B_N &= E_N
\end{aligned}
$$

a. Show that this system is equivalent to:

$$M_{11}B_1 = E_1 - M_{12}B_2 - \cdots - M_{1N}B_N$$
$$M_{22}B_2 = E_2 - M_{21}B_1 - \cdots - M_{2N}B_N$$
$$\vdots \qquad\qquad \vdots \qquad\qquad \vdots \qquad \Leftrightarrow \quad D*B = E - (M - D)*B.$$
$$M_{NN}B_N = E_N - M_{N1}B_1 - \cdots$$

b. Suppose that $M_{ii} \neq 0$ for $i = 1, \ldots, N$. Show that the system in part a is equivalent to:

$$B_1 = \frac{E_1}{M_{11}} - \frac{M_{12}}{M_{11}}B_2 - \cdots - \frac{M_{1N}}{M_{11}}B_N$$
$$B_2 = \frac{E_2}{M_{22}} - \frac{M_{21}}{M_{22}}B_1 - \cdots - \frac{M_{2N}}{M_{22}}B_N$$
$$\vdots \qquad\qquad \vdots \qquad\quad \vdots \qquad \Leftrightarrow \quad B = D^{-1}*E - \left(D^{-1}*M - I\right)*B.$$
$$B_N = \frac{E_N}{M_{NN}} - \frac{M_{N1}}{M_{NN}}B_1 - \cdots$$

c. Show that:

$$D_{ij}^{-1} = \begin{cases} \frac{1}{M_{ii}}, & i = j, \\ 0, & i \neq j, \end{cases}$$

$$D^{-1}*M - I = \begin{pmatrix} 0 & \frac{M_{12}}{M_{11}} & \cdots & \frac{M_{1N}}{M_{11}} \\ \frac{M_{21}}{M_{22}} & 0 & \cdots & \frac{M_{2N}}{M_{22}} \\ \vdots & \ddots & \ddots & \vdots \\ \frac{M_{N1}}{M_{NN}} & \cdots & \frac{M_{NN-1}}{M_{NN}} & 0 \end{pmatrix}.$$

d. Let $T(B) = D^{-1}*E - (D^{-1}*M - I)*B$. Show that Jacobi relaxation is equivalent to iterating the transformation $T(B)$.

e. Conclude from part d that Jacobi relaxation is equivalent to finding a fixed point of the transformation $T(B)$.

23.4. Let $M = \left(M_{ij}\right)$, $B = (B_1, \ldots, B_N)^T$, $E = (E_1, \ldots, E_N)^T$, and let $L$ be the lower triangular matrix defined by

$$L_{ij} = \begin{cases} M_{ij}, & i \geq j, \\ 0, & i < j. \end{cases}$$

Consider the system of linear equations:

$$M_{11}B_1 + M_{12}B_2 + \cdots + M_{1N}B_N = E_1$$
$$M_{21}B_1 + M_{22}B_2 + \cdots + M_{2N}B_N = E_2$$
$$\vdots \qquad\qquad\qquad \vdots \qquad\qquad \Leftrightarrow \quad M*B = E.$$
$$M_{N1}B_1 + M_{N2}B_2 + \cdots + M_{NN}B_N = E_N$$

a. Show that this system is equivalent to:

$$
\begin{aligned}
M_{11}B_1 &= E_1 - M_{12}B_2 - \cdots - M_{1N}B_N \\
M_{21}B_1 + M_{22}B_2 &= E_2 - M_{23}B_3 - \cdots - M_{2N}B_N \\
\vdots \quad & \qquad \vdots \\
M_{N1}B_1 + \cdots + M_{NN}B_N &= E_N
\end{aligned}
\quad \Leftrightarrow \quad L*B = E - (M - L)*B.
$$

b. Suppose that $M_{ii} \neq 0$ for $i = 1, \ldots, N$. Show that the system in part a is equivalent to:

$$
B = L^{-1}*E - \left(L^{-1}*M - I\right)*B.
$$

c. Show that the matrix $L^{-1}$—and hence the matrix $L^{-1}*M - I$—is easy to compute. In particular, show that $L^{-1}$ is lower triangular and that each entry can be computed by solving one linear equation in one unknown.

d. Let $T(B) = L^{-1}*E - (L^{-1}*M - I)*B$. Show that Gauss–Seidel relaxation is equivalent to iterating the transformation $T(B)$.

e. Conclude from part d that Gauss–Seidel relaxation is equivalent to finding a fixed point of the transformation $T(B)$.

---

## Programming Project

23.1. *Radiosity*

Implement radiosity in your favorite programming language using your favorite API.

a. Use your implementation to render several different scenes—apply both gathering and shooting.

b. Compare the scenes rendered by radiosity with the same scenes rendered using recursive ray tracing.

# Part IV

# Geometric Modeling:
# Freeform Curves and Surfaces

# Chapter 24

## Bezier Curves and Surfaces

*thou shalt be near unto me*

– Genesis 45:10

### 24.1    Interpolation and Approximation

Freeform curves and surfaces are smooth shapes often describing man-made objects. The hood of a car, the hull of a ship, and the fuselage of an airplane are all examples of freeform shapes. Freeform surfaces differ from the classical surfaces we encountered in earlier chapters such as spheres, cylinders, cones, and tori. Classical surfaces are typically easy to describe with a few simple parameters. A sphere can be represented by a center point and a scalar radius; a cone by a vertex point, a vertex angle, and an axis vector. The hood of a car or the hull of a ship is not so easy to describe with a few simple parameters. The goal of the next few chapters is to develop mathematical techniques for describing freeform curves and surfaces.

Scientists and engineers use freeform curves and surfaces to interpolate data and to approximate shape. But interpolation and approximation are not always compatible operations. Consider the data in Figure 24.1a. If we use a low degree polynomial as in Figure 24.1b to interpolate this data, the polynomial oscillates about the $x$-axis, even though there are no such oscillations in the data. Thus the shape of the interpolating polynomial does not reflect the shape of the data. Moreover, even providing more and more data points on the desired curve may not eliminate these unwanted oscillations.

The goal of approximation is to capture the shape of a desired curve or surface from a few data points without necessarily interpolating the points. Unlike interpolation, in approximation the data points themselves are not sacred. Rather the data points are *control points*; these points control the shape of the curve or surface and can be adjusted in order to provide a better representation of the desired shape. The curve in Figure 24.2b approximates the shape of the data in Figure 24.2a, even though the curve does not pass through all the data points. The curve in Figure 24.2b is called a *Bezier curve*. Bezier curves have many practical applications, ranging from the design of new fonts to the creation of mechanical components and assemblies for large scale industrial design and manufacture. The goal of this chapter is to develop some of the theory underlying Bezier curves and surfaces.

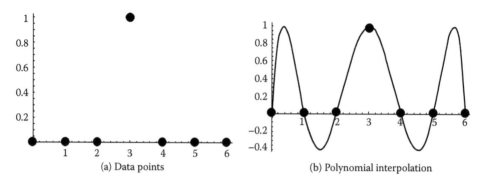

(a) Data points     (b) Polynomial interpolation

**FIGURE 24.1:** Polynomial interpolation. Notice that there are oscillations in the interpolating polynomial curve, even though there are no oscillations in the original data points.

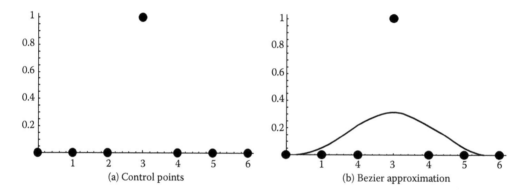

(a) Control points     (b) Bezier approximation

**FIGURE 24.2:** Polynomial approximation. A Bezier curve approximates the shape described by the control points, but the Bezier curve does not interpolate all the control points. The height of the curve can be adjusted by changing the height of the middle control point. Compare to Figure 24.1b.

## 24.2 The de Casteljau Evaluation Algorithm

The easiest curve to represent with control points is a straight line. Given two points $P_0, P_1$, the line $P(t)$ joining $P_0$ and $P_1$ can be expressed parametrically by setting

$$P(t) = P_0 + t(P_1 - P_0) = (1 - t)P_0 + tP_1. \tag{24.1}$$

It is easy to check that $P(t)$ represents a straight line, and that $P(0) = P_0$ and $P(1) = P_1$. Thus the curve $P(t)$ passes through the point $P_0$ at time $t = 0$ and through the point $P_1$ at time $t = 1$.

Suppose, however, that you want the straight line to pass through the point $P_0$ at time $t = a$ and through the point $P_1$ at time $t = b$. Then mimicking Equation 24.1, you might write

$$P(t) = (1 - f(t))P_0 + f(t)P_1 \tag{24.2}$$

with the requirement that

$$f(a) = 0 \quad \text{and} \quad f(b) = 1. \tag{24.3}$$

To find a simple explicit expression for the function $f(t)$, you would need to find the line in the coordinate plane interpolating the data $(a, 0)$ and $(b, 1)$. Using standard techniques from analytic geometry, you can write the equation of this line as

$$f(t) = \frac{(t-a)}{(b-a)}, \qquad (24.4)$$

and you can easily verify that $f(a) = 0$ and $f(b) = 1$. Substituting Equation 24.4 into Equation 24.2 yields

$$P(t) = \frac{b-t}{b-a} P_0 + \frac{t-a}{b-a} P_1. \qquad (24.5)$$

Equation 24.5 is called *linear interpolation* because the curve $P(t)$ interpolates the points $P_0, P_1$ with a straight line. You have encountered linear interpolation many times before in Computer Graphics; for example, Gouraud and Phong shading are both based on linear interpolation.

Equation 24.5 is so important that we are going to represent this equation by a simple graph. In Figure 24.3a, the control points $P_0, P_1$ are placed in the two nodes at the base of the diagram and the coefficients of the control points $P_0, P_1$ in Equation 24.5 are placed along the arrows emanating from these nodes. The values in the nodes are multiplied by the values along the arrows; these products are then added and the result is placed in the node at the apex of the diagram. Thus Figure 24.3a is a graphical representation of Equation 24.5. Figure 24.3b is the same as Figure 24.3a, except that to avoid cluttering the diagram, we have removed the normalizing constant $b - a$ in the denominator of the functions along the arrows. We can easily retrieve this normalizing constant, since the denominator is simply the sum of the numerators: $b - a = (b - t) + (t - a)$. Thus we shall interpret Figure 24.3b to mean Figure 24.3a, which in turn is equivalent to Equation 24.5.

A *Bezier curve* is a curve generated by an algorithm where the steps in Figure 24.3 are repeated over and over again. Figure 24.4a with three control points at the base represents a quadratic Bezier curve, and Figure 24.4b with four control points at the base represents a cubic Bezier curve. The corresponding curves are illustrated in Figure 24.5a and Figure 24.5b. The piecewise linear curve consisting of the lines connecting the control points with consecutive indices is called the *control polygon*. Notice how the Bezier curves in Figure 24.5 mimic the shape of their control polygons.

This algorithm for generating Bezier curves is called the *de Casteljau evaluation algorithm*. A curve generated by the de Casteljau algorithm with $n + 1$ control points at the base is called *a degree n Bezier curve*. Notice that in the de Casteljau algorithm, all the left pointing arrows are labeled with the function $t - a$ and all the right pointing arrows are labeled with the function $b - t$.

Intermediate nodes of the de Casteljau algorithm represent Bezier curves of lower degree. Thus the de Casteljau algorithm is a dynamic programming procedure for computing points on a Bezier curve. Usually, for reasons that will become clear in Section 24.4.2. Bezier curves are restricted to

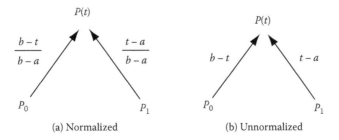

(a) Normalized          (b) Unnormalized

**FIGURE 24.3:** Linear Interpolation. Two diagrams representing the line in Equation 24.5.

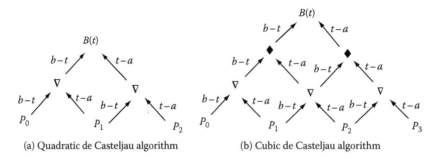

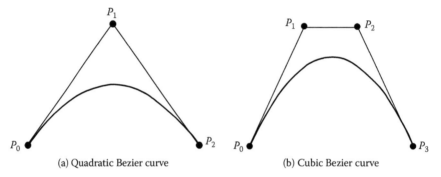

**FIGURE 24.4:**   The de Casteljau evaluation algorithm for (a) a quadratic Bezier curve and for (b) a cubic Bezier curve. The label on each edge must be normalized by dividing by $b - a$.

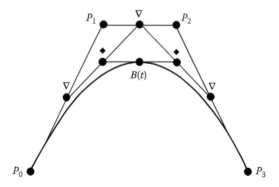

(a) Quadratic Bezier curve          (b) Cubic Bezier curve

**FIGURE 24.5:**   A quadratic Bezier curve (left) and a cubic Bezier curve (right). Notice how the shape of the Bezier curve (dark) mimics the shape of the control polygon (light).

the parameter interval $[a, b]$—that is, usually we shall insist that $a \le t \le b$. Typically, we shall take $a = 0$ and $b = 1$, though there are cases where we will need to take other values for $a, b$. Notice that when $a = 0$ and $b = 1$, no normalization is required because $b - a = 1 - 0 = 1$.

Since each node in the de Casteljau algorithm represents the equation of a straight line joining the points in the nodes immediately below to the left and the right, each node symbolizes a point on the line segment joining the two points whose arrows point into the node. Drawing all these line segments generates the trellis in Figure 24.6.

**FIGURE 24.6:**   Geometric construction algorithm for a point on a cubic Bezier curve based on a geometric interpretation of the de Casteljau evaluation algorithm. If the labels along the edges in the de Casteljau algorithm are $(b - t)$ and $(t - a)$, then at the parameter $t$ each line segment in the trellis is split in the ratio $(t - a)/(b - t)$.

## 24.3    The Bernstein Representation

Bezier curves are polynomial curves. The de Casteljau algorithm proceeds by adding and multiplying the linear functions $(b - t)/(b - a)$ and $(t - a)/(b - a)$ (see Figure 24.4). But adding and multiplying polynomials generates polynomials of higher degree. Therefore the de Casteljau algorithm generates polynomial curves.

We can find an explicit polynomial representation for Bezier curves. Let $B(t)$ denote the Bezier curve with control points $P_0, \dots, P_n$. Since $B(t)$ is a degree $n$ polynomial curve, we could try to express $B(t)$ relative to the standard polynomial basis $1, t, t^2, \dots, t^n$—that is, we could ask: what are the constant coefficients of the basis functions $1, t, t^2, \dots, t^n$ for the polynomial $B(t)$? Unfortunately, these coefficients are numerically unstable, so in practice these values are not very useful. A better, more insightful question to ask is: what are the polynomial coefficients of the control points $P_0, \dots, P_n$?

Let $B_k^n(t)$ denote the coefficient of the control point $P_k$ in the function $B(t)$. From the de Casteljau algorithm (Figure 24.4), it is easy to see that

$B_k^n(t) = $ *the sum over all paths from $P_k$ at the base to $B(t)$ at the apex of the graph,*

where

*a path $=$ the product of all the labels along the arrows in the path.*

Since $P_k$ lies in the $k$th position at the base of the diagram, to reach the apex a path must take exactly $k$ left turns and exactly $n - k$ right turns. But each left pointing arrow carries the label $(t - a)/(b - a)$ and each right pointing arrow carries the label $(b - t)/(b - a)$. Therefore, all the paths from $P_k$ at the base to $B(t)$ at the apex of the diagram generate the same product, so

$$B_k^n(t) = P(n, k) \frac{(t - a)^k (b - t)^{n-k}}{(b - a)^n},$$

where

$P(n, k) = $ *the number of paths from $P_k$ to $B(t)$.*

To find a closed formula for $P(n, k)$, observe that $P(n, k)$ satisfies the recurrence

$$P(n, k) = P(n - 1, k - 1) + P(n - 1, k),$$

because the only way to reach the $k$th position on the $n$th level is from either the $(k - 1)$st position on the $(n - 1)$st level or from the $k$th position on the $(n - 1)$st level (see Figure 24.7). Since $P(0, 0) = 1$, the values in Pascal's triangle (binomial coefficients) and the values in the path triangle are identical: $\binom{n}{k}$ and $P(n, k)$ start at the same value and satisfy the same recurrence.

Therefore

$$P(n, k) = \binom{n}{k} = \frac{n!}{k!(n - k)!},$$

so

$$B_k^n(t) = \binom{n}{k} \frac{(t - a)^k (b - t)^{n-k}}{(b - a)^n}.$$

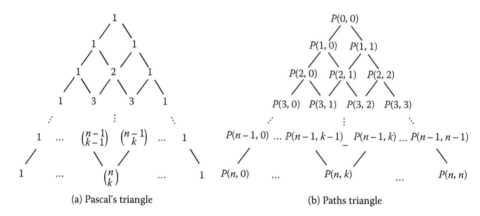

(a) Pascal's triangle       (b) Paths triangle

**FIGURE 24.7:** Pascal's triangle and the paths triangle start at the same value and satisfy the same recurrence. Therefore, $P(n,k) = \begin{pmatrix} n \\ k \end{pmatrix}$.

The functions $B_0^n(t), \ldots, B_n^n(t)$ are called the *Bernstein basis functions*. We shall show in Chapter 26, Section 26.3 that the Bernstein basis functions form a basis for the polynomials of degree $n$; every polynomial of degree $n$ can be expressed in terms of the Bernstein basis functions. From this perspective, the control points $P_0, \ldots, P_n$ of a Bezier curve $B(t)$ are simply the coefficients of the Bezier curve relative to the Bernstein basis $B_0^n(t), \ldots, B_n^n(t)$—that is,

$$
B(t) = \sum_{k=0}^{n} B_k^n(t) P_k,
$$

$$
B_k^n(t) = \begin{pmatrix} n \\ k \end{pmatrix} \frac{(t-a)^k (b-t)^{n-k}}{(b-a)^n}, \quad k = 0, \ldots, n.
$$

(24.6)

Equation 24.6 is called the *Bernstein representation* of the Bezier curve $B(t)$. The Bernstein basis functions $B_0^n(t), \ldots, B_n^n(t)$ are also called *blending functions*, since these functions blend the discrete control points $P_0, \ldots, P_n$ to form a smooth curve.

## 24.4   Geometric Properties of Bezier Curves

Bezier curves have the following key geometric features: they are affine invariant, lie in the convex hull of their control points, satisfy the variation diminishing property (i.e., they do not oscillate more than their control polygon), and interpolate their first and last control points.

There are two ways to derive these properties: either we can appeal to the de Casteljau evaluation algorithm or we can invoke the Bernstein representation of a Bezier curve. Here we shall derive most of these properties simply by reading these attributes off the graph representing the de Casteljau algorithm. For simplicity, we shall typically illustrate our arguments on cubic Bezier curves, but our proofs are completely general and apply to Bezier curves of arbitrary degree. Alternative proofs based on the Bernstein representation of a Bezier curve are provided in Exercises 24.4–24.7 at the end of this chapter.

### 24.4.1 Affine Invariance

A curve scheme is said to be *affine invariant* if applying an affine transformation to the control points transforms every point on the curve by the same affine transformation. For example, a curve scheme is *translation invariant* if translating each control point by a vector $v$ translates the entire curve by the vector $v$.

Translation invariance is an easy consequence of the de Casteljau algorithm. If we translate each control point by the vector $v$, then each node in the de Casteljau algorithm translates by the same vector $v$ because the coefficients along the two arrows entering each node sum to one (see Figure 24.8). More generally, an affine transformation can be represented by an affine matrix. If we multiply each control point by an affine transformation matrix $M$, then each node in the de Casteljau algorithm is multiplied by the same transformation matrix $M$ because matrix multiplication distributes through addition and commutes with scalar multiplication (see Figure 24.9).

Translation invariance is equivalent to asserting that the curve is independent of the choice of the origin of the coordinate system. Affine invariance insures that the curve is also independent of the choice of the coordinate axes. Both of these properties are essential for a good approximation scheme. In Computer Graphics, a curve should be completely determined by its control points; during design we do not want to worry about the location of the coordinate origin or the orientation of the coordinate axes. Bezier curves depend only on their control points; the location of the coordinate origin and the orientation of the coordinate axes affect neither the location nor the shape of a Bezier curve.

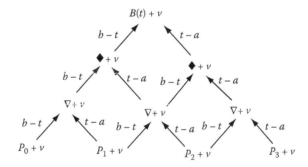

**FIGURE 24.8:** Translation invariance. Translating each control point by a vector $v$ translates each point on the curve by the same vector $v$.

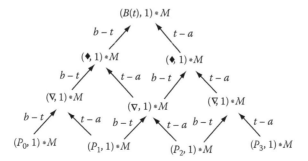

**FIGURE 24.9:** Affine invariance. Transforming each control point by an affine transformation matrix $M$ transforms every point on the curve by the same affine transformation matrix $M$.

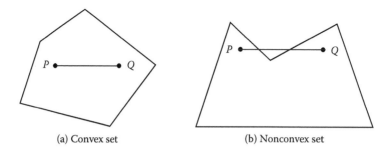

(a) Convex set　　　　　　　　(b) Nonconvex set

**FIGURE 24.10:**　In a convex set (a), the line segment joining any two points in the set lies entirely within the set. In a nonconvex set (b), part of the line segment joining two points in the set may lie outside the set.

### 24.4.2　Convex Hull Property

A set of points $S$ is said to be *convex* if whenever $P$ and $Q$ are points in $S$ the entire line segment from $P$ to $Q$ lies within $S$ (see Figure 24.10). The intersection $S$ of a collection of convex sets $\{S_i\}$ is a convex set because if $P$ and $Q$ are points in $S$, they must also be points in each of the sets $S_i$. Since, by assumption, the sets $S_i$ are convex, the entire line segment from $P$ to $Q$ lies in each set $S_i$. Hence the entire line segment from $P$ to $Q$ lies in the intersection $S$, so $S$ too is convex.

The *convex hull* of a collection of points $\{P_k\}$ is the intersection of all the convex sets containing the points $\{P_k\}$. Since the intersection of convex sets is a convex set, the convex hull is the smallest convex set containing the points $\{P_k\}$. For two points, the convex hull is the line segment joining the points. For three noncollinear points, the convex hull is the triangle whose vertices are the three points. The convex hull of a finite collection of points in the plane can be found mechanically by placing a nail at each point, stretching a rubber band so that its interior contains all the nails, and then releasing the rubber band. When the rubber band comes to rest on the nails, the interior of the rubber band is the convex hull of the points.

Bezier curves always lie in the convex hull of their control points. Once again, we can prove this assertion directly from the de Casteljau algorithm. First recall that, by convention, we always restrict the Bezier curve to the parameter interval $[a, b]$. With this convention, the points on the first level of the de Casteljau algorithm certainly lie in the convex hull of the control points on the zeroth level because, by construction, the points on the first level lie on the line segments joining adjacent control points (see Figure 24.6). For the same reason, the points on the $(n + 1)$st level of the de Casteljau algorithm lie in the convex hull of the points on the $n$th level of the de Casteljau algorithm. Hence, by induction, the point at the apex of the de Casteljau algorithm lies in the convex hull of the control points at the base of the de Casteljau algorithm. Thus each point on a Bezier curve lies in the convex hull of the control points.

The convex hull property is important in Computer Graphics because the convex hull property guarantees that if all the control points are visible on the graphics terminal, then the entire curve is visible as well. The restriction $a \le t \le b$ on the parameter $t$ is there precisely to guarantee the convex hull property.

### 24.4.3　Variation Diminishing Property

Intuitively, a curve is said to be *variation diminishing* if the curve does not oscillate more than its control points. In Section 24.1, we observed that interpolating polynomials may oscillate even if the data does not (see Figure 24.1). We abandoned interpolation in favor of approximation precisely in order to avoid unnecessary oscillations. Therefore we need to be sure that Bezier curves are variation diminishing.

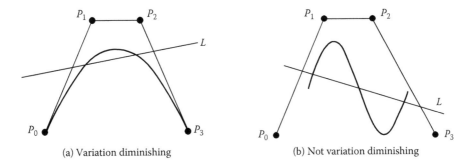

(a) Variation diminishing          (b) Not variation diminishing

**FIGURE 24.11:**  (a) A variation diminishing curve. An arbitrary line $L$ intersects both the curve and the control polygon twice. (b) A curve that is not variation diminishing. The line $L$ intersects the curve three times, but the control polygon only twice.

But how do we measure oscillations? In Figure 24.1, we considered oscillations with respect to the $x$-axis. The oscillations of the curve in Figure 24.1 are linked to the number of times this curve crosses the $x$-axis. But in affine invariant schemes, there is nothing special about the $x$-axis; any line will do. Therefore we say that a curve is *variation diminishing* if the number of intersections of the curve with each line in the plane (or each plane in three space) is less than or equal to the number of intersections of the line (or the plane) with the control polygon (see Figure 24.11). In this definition, we ignore lines that coincide with (or planes that contain) an edge of the control polygon.

Bezier curves are indeed variation diminishing, but it is not so clear how to derive this fact from the de Casteljau evaluation algorithm. Therefore we shall defer our proof of the variation diminishing property for Bezier curves till Chapter 25, Section 25.4, where we shall derive the variation diminishing property for Bezier curves from de Casteljau's subdivision algorithm.

### 24.4.4   Interpolation of the First and Last Control Points

Bezier curves do not generally interpolate all their control points. But Bezier curves always interpolate their first and last control points. In fact, if $B(t)$ represents the Bezier curve over the interval $[a, b]$ with control points $P_0, \ldots, P_n$, then $B(a) = P_0$ and $B(b) = P_n$. Thus $B(t)$ starts at the initial control point $P_0$ and ends at the final control point $P_n$.

As usual, we can establish these results by examining the de Casteljau algorithm. If we set $t = a$ in de the Casteljau algorithm, then all the right pointing arrows become one and all the left pointing arrows become zero (see Figure 24.12a). Now if we follow the de Casteljau algorithm from the base to the apex, then $P_0$ appears at each node along the left edge of Figure 24.12a. Therefore, at the

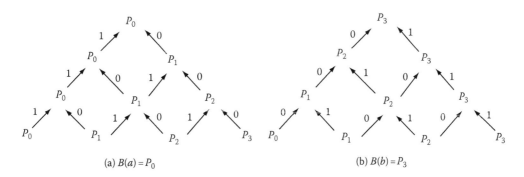

(a) $B(a) = P_0$                    (b) $B(b) = P_3$

**FIGURE 24.12:**   Interpolation of the first and last control points: $B(a) = P_0$ and $B(b) = P_3$.

apex, $B(a) = P_0$. Similarly, if we set $t = b$ in the de Casteljau algorithm, then all the right pointing arrows become zero and all the left pointing arrows become one (see Figure 24.12b). Now if we follow the algorithm from the base to the apex, then $P_n$ appears at each node along the right edge of Figure 24.12b. Therefore, at the apex, $B(b) = P_n$.

Interpolation at the end points is important because we often want to connect two Bezier curves. To insure that two Bezier curves join at their end points, all we need to do is to make sure that the initial control point of the second curve is the same as the final control point of the first curve. This device guarantees continuity. In Section 24.5.1, we shall develop techniques for guaranteeing higher order smoothness between adjacent Bezier curves, but before we can perform this analysis we need to know how to compute the derivative of a Bezier curve.

## 24.5   Differentiating the de Casteljau Algorithm

We can compute the derivative of a Bezier curve directly from its Bernstein representation. Let $B(t)$ be a Bezier curve with control points $P_0, \ldots, P_n$. Recall from Equation 24.6 that

$$B(t) = \sum_{k=0}^{n} B_k^n(t) P_k,$$

where

$$B_k^n(t) = \binom{n}{k} \frac{(t-a)^k (b-t)^{n-k}}{(b-a)^n}, \quad k = 0, \ldots, n.$$

Thus,

$$B'(t) = \sum_{k=0}^{n} \frac{dB_k^n(t)}{dt} P_k. \tag{24.7}$$

Therefore, to compute the derivative of a Bezier curve, we need only differentiate the Bernstein basis functions $B_k^n(t), k = 0, \ldots, n$. But it follows easily from the product rule that (see Exercise 24.8)

$$\frac{dB_k^n(t)}{dt} = n \left( \frac{B_{k-1}^{n-1}(t) - B_k^{n-1}(t)}{b-a} \right). \tag{24.8}$$

Inserting Equation 24.8 into Equation 24.7 and collecting the coefficients of $B_k^{n-1}(t)$ yields

$$B'(t) = n \sum_{k=0}^{n-1} B_k^{n-1}(t) \left( \frac{P_{k+1} - P_k}{b-a} \right). \tag{24.9}$$

Thus the derivative of the degree $n$ Bernstein polynomial with coefficients $P_0, \ldots, P_n$ is $n$ times the Bernstein polynomial of degree $n - 1$ with coefficients $\frac{P_1 - P_0}{b-a}, \ldots, \frac{P_n - P_{n-1}}{b-a}$.

One important consequence of this observation is that we can compute the derivative of a Bezier curve using the de Casteljau algorithm: simply place the coefficients $\frac{P_1 - P_0}{b-a}, \ldots, \frac{P_n - P_{n-1}}{b-a}$ at the base of the diagram, run $n - 1$ levels of the algorithm, and multiply the output by $n$ (see Figure 24.13).

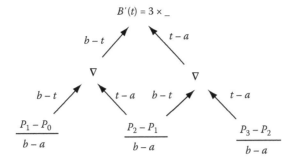

**FIGURE 24.13:** The derivative of a cubic Bezier curve with control points $\{P_k\}$ is, up to a constant multiple, a quadratic Bernstein polynomial with coefficients $\left\{\frac{P_{k+1} - P_k}{b - a}\right\}$.

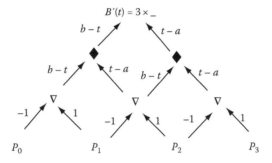

**FIGURE 24.14:** Computing the first derivative of a cubic Bezier curve with control points $P_0, P_1, P_2, P_3$ by differentiating the labels on the first level of the de Casteljau algorithm. To get the correct derivative $B'(t)$, we must multiply the output of this algorithm by $n = 3$ and we must remember to normalize the labels on each arrow—even the constants along the arrows on the first level—by dividing by $b - a$. Compare to Figure 24.13.

We can also compute the derivative of a Bezier curve by differentiating the de Casteljau algorithm. Differentiating the de Casteljau algorithm for a degree $n$ Bezier curve is actually quite easy: simply differentiate the labels—that is, replace $b - t \to -1$ and $t - a \to 1$—on the first level of the de Casteljau algorithm and multiply the output by $n$ (see Figure 24.14). The validity of this differentiation algorithm follows directly from Equation 24.9 because if we differentiate the labels on the first level and run the algorithm, then the values in the nodes on the first level become $\frac{P_1 - P_0}{b - a}, \ldots, \frac{P_n - P_{n-1}}{b - a}$ (remember the normalizing factor of $b - a$ in the denominator of every label), so the result of the algorithm is exactly the Bernstein polynomial of degree $n - 1$ with coefficients $\frac{P_1 - P_0}{b - a}, \ldots, \frac{P_n - P_{n-1}}{b - a}$.

In general, up to a constant multiple, the derivative of a Bezier curve with control points $\{P_k\}$ is a Bernstein polynomial of one lower degree with control vectors $\left\{\frac{P_{k+1} - P_k}{b - a}\right\}$. Therefore, by induction, we can express the $k$th derivative of a degree $n$ Bezier curve as a Bernstein polynomial of degree $n - k$. Moreover, to compute the $k$th derivative of a Bezier curve, we can simply differentiate the labels on the first $k$ levels the de Casteljau algorithm and multiply the output by $\frac{n!}{(n - k)!}$ (see Figure 24.15).

### 24.5.1 Smoothly Joining Two Bezier Curves

We prefaced our discussion of differentiation by saying that we wanted to be able to join two Bezier curves together smoothly. To do so, we need to calculate the derivatives at their end points.

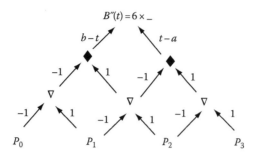

**FIGURE 24.15:** Computing the second derivative of a cubic Bezier curve by differentiating the labels on the bottom two levels of the de Casteljau algorithm. Here to get the correct second derivative $B''(t)$, we must multiply the output of this algorithm by $6 = 3 \times 2$. Also, as in Figure 24.14, we must remember to normalize the labels along all the arrows, even the constants along the arrows on the bottom two levels, by dividing by $b - a$.

Let $B(t)$ be a Bezier curve with control points $P_0, \ldots, P_n$. Substituting $t = a$ or $t = b$ into the differentiation algorithm and recalling the interpolation property at the end points, we see that:

$$B(a) = P_0, \qquad\qquad\qquad\qquad B(b) = P_n,$$
$$B'(a) = n(P_1 - P_0)/(b - a), \qquad B'(b) = n(P_n - P_{n-1})/(b - a),$$
$$B''(a) = n(n-1)(P_2 - 2P_1 + P_0)/(b-a)^2, \quad B''(b) = n(n-1)(P_n - 2P_{n-1} + P_{n-2})/(b-a)^2.$$

In general, it follows by induction on $k$ that the $k$th derivative at $t = a$ depends only on the first $k + 1$ control points, and the $k$th derivative at $t = b$ depends only on the last $k + 1$ control points.

Suppose then that we are given a Bezier curve $P(t)$ with control points $P_0, \ldots, P_n$ and we want to construct another Bezier curve $Q(t)$ with control points $Q_0, \ldots, Q_n$ so that $Q(t)$ meets $P(t)$ and matches the first $r$ derivatives of $P(t)$ at its end point. From the results in the previous paragraph, we find that the control points $Q_0, \ldots, Q_n$ must satisfy the following constraints:

$$r = 0: Q_0 = P_n,$$
$$r = 1: Q_1 - Q_0 = P_n - P_{n-1} \Rightarrow Q_1 = P_n + (P_n - P_{n-1}),$$
$$r = 2: Q_2 - 2Q_1 + Q_0 = P_n - 2P_{n-1} + P_{n-2} \Rightarrow Q_2 = P_{n-2} + 4(P_n - P_{n-1}),$$

and so on for higher values of $r$. Each additional derivative allows us to solve for one additional control point. Clearly we could go on in this manner solving for one control point at a time. An alternative approach for finding an explicit formula for the control points $\{Q_k\}$ that avoids all this tedious computation will be presented in Chapter 25, Section 25.5. Figure 24.16 illustrates two cubic Bezier curves that meet with matching first derivatives at their join.

### 24.5.2   Uniqueness of the Bezier Control Points

Another consequence of the differentiation algorithm for Bezier curves is the uniqueness of the Bezier control points of a degree $n$ Bezier curve over a fixed parameter interval. For suppose that $P_0, \ldots, P_n$ and $P_0^*, \ldots, P_n^*$ are two distinct sets of control points for the Bezier curve $B(t)$ over the parameter interval $[a, b]$. Substituting $t = a$ into the differentiation algorithm and recalling the interpolation property at the end points, we have the following formulas for the derivatives of $B(t)$ at $t = a$ in terms of $P_0, \ldots, P_n$ and $P_0^*, \ldots, P_n^*$:

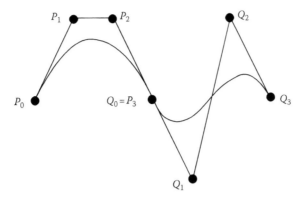

**FIGURE 24.16:** Two cubic Bezier curves—one with control points $P_0, P_1, P_2, P_3$ and the other with control points $Q_0, Q_1, Q_2, Q_3$—that meet with matching first derivatives at their join. Here $Q_0 = P_3$ and $Q_1 - Q_0 = P_3 - P_2$.

$$B(a) = P_0, \qquad\qquad\qquad\qquad B(a) = P_0^*,$$
$$B'(a) = n(P_1 - P_0)/(b-a), \qquad\qquad B'(a) = n(P_1^* - P_0^*)/(b-a),$$
$$B''(a) = n(n-1)(P_2 - 2P_1 + P_0)/(b-a)^2, \quad B''(a) = n(n-1)(P_2^* - 2P_1^* + P_0^*)/(b-a)^2.$$

and so on up to the $n$th derivative $B^{(n)}(a)$. In general, it follows by induction on $k$ that $B^{(k)}(a)$ depends only on the first $k+1$ control points $P_0, \ldots, P_k$ or $P_0^*, \ldots, P_k^*$, and the formulas for these derivatives are identical with respect to $P_0, \ldots, P_k$ and $P_0^*, \ldots, P_k^*$. From the formula for the zeroth derivative, we conclude that $P_0^* = P_0$; from this result and the formula for the first derivative, we conclude that $P_1^* = P_1$, and so on. Thus, from the formulas for first $n$ derivatives, we conclude that $P_k^* = P_k$, $k = 0, \ldots, n$. Hence the control points of a degree $n$ Bezier curve over a fixed parameter interval are unique.

## 24.6 Tensor Product Bezier Patches

Bezier technology—the de Casteljau algorithm and Bernstein polynomials—can be used to create surfaces as well as curves. A *tensor product Bezier patch* $B(s,t)$ is a rectangular parametric surface patch—the image of a planar rectangular domain $[a, b] \times [c, d]$—that approximates the shape of a rectangular array of control points $\{P_{ij}\}$, $i = 0, \ldots, m, j = 0, \ldots, n$. Connecting control points with adjacent indices by straight lines generates a control polyhedron that controls the shape of the Bezier patch in much the same way that a Bezier control polygon controls the shape of a Bezier curve (see Figure 24.17). In particular, dragging a control point pulls the surface patch in the same general direction as the control point.

To construct a tensor product Bezier patch $B(s,t)$ from a rectangular array of control points $\{P_{ij}\}$, $i = 0, \ldots, m, j = 0, \ldots, n$, we begin with the curves $P_0(t), \ldots, P_m(t)$, where $P_i(t)$ is the Bezier curve with control points $P_{i,0}, \ldots, P_{i,n}$. For each fixed value of $t$, let $B(s,t)$ be the Bezier curve for the control points $P_0(t), \ldots, P_m(t)$. Then as $s$ varies from $a$ to $b$ and $t$ varies from $c$ to $d$, the curves $B(s,t)$ sweep out a surface (see Figures 24.18 and 24.19). This surface is called a *tensor product Bezier patch of bidegree* $(m, n)$.

This construction suggests the following evaluation algorithm for tensor product Bezier patches: first use the de Casteljau algorithm $m + 1$ times to compute the points at the parameter $t$ along the

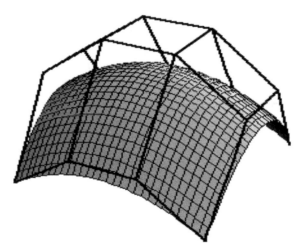

**FIGURE 24.17: (See color insert following page 354.)**   A bicubic tensor product Bezier surface with its control polyhedron, formed by connecting control points with adjacent indices.

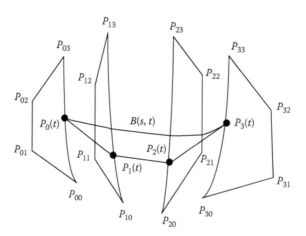

**FIGURE 24.18:**   Construction of points on a bicubic tensor product Bezier surface $B(s,t)$. First the Bezier curves $P_i(t)$, $i = 0, \ldots, 3$ are constructed from the control points $P_{i0}, P_{i1}, P_{i2}, P_{i3}$. Then for a fixed value of $t$, the Bezier curve $B(s,t)$ is constructed using the points $P_0(t), P_1(t), P_2(t), P_3(t)$ as control points. As $s$ varies from $a$ to $b$ and $t$ varies from $c$ to $d$, these curves sweep out a surface patch (see Figure 24.19).

degree $n$ Bezier curves $P_0(t), \ldots, P_m(t)$; then use the de Casteljau algorithm one more time to compute the point at the parameter $s$ along the degree $m$ Bezier curve with control points $P_0(t), \ldots, P_m(t)$ (see Figure 24.20).

Alternatively, instead of starting with the Bezier curves $P_0(t), \ldots, P_m(t)$, we could begin with the Bezier curves $P_0^*(s), \ldots, P_n^*(s)$, where $P_j^*(s)$ is the degree $m$ Bezier curve with control points $P_{0,j}, \ldots, P_{m,j}$. Now for each fixed value of $s$, let $B^*(s,t)$ be the Bezier curve for the control points $P_0^*(s), \ldots, P_n^*(s)$. Again as $s$ varies from $a$ to $b$ and $t$ varies from $c$ to $d$, the curves $B^*(s,t)$ sweep out a surface, and we can use the de Casteljau algorithm to evaluate points along this surface (see Figure 24.21).

The surface $B^*(s,t)$ is exactly the same as the surface $B(s,t)$. Indeed both of these surfaces have the same Bernstein representation. To compute the Bernstein representation of the Bezier patch

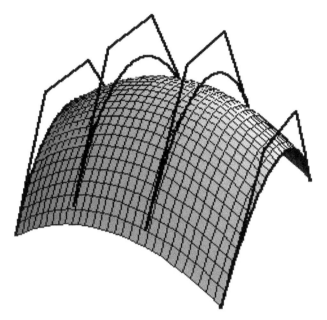

**FIGURE 24.19: (See color insert following page 354.)** The bicubic Bezier patch in Figure 24.17 along with its cubic Bezier control curves $P_0(t), \ldots, P_3(t)$. Notice that only the boundary control curves are interpolated by the surface.

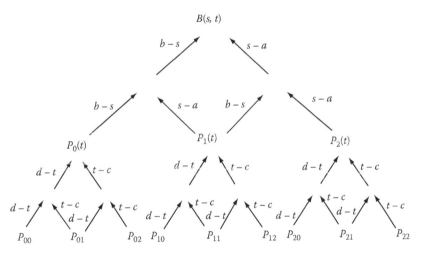

**FIGURE 24.20:** The de Casteljau evaluation algorithm for a biquadratic Bezier patch. The three lower triangles represent Bezier curves in the $t$ direction; the upper triangle blends these curves in the $s$ direction.

$B(s, t)$, recall that for a fixed value of $t$, $B(s, t)$ is the Bezier curve with control points $P_0(t), \ldots, P_m(t)$. Therefore,

$$B(s, t) = \sum_{i=0}^{m} B_i^m(s) P_i(t). \tag{24.10}$$

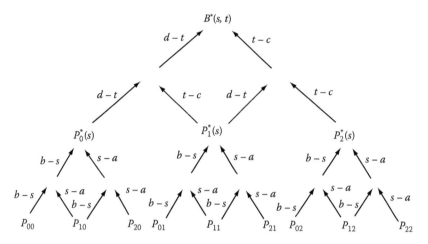

**FIGURE 24.21:**    An alternative version of the de Casteljau evaluation algorithm for a biquadratic Bezier patch with the same control points as in Figure 24.20. The three lower triangles represent Bezier curves in the $s$ direction, and the upper triangle blends these curves in the $t$ direction. Compare to Figure 24.20.

Moreover, $P_i(t)$ is the Bezier curve with control points $P_{i,0}, \ldots P_{i,n}$, so

$$P_i(t) = \sum_{j=0}^{n} B_j^n(t) P_{i,j}. \tag{24.11}$$

Substituting Equation 24.11 into Equation 24.10 yields the Bernstein representation:

$$B(s, t) = \sum_{i=0}^{m} \sum_{j=0}^{n} B_i^m(s) B_j^n(t) P_{i,j}. \tag{24.12}$$

Similarly,

$$B^*(s, t) = \sum_{j=0}^{n} B_j^n(t) P_j^*(s), \tag{24.13}$$

where

$$P_j^*(s) = \sum_{i=0}^{m} B_i^m(s) P_{i,j}, \tag{24.14}$$

so

$$B^*(s, t) = \sum_{j=0}^{n} \sum_{i=0}^{m} B_i^m(s) B_j^n(t) P_{i,j}. \tag{24.15}$$

Hence, $B^*(s, t) = B(s, t)$. Thus it does not matter which version of the de Casteljau algorithm we apply; both constructions generate the same surface. However, if $m < n$, then the de Casteljau

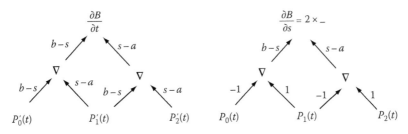

**FIGURE 24.22:** Computing the partial derivatives of a biquadratic Bezier patch by applying the procedure for differentiating the de Casteljau algorithm for Bezier curves. To find $\partial B/\partial t$ (left), simply differentiate the de Casteljau algorithm for each of the Bezier curves $P_0(t), P_1(t), P_2(t)$. To find $\partial B/\partial s$ (right), simply differentiate the first level of the de Casteljau algorithm in $s$ and multiply the result by the degree in $s$.

algorithm for $B(s,t)$ is more efficient, whereas if $n < m$, then the de Casteljau algorithm for $B^*(s,t)$ is more efficient.

Tensor product Bezier patches inherit many of the characteristic properties of Bezier curves: they are affine invariant, lie in the convex hull of their control points, and interpolate their corner control points (see Exercises 24.12 and 24.13). These properties follow easily from the de Casteljau algorithm for Bezier patches (Figures 24.20 and 24.21) and the corresponding properties of Bezier curves. Moreover, the boundaries of a tensor product Bezier patch are the Bezier curves determined by their boundary control points, since

$$\begin{aligned} B(a,t) &= P_0(t) \\ B(b,t) &= P_m(t) \end{aligned} \quad \text{and} \quad \begin{aligned} B(s,c) &= P_0^*(s) \\ B(s,d) &= P_n^*(s). \end{aligned}$$

It follows that although tensor product Bezier patches do not generally interpolate their control points, they always interpolate the four corner control points $P_{00}, P_{m0}, P_{0n}, P_{mn}$.

There is no known analogue of the variation diminishing property for tensor product Bezier patches. Thus although Bezier patches typically follow the shape of their control polyhedra, there is no theorem which guarantees that Bezier surfaces do not oscillate more than their control points.

To compute the partial derivatives of a Bezier patch, we can apply our procedure for differentiating the de Casteljau algorithm for Bezier curves. Consider Figure 24.20. We can compute $\partial B/\partial t$ simply by differentiating the de Casteljau algorithm for each of the Bezier curves $P_0(t), \ldots, P_m(t)$ at the base of Figure 24.20. Similarly, we can compute $\partial B/\partial s$ by differentiating the first upper $s$ level of Figure 24.20 and multiplying the result by the degree in $s$ (see Figure 24.22). Symmetric results hold for differentiating the algorithm in Figure 24.21: simply reverse the roles of $s$ and $t$. The normal vector $N$ to a Bezier patch $B(s,t)$ is given by setting $N = \frac{\partial B}{\partial s} \times \frac{\partial B}{\partial t}$.

## 24.7 Summary

Beizer curves and surfaces are used to approximate freeform shapes described by a finite array of control points. The fundamental algorithm for Bezier curves and surfaces is the de Casteljau evaluation algorithm, an algorithm based on repeated linear interpolation (Figure 24.23a). The decomposition of evaluation into successive identical linear interpolation steps makes the evaluation algorithm easy to differentiate (Figure 24.23b), allowing us to compute derivatives as well as points

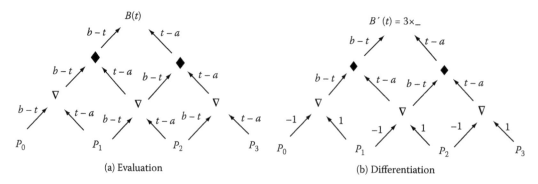

(a) Evaluation                                      (b) Differentiation

**FIGURE 24.23:**   The de Casteljau algorithm for (a) evaluation and (b) differentiation.

along a Bezier curve using a variant of the de Casteljau algorithm. We applied the differentiation algorithm to derive constraints on the control points which guarantee that two Bezier curves meet smoothly at their join. We also used the differentiation algorithm to establish the uniqueness of the degree $n$ Bezier control points over a fixed parameter interval.

The Bernstein representation provides an alternative explicit polynomial representation for Bezier curves. If $B(t)$ is the Bezier curve over the interval $[a, b]$ with control points $P_0, \ldots, P_n$, then the Bernstein representation is given by setting

$$B(t) = \sum_{k=0}^{n} B_k^n(t)P_k, \quad a \le t \le b,$$

$$B_k^n(t) = \binom{n}{k} \frac{(t-a)^k(b-t)^{n-k}}{(b-a)^n}, \quad k = 0, \ldots, n.$$

(24.16)

Bezier curves have the following important geometric properties:

1. Affine invariance

2. Convex hull property

3. Variation diminishing property

4. Interpolate their first and last control points

Tensor product Bezier patches are an extension of the Bezier representation from curves to surfaces. Bezier patches can be evaluated either by extending the de Casteljau algorithm to surfaces (Figure 24.24) or by invoking the Bernstein representation in two variables (Equation 24.17).

The Bernstein representation of a tensor product Bezier surface $B(s, t)$ of bidegree $(m, n)$ with control points $\{P_{ij}\}$ and domain $[a, b] \times [c, d]$ is given by

$$B(s, t) = \sum_{i=0}^{m} \sum_{j=0}^{n} B_i^m(s)B_j^n(t)P_{i,j},$$

(24.17)

where $B_i^m(s), B_j^n(t)$ are the Bernstein basis functions over the intervals $[a, b]$ and $[c, d]$.

Bezier patches are affine invariant, lie in the convex hull of their control points, and interpolate the boundary Bezier curves defined by their boundary control points. There is, however, no known analogue of the variation diminishing property for Bezier surfaces.

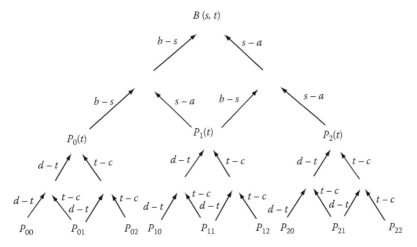

**FIGURE 24.24:** The de Casteljau evaluation algorithm for a biquadratic Bezier patch. The three lower triangles represent Bezier curves in the $t$ direction; the upper triangle blends these curves in the $s$ direction.

## Exercises

24.1. Give an example to show that Bezier curves depend on the order of their control points $P_0, \ldots, P_n$—that is, if we change the order but not the location of the control points, we may generate a different Bezier curve.

24.2. Prove that the Bezier curve for the control points $P_n, \ldots, P_0$ is the same as the Bezier curve with the control points $P_0, \ldots, P_n$ but with opposite orientation.

24.3. Show that a Bezier curve collapses to a single point $P$ if and only if all the Bezier control points are located at $P$.

24.4. Consider the Bernstein basis functions

$$B_k^n(t) = \binom{n}{k} \frac{(t-a)^k (b-t)^{n-k}}{(b-a)^n}, \quad k = 0, \ldots, n.$$

Show that:

a. $\displaystyle\sum_{k=0}^{n} B_k^n(t) \equiv 1$.

b. $0 \leq B_k^n(t) \leq 1$ for $a \leq t \leq b$.

c. $B_k^n(a) = \delta_{k,0}$ and $B_k^n(b) = \delta_{k,n}$.

24.5. Using the results in Exercise 24.4, show that:

a. Bezier curves are translation invariant.

b. Bezier curves interpolate their first and last control points.

24.6. Prove that

$$Convex\ Hull(P_0, \dots, P_n) = \left\{ \sum_{k=0}^{n} c_k P_k \mid \sum_{k=0}^{n} c_k \equiv 1 \quad and \quad c_k \geq 0 \right\}.$$

24.7. Using the results of Exercises 24.4 and 24.6, show that Bezier curves lie in the convex hull of their control points.

24.8. Consider a Bezier curve

$$B(t) = \sum_{k=0}^{n} B_k^n(t) P_k,$$

where

$$B_k^n(t) = \binom{n}{k} \frac{(t-a)^k (b-t)^{n-k}}{(b-a)^n}, \quad k = 0, \dots, n$$

are the Bernstein basis functions. Show that:

a. $\frac{dB_k^n(t)}{dt} = n \left( \frac{B_{k-1}^{n-1}(t) - B_k^{n-1}(t)}{b-a} \right)$.

b. $B'(t) = n \sum_{k=0}^{n-1} B_k^{n-1}(t) \left( \frac{P_{k+1} - P_k}{b-a} \right)$.

24.9. Let $P_{i_0 \cdots i_n}(t)$ denote the degree $n$ polynomial curve that interpolates the control points $P_{i_0}, \dots, P_{i_n}$ at the parameter values $t_{i_0}, \dots, t_{i_n}$—that is, $P_{i_0 \cdots i_n}(t_{i_k}) = P_{i_k}, k = 0, \dots, n$. Show that:

a. $P_{01}(t) = \frac{t_1 - t}{t_1 - t_0} P_0 + \frac{t - t_0}{t_1 - t_0} P_1$.

b. $P_{012}(t)$ and $P_{0123}(t)$ can be built using the dynamic programming algorithm (Neville's algorithm) illustrated in Figure 24.25.

c. Explain how to extend the dynamic programming algorithm in Figure 24.25 to $P_{0 \cdots n}(t)$.

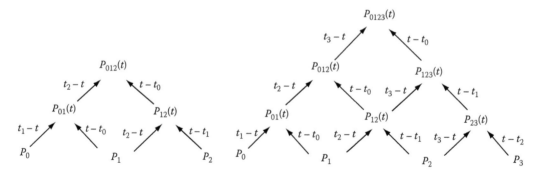

**FIGURE 24.25:** Neville's algorithm for computing points on the polynomial curve that interpolates the points at the base of the diagram. The value at each node must be normalized in the usual manner by dividing by the sum of the labels along the arrows that enter the node.

24.10. Consider a Bezier curve with control points $P_0, \ldots, P_n$. Show that:

    a. The line segment $P_0 P_1$ is tangent to the curve at $P_0$.

    b. The line segment $P_{n-1} P_n$ is tangent to the curve at $P_n$.

24.11. Given point and derivative data $(R_0, v_0), \ldots, (R_n, v_n)$, explain how to place Bezier control points to generate a smooth piecewise cubic curve to interpolate this data.

24.12. Show that tensor product Bezier patches:

    a. Are affine invariant.

    b. Lie in the convex hull of their control points.

24.13. Show that every tensor product Bezier patch interpolates:

    a. The four Bezier curves defined by the boundary control points.

    b. The four corner control points.

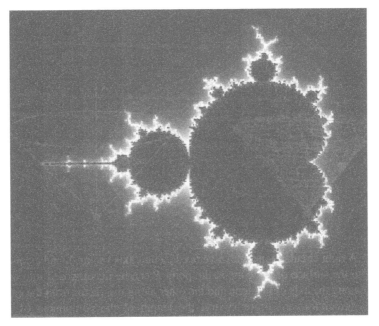

**FIGURE 15.2:** The Mandelbrot set, generating by iterating the expression $z_{n+1}(z_0) = z_n^2(z_0) + z_0$.

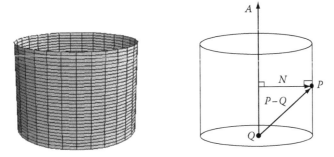

**FIGURE 19.3:** A right circular cylinder with an axis line $L$ determined by a point $Q$ and a unit direction vector $A$. The normal vector $N$ to the surface of the cylinder at the point $P$ is the orthogonal component of the vector $P - Q$ with respect to the axis vector $A$.

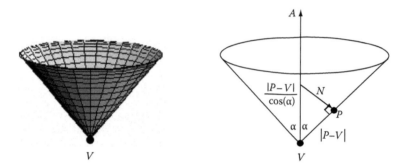

**FIGURE 19.4:** A right circular cone with a vertex $V$, a unit axis vector $A$, and a cone angle $\alpha$. The normal vector $N$ to the surface of the cone at the point $P$ can be decomposed into the sum of two components: one along the axis of the cone and the other along the vector from the cone vertex $V$ to the point $P$. By straightforward trigonometry, the length of the component of $N$ along $A$ is $|P - V|/\cos(\alpha)$.

**FIGURE 19.5:** The ellipsoid is the image of the sphere under nonuniform scaling.

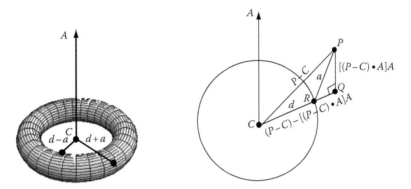

**FIGURE 19.6:** The torus. The circle at a distance $d$ from the center of the torus in the plane perpendicular to the axis $A$ is the generator for the torus. The radius of the inner tube is $a$.

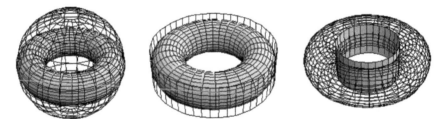

**FIGURE 19.7:** A sphere (left) and a cylinder (center) bounding a torus on the outside, and a cylinder (right) surrounding the hole to the torus on the inside. Notice that the cylinder is a tighter fit than the sphere to the torus on the outside.

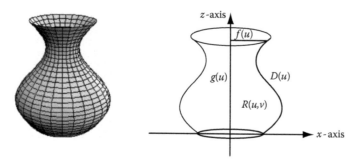

**FIGURE 19.8:** A surface of revolution $R(u, v)$ with a directrix $D(u)$ in the $xz$-plane. The axis of revolution is the $z$-axis.

**FIGURE 20.1:** Primitives for a solid modeling system: a box, a sphere, a cylinder, a cone, and a torus.

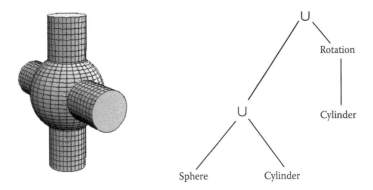

**FIGURE 20.2:** A spherical tank with two cylindrical pipes (left), modeled by a CSG tree consisting of the union of a sphere and two cylinders (right).

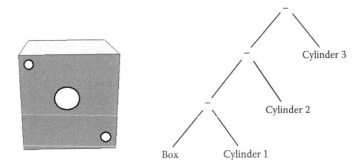

**FIGURE 20.3:** A solid block with three cylindrical holes (left), modeled by a CSG tree that subtracts three cylinders from a box (right).

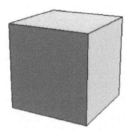 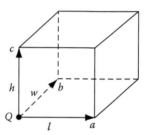

**FIGURE 20.8:** A box with a corner at $Q$, edges parallel to the unit vectors $a, b, c$, with length, width, and height given by the scalars $l, w, h$.

 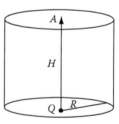

**FIGURE 20.9:** A bounded cylinder of radius $R$ and height $H$ with an axis line $L$ determined by a base point $Q$ and a unit direction vector $A$.

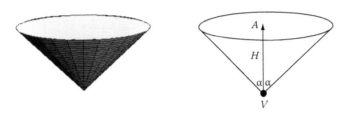

**FIGURE 20.10:** A bounded cone with vertex $V$, cone angle $\alpha$, and height $H$, with an axis line parallel to the unit direction vector $A$.

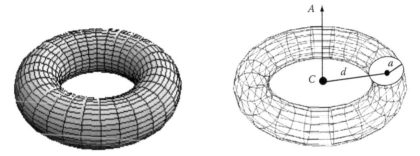

**FIGURE 20.11:** A torus with center $C$ and axis parallel to the unit vector $A$, with a generator of radius $d$ and a tube radius $a$.

**FIGURE 23.4:** A hemi-cube. The lengths of the sides are twice the height of the hemi-cube.

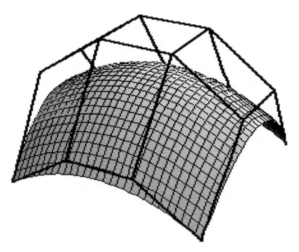

**FIGURE 24.17:** A bicubic tensor product Bezier surface with its control polyhedron, formed by connecting control points with adjacent indices.

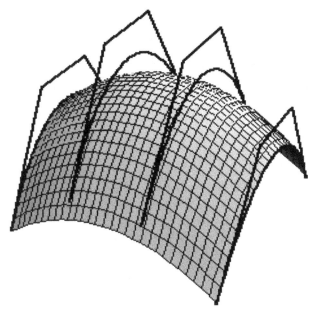

**FIGURE 24.19:** The bicubic Bezier patch in Figure 24.17 along with its cubic Bezier control curves $P_0(t), \ldots, P_3(t)$. Notice that only the boundary control curves are interpolated by the surface.

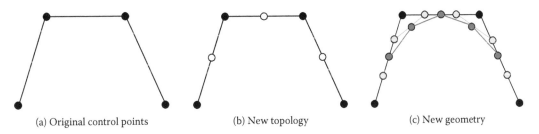

(a) Original control points      (b) New topology      (c) New geometry

**FIGURE 30.11:** One level of the Lane–Riesenfeld algorithm for uniform cubic B-spline curves. In (a), a segment of the original control polygon (black) is illustrated. In the topological stage (b), new control points (yellow) are introduced at the midpoints of the edges of the control polygon. Notice that there are now two types of control points: edge points (yellow) and vertex points (black). In the geometric stage (c), the control points (black and yellow) are repositioned to the midpoints (blue) of the midpoints (green) of these control points. Edge points (yellow) are relocated along the same edge (in fact, to the same position along the edge), but vertex points (black) are relocated to new positions off the original control polygon. The new control polygon is illustrated in blue.

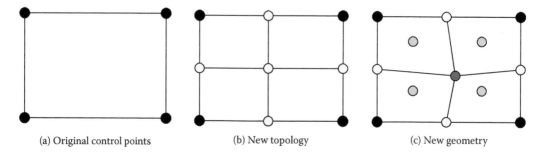

| (a) Original control points | (b) New topology | (c) New geometry |

**FIGURE 30.12:** One level of the Lane–Riesenfeld algorithm for bicubic B-spline surfaces. In (a), we illustrate one face of the control polyhedron. In the topological stage (b), new control points (yellow) are inserted at the midpoints of the edges and at the centroids of the faces of the control polyhedron; each edge is then split into two edges and each face is split into four faces. In the geometric stage (c), each control point—yellow or black—is repositioned to the centroid (blue) of the centroids (green) of the faces adjacent to the control point. Thus, after the geometric stage, each black and each yellow control point is repositioned based on the location of the green centroids of adjacent faces, but the topology of the polyhedron is not changed. The green centroids are used only to reposition vertices and are not themselves vertices of the refined polyhedron.

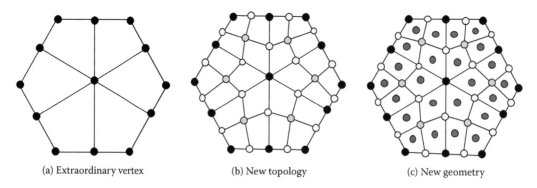

| (a) Extraordinary vertex | (b) New topology | (c) New geometry |

**FIGURE 30.13:** One level of centroid averaging. In (a), six faces of a quadrilateral mesh surround an extraordinary vertex. In the topological stage (b), new vertices are inserted at the midpoints (yellow) of the edges and at the centroids (green) of the faces, and the centroid of each face is joined to the midpoints of the surrounding edges, splitting each face into four new faces. In the geometric stage (c), each vertex—yellow, green, or black—is repositioned to the centroid of the centroids (blue) of the faces adjacent to the vertex, but the topology of the mesh is not changed. Notice that the extraordinary vertex at the center remains an extraordinary vertex and is repositioned as the centroid of six adjacent centroids. All the other vertices—black, yellow, and green—are ordinary vertices and their new location is computed as the centroid of four blue points, just as in the Lane–Riesenfeld algorithm for bicubic B-splines.

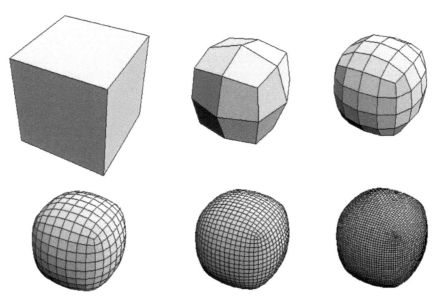

**FIGURE 30.14:** Five levels of centroid averaging applied to a cube. The corners are rounded, but the limit shape is not a perfect sphere. Notice the extraordinary vertices of valence three at the corners of the cube.

**FIGURE 30.15:** Centroid averaging applied to two cubes connected by a thin rectangular rod. The original quadrilateral mesh is on the left; the first and third levels of subdivision are illustrated in the middle and on the right.

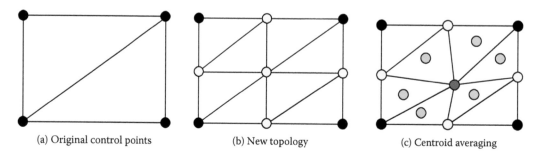

| (a) Original control points | (b) New topology | (c) Centroid averaging |

**FIGURE 30.22:** One level of subdivision for the three-direction quartic box spline. In (a), we illustrate two faces of the original mesh. In the topological stage (b), new control points (yellow) are inserted at the midpoints of the edges, and new edges and faces are added to the mesh by connecting vertices along adjacent edges. Thus each edge is split into two edges and each face is split into four faces. In the geometric stage (c), each control point—yellow or black—is repositioned to the centroid (blue) of the weighted centroids (green) of the faces adjacent to the control point, but the topology of the mesh is not changed. Note that since the green centroids are weighted centroids, we cannot reuse the same centroids to reposition different control points; rather we must recompute these centroids for each vertex. Compare to Figure 30.12 for bicubic B-splines.

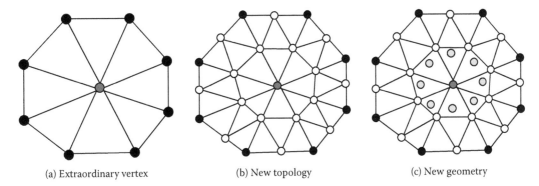

(a) Extraordinary vertex      (b) New topology      (c) New geometry

**FIGURE 30.23:** One level of centroid averaging. In (a), we illustrate eight faces of a triangular mesh surrounding an extraordinary vertex (blue). In the topological stage (b), new vertices (yellow) are inserted at the midpoints of the edges of the triangular mesh, and new edges and faces are added to the mesh by joining the midpoints of adjacent edges. In the geometric stage (c), each vertex—blue, yellow, or black—is repositioned to the centroid of the weighted centroids (green) of the faces adjacent to the vertex, but the topology of the mesh is not changed. Notice that all the new vertices (yellow) are ordinary vertices with valence six. Also, the extraordinary vertex (blue) at the center with valence eight remains an extraordinary vertex with valence eight and is repositioned to the centroid of eight adjacent weighted centroids (green). Since the green centroids are weighted centroids, we cannot reuse the same centroids to reposition different vertices; rather we must recompute these green centroids for each vertex. Compare to Figure 30.13 for quadrilateral meshes.

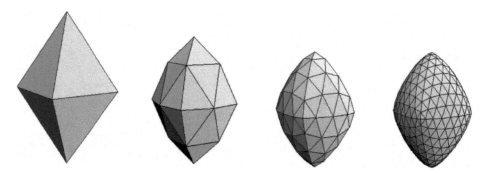

**FIGURE 30.24:** Three levels of centroid averaging applied to an octahedron (left). Notice the extraordinary vertices of valence four at the corners of the octahedron.

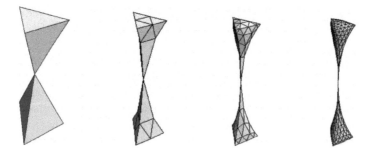

**FIGURE 30.25:** Centroid averaging applied to two tetrahedra meeting at a vertex. The original triangular mesh is on the left; the first three levels of subdivision are illustrated successively on the right.

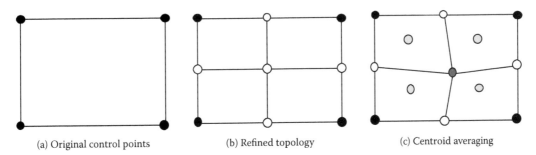

(a) Original control points          (b) Refined topology          (c) Centroid averaging

**FIGURE 30.29:** Centroid averaging for bicubic B-splines. First (b) refine the topology by placing new edge vertices (yellow) at the center of each edge and new face vertices (yellow) at the centroid of each face; then connect these vertices with edges to split each face into four faces. The new control point (blue) in (c) is repositioned to the centroid of the centroids (green) of the adjacent faces.

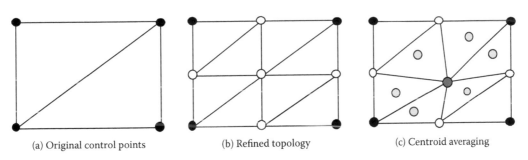

(a) Original control points          (b) Refined topology          (c) Centroid averaging

**FIGURE 30.30:** Centroid averaging for three-direction quartic box splines. First (b) refine the topology by inserting a new edge vertex (yellow) at the center of each edge; then connect vertices on adjacent edges to split each face into four faces. The new control point (blue) in (c) is repositioned to the centroid of the weighted centroids (green) of the adjacent faces.

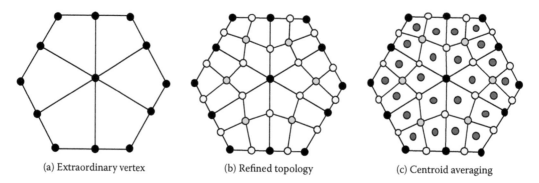

(a) Extraordinary vertex        (b) Refined topology        (c) Centroid averaging

**FIGURE 30.33:**  Quadrilateral meshes. Each vertex is repositioned to (c) the centroid of the centroids (blue) of the adjacent faces after (b) refining the topology by splitting each quadrilateral face into four quadrilateral faces.

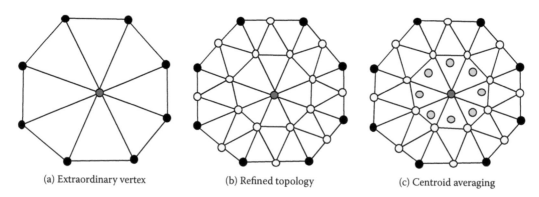

(a) Extraordinary vertex        (b) Refined topology        (c) Centroid averaging

**FIGURE 30.34:**  Triangular meshes. Each vertex is repositioned to (c) the centroid of the weighted centroids (green) of the adjacent faces after (b) refining the topology by splitting each triangular face into four triangular faces.

# Chapter 25

## Bezier Subdivision

*And he took unto him all these, and divided them in the midst, and laid each piece one against another:*

– Genesis 15:10

## 25.1 Divide and Conquer

If we are going to build useful computer models of freeform shapes using Bezier curves and surfaces, we need procedures to do more than simply evaluate points along these curves and surfaces. We need algorithms to analyze their geometry. For example, to ray trace a Bezier patch, we need algorithms both for finding surface normals and for computing ray–surface intersections. In Chapter 24 we developed an algorithm for differentiating Bezier curves and surfaces, which allows us to find the surface normal of a Bezier patch at any parameter value by taking the cross product of the partial derivatives of the patch. Here we shall develop an algorithm that will permit us to compute the intersection of an arbitrary ray with a Bezier patch.

The method we shall employ is a divide and conquer technique called *Bezier subdivision*. The main idea behind Bezier subdivision is that although globally Bezier curves and surfaces represent curved shapes, if we divide these curves and surfaces into a collection of small enough segments, each segment will be almost flat. Flat shapes like lines and planes are simple to analyze. For example, we can easily ray trace a planar polygon. Thus we shall analyze Bezier curves and surfaces by dividing these curves and surfaces into lots of small flat segments, analyzing the individual flat segments, and then recombining the results to build an understanding of the global geometry of the entire curve or surface.

Bezier subdivision is a powerful tool with many applications. In this chapter we shall use Bezier subdivision to develop fast, robust algorithms for rendering and intersecting Bezier curves and surfaces, We shall also use Bezier subdivision to prove the variation diminishing property for Bezier curves and to show how to connect two Bezier curves smoothly at their join.

## 25.2 The de Casteljau Subdivision Algorithm

The control points of a Bezier curve describe a polynomial curve with respect to a fixed parameter interval $[a, b]$. Sometimes, however, we are interested only in the part of the polynomial curve in a subinterval of $[a, b]$. For example, when rendering a Bezier curve, we may need to clip the curve to a window on the graphics terminal. Since any segment of a Bezier curve is itself a

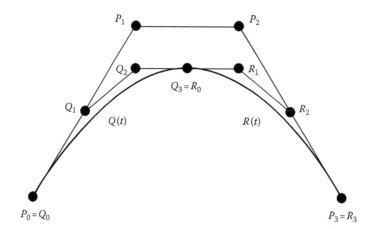

**FIGURE 25.1:** Bezier subdivision for a cubic Bezier curve. The Bezier curve $P(t)$ with control points $P_0, P_1, P_2, P_3$ is split into two Bezier curves: the left Bezier segment $Q(t)$ has control points $Q_0, Q_1, Q_2, Q_3$; the right Bezier segment $R(t)$ has control points $R_0, R_1, R_2, R_3$.

polynomial curve, we should be able to represent each segment of a Bezier curve as a Bezier curve with a new set of control points. Splitting a Bezier curve into smaller pieces is also useful as a divide and conquer strategy for rendering and intersection algorithms. The process of splitting a Bezier curve into two or more Bezier curves that represent the exact same curve is called *Bezier subdivision* (see Figure 25.1).

The basic problem in Bezier subdivision is: given a collection of control points $P_0, \ldots, P_n$ that describe a Bezier curve over the interval $[a, b]$, find the control points $Q_0, \ldots, Q_n$ and $R_0, \ldots, R_n$ that describe the segments of the same Bezier curve over the intervals $[a, c]$ and $[c, b]$. Remarkably the solution to this problem is provided by the same de Casteljau algorithm that we use to evaluate points on a Bezier curve.

*The de Casteljau Subdivision Algorithm*

Let $P(t)$ be a Bezier curve over the interval $[a, b]$ with control points $P_0, \ldots, P_n$. To subdivide $P(t)$ at $t = c$, run the de Casteljau evaluation algorithm for $P(t)$ at $t = c$. The points $Q_0, \ldots, Q_n$ that emerge along the left lateral edge of the de Casteljau diagram are the Bezier control points for the segment of the curve from $t = a$ to $t = c$, and the points $R_0, \ldots, R_n$ that emerge along the right lateral edge of the de Casteljau diagram are the Bezier control points for the segment of the curve from $t = c$ to $t = b$ (see Figure 25.2).

Notice that when $c = (a + b)/2$ is the midpoint of the parameter interval $[a, b]$, then $(b - c)/(b - a) = 1/2 = (c - a)/(b - a)$, so the labels along all the edges in the de Casteljau subdivision algorithm evaluate to $1/2$. Thus for midpoint subdivision the de Casteljau subdivision algorithm is independent of the parameter interval (see Figure 25.3).

We shall defer the proof of the de Casteljau subdivision algorithm till Chapter 26, Section 26.3.1. In this chapter we will study the consequences of this subdivision algorithm. Figure 25.4 provides a geometric interpretation for the de Casteljau subdivision algorithm.

When we subdivide a Bezier curve, the new control polygons appear closer to the Bezier curve than the initial control polygon (see Figures 25.1 and 25.4). Suppose we continue recursively subdividing each Bezier segment. Then in the limit, these control polygons converge to a smooth curve (see Figure 25.5). We shall now show that this limit curve is, in fact, the Bezier curve for the original control polygon. For convenience, we will restrict our attention to recursive subdivision at the midpoint of the parameter interval, though the results are much the same for any value that splits the parameter intervals in a constant ratio.

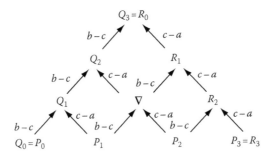

**FIGURE 25.2:** The de Casteljau subdivision algorithm for a cubic Bezier curve with control points $P_0, P_1, P_2, P_3$. The points $Q_0, Q_1, Q_2, Q_3$ that emerge along the left lateral edge of the triangle are the Bezier control points for the segment of the original curve from $t = a$ to $t = c$; the points $R_0, R_1, R_2, R_3$ that emerge along the right lateral edge of the triangle are the Bezier control points for the segment of the original curve from $t = c$ to $t = b$.

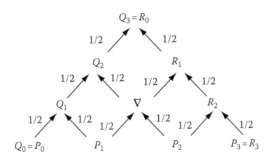

**FIGURE 25.3:** The de Casteljau subdivision algorithm for a cubic Bezier curve at the midpoint of the parameter interval. The labels along all the edges evaluate to $1/2$, so midpoint subdivision does not depend on the choice of the parameter interval.

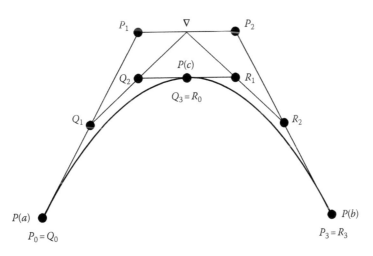

**FIGURE 25.4:** Geometric interpretation of the de Casteljau subdivision algorithm for a cubic Bezier curve $P(t)$ with control points $P_0, P_1, P_2, P_3$. The points $Q_0, Q_1, Q_2, Q_3$ are the Bezier control points of the segment of the original curve from $t = a$ to $t = c$, and the points $R_0, R_1, R_2, R_3$ are the Bezier control points of the segment of the original curve from $t = c$ to $t = b$.

**FIGURE 25.5:** Control polygons converge to Bezier curves under recursive subdivision. Here the original control polygons (left) and the first three levels of subdivision (right) are illustrated.

**Theorem 25.1:** *The control polygons generated by recursive subdivision converge to the Bezier curve for the original control polygon.*

**Proof:** Suppose that the maximum distance between any two adjacent control points is $d$. By construction, the points on any level of the de Casteljau algorithm evaluated at the midpoint of the parameter domain lie at the midpoints of the edges of the polygons generated by the previous level because the labels along all the edges evaluate to $1/2$ (see Figure 25.3). Therefore it follows easily by induction that adjacent points on any level of the de Casteljau diagram are no further than $d$ units apart (see Figure 25.6).

By the same midpoint argument, as we proceed up the de Casteljau diagram adjacent points along the left or right lateral edge of the triangle can be no further than $d/2$ units apart. Hence as we apply recursive subdivision, the distance between the control points of any single control polygon must converge to zero. Since the first and last control points of a Bezier control polygon always lie on the Bezier curve, these control polygons must converge to points along the Bezier curve for the original control polygon.

We can also subdivide tensor product Bezier patches. Recall from Chapter 24, Section 24.6 that a Bezier surface $B(s,t)$ is defined by applying the de Casteljau algorithm in the parameter $s$ to a

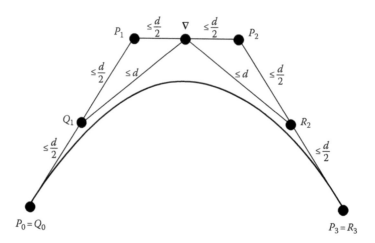

**FIGURE 25.6:** One level of the de Casteljau algorithm for a cubic Bezier curve evaluated at the midpoint of the parameter domain. If adjacent control points $P_0, P_1, P_2, P_3$ are no further than $d$ units apart, then adjacent points $Q_1, \nabla, R_2$ on the first level of the de Casteljau algorithm can also be no further than $d$ units apart.

collection of control points defined by the Bezier curves $P_0(t), \ldots, P_m(t)$ in the parameter $t$. To subdivide the Bezier patch along the parameter line $t = t_0$, simply subdivide each of the curves $P_0(t), \ldots, P_m(t)$ at $t = t_0$. Similarly, to subdivide the Bezier patch along a parameter line $s = s_0$, recall that the same Bezier surface $B(s, t)$ is defined by applying the de Casteljau algorithm in the parameter $t$ to a collection of control points defined by the Bezier curves $P_0^*(s), \ldots, P_n^*(s)$ in the parameter $s$. Thus to subdivide the Bezier patch along the parameter line $s = s_0$, simply subdivide each of the curves $P_0^*(s), \ldots, P_n^*(s)$ at $s = s_0$. Now we have the following result.

**Theorem 25.2:** *The control polyhedra generated by recursive subdivision converge to the tensor product Bezier patch for the original control polyhedron provided that the subdivision is done in both the s and t directions.*

The proof of Theorem 25.2 for Bezier surfaces is much the same as the proof of Theorem 25.1 for Bezier curves, so we shall not repeat the proof here. Notice, however, that we must be careful to subdivide in both the $s$ and the $t$ directions. If we subdivide only along one parameter direction, convergence is not assured because in the limit each of the control polyhedra will not necessarily shrink to a single point. Thus to guarantee convergence it is best to alternate subdividing in the $s$ and $t$ parameter directions.

---

## 25.3  Rendering and Intersection Algorithms

Subdivision is our main mathematical tool for analyzing Bezier curves and surfaces. Here we shall apply recursive subdivision to develop fast, robust algorithms for rendering and intersecting Bezier curves and surfaces.

### 25.3.1  Rendering and Intersecting Bezier Curves

To render a Bezier curve, we could begin by applying the de Casteljau evaluation algorithm to compute lots of points along the curve. We could then display a dense collection of points on the curve or we could connect consecutive points on the curve with straight lines and display these line segments.

But Bezier curves are smooth curves. If we were to apply this approach to rendering Bezier curves, how many points would we need to compute in order for the curve to appear smooth? Some parts of a Bezier curve may be relatively flat, so in these segments we would need to compute only a few points and connect these points with straight lines, whereas other parts of a Bezier curve may be highly bent and in these locations we would need to compute quite a lot of points for the line segments approximating the curve to appear smooth (see Figure 25.7). The de Casteljau evaluation algorithm allows us to compute lots of points along a Bezier curve. But if we want an accurate portrayal of a Bezier curve, it is not clear how many points to compute or where on the curve to concentrate these computations.

Recursive subdivision solves both of these problems: only a small amount of subdivision is required on segments where the curve is relatively flat, while additional subdivision can be performed on segments where the curve is highly bent. The convergence of recursive subdivision is quite fast and the output of recursive subdivision appears smooth (see Figure 25.5). This rapid convergence and smooth appearance leads to the following recursive subdivision algorithm for rendering Bezier curves.

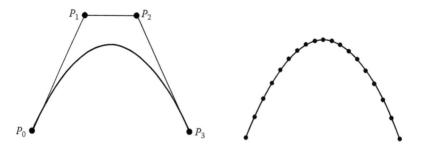

**FIGURE 25.7:** A cubic Bezier curve that is relatively flat near the end points but highly curved toward the center (left). Thus near the end points we need to compute only a few points and connect these points with straight lines, whereas near the center we need to compute quite a lot of points for the line segments approximating the curve to appear smooth (right).

*Rendering Algorithm—Bezier Curves*

If the Bezier curve can be approximated to within tolerance by the straight line segment joining its first and last control points, then draw either this line segment or the control polygon. Otherwise *subdivide* the Bezier curve (at the midpoint of the parameter interval) and render the segments recursively.

To intersect two Bezier curves, we could use the rendering algorithm to generate a piecewise linear approximation for each of the two curves and then intersect all these line segments. But this approach would be highly inefficient because most of the short line segments would not intersect. We can avoid these needless computations by combining recursive subdivision with the convex hull property in order to avoid trying to compute intersections for those parts of the curve that fail to intersect.

*Intersection Algorithm—Bezier Curves*

If the convex hulls of the control points of two Bezier curves fail to intersect, then the curves themselves do not intersect.
Otherwise:
   If each Bezier curve can be approximated by the straight line segment joining its first and last control points, then intersect these line segments.
   Otherwise *subdivide* the two Bezier curves and intersect the pieces recursively.

To determine whether a Bezier curve can be approximated to within tolerance by the straight line segment joining its first and last control points, it is sufficient, by the convex hull property, to test whether all the interior control points lie within tolerance of this line segment. But recall from Chapter 11, Section 11.4.1.2 that the distance between a point $P$ and a line $L$ determined by a point $Q$ and a direction vector $v$ is given by

$$dist^2(P,L) = |P - Q|^2 - \frac{((P - Q) \cdot v)^2}{v \cdot v}. \tag{25.1}$$

Therefore to test whether the control points $P_k$, $k = 1, \ldots, n - 1$ lie within some tolerance $\varepsilon$ of the line determined by the first and last control points $P_0, P_n$, we set $v = P_n - P_0$ in Equation 25.1 and test whether or not

$$|P_k - P_0|^2 - \frac{((P_k - P_0) \cdot (P_n - P_0))^2}{(P_n - P_0) \cdot (P_n - P_0)} < \varepsilon^2. \tag{25.2}$$

Be careful. It may happen that a control point $P_k$ is close to the line $L$ determined by the first and last control points $P_0, P_n$ even though the orthogonal projection of $P_k$ onto $L$ does not lie inside the line segment $P_0P_n$—that is, $P_k$ may be close to the line $L$ even though it is not close to the line segment $P_0P_n$. To be sure that $P_k$ lies close to the line segment $P_0P_n$, you need only check that in addition to Equation 25.2

$$0 \le (P_k - P_0) \cdot (P_n - P_0) \le |P_n - P_0|^2. \tag{25.3}$$

It is relatively easy to test whether or not a Bezier curve can be approximated to within some tolerance by a straight line segment. On the other hand, finding and intersecting the convex hulls of two Bezier curves can be quite difficult and time consuming. In practice, the convex hulls in the intersection algorithm are typically replaced by bounding boxes which are much easier to compute and intersect than the actual convex hulls: simply take the minimum and maximum $x$ and $y$ coordinates of the control points. Since the subdivision algorithm converges rapidly, not much time is lost by replacing convex hulls with bounding boxes.

To complete the intersection algorithm for Bezier curves, we need to be able to intersect two line segments:

$$L_1(s) = (1 - s)P_0 + sP_1 = P_0 + s(P_1 - P_0), \quad 0 \le s \le 1,$$
$$L_2(t) = (1 - t)Q_0 + tQ_1 = Q_0 + t(Q_1 - Q_0), \quad 0 \le t \le 1.$$

Two infinite lines intersect when $L_1(s) = L_2(t)$—that is, when

$$P_0 + s(P_1 - P_0) = Q_0 + t(Q_1 - Q_0),$$

or equivalently when

$$s(P_1 - P_0) - t(Q_1 - Q_0) = (Q_0 - P_0).$$

Dotting both sides first with $P_1 - P_0$ and then with $Q_1 - Q_0$ generates two linear equations in two unknowns:

$$\begin{aligned} s\{(P_1 - P_0) \cdot (P_1 - P_0)\} - t\{(Q_1 - Q_0) \cdot (P_1 - P_0)\} &= (Q_0 - P_0) \cdot (P_1 - P_0), \\ s\{(P_1 - P_0) \cdot (Q_1 - Q_0)\} - t\{(Q_1 - Q_0) \cdot (Q_1 - Q_0)\} &= (Q_0 - P_0) \cdot (Q_1 - Q_0), \end{aligned} \tag{25.4}$$

which are easy to solve for the parameters $s, t$ (see Chapter 11, Section 11.5.1). The intersection point lies on the two line segments if and only $0 \le s, t \le 1$. In this case we can compute either $L_1(s)$ or $L_2(t)$ to find the actual intersection point.

## 25.3.2 Rendering and Intersecting Bezier Surfaces

The rendering and intersection algorithms for Bezier curves can be extended to rendering and intersection algorithms for Bezier surfaces. Line are replaced by planes, line segments are replaced by triangles, and recursive subdivision for Bezier curves is replaced by recursive subdivision for Bezier surfaces. To render a Bezier patch, we must first polygonalize the patch, so we begin with an algorithm to approximate a Bezier patch by a collection of triangles.

*Triangulation Algorithm—Bezier Surfaces*

If the Bezier patch can be approximated to within tolerance by two triangles each determined by three of its four corner control points and if the four Bezier boundaries of the Bezier patch can be approximated by straight line segments joining the four corner control points, then triangulate the Bezier patch by the two triangles determined by the four corner control points.

Otherwise *subdivide* the Bezier surface (at the midpoint of the parameter interval in *s* or *t*) and triangulate the segments recursively.

### Rendering Algorithm—Bezier Surfaces

Triangulate the Bezier patch.

Apply your favorite shading and hidden surface algorithms to render the triangulated patch.

### Ray Tracing Algorithm—Bezier Surfaces

If the ray does not intersect the convex hull of the control points of the Bezier patch, then the ray and the patch do not intersect.

Otherwise:

If the Bezier patch can be approximated to within tolerance by two triangles each determined by three of its four corner control points and if the four Bezier boundaries of the Bezier patch can be approximated by straight line segments joining the four corner control points, then intersect the ray with the two triangles determined by the four corner control points.

Otherwise *subdivide* the Bezier patch and ray trace the segments recursively.

Keep the intersection closest to the eye.

We already know from Section 25.3.1 how to test if a boundary Bezier curve can be approximated by a straight line joining two of the boundary control points. We need to perform this test in the triangulation algorithm to be sure that after subdivision cracks do not appear between adjacent triangles. If a boundary is not a straight line, then subdividing on one side of the boundary but not the other may introduce a tear between adjacent triangles. But when the boundaries are flat, such cracks will not appear.

To determine whether a Bezier surface can be approximated to within some tolerance by a pair of triangles determined by its four corner control points, it is sufficient, by the convex hull property, to test whether each control point lies within some tolerance of at least one of the two triangles. Now recall from Chapter 11, Section 11.4.1.3 that the distance between a point $P$ and a plane $S$ determined by a point $Q$ and a unit normal vector $N$ is given by

$$dist(P, S) = |(P - Q) \cdot N|. \tag{25.5}$$

Therefore to test whether a control point $P_{ij}$ lies within some tolerance $\varepsilon$ of the plane determined by the three corner control points $P_{00}, P_{0n}, P_{m0}$, we set $N = \frac{(P_{0n}-P_{00})\times(P_{m0}-P_{00})}{|(P_{0n}-P_{00})\times(P_{m0}-P_{00})|}$ in Equation 25.5 and test whether or not

$$|(P_{ij} - P_{00}) \cdot N| < \varepsilon. \tag{25.6}$$

If this test fails, we apply the analogous test with the plane determined by the control points $P_{mn}, P_{0n}, P_{m0}$. Again we must be careful. It may happen that a control point $P_{ij}$ lies close to the plane $S$ of a triangle $\Delta$ even though the orthogonal projection of the point $P_{ij}$ onto the plane $S$ does not lie inside the triangle $\Delta$—that is, $P_{ij}$ may be close to the plane $S$ even though $P_{ij}$ is not close to the triangle $\Delta$. To be sure that this is not the case, you need only check that the orthogonal projection $R_{ij} = P_{ij} - \{(P_{ij} - Q) \cdot N\}N$ lies inside $\Delta$. Now recall from Chapter 19, Section 19.3 that a point $R$ lies inside a triangle with vertices $P_0, P_1, P_2$ if and only if

$$\det(N, P_{i+1} - P_i, R - P_i) \geq 0, \quad i = 0, 1, 2,$$

where $N$ is the normal to the plane of the triangle. These tests should be carried out first with $\Delta P_{00}P_{0n}P_{m0}$ and only if these tests fail should the tests then be performed with $\Delta P_{mn}P_{0n}P_{m0}$.

Notice that in the ray tracing algorithm we do not simply triangulate the patch and then ray trace the triangulated approximation. Rather we first perform a convex hull test in order to eliminate as many subpatches as possible before we perform intersection calculations.

Finding the convex hull of a Bezier patch can be quite difficult and time consuming. In practice for surfaces, as with curves, the convex hull is replaced by a bounding box. Again since the subdivision algorithm converges rapidly, not much time is lost in the ray tracing algorithm by replacing convex hulls with bounding boxes.

To complete the ray tracing algorithm for Bezier patches, we need to be able to intersect a line with a planar polygon. Here we can use the algorithm described in Chapter 19, Section 19.3.

Finally, a Bezier patch may self-shadow—that is, a light ray may not be able to reach part of a Bezier patch because the light has already struck another part of the patch. Therefore at the end of the ray tracing algorithm we discard all the intersections of the ray and the patch, except for the intersection closest to the eye, and we fire rays from this point to each light source to check for self shadowing.

## 25.4   The Variation Diminishing Property of Bezier Curves

We introduced Bezier curves because, unlike polynomial interpolation, Bezier approximation does not introduce oscillations not already present in the data. This property of Bezier curves is called the *variation diminishing property*. Recall from Chapter 24, Section 24.4.3 that a curve $B(t)$ for a control polygon $P$ is said to be *variation diminishing* if for every line $L$

the number of intersections of $B(t)$ and $L \leq$ the number of intersections of $P$ and $L$.

We are now going to use Bezier subdivision to prove the variation diminishing property of Bezier curves.

One way to introduce variation diminishing schemes is by corner cutting. Start with a polygon $P$ and form another polygon $Q$ by cutting a corner off $P$ (see Figure 25.8). Then it is easy to see that every line $L$ intersects $P$ at least as often as $L$ intersects $Q$. Thus $Q$ is variation diminishing with respect to $P$.

Now look at the geometric interpretation of the de Casteljau subdivision algorithm in Figure 25.4. Evidently, the de Casteljau subdivision algorithm is a corner cutting procedure: first the corners at $P_1, P_2$ are cut off, then the corner at $\Delta$ is cut off (see Figure 25.9).

Although we have illustrated only the cubic case in Figure 25.9, it is easy to verify that the de Casteljau subdivision algorithm is a corner cutting procedure in all degrees. This observation leads to the following result.

**Theorem 25.3:** *Bezier curves are variation diminishing.*

**Proof:** Since recursive subdivision is a corner cutting procedure, the limit curve must be variation diminishing with respect to the original control polygon. But by Theorem 25.1, the Bezier curve is the limit curve generated by recursive subdivision, so Bezier curves are variation diminishing.

By the way, there is no known analogue of the variation diminishing property for Bezier surfaces because for Bezier surfaces subdivision is not a corner cutting procedure.

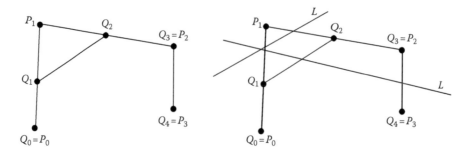

**FIGURE 25.8:** Corner cutting. The polygon with vertices $Q = \{Q_i\}$ is generated from the polygon with vertices $P = \{P_j\}$ by cutting off the corner at $P_1$ (left). Every line $L$ intersects $P$ at least as often as $L$ intersects $Q$ because if $L$ intersects the line segment $Q_1 Q_2$, then $L$ must intersect either $P_0 P_1$ or $P_1 P_2$ (right).

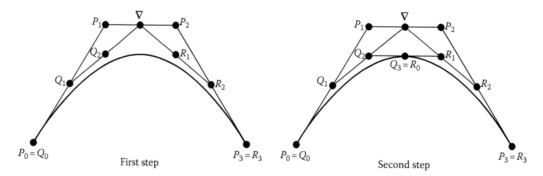

**FIGURE 25.9:** The de Casteljau subdivision algorithm as a sequence of corner cuts. In the first stage corners are cut off at $P_1$ and $P_2$; in the second stage the corner is cut off at $\Delta$. Compare to Figure 25.4.

## 25.5   Joining Bezier Curves Smoothly

In Chapter 24, Section 24.5.1, we used our procedure for differentiating the de Casteljau evaluation algorithm for Bezier curves to derive constraints on the location of the control points to insure that two Bezier curves meet smoothly where they join. Given a Bezier curve $P(t)$ with control points $P_0, \ldots, P_n$, we derived the following constraints for the location of the control points $Q_0, \ldots, Q_n$ of another Bezier curve $Q(t)$ that meets the first curve and matches the first $k$ derivatives of $P(t)$ where they join.

$$
\begin{aligned}
k = 0 &\implies Q_0 = P_n, \\
k = 1 &\implies Q_1 = P_n + (P_n - P_{n-1}), \\
k = 2 &\implies Q_2 = P_{n-2} + 4(P_n - P_{n-1}).
\end{aligned}
\tag{25.7}
$$

We observed that each derivative generates one new constraint on one additional control point. These constraints are easy to solve, but rather cumbersome to write down. Here we show how to use subdivision to determine the location of the points $Q_0, \ldots, Q_k$ that guarantee $k$-fold continuity at the join directly, without solving a system of linear equations.

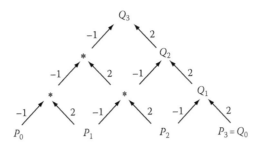

**FIGURE 25.10:**   The de Casteljau subdivision algorithm at $c = 2b - a$ for cubic Bezier curves over the interval $[a, b]$. The first $k + 1$ points $Q_0, \ldots, Q_k$ that emerge on the right lateral edge of the diagram are the control points that guarantee $k$-fold continuity at the join. Compare these points to the values in Equation 25.7.

We know that the location of the points $Q_0, \ldots, Q_k$ is unique. If we could find a curve $Q(t)$ that meets the original curve $P(t)$ smoothly at the join, then the control points of $Q(t)$ would necessarily give the location of the points $Q_0, \ldots, Q_k$. But we certainly do know such a curve, for $P(t)$ meets itself smoothly at the join. Suppose that $P(t)$ is parameterized over the interval $[a, b]$. Let $Q(t)$ be $P(t)$ over the interval $[b, 2b - a]$—that is, the interval starting at $b$ and with the same length as $[a, b]$. Then $P(t)$ and $Q(t)$ surely meet smoothly at $t = b$. All we need now are the Bezier control points of $Q(t)$.

To find the Bezier control points of $P(t)$ over the intervals $[a, c]$ and $[c, b]$, we can subdivide the Bezier curve $P(t)$ at $t = c$. Nothing in our subdivision algorithm requires that $a \leq c \leq b$. If we take $c = 2b - a$, then the de Casteljau subdivision algorithm will generate the Bezier control points for the curve $P(t)$ over the intervals $[a, 2b - a]$ and $[2b - a, b]$. By the symmetry property of Bezier curves (see Chapter 24, Exercise 24.2) the Bezier control points for the interval $[2b - a, b]$ are the same as the Bezier control points for the interval $[b, 2b - a]$ but in reverse order. Thus we can read off the control points $Q_0, \ldots, Q_k$ from the right lateral edge of the de Casteljau subdivision algorithm for $c = 2b - a$. But in this algorithm the labels on all the right pointing arrows are $-1$ because $(b - (2b - a))/(b - a) = (a - b)/(b - a) = -1$ and the labels on all the left pointing arrows are 2 because $((2b - a) - a)/(b - a) = 2(b - a)/(b - a) = 2$. Notice that these labels are independent of the interval $[a, b]$; hence this algorithm is independent of the parameter interval. We illustrate this algorithm in Figure 25.10 for cubic Bezier curves.

## 25.6   Summary

Subdivision is our main technical tool for analyzing Bezier curves and surfaces. In this chapter we used Bezier subdivision to:

i. Develop a divide and conquer strategy for rendering and intersecting Bezier curves and surfaces.

ii. Prove the variation diminishing property for Bezier curves.

iii. Find the control points which guarantee that two Bezier curves connect with $k$-fold smoothness at their join.

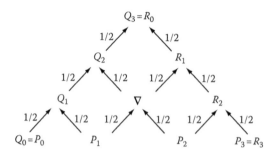

**FIGURE 25.11:** The de Casteljau subdivision algorithm for a cubic Bezier curve at the midpoint of the parameter interval. Midpoint subdivision is independent of the choice of the parameter domain.

The basic problem in Bezier subdivision is: given a collection of control points $P_0, \ldots, P_n$ that represent a Bezier curve over the parameter interval $[a, b]$, find the control points $Q_0, \ldots, Q_n$ and $R_0, \ldots, R_n$ that represent the segments of the same Bezier curve over the parameter intervals $[a, c]$ and $[c, b]$.

De Casteljau's evaluation algorithm for Bezier curves is also a subdivision procedure. The subdivision control points $Q_0, \ldots, Q_n$ and $R_0, \ldots, R_n$ emerge along the left and right lateral edges of the de Casteljau evaluation algorithm at $t = c$. Bezier subdivision is usually applied at the midpoint of the parameter domain, where the labels along all the arrows are $1/2$ (see Figure 25.11).

---

## Exercises

25.1. Let $P(t)$ be a Bezier curve defined over the interval $[a, b]$.

    a. Show that in the de Casteljau subdivision algorithm at $c = (1 - r)a + rb$ the labels on all the right pointing arrows are $1 - r$ and the labels on all the left pointing arrows are $r$ (see Figure 25.12).

    b. Conclude from part a that if we subdivide at a fixed ratio of the parameter domain, then the de Casteljau subdivision algorithm is independent of the parameter interval.

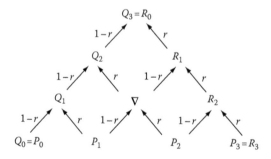

**FIGURE 25.12:** The de Casteljau subdivision algorithm at the parameter value $c = (1 - r)a + rb$ for a cubic Bezier curve defined over the interval $[a, b]$. The labels on all the right pointing arrows are $1 - r$ and the labels on all the left pointing arrows are $r$.

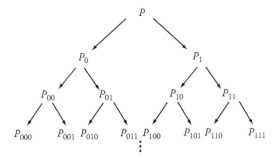

**FIGURE 25.13:** The binary tree of control polygons generated by recursive subdivision of a Bezier curve over the interval $[0, 1]$ at the parameter $t = 1/2$.

25.2. Consider a Bezier curve defined over the interval $[0, 1]$. Bezier subdivision at $t = 1/2$ generates a binary tree whose nodes are control polygons. Denote the original control polygon by $P$, and let this polygon be the root of the tree. Let $P_0$—the left child of $P$—denote the control polygon for the left segment of the Bezier curve (from $t = 0$ to $t = 1/2$), and let $P_1$—the right child of $P$—denote the control polygon for the right portion of the curve (from $t = 1/2$ to $t = 1$). Continue to build this binary tree recursively in this fashion. Thus if $P_b$ is a node in the tree, then $P_b$ represents the control polygon for a portion of the curve, and $P_{b0}$—the left child of $P_b$—represents the control polygon for the left half of the Bezier segment represented by $P_b$, while $P_{b1}$—the right child of $P_b$—represents the control polygon for the right half of the Bezier segment represented by $P_b$ (see Figure 25.13).

   a. Prove that $P_{b_1 \dots b_n}$ is the control polygon for the original Bezier curve from $t = b$ to $t = b + 2^{-n}$, where $b$ is the binary fraction represented by $0.b_1 \cdots b_n$.

   b. Prove that the sequence of control polygons $P_{b_1}, P_{b_1 b_2}, \dots, P_{b_1 \dots b_n}, \dots$ converges to the point on the Bezier curve at the parameter value $b = Lim_{n \to \infty} 0.b_1 \dots b_n$.

25.3. Develop an algorithm to intersect a Bezier curve with a Bezier surface based on recursive subdivision.

25.4. Develop an algorithm to intersect two Bezier patches based on recursive subdivision.

25.5. Prove Theorem 25.2: that the control polyhedra generated by recursive subdivision converge to the tensor product Bezier patch $B(s, t)$ for the original control polyhedron provided that the subdivision is done in both the $s$ and $t$ directions.

25.6. Let $P(t)$ be a Bezier curve with control points $P_0, \dots, P_n$.

   a. Prove that the arc length of a Bezier curve is greater than or equal to the length of the line segment joining the first and last control points and less than or equal to the perimeter of the control polygon—that is

$$|P_n - P_0| \leq arclength\ (P(t)) \leq \sum_{k=0}^{n-1} |P_{k+1} - P_k|.$$

   b. Explain how to use the result of part a to compute the arc length of a Bezier curve to within any desired tolerance.

---

## Programming Projects

25.1. *Bezier Curves and Surfaces*

Implement a modeling system based on Bezier curves and surfaces in your favorite programming language using your favorite API.

a. Include algorithms for rendering Bezier curves and surfaces based on recursive subdivision.

b. Incorporate the ability to move control points interactively and have the Bezier curve or surface adjust in real time.

c. Create a new font using Bezier curves.

d. Build some interesting freeform shapes using Bezier patches.

25.2. *Root Finding Algorithm for Polynomials in Bezier Form.*

A. Implement the following root finding algorithm for polynomials of arbitrary degree $n$.

*Input:* $c_0, \ldots, c_n =$ Bezier coefficients of $P(t)$ over the interval $[a, b]$.

*Root Finding Algorithm*

1. If $c_i = 0, i = 0, \ldots, j - 1$, then

   a. there is a multiple root of order $j$ at $t = a$; set:

   $$c_k = \frac{n \ldots (n - j + 1)}{(k + j) \ldots (k + 1)(b - a)^j} c_{j+k}, \quad k = 0, \ldots, n - j,$$

   $$n = n - j.$$

2. If $c_{n-i} = 0, i = 0, \ldots, j - 1$, then

   a. there is a multiple root of order $j$ at $t = b$; set:

   $$c_k = \frac{n \ldots (n - j + 1)}{(n - k) \ldots (n - j - k + 1)(b - a)^j} c_k, \quad k = 0, \ldots, n - j,$$

   $$n = n - j.$$

3. If $c_k > 0$ for all $k$ or if $c_k < 0$ for all $k$, then there is no root in the interval $[a, b]$. STOP.

4. If $c_k \geq 0$ for $0 \leq k < i$ and $c_k \leq 0$ for $i \leq k \leq n$ (one sign change), then:

   a. There is exactly one root $r$ in the interval $[a, b]$.

   b. If $b - a < \varepsilon$, set $r = (a + b)/2$.

   Otherwise subdivide the interval at the midpoint and search for the root in each subinterval recursively.

5. Otherwise subdivide the interval at the midpoint and find the roots in each subinterval recursively.

B. Prove that this algorithm finds all the roots of the given polynomial in the interval $[a, b]$.

C. Compare the speed and accuracy of this root finding algorithm to Newton's method (see Chapter 7, Section 7.3.1).

25.3. *De Casteljau's Subdivision Algorithm for Planar Bezier Curves*

Implement the de Casteljau subdivision algorithm for planar Bezier curves defined over the parameter interval [0,1] in your favorite programming language using your favorite API.

a. Verify that the control polygons generated by recursive subdivision converge to the Bezier curve for the original control polygon for any subdivision parameter $0 < r < 1$.

b. Investigate what happens if the subdivision parameter $r = 0$ or $r = 1$.

c. Investigate what happens if the subdivision parameter $r < 0$ or $r > 1$.

d. Represent the control points by complex numbers and points on the Bezier curve by complex numbers. Use complex multiplication in the implementation of the de Casteljau subdivision algorithm, and investigate what happens if the subdivision parameter is also a complex number.

    i. Investigate the curves generated by subdivision at the complex parameter $r = (1 + i)/2$.

    ii. For which complex numbers does the algorithm converge?

    iii. What kinds of shapes are generated by subdivision at complex parameters?

    iv. Investigate what kinds of fractals can you generate by various combinations of left and right subdivision at complex parameter values. Compare your results to Chapter 15, Programming Project 15.1.

# Chapter 26

---

## Blossoming

*blossom abundantly, and rejoice even with joy and singing*

– Isaiah 35:2

---

## 26.1 Motivation

Linear functions are simple; polynomials are complicated. If $L(t) = at$, then clearly

$$L(\mu s + \lambda t) = \mu L(s) + \lambda L(t).$$

More generally, if $L(t) = at + b$, then it is easy to verify that

$$L((1 - \lambda)s + \lambda t) = (1 - \lambda)L(s) + \lambda L(t).$$

In the first case $L(t)$ preserves linear combinations; in the second case $L(t)$ preserves affine combinations. But if $P(t) = a_n t^n + \cdots + a_0$ is a polynomial of degree $n > 1$, then

$$P(\mu s + \lambda t) \neq \mu P(s) + \lambda P(t),$$

$$P((1 - \lambda)s + \lambda t) \neq (1 - \lambda)P(s) + \lambda P(t).$$

Thus arbitrary polynomials preserve neither linear nor affine combinations. The key idea behind blossoming is to replace a complicated polynomial function $P(t)$ in one variable by a simple polynomial function $p(u_1, \ldots, u_n)$ in many variables that is either linear or affine in each variable. The function $p(u_1, \ldots, u_n)$ is called the *blossom* of $P(t)$, and converting from $P(t)$ to $p(u_1, \ldots, u_n)$ is called *blossoming*.

Blossoming is intimately linked to Bezier curves. We shall see in Section 26.3 that the Bezier control points of a polynomial curve are given by the blossom of the curve evaluated at the end points of the parameter interval. Moreover, there is an algorithm for evaluating the blossom recursively that closely mimics the de Casteljau evaluation algorithm for Bezier curves. In this chapter we shall apply the blossom to derive two standard algorithms for Bezier curves: the de Casteljau subdivision algorithm and the procedure for differentiating the de Casteljau evaluation algorithm. We shall also derive algorithms for converting between the Bernstein and monomial representations of a polynomial. Other properties of Bezier curves such as degree elevation are derived using blossoming in the exercises at the end of this chapter.

## 26.2   The Blossom

The *blossom* of a degree $n$ polynomial $P(t)$ is the unique symmetric multiaffine function $p(u_1, \ldots, u_n)$ that reduces to $P(t)$ along the diagonal. That is, $p(u_1, \ldots, u_n)$ is the unique multivariate polynomial satisfying the following three axioms:

*Blossoming Axioms*

1. Symmetry

   $p(u_1, \ldots, u_n) = p(u_{\sigma(1)}, \ldots, u_{\sigma(n)})$ for every permutation $\sigma$ of $\{1, \ldots, n\}$.

2. Multiaffine

   $p(u_1, \ldots, (1 - \alpha)u_k + \alpha v_k, \ldots, u_n) = (1 - \alpha)\, p(u_1, \ldots, u_k, \ldots, u_n) + \alpha p(u_1, \ldots, v_k, \ldots, u_n).$

3. Diagonal

   $p(t, \ldots, t) = P(t).$

The symmetry property says that the order of the parameters $u_1, \ldots, u_n$ does not matter when we evaluate the blossom $p(u_1, \ldots, u_n)$. The multiaffine property is just a fancy way to say that $p(u_1, \ldots, u_n)$ is degree one in each variable (see Exercise 26.4). The diagonal property connects the blossom back to the original polynomial. At first, these three axioms may seem very abstract and complicated, but we shall see shortly that blossoming is both explicit and straightforward.

We are interested in blossoming because of the following key property, which we shall derive in Section 26.3, relating the blossom of a polynomial to its Bezier control points.

*Dual Functional Property*

Let $P(t)$ be a Bezier curve over the interval $[a, b]$ with control points $P_0, \ldots, P_n$. Then,

$$P_k = p(\underbrace{a, \ldots, a}_{n-k}, \underbrace{b, \ldots, b}_{k}), \quad k = 0, \ldots, n.$$

We have yet to establish the existence and uniqueness of a function satisfying the three blossoming axioms. But before we proceed to prove both existence and uniqueness, let us get a better feel for the blossom by computing a few simple examples.

### Example 26.1   Cubic Polynomials

Consider the monomials $1, t, t^2, t^3$ as cubic polynomials. It is easy to blossom these monomials, since in each case it is easy to verify that the associated function $p(u_1, u_2, u_3)$ given below is symmetric, multiaffine, and reduces to the required monomial along the diagonal:

$$P(t) = 1 \Rightarrow p(u_1, u_2, u_3) = 1,$$

$$P(t) = t \Rightarrow p(u_1, u_2, u_3) = \frac{u_1 + u_2 + u_3}{3},$$

$$P(t) = t^2 \Rightarrow p(u_1, u_2, u_3) = \frac{u_1 u_2 + u_2 u_3 + u_3 u_1}{3},$$

$$P(t) = t^3 \Rightarrow p(u_1, u_2, u_3) = u_1 u_2 u_3.$$

Notice that the functions on the right-hand side are, up to a constant multiple, simply the elementary symmetric functions in three variables. Notice too that we must decide the degree of each monomial before we blossom, since the degree of the monomial tells us how many variables must appear in the blossom. Using these results, we can blossom any cubic polynomial, since

$$P(t) = a_3 t^3 + a_2 t^2 + a_1 t + a_0,$$

$$p(u_1, u_2, u_3) = a_3 u_1 u_2 u_3 + a_2 \frac{u_1 u_2 + u_2 u_3 + u_3 u_1}{3} + a_1 \frac{u_1 + u_2 + u_3}{3} + a_0.$$

Similar techniques can be applied to blossom polynomials of any arbitrary degree $n$ by first blossoming the monomials $M_k^n(t) = t^k, k = 0, \ldots, n$, using elementary symmetric functions in $n$ variables, and then applying linearity. This observation leads to the following results.

**Proposition 26.1   (Blossoming Monomials)**
    *The blossom of the monomial $M_k^n(t) = t^k$ (considered as a polynomial of degree n) is given by*

$$m_k^n(u_1, \ldots, u_n) = \frac{\sum u_{i_1} \cdots u_{i_k}}{\binom{n}{k}}, \tag{26.1}$$

*where the sum is taken over all subsets $\{i_1, \ldots, i_k\}$ of $\{1, \ldots, n\}$.*

**Proof:** To show that $m_k^n(u_1, \ldots, u_n)$ is the blossom of $M_k^n(t)$, we need to verify that the three blossoming axioms are satisfied. By construction $m_k^n(u_1, \ldots, u_n)$ is symmetric, since the sum on the right-hand side of Equation 26.1 is taken over all subsets $\{i_1, \ldots, i_k\}$ of $\{1, \ldots, n\}$. Also the function on the right-hand side of Equation 26.1 is multiaffine, since each variable appears in each term to at most the first power. Finally, since the number of subsets of $\{1, \ldots, n\}$ with $k$ elements is $\binom{n}{k}$, along the diagonal

$$m_k^n(t, \ldots, t) = \frac{\binom{n}{k} t^k}{\binom{n}{k}} = M_k^n(t).$$

**Theorem 26.1   (Existence of the Blossom)**
    *For every degree n polynomial $P(t)$, there exists a symmetric multiaffine function $p(u_1, \ldots, u_n)$ that reduces to $P(t)$ along the diagonal. That is, there exists a blossom $p(u_1, \ldots, u_n)$ for every polynomial $P(t)$.*

**Proof:** By Proposition 26.1, the blossom exists for the monomials $t^k, k = 0, \ldots, n$. Therefore since every polynomial is the sum of monomials and since the blossom of the sum is the sum of the blossoms (see Exercise 26.1), a blossom $p(u_1, \cdots, u_n)$ exists for every polynomial $P(t)$.

---

## 26.3   Blossoming and the de Casteljau Algorithm

A word about notation before we proceed. Throughout this chapter, especially in the diagrams, we shall adopt the multiplicative notation $u_1 \cdots u_n$ to represent the blossom value $p(u_1, \ldots, u_n)$. Though an abuse of notation, this multiplicative representation is highly suggestive. For example, multiplication is commutative and the blossom is symmetric

$$u_1 \cdots u_n = u_{\sigma(1)} \cdots u_{\sigma(n)} \quad \leftrightarrow \quad p(u_1, \ldots, u_n) = p(u_{\sigma(1)}, \ldots, u_{\sigma(n)}).$$

Moreover, multiplication distributes through addition and the blossom is multiaffine. Thus

$$u = \frac{b-u}{b-a}a + \frac{u-a}{b-a}b \quad \Rightarrow \quad u_1 \cdots u_n u = \frac{b-u}{b-a}u_1 \cdots u_n a + \frac{u-a}{b-a}u_1 \cdots u_n b,$$

and similarly

$$u = \frac{b-u}{b-a}a + \frac{u-a}{b-a}b \quad \Rightarrow \quad p(u_1,\ldots,u_n,u) = \frac{b-u}{b-a}p(u_1,\ldots,u_n,a) + \frac{u-a}{b-a}p(u_1,\ldots,u_n,b).$$

Since symmetry and multiaffinity are the main properties featured in the diagrams, the same diagrams make sense both for multiplication and for blossoming. Thus this multiplicative notation for the blossom is both natural and evocative. In addition, these similarities between multiplication and blossoming suggest that corresponding to identities for multiplication we should expect analogous identities for the blossom. In fact, we shall see an important example of such an analogous identity in Section 26.4.2, where we introduce an analogue of the binomial theorem for the blossom.

Using this multiplicative notation, Figure 26.1 shows how to compute values of $p(t,\ldots,t)$ from the blossom values $p(\underbrace{a,\ldots,a}_{n-k},\underbrace{b,\ldots,b}_{k})$, $k = 0,\ldots,n$, recursively by applying the multiaffine and symmetry properties at each node. For example, since

$$t = \frac{b-t}{b-a}a + \frac{t-a}{b-a}b,$$

it follows by the multiaffine property that

$$p(a,b,t) = p\left(a,b,\frac{b-t}{b-a}a + \frac{t-a}{b-a}b\right) = \frac{b-t}{b-a}p(a,b,a) + \frac{t-a}{b-a}p(a,b,b).$$

Similar identities hold at all the other nodes in Figure 26.1.

Now compare the blossoming algorithm in Figure 26.1 to the de Casteljau algorithm in Figure 26.2. Clearly Figure 26.1 is the de Casteljau algorithm for $p(t,\ldots,t) = P(t)$ with control

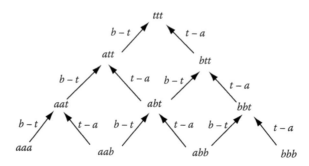

**FIGURE 26.1:** Computing $p(t,\ldots,t)$ from the blossom values $p(\underbrace{a,\ldots,a}_{n-k},\underbrace{b,\ldots,b}_{k})$, $k = 0,\ldots,n$. Here we illustrate the cubic case, and we adopt the multiplicative notation $uvw$ for $p(u,v,w)$. As usual the label on each edge must be normalized by dividing by $b-a$, so that the labels along the two arrows entering each node sum to one.

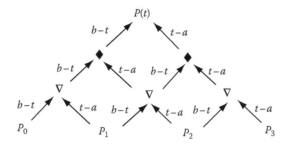

**FIGURE 26.2:** The de Casteljau algorithm for a cubic Bezier curve $P(t)$ on the interval $[a, b]$. Compare to Figure 26.1.

points $P_k = p(\underbrace{a, \ldots, a}_{n-k}, \underbrace{b, \ldots, b}_{k})$. This observation has several important consequences, which we summarize in Theorems 26.2 through 26.4.

### Theorem 26.2 (Every Polynomial Curve is a Bezier Curve)

*Every polynomial can be expressed in Bezier form. That is, for every polynomial $P(t)$ and every interval $[a, b]$ there exist control points $P_0, \ldots, P_n$ such that over the interval $[a, b]$ the polynomial $P(t)$ is generated by the de Casteljau algorithm with control point $P_0, \ldots, P_n$.*

**Proof:** This result follows immediately from Figure 26.1, since Figure 26.1 is the de Casteljau algorithm for $p(t, \ldots, t) = P(t)$ with control points $P_k = p(\underbrace{a, \ldots, a}_{n-k}, \underbrace{b, \ldots, b}_{k}), k = 0, \ldots, n$.

### Theorem 26.3 (Dual Functional Property of the Blossom)

*Let $P(t)$ be a Bezier curve defined over the interval $[a, b]$, and let $p(u_1, \ldots, u_n)$ be the blossom of $P(t)$. Then*

$$p(\underbrace{a, \ldots, a}_{n-k}, \underbrace{b, \ldots, b}_{k}) = P_k = kth \ Bezier \ Control \ Point \ of \ P(t). \qquad (26.2)$$

**Proof:** Again this result follows immediately from Figure 26.1.

### Theorem 26.4 (Uniqueness of the Blossom)

*Let $P(t)$ be a polynomial of degree n. Then the blossom $p(u_1, \ldots, u_n)$ of $P(t)$ is unique. That is, for each polynomial $P(t)$ there is only one function $p(u_1, \ldots, u_n)$ that is symmetric, multiaffine, and reduces to $P(t)$ along the diagonal.*

**Proof:** For any polynomial $P(t)$, Figure 26.3 shows how to compute an arbitrary blossom value $p(u_1, \ldots, u_n)$ from the blossom values $p(\underbrace{a, \ldots, a}_{n-k}, \underbrace{b, \ldots, b}_{k}), k = 0, \ldots, n$ by substituting $u_j$ for $t$ on the $j$th level of the de Casteljau algorithm, $j = 1, \ldots, n$.

Thus every blossom value $p(u_1, \ldots, u_n)$ is completely determined by the blossom values $p(\underbrace{a, \ldots, a}_{n-k}, \underbrace{b, \ldots, b}_{k}), k = 0, \ldots, n$. Now suppose that the polynomial $P(t)$ has two blossoms $p(u_1, \ldots, u_n)$ and $q(u_1, \ldots, u_n)$. Then by Theorem 26.3

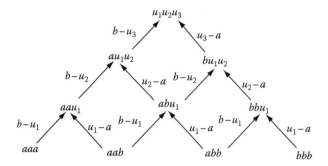

**FIGURE 26.3:** Computing an arbitrary blossom value $p(u_1, \ldots, u_n)$ from the blossom values $p(\underbrace{a, \ldots, a}_{n-k}, \underbrace{b, \ldots, b}_{k})$, $k = 0, \ldots, n$. Here again we illustrate the cubic case, and we adopt the multiplicative notation $uvw$ for $p(u, v, w)$. As usual the label on each edge must be normalized by dividing by $b - a$.

$$p(\underbrace{a, \ldots, a}_{n-k}, \underbrace{b, \ldots, b}_{k}) = q(\underbrace{a, \ldots, a}_{n-k}, \underbrace{b, \ldots, b}_{k}), \quad k = 0, \ldots, n,$$

since both sides represent the Bezier control points of $P(t)$ and these control points are unique (see Chapter 24, Section 24.5.2). But we have just seen in Figure 26.3 that any arbitrary blossom value is completely determined by these particular $n + 1$ blossom values. Therefore

$$p(u_1, \ldots, u_n) = q(u_1, \ldots, u_n),$$

so the blossom of $P(t)$ is unique.

### 26.3.1   Bezier Subdivision from Blossoming

The de Casteljau algorithm for subdividing Bezier curves is easy to derive using the dual functional property (Equation 26.2) of the blossom. Let $P(t)$ be a Bezier curve over the interval $[a, b]$ with control points $P_0, \ldots, P_n$. By the dual functional property,

$$P_k = p(\underbrace{a, \ldots, a}_{n-k}, \underbrace{b, \ldots, b}_{k}), \quad k = 0, \ldots, n.$$

To subdivide this Bezier curve at the parameter $t$, we must find the Bezier control points for $P(t)$ over the intervals $[a, t]$ and $[t, b]$. Again by the dual functional property, these control points $Q_0, \ldots, Q_n$ and $R_0, \ldots, R_n$ are

$$Q_k = p(\underbrace{a, \ldots, a}_{n-k}, \underbrace{t, \ldots, t}_{k}), \quad k = 0, \ldots, n,$$

$$R_k = p(\underbrace{t, \ldots, t}_{n-k}, \underbrace{b, \ldots, b}_{k}), \quad k = 0, \ldots, n.$$

But look at Figure 26.1. If we interpret every triple $uvw$ as the blossom value $p(u, v, w)$, then the control points $Q_k$ and $R_k$ emerge along the left and right lateral edges of the triangle. Generalized to arbitrary degree, this observation is precisely the de Casteljau subdivision algorithm of Chapter 25. Notice how blossoming simplifies the analysis of Bezier curves by providing natural labels for all the interior nodes in the de Casteljau algorithm. Blossoming is a beautiful, powerful, clever, almost magical idea.

## 26.4 Differentiation and the Homogeneous Blossom

Points along a polynomial curve can be represented in terms of the blossom of the curve by invoking the diagonal property. But what about derivatives? A polynomial curve takes on values that are points in an affine space; so too does its blossom. Indeed, only affine combinations of blossom values are permitted. But derivatives represent tangent vectors not points. To represent derivatives, we need a variant of the blossom that takes on values in a vector space, rather than an affine space. Here we shall construct such a blossom by applying the technique of *homogenization*. Homogenization lifts the domain (and the range) from an affine space to a vector space, so we are no longer restricted to affine combinations, but can exploit instead arbitrary linear combinations.

### 26.4.1 Homogenization and the Homogeneous Blossom

Monomials are the simplest polynomials. If $P(t) = t^n$, then $P(ct) = c^n t^n = c^n P(t)$. Monomials are not linear—$P(ct) \neq cP(t)$—but $P(ct) = c^n P(t)$ is almost as good. On the other hand, if $P(t) = a_n t^n + \cdots + a_0$, $n > 1$, then $P(ct) \neq c^n P(t)$ because different terms have different degrees. The key idea behind homogenization is to introduce a new variable to homogenize the polynomial so that all the terms have the same degree.

To homogenize a degree $n$ polynomial $P(t) = \sum_{k=0}^{n} a_k t^k$, we multiply each term $t^k$ by $w^{n-k}$. Homogenization creates a new polynomial in two variables, $P(t, w) = \sum_{k=0}^{n} a_k t^k w^{n-k}$, which is homogeneous of degree $n$—that is, each term has the same total degree $n$. Therefore $P(ct, cw) = c^n P(t, w)$. We can easily recover $P(t)$ from $P(t, w)$ because $P(t) = P(t, 1)$. Constructing $P(t, w)$ from $P(t)$ is called *homogenization*; recovering $P(t)$ from $P(t, w)$ is called *dehomogenization*.

To homogenize the blossom $p(u_1, \ldots, u_n)$, we homogenize with respect to each variable independently. Thus the homogeneous version of $p(u_1, \ldots, u_n)$ is another polynomial $p((u_1, v_1), \ldots, (u_n, v_n))$ that is homogeneous with respect to each pair of variables $(u_k, v_k)$. In every term of $p(u_1, \ldots, u_n)$ each variable $u_k$ appears to at most the first power, so every term of the homogeneous polynomial $p((u_1, v_1), \ldots, (u_n, v_n))$ has as a factor either $u_k$ or $v_k$ but not both. Since $p((u_1, v_1), \ldots, (u_n, v_n))$ is homogeneous of degree one in each pair of variables $(u_k, v_k)$,

$$p((u_1, v_1), \ldots, c(u_k, v_k), \ldots, (u_n, v_n)) = cp((u_1, v_1), \ldots, (u_k, v_k), \ldots, (u_n, v_n)).$$

Again we can dehomogenize $p((u_1, v_1), \ldots, (u_n, v_n))$ by setting $v_k = 1, k = 1, \ldots, n$. Thus $p((u_1, 1), \ldots, (u_n, 1)) = p(u_1, \ldots, u_n)$.

We can also blossom the homogenization of $P(t)$. We define the blossom of a homogeneous polynomial $P(t, w)$ to be the unique symmetric, multilinear polynomial $p((u_1, v_1), \ldots, (u_n, v_n))$ that reduces to $P(t, w)$ along the diagonal. Thus the homogeneous blossom satisfies the following axioms.

*Homogeneous Blossoming Axioms*

1. Symmetry

$$p((u_1, v_1), \ldots, (u_k, v_k), \ldots, (u_n, v_n)) = p((u_{\sigma(1)}, v_{\sigma(1)}), \ldots, (u_{\sigma(n)}, v_{\sigma(n)}))$$ for every permutation $\sigma$ of $\{1, \ldots, n\}$.

2. Multilinear

$$p((u_1, v_1), \ldots, (u_k, v_k) + (r_k, s_k), \ldots, (u_n, v_n))$$
$$= p((u_1, v_1), \ldots, (u_k, v_k), \ldots, (u_n, v_n)) + p((u_1, v_1), \ldots, (r_k, s_k), \ldots, (u_n, v_n)),$$
$$p((u_1, v_1), \ldots, c(u_k, v_k), \ldots, (u_n, v_n)) = cp((u_1, v_1), \ldots, (u_k, v_k), \ldots, (u_n, v_n)).$$

3. Diagonal

$$p((t, w), \ldots, (t, w)) = P(t, w).$$

Geometrically, the variables $t, u_1, \ldots, u_n$ represent parameters in a one-dimensional affine space. The homogenizing parameters $w, v_1, \ldots, v_n$ play the role of mass, lifting the variables $t, u_1, \ldots, u_n$ from a one-dimensional affine space to a two-dimensional vector space, where we can perform linear algebra.

Thus in the homogeneous blossom, we have replaced the multiaffine property by the multilinear property:

$$p((u_1, v_1), \ldots, (u_k, v_k) + (r_k, s_k), \ldots, (u_n, v_n))$$
$$= p((u_1, v_1), \ldots, (u_k, v_k), \ldots, (u_n, v_n)) + p((u_1, v_1), \ldots, (r_k, s_k), \ldots, (u_n, v_n)),$$
$$p((u_1, v_1), \ldots, c(u_k, v_k), \ldots, (u_n, v_n)) = cp((u_1, v_1), \ldots, (u_k, v_k), \ldots, (u_n, v_n)).$$

This property is equivalent to the fact that for each parameter pair $(u_i, v_i)$, either $u_i$ or $v_i$ but not both appear in each term to the first power (see Exercise 26.5).

Now starting with any polynomial $P(t)$, we can blossom and then homogenize or we can homogenize and then blossom. Figure 26.4 illustrates how this works in practice for the monomials $1, t, t^2, t^3$ considered as cubic polynomials. Notice that if we blossom and then homogenize, we get the same result as when we homogenize and then blossom. We formalize this result in Theorem 26.5.

**Theorem 26.5** *Blossoming and homogenization commute.*

**Proof:** This result follows immediately from Figure 26.5, since it is straightforward to verify that the functions in the right-hand column are symmetric, multiaffine on top, multilinear on bottom, and reduce to the corresponding functions in the left-hand column along the diagonal.

Notice that Figure 26.5 also establishes the existence of the homogeneous blossom, since the expression in the lower right-hand corner of the diagram is symmetric, multilinear and reduces to the expression on the bottom left along the diagonal. Uniqueness follows from dehomogenization: if we dehomogenize the homogeneous blossom, then we get the multiaffine blossom. Since the multiaffine blossom is unique, rehomogenization shows that the homogeneous blossom is also unique.

We constructed the multiaffine blossom in Figure 26.3 by blossoming the de Casteljau algorithm in Figures 26.1 and 26.2, replacing $t$ by a different parameter $u_k$ on the $k$th level of the algorithm. We can do the same for the multilinear blossom. Begin by homogenizing the de

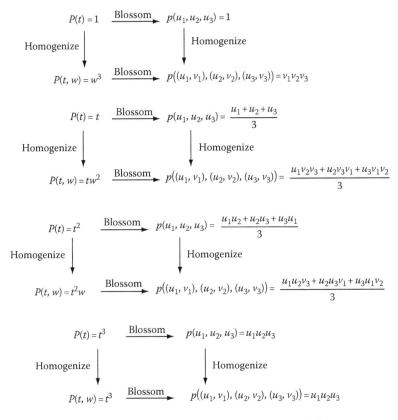

**FIGURE 26.4:** Blossoming and homogenizing the monomials $1, t, t^2, t^3$. Observe that blossoming and homogenization commute.

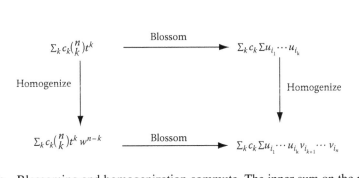

**FIGURE 26.5:** Blossoming and homogenization commute. The inner sum on the right-hand side is taken over all subsets $\{i_1, \ldots, i_k\}$ of $\{1, \ldots, n\}$.

Casteljau algorithm. This amounts to replacing $b - t \to bw - t$ and $t - a \to t - aw$ to insure that each term has the same total degree (see Figure 26.6). To blossom, we now replace the pair $(t, w)$ by the pair $(u_k, v_k)$ on $k$th level of the algorithm (see Figure 26.7). This process generates a symmetric, multilinear function that reduces to the homogeneous curve when we replace each pair $(u_k, v_k)$ by $(t, w)$. This function is multilinear rather than multiaffine because it is linear in $(u_k, v_k)$ on the $k$th level of the algorithm. Thus $u_k$ or $v_k$, but not both, appears in every term to the first power. Notice, by the way, that if we blossom first and then homogenize, we get exactly the same diagram. Blossoming first gives us Figure 26.3; homogenizing this diagram generates Figure 26.7.

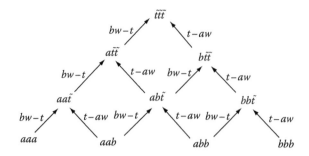

**FIGURE 26.6:** The homogeneous version of the de Casteljau algorithm for cubic Bezier curves. This diagram is generated from the de Casteljau algorithm (Figure 26.1) by replacing $b - t \to bw - t$ and $t - a \to t - aw$ along the edges of the triangle. Here we use the symbol "~" to denote homogeneous values, so $\tilde{t} = (t, w)$ while $a = (a, 1)$. Notice that

$$(t, w) = \frac{bw - t}{b - a}(a, 1) + \frac{t - aw}{b - a}(b, 1).$$

Therefore, by the linearity of the homogenous blossom,

$$p(a, b, \tilde{t}) = p\left(a, b, \frac{bw - t}{b - a}a + \frac{t - aw}{b - a}b\right) = \frac{bw - t}{b - a}p(a, b, a) + \frac{t - aw}{b - a}p(a, b, b),$$

and similar identities hold at all the other nodes in this diagram.

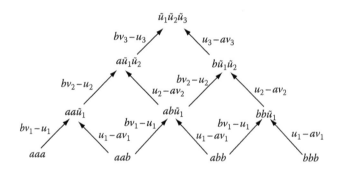

**FIGURE 26.7:** The homogeneous blossom of a cubic Bezier curve. This diagram is generated by blossoming the homogeneous version of the de Casteljau algorithm (Figure 26.6)—that is, by replacing the pair $(t, w)$ with the pair $(u_k, v_k)$ on $k$th level of the algorithm. This diagram can also be generated from the multiaffine blossom (Figure 26.3) by homogenizing the functions along the edges. Thus once again we see that blossoming and then homogenizing is equivalent to homogenizing and then blossoming. As in Figure 26.6, we use the symbol "~" to denote homogeneous values, so $\tilde{u} = (u, v)$ while $a = (a, 1)$. Notice that

$$(u, v) = \frac{bv - u}{b - a}(a, 1) + \frac{u - av}{b - a}(b, 1).$$

Therefore, by the linearity of the homogenous blossom,

$$p(a, b, \tilde{u}_1) = p\left(a, b, \frac{bv_1 - u_1}{b - a}a + \frac{u_1 - av_1}{b - a}b\right) = \frac{bv_1 - u_1}{b - a}p(a, b, a) + \frac{u_1 - av_1}{b - a}p(a, b, b),$$

and similar identities hold at all the other nodes in this diagram.

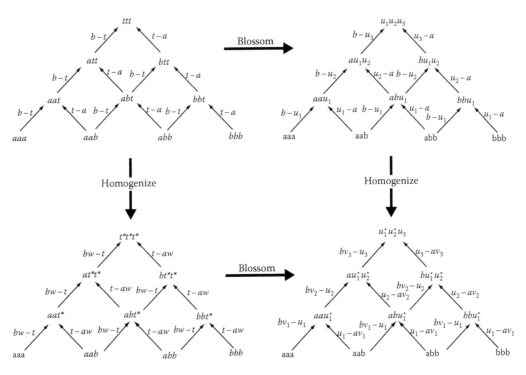

**FIGURE 26.8:** Four versions of the de Casteljau algorithm for cubic Bezier curves: the standard version (upper left), the homogeneous version (lower left), the blossomed version (upper right), and the homogeneous blossom (lower right). Blossoming and homogenization commute, since we get the same result by traversing right and then down or down and then right. Here we use the symbol "*" to denote homogeneous values, so $t^* = (t, w)$ and $u^* = (u, v)$ whereas $a = (a, 1)$ and $b = (b, 1)$.

We illustrate all four variants of the de Casteljau algorithm—original, blossomed, homogenized, and blossomed and homogenized—in Figure 26.8.

### 26.4.2 Differentiating the de Casteljau Algorithm

We built the homogeneous blossom to deal specifically with differentiation. We are now going to use a variant of Taylor's theorem along with the binomial expansion of the homogenous blossom to derive a formula for the derivatives of a polynomial in terms of the values of its homogeneous blossom.

**Theorem 26.6 (Taylor's Theorem)**
*Let $P(t)$ be a polynomial of degree $n$. Then*

$$P(t + h) = \sum_{k=0}^{n} \frac{P^{(k)}(t)}{k!} h^k. \tag{26.3}$$

**Proof:** Recall from calculus that the standard version of Taylor's theorem is

$$P(x) = P(t) + P'(t)(x - t) + \frac{P''(t)}{2!}(x - t)^2 + \cdots + \frac{P^{(n)}(t)}{n!}(x - t)^n.$$

Setting $x = t + h$ yields

$$P(t + h) = P(t) + P'(t)h + \frac{P''(t)}{2!}h^2 + \cdots + \frac{P^{(n)}(t)}{n!}h^n.$$

We can also compute $P(t + h)$ using the diagonal property of the homogeneous blossom. Let $t = (t, 1)$, and $\delta = (1, 0)$. Then

$$t + h \equiv (t + h, 1) = (t, 1) + h(1, 0) = t + h\delta.$$

Therefore, by the diagonal property of the homogeneous blossom,

$$P(t + h) = p(t + h\delta, \ldots, t + h\delta).$$

We can expand $p(t + h\delta, \ldots, t + h\delta)$ by the multilinear property to get a formula for $p(t + h\delta, \ldots, t + h\delta)$ similar in form to the binomial theorem.

**Theorem 26.7    (Binomial Expansion)**
  *Let $p(u_1, \ldots, u_n)$ be a symmetric multiaffine function. Then*

$$p(t + h\delta, \ldots, t + h\delta) = \sum_{k=0}^{n} \binom{n}{k} p(\underbrace{t, \ldots, t}_{n-k}, \underbrace{\delta, \ldots, \delta}_{k})h^k, \tag{26.4}$$

*where $t = (t, 1)$ and $\delta = (1, 0)$.*

**Proof:** The proof is an inductive argument, similar to the inductive proof of the binomial theorem. We illustrate the case $n = 2$ and leave the general case as an exercise (see Exercise 26.8). To prove the result for $n = 2$, just expand by linearity:

$$\begin{aligned}
p(t + h\delta, t + h\delta) &= p(t, t + h\delta) + p(h\delta, t + h\delta) \\
&= p(t, t) + p(t, h\delta) + p(h\delta, t) + p(h\delta, h\delta) \\
&= p(t, t) + 2p(t, h\delta) + p(h\delta, h\delta) \\
&= p(t, t) + 2p(t, \delta)h + p(\delta, \delta)h^2.
\end{aligned}$$

**Theorem 26.8** *Let $P$ be a polynomial of degree $n$, and let $p$ be the homogenous blossom of $P$. Then*

$$P^{(k)}(t) = \frac{n!}{(n-k)!}p(\underbrace{t, \ldots, t}_{n-k}, \underbrace{\delta, \ldots, \delta}_{k}), \tag{26.5}$$

*where $t = (t, 1)$ and $\delta = (1, 0)$.*

**Proof:** By the diagonal property of the homogeneous blossom

$$P(t + h) = p(t + h\delta, \ldots, t + h\delta),$$

and this equality is valid for all values of $h$. Comparing the coefficients of $h^k$ in Equations 26.3 and 26.4 for $P(t + h)$ and $p(t + h\delta, \ldots, t + h\delta)$ yields

$$P^{(k)}(t) = \frac{n!}{(n-k)!} p(\underbrace{t, \ldots, t}_{n-k}, \underbrace{\delta, \ldots, \delta}_{k}),$$

Thus to find the $k$th derivative of a polynomial $P(t)$, we need only compute the blossom value $p(\underbrace{t, \ldots, t}_{n-k}, \underbrace{\delta, \ldots, \delta}_{k})$ and multiply by a constant. But we have already shown that by blossoming and homogenizing the de Casteljau algorithm, we can compute the homogenous blossom $p((u_1, v_1), \ldots, (u_n, v_n))$ for any values of the parameters $(u_1, v_1), \ldots, (u_n, v_n)$ (see Figure 26.7). We illustrate the algorithms for computing $p(t, \ldots, t, \delta)$ and $p(t, \ldots, t, \delta, \delta)$ in Figures 26.9 and 26.10.

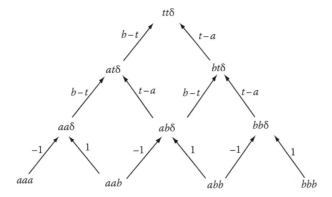

**FIGURE 26.9:** The first derivative of a cubic Bezier curve. Compare to Figure 26.7 with $(u_1, v_1) = \delta = (1, 0)$, and $(u_2, v_2) = (u_3, v_3) = (t, 1)$. To get the derivative, we must multiply the output at the apex of the triangle by $n = 3$, and we must remember to normalize the labels on each row—even the constants along the arrows on the first level—by dividing by $b-a$. Compare to Chapter 24, Figure 24.14.

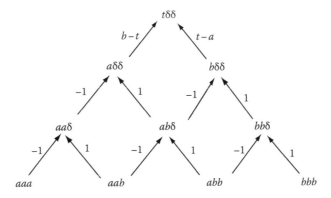

**FIGURE 26.10:** The second derivative of a cubic Bezier curve. Compare to Figure 26.7 with $(u_1, v_1) = (u_2, v_2) = \delta = (1, 0)$, and $(u_3, v_3) = (t, 1)$. To get the derivative, we must multiply the output at the apex of the triangle by $n! = 6$, and we must remember to normalize the labels on each row—even the constants along the arrows on the first two levels—by dividing by $b-a$. Compare to Chapter 24, Figure 24.15.

Notice that if we homogenize the de Casteljau algorithm and then we replace $(t, w) \to (1, 0)$ on any level of the de Casteljau algorithm, the effect is to replace $bw - t \to -1$ and $t - aw \to +1$; that is, the effect is to differentiate one level of the de Casteljau algorithm. Now if we replace $(t, w) \to (t, 1)$ on the remaining levels of the algorithm—that is, if we dehomogenize the remaining levels—then, up to a constant multiple, we get the derivative of the original Bezier curve (see Figure 26.9). If on $k$ levels of the homogeneous de Casteljau algorithm we replace $(t, w) \to (1, 0)$ and on the remaining $n - k$ levels we replace $(t, w) \to (t, 1)$, then, up to a constant multiple, we obtain the $k$th derivative of the original Bezier curve. This algorithm is precisely the derivative algorithm we introduced in Chapter 24, Section 24.5. Blossoming provides an alternative proof as well as a more general algorithm.

### 26.4.3   Conversion Algorithms between Monomial and Bezier Form

We can also use blossoming to develop simple algorithms to convert between monomial and Bezier form. That is, given the monomial coefficients of a polynomial, we can find the Bezier control points, and conversely given the Bezier control points we can find the monomial coefficients. The key observation here is that the monomial coefficients are derivatives, at least up to constant multiples, and we have just seen that derivatives can be represented, at least up to constant multiples, by values of the homogeneous blossom.

To see the connection between monomial coefficients, derivatives, and blossom values, suppose

$$P(t) = a_0 + a_1 t + \cdots + a_k t^k + \cdots + a_n t^n.$$

Recall that by Taylor's theorem

$$P(t) = P(0) + P'(0)t + \cdots + \frac{P^{(k)}(0)}{k!} t^k + \cdots + \frac{P^{(n)}(0)}{n!} t^n.$$

Comparing the coefficients of $t^k$ and invoking Theorem 26.8 yields

$$a_k = \frac{P^{(k)}(0)}{k!} = \binom{n}{k} p(\underbrace{0, \ldots, 0}_{n-k}, \underbrace{\delta, \ldots, \delta}_{k}), \quad k = 0, \ldots, n, \tag{26.6}$$

or equivalently

$$p(\underbrace{0, \ldots, 0}_{n-k}, \underbrace{\delta, \ldots, \delta}_{k}) = \frac{a_k}{\binom{n}{k}},$$

where $\delta = (1, 0)$. Equation 26.6 is reminiscent of Equation 26.2 for the interval $[0, 1]$

$$P_k = p(\underbrace{0, \ldots, 0}_{n-k}, \underbrace{1, \ldots, 1}_{k}), \quad k = 0, \ldots, n,$$

which gives the Bezier control points of $P(t)$ over the interval $[0, 1]$. Thus to convert between monomial and Bezier form, we need only convert from one set of blossom values to another. We illustrate these conversion algorithms for cubic curves in Figure 26.11; these algorithms generalize to arbitrary degree in the obvious manner. Notice that to convert between monomial and Bezier form, the monomial coefficients must be normalized by multiplying or dividing by binomial coefficients.

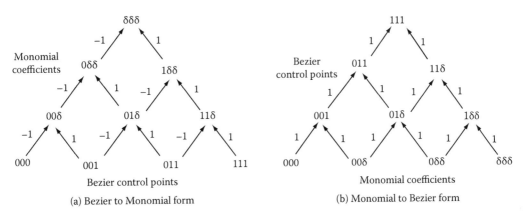

(a) Bezier to Monomial form  (b) Monomial to Bezier form

**FIGURE 26.11:** Conversion between cubic Bezier and cubic monomial form. In (a) we convert from Bezier to monomial form; in (b) we convert from monomial to Bezier form. Notice that the labels along the edges in these diagrams do not need to be normalized because $1 = (1,1) = (0,1) + (1,0) = 0 + \delta$. Thus we subtract at each node to get from Bezier control points to monomial coefficients, and we add at each node to get from monomial coefficients to Bezier control points. These algorithms work in arbitrary degree. Notice that the diagram for the algorithm to covert from Bezier to monomial form is the same as the diagram for the algorithm to compute the $n$th derivative of a Bezier curve over the interval $[0, 1]$.

## 26.5  Summary

Blossoming is a potent tool for deriving properties of Bezier curves. In this chapter we applied blossoming to derive the de Casteljau subdivision algorithm and the de Casteljau differentiation algorithm for Bezier curves, as well as algorithms to convert between monomial and Bezier form. B-splines can also be conveniently investigated from the perspective of blossoming; we shall investigate B-splines in Chapter 27.

Blossoming is related to Bezier curves because each node in the de Casteljau algorithm has an interpretation in terms of a blossom value. There are also several important variants of the de Casteljau algorithm that can be generated by straightforward substitutions.

*Blossoming*

$$t \rightarrow u_k \quad \text{on the } k\text{th level.}$$

*Homogenization*

$$\begin{aligned} t - a &\rightarrow t - aw \\ b - t &\rightarrow bw - t \end{aligned} \quad \text{on every level.}$$

*Differentiation (kth derivative)*

$$\begin{aligned} t - a &\rightarrow 1 \\ b - t &\rightarrow -1 \end{aligned} \quad \text{on } k \text{ levels and multiply the output by } \frac{n!}{(n-k)!}.$$

We illustrate the blossoming interpretation of the de Casteljau algorithm along with these three variants of the de Casteljau algorithm for cubic Bezier curves in Figures 26.12 through 26.15. As usual, we use the multiplicative notation $uvw$ to represent the blossom value $p(u, v, w)$.

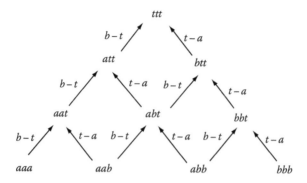

**FIGURE 26.12:** The de Casteljau evaluation algorithm for cubic Bezier curves with each node interpreted as a blossom value.

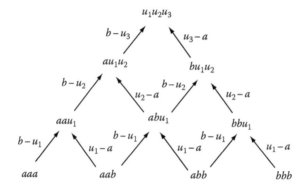

**FIGURE 26.13:** The blossomed version of the de Casteljau algorithm for cubic Bezier curves.

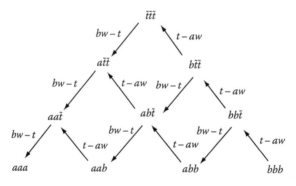

**FIGURE 26.14:** The homogeneous version of the de Casteljau algorithm for cubic Bezier curves: $\tilde{t} = (t, w)$.

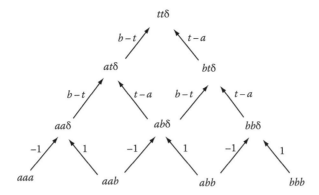

**FIGURE 26.15:** The differentiated version of the de Casteljau algorithm for cubic Bezier curves: $\delta = (1, 0)$.

We close with a summary of the primary properties of the blossom. The first three properties are the blossoming axioms—the three properties that uniquely characterize the blossom. Property *iv* is the key connection between blossoming and Bezier curves, and Property *v* is the connection between blossoming and differentiation.

*Primary Properties of the Blossom*

   i. Symmetry

$$p(u_1, \ldots, u_n) = p(u_{\sigma(1)}, \ldots, u_{\sigma(n)}) \text{ for every permutation } \sigma \text{ of} \{1, \ldots, n\}.$$

  ii. Multiaffine

$$p(u_1, \ldots, (1 - \alpha)u_k + \alpha v_k, \ldots, u_n) = (1 - \alpha)p(u_1, \ldots, u_k, \ldots, u_n) + \alpha p(u_1, \ldots, v_k, \ldots, u_n).$$

 iii. Diagonal

$$P(t) = p(\underbrace{t, \ldots, t}_{n})$$

 iv. Dual functional

$$P_k = p(\underbrace{a, \ldots, a}_{n-k}, \underbrace{b, \ldots, b}_{k}) = k\text{th Bezier control point}$$

  v. Differentiation

$$P^{(k)}(t) = \frac{n!}{(n-k)!} p(\underbrace{t, \ldots, t}_{n-k}, \underbrace{\delta, \ldots, \delta}_{k}) \quad \delta = (1, 0)$$

## Exercises

Hint: In the following exercises, you should use the uniqueness property of the blossom along with the axioms for the blossom to prove the required results. *Avoid using explicit formulas for blossoming monomials*, since these formulas will only make the proofs much more difficult.

26.1. Let $P(t)$ and $Q(t)$ be polynomials of degree $n$. Prove that:

a. If $R(t) = P(t) + Q(t)$, then $r(u_1, \ldots, u_n) = p(u_1, \ldots, u_n) + q(u_1, \ldots, u_n)$.

b. If $S(t) = cP(t)$, then $s(u_1, \ldots, u_n) = cp(u_1, \ldots, u_n)$.

26.2. Let $P(t) = (t - a)^n$. Show that $p(u_1, \ldots, u_n) = (u_1 - a) \cdots (u_n - a)$.

26.3. Let $P_k^n(t) = \dfrac{(t - t_k)^n}{\prod_{j \neq k} (t_j - t_k)}$, $k = 0, \ldots, n$, and let $P(t) = \sum_{k=0}^{n} P_k^n(t) P_k$. Using the result of Exercise 26.2, show that $P_k = p(t_0, \ldots, t_{k-1}, t_{k+1}, \ldots, t_n)$.

26.4. Let $p(u_1, \ldots, u_n)$ be a polynomial in which each variable appears to at most the first power. Show that $p(u_1, \ldots, u_n)$ is multiaffine.

26.5. Let $p((u_1, v_1), \ldots, (u_n, v_n))$ be a polynomial in which either $u_k$ or $v_k$, but not both, appears in every term to the first power for $k = 1, \ldots, n$. Show that $p((u_1, v_1), \ldots, (u_n, v_n))$ is multilinear.

26.6. Let $u_1 \cdots u_n$ denote the polynomial $(u_1 - t) \cdots (u_n - t)$. Show that with this interpretation the algorithms represented by Figures 26.1 and 26.3 remain valid.

26.7. Let $P(t, w)$ and $Q(t, w)$ be two homogeneous polynomials of degree $n$. Show that:

a. $\dfrac{P(ct, cw)}{Q(ct, cw)} = \dfrac{P(t, w)}{Q(t, w)}$

b. $P(t, w) = w^n P(t/w, 1)$.

26.8. In this exercise, you will complete the proof of Theorem 26.7.

a. Use induction to prove the binomial theorem

$$(a + b)^n = \sum_{k=0}^{n} \binom{n}{k} a^k b^{n-k}.$$

b. Mimic the inductive proof of the binomial theorem in part a to prove that

$$p(t + h\delta, \ldots, t + h\delta) = \sum_{k=0}^{n} \binom{n}{k} p(\underbrace{t, \ldots, t}_{n-k}, \underbrace{\delta, \ldots, \delta}_{k}) h^k.$$

26.9. Let $P(t)$ be a polynomial of degree $n$ over the interval $[a, b]$ with Bernstein coefficients $P_0, \ldots, P_n$. Let $G(P)(t) = (t, P(t))$ denote the graph of the polynomial $P(t)$. Show that the Bezier control points of $G(P)(t)$ over the interval $[a, b]$ are given by

$$Q_k = \left(a + \frac{k(b-a)}{n}, P_k\right), \quad k = 0, \ldots, n.$$

26.10. Let $M_k^n(t, a) = (t - a)^k, k = 0, \ldots, n$ denote the monomial basis at $x = a$. Develop algorithms to convert between:

    a. The monomial basis at $x = a$ and the Bernstein basis over the interval $[a, b]$.

    b. The monomial basis at $x = 0$ and the monomial basis at $x = a$.

    c. The Bernstein basis over the interval $[a, b]$ and the monomial basis at $x = 0$.

26.11. Using the results of Exercises 26.9, 26.10c and the intersection algorithm for Bezier curves in Chapter 25. Section 25.3.1 develop an algorithm for finding the roots of a polynomial lying in the interval $[a, b]$.

26.12. Let $B_k^n(t)$ denote the Bernstein basis functions over the interval $[a, b]$ and let $L(t) = ct + d$ be a linear function.

    a. Using the dual functional property, show that:

        i. $\displaystyle\sum_{k=0}^{n} B_k^n(t) \equiv 1.$

        ii. $\displaystyle\sum_{k=0}^{n} (a + \tfrac{k}{n}(b - a))B_k^n(t) \equiv t.$

    b. Using the result of part a, show that:

$$\sum_{k=0}^{n} L\left(a + \frac{k}{n}(b - a)\right) B_k^n(t) \equiv L(t).$$

26.13. Let $B_k^n(t)$ denote the Bernstein basis functions over the interval $[0, 1]$. Using the dual functional property, prove the following identities:

    a. $\displaystyle\sum_{k=0}^{n} \binom{k}{j} B_k^n(t) \equiv \binom{n}{j} t^j, \quad j = 0, \ldots, n.$

    b. $\displaystyle\sum_{k=0}^{n} (-1)^k B_k^n(t) \equiv (1 - 2t)^n.$

    c. $\displaystyle\sum_{k=0}^{n} \frac{(-1)^k}{\binom{n}{k}} B_{n-k}^n(x) B_k^n(t) \equiv (x - t)^n.$

26.14. Let $P(t)$ be a Bezier curve of degree $n$ over the interval $[a, b]$. Every polynomial of degree $n$ is also a polynomial of degree $n + 1$. Let $p_n(u_1, \ldots, u_n)$ denote the blossom of $P(t)$ as a polynomial of degree $n$ and let $p_{n+1}(u_1, \ldots, u_{n+1})$ denote the blossom of $P(t)$ as a polynomial of degree $n + 1$. In addition, let $P_0, \ldots, P_n$ be the Bezier control points for $P(t)$ as a polynomial of degree $n$ and let $Q_0, \ldots, Q_{n+1}$ be the Bezier control points for $P(t)$ as a polynomial of degree $n + 1$. Show that:

    a. $\displaystyle p_{n+1}(u_1, \ldots, u_{n+1}) = \sum_{k=1}^{n+1} \frac{p_n(u_1, \ldots, u_{k-1}, u_{k+1}, \ldots, u_{n+1})}{n + 1}.$

    b. $\displaystyle Q_k = \frac{k}{n+1} P_{k-1} + \frac{n+1-k}{n+1} P_k, \quad k = 0, \ldots, n + 1.$

# Chapter 27

---

## B-Spline Curves and Surfaces

*He hath made every thing beautiful in his time*

– Ecclesiastes 3:11

---

## 27.1 Motivation

Bezier curves and surfaces are polynomials that approximate the shape defined by their control points. Nevertheless, Bezier curves and surfaces are not suitable for large-scale industrial design and manufacture. A freeform shape such as the hull of a ship or the fuselage of an airplane or the body of a car cannot be represented accurately by a single polynomial. Moreover, Bezier control points have global effects (see Figure 27.1); adjusting a Bezier control point near the rear of a car would affect the shape near the front of the car. Designers require local control. Adjusting a control point should change the shape of the curve or surface only near the control point; fixing an undesirable artifact in one location should not create a new artifact in a far away location.

One possible alternative is to employ piecewise Bezier curves and surfaces. But piecewise Bezier curves and surfaces also have drawbacks. The control points for piecewise Bezier curves and surfaces that meet smoothly at their join must satisfy rigid constraints. The burden of maintaining these constraints would fall naturally to the designer, who may have little or no knowledge of the mathematics underlying Bezier curves and surfaces.

B-spline techniques overcome these shortcomings of Bezier methods. B-spline curves and surfaces are inherently piecewise polynomials. Moreover, with B-splines there are no constraints on the location of the control points; B-splines remain smooth no matter where the designers place their control points. B-splines also provide local control. Moving a control point of a B-spline curve or surface affects the shape only in a neighborhood of the control point. Parts of the shape far from a B-spline control point are not influenced by the control point (see Figure 27.2).

In this chapter we are going to study B-spline curves and surfaces. There are many analogies between Bezier and B-spline techniques. Bezier curves and surfaces are polynomials; B-spline curves and surfaces are piecewise polynomials. Bezier curves and surfaces are generated from the de Casteljau algorithm; B-spline curves and surfaces are generated from the de Boor algorithm. Bezier curves and surfaces can be analyzed using subdivision; B-spline curves and surfaces can be analyzed using knot insertion. The tool that unifies these two theories is blossoming, so if you do not fully understand blossoming, now would be a good time to review Chapter 26.

**FIGURE 27.1:** Quartic Bezier curves with five control points. Changing the initial control point of the curve on the left alters the entire curve—see the curve on the right.

**FIGURE 27.2:** Quadratic B-spline curves with three polynomial segments and five control points. Changing the initial control point of the curve on the left does not alter the second or third segments of the curve—see the curve on the right. Compare to the Bezier curves in Figure 27.1.

## 27.2 Blossoming and the Local de Boor Algorithm

The blossoming interpretation of the de Casteljau algorithm (Figure 27.3) follows easily from the multiaffine property of the blossom (Figure 27.4). The multiaffine property permits us to proceed to the next level in the diagram whenever the blossom values in two adjacent nodes agree in all but one parameter. These common parameters ascend to a node in the next level of the diagram, where

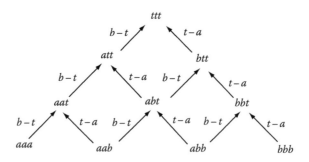

**FIGURE 27.3:** The blossoming interpretation of the de Casteljau algorithm (unnormalized) for cubic Bezier curves. Each node in the diagram has an interpretation in terms of blossom values. As usual, we adopt the multiplicative notation $uvw$ for the blossom $p(u, v, w)$.

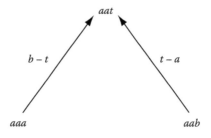

**FIGURE 27.4:** The multiaffine property (unnormalized) of the blossom for cubic polynomials: $t = \frac{b-t}{b-a}a + \frac{t-a}{b-a}b \Rightarrow aat = \frac{b-t}{b-a}aaa + \frac{t-a}{b-a}aab.$

they are joined by one new parameter. The base case of the de Casteljau algorithm, however, is somewhat special, since at the base of the de Casteljau diagram all the blossom parameters are either $a$ or $b$.

We do not need this special base case to generate these diagrams; all we require is that the blossom values in adjacent nodes agree in all but one parameter (see Figure 27.5). If we start with $n$ arbitrary parameters and we change one parameter at a time as we move along the base of the diagram, then we arrive at the algorithm in Figure 27.6. This algorithm is the *de Boor algorithm* for one segment of a degree $n$ B-spline curve.

The de Boor algorithm for polynomials of degree $n$ is generated from $2n$ constants $t_1, \ldots, t_{2n}$. The parameters $t_1, \ldots, t_{2n}$ are called *knots*. These knots are not completely arbitrary because their differences appear in the denominators of the labels along the edges. For example, in Figure 27.6, $t_4 - t_1$ is the denominator along the lower left edge and $t_6 - t_3$ is the denominator along the lower right edge. To avoid zeroes in the denominators, we shall insist that the knots $t_1, \ldots, t_{2n}$ are increasing—in particular we shall insist that $t_1 \leq \cdots \leq t_n < t_{n+1} \leq \cdots \leq t_{2n}$. Also, as with Bezier curves, we shall want B-spline curves to lie in the convex hull of their control points. This constraint requires that the labels along the edges must be nonnegative for all values of $t$. We ensure this outcome by restricting $t$ so that $t_n \leq t \leq t_{n+1}$. Notice that the de Casteljau algorithm for Bezier curves of degree $n$ over the interval $[a,b]$ is the special case of the de Boor algorithm where $t_1 = \cdots = t_n = a$ and $t_{n+1} = \cdots = t_{2n} = b$.

We can blossom and homogenize the de Boor algorithm in much the same way that we blossom and homogenize the de Casteljau algorithm. To blossom the de Boor algorithm, replace the parameter $t$ on the $k$th level of the de Boor algorithm by the parameter $u_k$, $k = 1, \ldots, n$ (see Figure 27.7). To homogenize the de Boor algorithm, homogenize the labels along all the arrows—that is, replace $t - t_j \rightarrow t - t_j w$ and $t_k - t \rightarrow t_k w - t$ (see Figure 27.8). We shall return to the blossomed version of

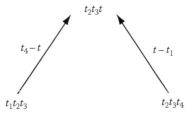

**FIGURE 27.5:** The multiaffine property (unnormalized) of the blossom for a cubic polynomial: $t = \frac{t_4 - t}{t_4 - t_1} t_1 + \frac{t - t_1}{t_4 - t_1} t_4 \Rightarrow t_2 t_3 t = \frac{t_4 - t}{t_4 - t_1} t_1 t_2 t_3 + \frac{t - t_1}{t_4 - t_1} t_2 t_3 t_4$.

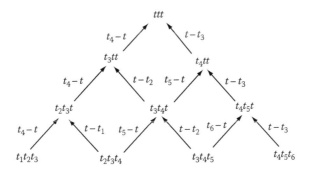

**FIGURE 27.6:** The de Boor algorithm (unnormalized) for one segment of a cubic B-spline curve. As usual, we adopt the multiplicative notation $uvw$ for the blossom $p(u, v, w)$. Compare to the de Casteljau algorithm in Figure 27.3.

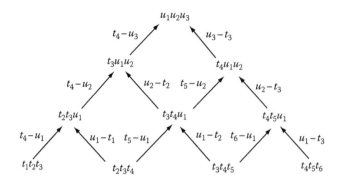

**FIGURE 27.7:** Blossoming the de Boor algorithm by replacing the parameter $t$ on the $k$th level of the de Boor algorithm by the parameter $u_k$, $k = 1, \ldots, n$. Here we illustrate the cubic case ($n = 3$). As usual, we adopt the multiplicative notation $uvw$ for the blossom $p(u, v, w)$. Compare to Figure 27.6.

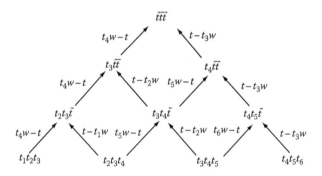

**FIGURE 27.8:** Homogenizing de Boor algorithm by replacing the labels $t - t_j \rightarrow t - t_j w$ and $t_k - t \rightarrow t_k w - t$. Again we illustrate the cubic case, and, as usual, we adopt the multiplicative notation $uvw$ for the blossom $p(u, v, w)$. Here $t = (t, 1)$ and $\tilde{t} = (t, w)$. Compare to Figure 27.6.

the de Boor algorithm in Chapter 28, Section 28.3.2 when we study the Oslo algorithm for knot insertion. In this chapter we shall use the homogeneous version of the de Boor algorithm in Section 27.4 to differentiate B-spline curves:

The control points for one segment of a B-spline curve are unique. To establish uniqueness, fix the knots $t_1, \ldots, t_{2n}$ and let

$B_{k,n}(t) =$ the sum over all paths from the $k$th position at the base to the apex of the de Boor diagram, $k = 0, \ldots, n$.

Since every polynomial $P(t)$ has a blossom $p(u_1, \ldots, u_n)$, every polynomial has a representation in terms of the functions $B_{0,n}(t), \ldots, B_{n,n}(t)$. In fact, by construction

$$P(t) = \sum_{k=0}^{n} B_{k,n}(t) p(t_{k+1}, \ldots, t_{k+n}). \tag{27.1}$$

Thus the polynomials $B_{0,n}(t), \ldots, B_{n,n}(t)$ span the space of polynomials of degree $n$. But the polynomials of degree $n$ form a vector space of dimension $n + 1$, since every polynomial of degree $n$ can be expressed uniquely as a linear combination of the monomials $1, t, \ldots, t^n$. Thus it follows from linear algebra that every spanning set of polynomials of degree $n$ of order $n + 1$ is a basis for the polynomials of degree $n$. In particular, $B_{0,n}(t), \ldots, B_{n,n}(t)$ is a spanning set for polynomials of degree $n$ of order $n + 1$, so $B_{0,n}(t), \ldots, B_{n,n}(t)$ is a basis for the polynomials of degree $n$. Therefore the control points of a B-spline segment are unique. Moreover by Equation 27.1, we have the following generalization of the dual functional property from Bezier curves to B-splines curves:

*Dual Functional Property*

Let $P(t)$ be one segment of a B-spline curve with knots $t_1, \ldots, t_{2n}$ and control points $P_0, \ldots, P_n$. Then

$$P_k = p(t_{k+1}, \ldots, t_{k+n}), \quad k = 0, \ldots, n. \tag{27.2}$$

## 27.3 B-Spline Curves and the Global de Boor Algorithm

So far we have discussed only one polynomial segment of a B-spline curve. But a B-spline curve is not just a single polynomial segment; a B-spline curve is a piecewise polynomial made up of many polynomial segments. It is time now to usher in these additional polynomial segments.

For each additional B-spline segment, we introduce one additional knot. We then shift all the indices in the de Boor algorithm by one. For example, introducing one new knot and shifting all the indices in Figure 27.6 by one generates Figure 27.9.

Figures 27.6 and 27.9 have a lot in common. In fact, the entire diagram below the node $t_4tt$ (see Figure 27.10) is common to both figures. Thus we can paste these two figures together to form Figure 27.11. Additional diagrams can be attached to the left and to the right to join arbitrarily many curve segments.

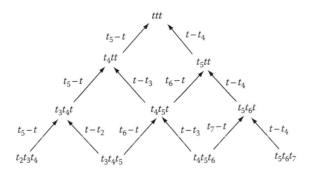

**FIGURE 27.9:** The de Boor algorithm (unnormalized) for another segment of a cubic B-spline curve. Compare to Figure 27.6.

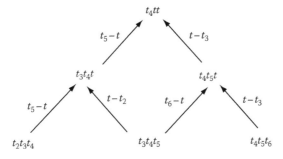

**FIGURE 27.10:** The diagram below the node $t_4tt$ is common to Figures 27.6 and 27.9.

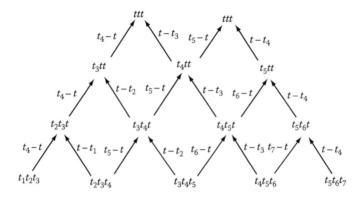

**FIGURE 27.11:** Overlaying the local de Boor algorithms (unnormalized) for two segments of a cubic B-spline curve.

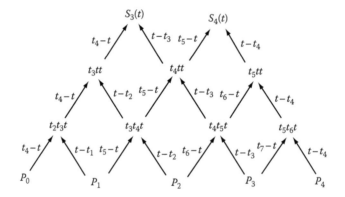

**FIGURE 27.12:** The de Boor algorithm for two segments of a cubic B-spline curve. The control points are placed at the base of the diagram and the values of the B-spline curve emerge at the apexes of the diagram. When $t_3 \leq t \leq t_4$, we compute the value at the apex on the left ($S_3(t)$); when $t_4 \leq t \leq t_5$, we compute the value at the apex on the right ($S_4(t)$).

In Figure 27.11 the knots $t_1, \ldots, t_7$ are fixed. The control points are placed at the base of the diagram and the values of the B-spline curve emerge at the apexes of the diagram (see Figure 27.12). When $t_3 \leq t \leq t_4$, we compute the value at the apex on the left ($S_3(t)$); when $t_4 \leq t \leq t_5$, we compute the value at the apex on the right ($S_4(t)$).

Notice that in Figure 27.11 we have overloaded our notation. We are using the blossom value $t_2 t_3 t_4$ to denote the control point $P_1$. But which polynomial are we blossoming, $S_3(t)$ or $S_4(t)$? The answer is both! By the dual functional property,

$$P_k = s_3(t_{k+1}, t_{k+2}, t_{k+3}), \quad k = 0, \ldots, 3,$$

and

$$P_k = s_4(t_{k+1}, t_{k+2}, t_{k+3}), \quad k = 1, \ldots, 4.$$

Thus $P_1 = s_3(t_2, t_3, t_4)$ and $P_1 = s_4(t_2, t_3, t_4)$, so $s_3(t_2, t_3, t_4) = s_4(t_2, t_3, t_4)$. Therefore we can safely use $t_2 t_3 t_4$ to denote $P_1$; in fact, any polynomial segment containing the control point $P_1$ can be used to evaluate the blossom at $(t_2, t_3, t_4)$. The same observation holds for each of the blossom values in

Figure 27.11 where the two diagrams overlap. We can choose to evaluate either $s_3$ or $s_4$; the values of these blossoms are necessarily identical at all the nodes common to both triangles.

---

## 27.4   Smoothness

The diagrams of the local de Boor algorithms for adjacent B-spline segments fit together nicely. We shall now use these diagrams to show that the corresponding B-spline curve segments meet smoothly at their join.

By convention, the parameter $t$ for the left segment of Figure 27.11, which we denote by $S_3(t)$ in Figure 27.12, ranges over the interval $[t_3, t_4]$, and the parameter $t$ for the right segment in Figure 27.11, which we denote by $S_4(t)$ in Figure 27.12, ranges over the interval $[t_4, t_5]$. To show that these segments meet continuously at their join, we need to investigate what happens at $t = t_4$. When $t = t_4$, the node labeled $t_4tt$ becomes $t_4t_4t_4$. Since this node is common to both triangles, the value of the blossom at this node must be the same for both curve segments. Thus

$$s_3(t_4, t_4, t_4) = s_4(t_4, t_4, t_4).$$

But by the diagonal property of the blossom

$$S_3(t_4) = s_3(t_4, t_4, t_4) \quad \text{and} \quad S_4(t_4) = s_4(t_4, t_4, t_4).$$

Therefore

$$S_3(t_4) = S_4(t_4),$$

so the two curve segments join continuously at $t = t_4$.

To investigate smoothness, we need to compute derivatives. Recall from Chapter 26, Section 26.4.2 that for any polynomial $P(t)$,

$$P^{(k)}(t) = \frac{n!}{(n-k)!} p(\underbrace{t, \ldots, t}_{n-k}, \underbrace{\delta, \ldots, \delta}_{k}). \tag{27.3}$$

Thus, up to constant multiples, derivatives are given by the homogeneous blossom evaluated at $t = (t, 1)$ and $\delta = (1, 0)$. Using this result together with Figure 27.8 for the homogenous blossom, we can differentiate the de Boor algorithm $k$ times by replacing $(t, w) \rightarrow \delta = (1, 0)$ on $k$ levels of the algorithm. Thus we can differentiate the de Boor algorithm in the same way that we differentiated the de Casteljau algorithm: by differentiating the labels along the edges. Figures 27.13 and 27.14 illustrate the diagrams for the first and second derivatives of the de Boor algorithm for two segments of a cubic B-spline curve.

In Figure 27.13, at $t = t_4$ the node labeled $t_4t\delta$ becomes $t_4t_4\delta$. Since this node is common to both triangles, the value of the blossom at this node must be the same for both curve segments. Thus

$$s_3(t_4, t_4, \delta) = s_4(t_4, t_4, \delta).$$

We conclude from Equation 27.3 that

$$S_3'(t_4) = S_4'(t_4),$$

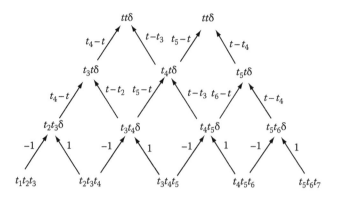

**FIGURE 27.13:** The first derivative of the de Boor algorithm for two segments of a cubic B-spline curve computed by differentiating the labels on the first level of the de Boor algorithm.

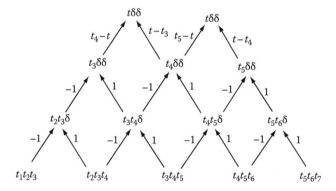

**FIGURE 27.14:** The second derivative of the de Boor algorithm for two segments of a cubic B-spline curve computed by differentiating the labels on the first two levels of the de Boor algorithm.

so the two curve segments join smoothly at $t = t_4$. Similarly in Figure 27.14, at $t = t_4$ the node labeled $t_4\delta\delta$ is common to both triangles, so the value of the blossom at this node must be the same for both curve segments. Thus

$$s_3(t_4, \delta, \delta) = s_4(t_4, \delta, \delta).$$

Again we conclude from Equation 27.3 that

$$S_3''(t_4) = S_4''(t_4),$$

so the two curve segments join with two continuous derivatives at $t = t_4$. Similar arguments can be used to show that two adjacent B-spline curve segments of degree $n$ meet with $n - 1$ continuous derivatives at their join.

If a knot is repeated, then the order of smoothness is reduced by the multiplicity of the knot. Consider, for example, a cubic B-spline curve with a double knot at $t_4 = t_5$. Since the segment of the curve between $t_4$ and $t_5$ collapses to a single point, we remove this segment from the de Boor algorithm. Now the de Boor algorithm for the segment of the curve in the interval $[t_3, t_4]$ overlaps the de Boor algorithm for the segment of the curve for the interval $[t_5, t_6]$ only on the first level of the diagram (see Figure 27.15).

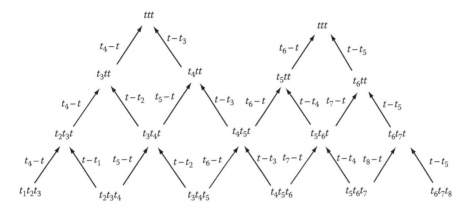

**FIGURE 27.15:** The de Boor algorithm for two segments of a cubic B-spline curve with a double knot at $t_4 = t_5$.

The arguments that we gave in the previous paragraphs for the continuity of the curve segments and their first derivatives still apply at the double knot $t_4 = t_5$, since the two triangles in the de Boor algorithm overlap at the node $t_4 t_5 t$. But the second derivatives of these curve segments will no longer match: their corresponding blossoms need no longer agree at the blossom value $t_4 \delta \delta$ because the two triangles no longer overlap at the node $t_4 t t$. Similar arguments apply to B-spline curves of higher degree and knots with higher order multiplicity.

## 27.5  Labeling and Locality in the Global de Boor Algorithm

The de Boor algorithm is more complicated than the de Casteljau algorithm because the de Boor algorithm depends on the knots as well as on the control points. Moreover the de Boor algorithm defines many polynomial segments, whereas the de Casteljau algorithm defines only a single polynomial segment. For these reasons, the labels along the edges of the de Boor algorithm are more complicated than the labels along the edges in the de Casteljau algorithm. Therefore, before we proceed further in our study of B-splines, we shall pause here briefly to clarify the labeling along the edges in the de Boor algorithm.

Actually the labeling scheme in the de Boor algorithm is quite straightforward and easy to remember. Consider a B-spline curve of degree $n$ with control points $P_0, \ldots, P_m$ and knots $t_1, \ldots, t_{m+n}$. Then the label exiting $P_k$ to the upper left is $t - t_k$ and the label exiting $P_k$ to the upper right is $t_{k+n+1} - t$ (see Figure 27.16). All the other labels in the de Boor algorithm can be retrieved from these labels on the first level and the following *in–out property* of the de Boor algorithm:

*In–Out Property*
    *Any label that enters a node in the de Boor algorithm, exits the node in the same direction.*

To understand the in–out property, consider Figure 27.12. Here, for example, the label $t - t_3$ enters the node $t_4 t_5 t$ from the lower right and exits the node $t_4 t_5 t$ to the upper left. Similar observations hold for each of the labels on the edges in Figure 27.12. Notice that the in–out property applies only to the numerators of the labels, not to the denominators. The denominators, however, are easily retrieved from the numerators because any two labels that enter a node must sum to one.

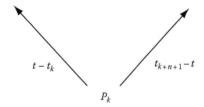

**FIGURE 27.16:**   Labels along the edges in the first level of the de Boor algorithm for a B-spline curve of degree $n$. The label exiting $P_k$ to the upper left is $t - t_k$; the label exiting $P_k$ to the upper right is $t_{k+n+1} - t$.

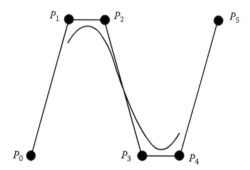

**FIGURE 27.17:**   A cubic B-spline curve with three polynomial segments. The knots are located at the integers $1, \ldots, 8$. The three polynomials segments lie over the three intervals: [3,4], [4,5], [5,6].

Now, by construction (see Figure 27.17), the segment of a degree $n$ B-spline curve over the parameter interval $[t_k, t_{k+1}]$:

- Has $n + 1$ control points—$P_{k-n}, \ldots, P_k$.

- Depends on $2n$ knots—$t_{k-n+1}, \ldots, t_{k+n}$.

Therefore B-splines exhibit the following locality:

- $P_k$ influences only the $n + 1$ curve segments over the interval $[t_k, t_{k+n+1}]$.

- $t_k$ influences solely the $2n$ curve segments over the interval $[t_{k-n}, t_{k+n}]$.

---

## 27.6   Every Spline is a B-Spline

B-spline curves are piecewise polynomials that join smoothly at the knots. Conversely, we shall now show that every piecewise polynomial curve where the polynomial segments meet smoothly at the joins has a B-spline representation. Thus every spline curve is a B-spline curve. This theorem is analogous to the proposition that every polynomial curve is a Bezier curve (see Chapter 26, Section 26.3). To prove this result for splines, we begin with the following technical lemma.

**Lemma 27.1:** *Let $P(t)$ and $Q(t)$ be polynomials of degree n. If $P^{(j)}(\tau) = Q^{(j)}(\tau), 0 \leq j \leq n - 1$, then* $p(\underbrace{\tau, \ldots, \tau}_{n-j}, u_1, \ldots, u_j) = q(\underbrace{\tau, \ldots, \tau}_{n-j}, u_1, \ldots, u_j)$ *for every set of parameters $u_1, \ldots, u_j, 0 \leq j \leq n - 1$.*

**Proof:** Recall from Equation 27.3 that

$$P^{(j)}(\tau) = \frac{n!}{(n-j)!} p(\underbrace{\tau, \ldots, \tau}_{n-j}, \underbrace{\delta, \ldots, \delta}_{j}),$$

$$Q^{(j)}(\tau) = \frac{n!}{(n-j)!} q(\underbrace{\tau, \ldots, \tau}_{n-j}, \underbrace{\delta, \ldots, \delta}_{j}).$$

Therefore,

$$P^{(j)}(\tau) = Q^{(j)}(\tau) \Rightarrow p(\underbrace{\tau, \ldots, \tau}_{n-j}, \underbrace{\delta, \ldots, \delta}_{j}) = q(\underbrace{\tau, \ldots, \tau}_{n-j}, \underbrace{\delta, \ldots, \delta}_{j}).$$

Now starting with the $n$ blossom values

$$p(\tau, \ldots, \tau), p(\tau, \ldots, \tau, \delta), \ldots, p(\tau, \underbrace{\delta, \ldots, \delta}_{n-1}),$$

along the base of a triangle and running the blossomed version of the homogeneous de Boor algorithm, we see that the $n$ blossom values

$$p(\tau, \ldots, \tau), p(\tau, \ldots, \tau, u_1), \ldots, p(\tau, u_1, \ldots, u_{n-1}),$$

emerge along the left lateral edge of the triangle (see Figure 27.18).

But the same algorithm (Figure 27.18) starting with the $n$ blossom values

$$q(\tau, \ldots, \tau), q(\tau, \ldots, \tau, \delta), \ldots, q(\tau, \underbrace{\delta, \ldots, \delta}_{n-1}),$$

generates the $n$ blossom values

$$q(\tau, \ldots, \tau), q(\tau, \ldots, \tau, u_1), \ldots, q(\tau, u_1, \ldots, u_{n-1}).$$

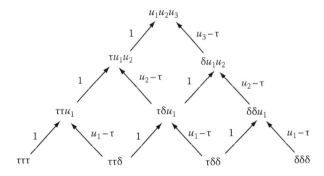

**FIGURE 27.18:** Computing the blossom values $\tau\tau\tau$, $\tau\tau u_1$, $\tau u_1 u_2$, $u_1 u_2 u_3$ from the blossom values $\tau\tau\tau$, $\tau\tau\delta$, $\tau\delta\delta$, $\delta\delta\delta$. Notice that the first three blossom values in the first set depend only on the first three blossom values in the second set. Here we use the homogeneous blossom and the identity $u_k = (u_k - \tau)\delta + \tau$, so no normalization is necessary in the labels along the edges.

Since by assumption the input to these two algorithms is the same, the output must also be the same. Therefore

$$p(\underbrace{\tau,\ldots,\tau}_{n-j},u_1,\ldots,u_j) = q(\underbrace{\tau,\ldots,\tau}_{n-j},u_1,\ldots,u_j) \text{ for every set of parameters } u_1,\ldots,u_j,\ 0 \le j \le n-1.$$

**Theorem 6.2:   (Every Spline is a B-Spline)**

*Every spline curve can be represented as a B-spline curve. That is, for every spline $S(t)$ of degree $n$ with knots $t_1,\ldots,t_{m+n}$, there exist control points $P_0,\ldots,P_m$ such that the spline $S(t)$ is generated by the de Boor algorithm with control point $P_0,\ldots,P_m$ and knots $t_1,\ldots,t_{m+n}$.*

**Proof:**  Locally every spline is a polynomial. Therefore every spline has a local blossom, so every spline has a local de Boor algorithm (Figure 27.6). It remains only to show that the local de Boor algorithms for adjacent polynomial segments have overlapping control points as in Figures 27.11 and 27.12. Consider a knot $t_k$. By Lemma 27.1, the first $n-1$ derivatives of a degree $n$ polynomial at the parameter $t = t_k$ completely determine its blossom at the values $t_k u_1 \cdots u_{n-1}$. Therefore the local de Boor algorithms for adjacent polynomial segments have matching overlapping control points. Indeed by the dual functional property, the control points for the polynomial segment $P_k(t)$ over the interval $[t_{k-1}, t_k]$ are

$$p_k(t_{k-n},\ldots,t_{k-1}), p_k(t_{k-n+1},\ldots,t_k),\ldots,p_k(t_k,\ldots,t_{k+n-1}),$$

and the control points for the polynomial segment $P_{k+1}(t)$ over the interval $[t_k, t_{k+1}]$ are

$$p_{k+1}(t_{k-n+1},\ldots,t_k),\ldots,p_{k+1}(t_k,\ldots,t_{k+n-1}), p_{k+1}(t_{k+1},\ldots,t_{k+n}).$$

Since, by assumption,

$$P_k^{(j)}(t_k) = P_{k+1}^{(j)}(t_k), \quad j = 0,\ldots,n-1,$$

it follows by Lemma 27.1 that

$$p_k(t_j,\ldots,t_k,\ldots,t_{j+n-1}) = p_{k+1}(t_j,\ldots,t_k,\ldots,t_{j+n-1}), \quad j = k-n+1,\ldots,k.$$

Therefore adjacent segments have overlapping control points.

## 27.7   Geometric Properties of B-Spline Curves

B-spline curves share several properties with Bezier curves. B-spline curves are affine invariant, lie in the convex hull of their control points, and satisfy the variation diminishing property. B-spline curves also interpolate control points that correspond to knots where the multiplicity of the knot is equal to the degree of the spline.

The proofs of affine invariance and the convex hull property are identical to the proofs of the corresponding results for Bezier curves: simply replace the de Casteljau algorithm with the de Boor algorithm (see Exercises 27.1 and 27.2). Notice that B-spline curves satisfy a somewhat stronger local convex hull property: each B-spline segment lies in the convex hull of the control points for the corresponding segment because the parameter is restricted so that each label in the diagram for the local de Boor algorithm is nonnegative.

B-spline curves do not, in general, interpolate any of their control points (see Figure 27.17). But B-spline curves do interpolate control points that correspond to knots where the multiplicity of the knot is equal to the degree of the spline. To establish this result, recall that by the dual functional property—Equation 27.2—if $P(t)$ is a B-spline curve, then

$$P_k = p(t_{k+1}, \ldots, t_{k+n}).$$

Therefore if $t_{k+1} = \cdots = t_{k+n}$, then by the diagonal property of the blossom

$$P_k = p(t_{k+1}, \ldots, t_{k+n}) = p(t_{k+1}, \ldots, t_{k+1}) = P(t_{k+1}).$$

Thus we can force interpolation at any control point by increasing the multiplicity of the corresponding knot. Notice, however, that the price we pay for interpolation is to reduce the smoothness of the curve at the corresponding knot. Often multiple knots are used to force interpolation at the end points of the spline, where the smoothness of adjacent segments is not an issue.

We say that a curve is *variation diminishing* if the number of intersections of the curve with each line in the plane (or each plane in three-space) is less than or equal to the number of intersections of the line (or the plane) with the control polygon. The variation diminishing property is important because this property guarantees that the curve does not oscillate more than its control polygon. B-spline curves are known to be variation diminishing, but it is not so clear how to derive this result directly from the de Boor algorithm. For Bezier curves, we derived the variation diminishing property using subdivision. For B-spline curves, the analogue of subdivision is knot insertion, so we shall postpone our derivation of the variation diminishing property for B-spline curves till Chapter 28 where we will discuss knot insertion procedures for B-spline curves and surfaces.

## 27.8   Tensor Product B-Spline Surfaces

A *tensor product B-spline surface* is a rectangular piecewise polynomial surface, where the rectangular polynomial pieces meet smoothly along their common boundaries. The construction of a tensor product B-spline surface is similar to the construction of a tensor product Bezier patch (see Chapter 24, Section 24.6): simply replace the de Casteljau algorithm for Bezier curves by the de Boor algorithm for B-splines. The only novelty here is the knots.

To build a tensor product B-spline surface $B(s, t)$, we need a rectangular array of control points $\{P_{ij}\}$, $i = 0, \ldots, \mu$, $j = 0, \ldots, \nu$ and two knot sequences $s_1, \ldots, s_{m+\mu}$ and $t_1, \ldots, t_{n+\nu}$. Let $P_i(t), i = 0, \ldots, \mu$, be the degree $n$ B-spline curve with control points $P_{i,0}, \ldots, P_{i,\nu}$ and knots $t_1, \ldots, t_{n+\nu}$. For each fixed value of $t$, let $B(s, t)$ be the degree $m$ B-spline curve for the control points $P_0(t), \ldots, P_\mu(t)$ and knots $s_1, \ldots, s_{m+\mu}$. Then as $s$ varies from $s_m$ to $s_{\mu+1}$ and $t$ varies from $t_n$ to $t_{\nu+1}$, the curves $B(s, t)$ sweep out a surface. This surface is called a *tensor product B-spline surface of bidegree* $(m, n)$.

This construction suggests the following evaluation algorithm for tensor product B-spline surfaces: first use the de Boor algorithm $\mu + 1$ times to compute the points at the parameter $t$ along the degree $n$ B-spline curves $P_0(t), \ldots, P_\mu(t)$; then use the de Boor algorithm one more time to compute the point at the parameter $s$ along the degree $m$ B-spline curve with control points $P_0(t), \ldots, P_\mu(t)$ (see Figure 27.19).

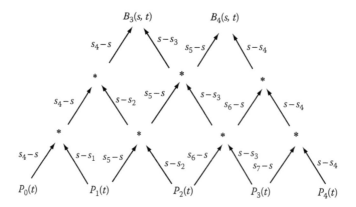

**FIGURE 27.19:** The de Boor algorithm for a bicubic B-spline surface. The functions $P_0(t), \ldots, P_4(t)$ represent cubic B-spline curves in the $t$ direction; the upper triangles represent the de Boor algorithm in $s$ direction.

Alternatively, instead of starting with the B-spline curves $P_0(t), \ldots, P_\mu(t)$, we could begin with the B-spline curves $P_0^*(s), \ldots, P_\nu^*(s)$, where $P_j^*(s)$ is the degree $m$ B-spline curve with control points $P_{0,j}, \ldots, P_{\mu,j}$ and knots $s_1, \ldots, s_{m+\mu}$. Now for each fixed value of $s$, let $B^*(s, t)$ be the B-spline curve for the control points $P_0^*(s), \ldots, P_\nu^*(s)$ and knots $t_1, \ldots, t_{n+\nu}$. Again as $s$ varies from $s_m$ to $s_{\mu+1}$ and $t$ varies from $t_n$ to $t_{\nu+1}$, the curves $B^*(s, t)$ sweep out a surface, and we can use the de Boor algorithm to evaluate points along this surface. As with Bezier patches, the surface $B^*(s, t)$ is exactly the same as the surface $B(s, t)$ (see Exercise 27.9).

Tensor product B-spline surfaces inherit many of the characteristic properties of B-spline curves: they are affine invariant, lie in the convex hull of their control points, and interpolate control points where the multiplicity of the knots is equal to the degree of the surface. Also, the polynomial patches join smoothly along common boundaries, These properties follow easily from the de Boor algorithm for tensor product B-spline surfaces and the corresponding properties of B-spline curves (see Exercise 27.10).

To compute the partial derivatives of a B-spline surface, we can apply our procedure for differentiating the de Boor algorithm for B-spline curves. Consider Figure 27.19. We can compute $\partial B / \partial t$ simply by differentiating the de Boor algorithm for each of the B-spline curves $P_0(t), \ldots, P_\mu(t)$ at the base of the diagram. Similarly, we can compute $\partial B / \partial s$ by differentiating the first $s$ level of the diagram and multiplying the result by the degree in $s$ (see Figure 27.20). Symmetric results hold for if we use the surface $B^*(s, t)$ in place of the surface $B(s, t)$: simply reverse the roles of $s$ and $t$. The normal vector $N$ to a B-spline surface $B(s, t)$ is given by setting $N = \frac{\partial B}{\partial s} \times \frac{\partial B}{\partial t}$.

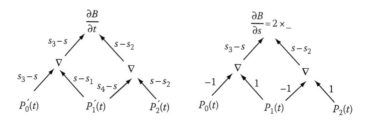

**FIGURE 27.20:** Computing the partial derivatives for a biquadratic B-spline surface by applying the procedure for differentiating the de Boor algorithm for B-spline curves. To find $\partial B / \partial t$ (left), simply differentiate the de Boor algorithm for each of the B-spline curves $P_0(t)$, $P_1(t)$, $P_2(t)$. To find $\partial B / \partial s$ (right), simply differentiate the first level of the de Boor algorithm in $s$ and multiply the result by the degree in $s$.

## 27.9   Non-Uniform Rational B-Splines (NURBS)

B-splines are piecewise polynomial curves and surfaces. Polynomials and piecewise polynomials have many advantages: they are easy to represent and simple to compute. Nevertheless, polynomials and piecewise polynomials have one severe disadvantage: these functions cannot exactly represent many of the simplest bounded curves and surfaces. For example, there is no exact representation for the circle or the sphere with polynomials or piecewise polynomials. Polynomials are unbounded functions: if $P(t)$ is a polynomial, then as $t \to \infty$ the function $P(t) \to \pm\infty$. Thus polynomials cannot exactly represent bounded curves such as circles or bounded surfaces such as spheres.

To overcome this deficiency, we shall use *rational functions*—that is, functions that are the ratio of two polynomials. For example, it is easy to verify that the parametric equations

$$x(t) = \frac{1 - t^2}{1 + t^2}, \quad y(t) = \frac{2t}{1 + t^2}, \tag{27.4}$$

represent a circle, since $x^2(t) + y^2(t) \equiv 1$. Similarly, the parametric equations

$$x(s, t) = \frac{1 - s^2}{1 + s^2} \frac{1 - t^2}{1 + t^2}, \quad y(s, t) = \frac{1 - s^2}{1 + s^2} \frac{2t}{1 + t^2}, \quad z(s, t) = \frac{2s}{1 + s^2}, \tag{27.5}$$

represent a sphere, since $x^2(s, t) + y^2(s, t) + z^2(s, t) \equiv 1$.

A *rational B-spline* is a function that is the ratio of two B-splines. A NURB is a rational B-spline where the knots are not evenly spaced. Fortunately, we do not need to invent any new algorithms to represent rational B-spline curves and surfaces; we can still use the de Boor algorithm. The only change we need to make to generate rational functions is to use mass-points as our control points. Initially a rational B-spline is a curve or surface in the vector space of mass-points. To generate the corresponding curve or surface in affine space, we simply divide by the mass. For example, to construct a rational B-spline curve, we start with a collection of mass-points $(m_k P_k, m_k)$ and apply the de Boor algorithm with respect to a knot sequence $(t_j)$ to generate a B-spline curve $(P(t), m(t))$ in the vector space of mass-points. We then divide by the mass $m(t)$ to get the rational B-spline curve $R(t) = P(t)/m(t)$.

Rational B-spline curves and surfaces inherit many of the properties of standard B-spline curves and surfaces: they are affine invariant, lie in the convex hull of their control points when the weights are nonnegative, and interpolate control points where the multiplicity of the knot is equal to the degree of the curve or surface. Also, the rational segments join smoothly along their joins. These properties follow easily from the de Boor algorithm for rational B-splines and the fact that the projection into affine space is simply division which does not adversely affect any of these properties.

Rational Bezier curves and surfaces are special cases of rational B-spline curves and surfaces. Again we use mass-points to represent the control points, but to generate rational Bezier curves and surfaces, we replace the de Boor algorithm by the de Casteljau algorithm. Using blossoming and the dual functional property for the denominator as well as for each coordinate in the numerator, we find (see Exercise 27.11) that the rational quadratic Bezier control points for the quarter circle represented by Equation 27.4 with $0 \le t \le 1$ are given by the mass-points:

$$(m_0 P_0, m_0) = (1, 0, 1), \quad (m_1 P_1, m_1) = (1, 1, 1), \quad (m_2 P_2, m_2) = (0, 2, 2).$$

Similar techniques can be used to find the rational biquadratic Bezier control points and weights for the part of the sphere represented by Equation 27.5 with $0 \le s, t \le 1$ (see Exercise 27.13).

## 27.10 Summary

B-spline curves and surfaces are important in Geometric Modeling because they allow us to model complicated freeform curves and surfaces with piecewise polynomials of low degree. B-splines also admit local control; moving a control point affects a B-spline curve or surface only in a local neighborhood of the control point. Here we summarize the main properties of B-spline curves and surfaces.

The fundamental algorithm for B-spline curves and surfaces is the de Boor evaluation algorithm. The labels along the edges in the de Boor algorithm for B-spline curves (Figure 27.22) can be retrieved from the labels along the first level (Figure 27.21) and the in–out property.

*In–Out Property*

- *Any label that enters a node in the de Boor algorithm, exits the node in the same direction.*

The segment of a degree $n$ B-spline curve over the parameter interval $[t_k, t_{k+1}]$:

- Has $n + 1$ control points—$P_{k-n}, \ldots, P_k$.

- Depends on $2n$ knots—$t_{k-n+1}, \ldots, t_{k+n}$.

Therefore B-spline curves admit local control.

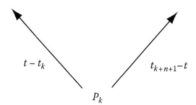

**FIGURE 27.21:** Labels along the edges in the first level of the de Boor algorithm for a B-spline curve of degree $n$.

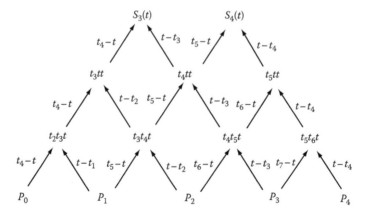

**FIGURE 27.22:** The de Boor algorithm for two segments of a cubic B-spline curve. The labels along the edges on first level are consistent with the labels in Figure 27.21, and the labels on subsequent levels satisfy the in–out property.

*Local Control*

- $P_k$ influences only the $n + 1$ curve segments over the interval $[t_k, t_{k+n+1}]$.

- $t_k$ influence solely the $2n$ curve segments over the interval $[t_{k-n}, t_{k+n}]$.

Knots can have multiplicity greater than one.

- *The continuity at a knot = the degree of the spline – the multiplicity of the knot.*

B-spline curves have the following geometric properties:

- Affine invariance.

- Convex hull property.

- Variation diminishing property.

- Interpolate control points that correspond to knots where the multiplicity of the knot is equal to the degree of the spline.

The dual functional property of the blossom provides a direct relationship between the control points and the knots of a B-spline curve.

*Dual Functional Property of the Blossom*
*Let $P(t)$ be a degree $n$ B-spline curve with knots $\{t_i\}$ and control points $\{P_j\}$. Then*

$$P_k = p(t_{k+1}, \ldots, t_{k+n}).$$

Tensor product B-spline surfaces are an extension of the B-spline representation from curves to surfaces. Tensor product B-spline surfaces are affine invariant, lie in the convex hull of their control points, and interpolate control points where the multiplicity of the knot is equal to the degree of the surface. Also, the polynomial patches join smoothly along common boundaries.

NURBS are the ratios of two B-splines. The control points of rational B-splines are mass-points rather than points in affine space. Rational B-splines are useful for representing bounded curves and surfaces such as circles and spheres, which cannot be represented exactly either by polynomials or by piecewise polynomials.

---

## Exercises

27.1. Prove that B-spline curves are affine invariant.

27.2. Prove that each B-spline curve segment lies in the convex hull of its control points.

27.3. Show that a B-spline curve collapses to a single point $P$ if and only if all the B-spline control points are located at $P$.

27.4. Show that the functions:

$$\frac{(t - t_0)^n}{(t_1 - t_0) \ldots (t_n - t_0)}, \ldots, \frac{(t - t_n)^n}{(t_1 - t_n) \ldots (t_{n-1} - t_n)}$$

are the basis functions corresponding to the local de Boor algorithm for the knot sequence $t_1, \ldots, t_n, t_0, \ldots, t_{n-1}$. (Note that here the knots are not increasing, but the local de Boor algorithm is still valid.)

27.5. Let $P_0, \ldots, P_m$ be the control points for a degree $n$ B-spline curve $S(t)$ with knots $t_1, \ldots, t_{m+n}$. Show that the derivative $S'(t)$ can be computed from the de Boor algorithm for a B-spline curve of degree $n - 1$ with the same knots and input

$$Q_k = \frac{P_{k+1} - P_k}{t_{k+n+1} - t_{k+1}}, \quad k = 0, \ldots, m - 1.$$

27.6. Let $N_{k,n}(t)$ denote the output of the global de Boor algorithm of degree $n$ for the knot sequence $t_1, \ldots, t_{m+n}$ when the control points are given by $P_j = \delta_{j,k}$. The functions $\{N_{k,n}(t)\}$ are called the *degree $n$ B-spline basis functions*.

a. Graph the basis functions $N_{0,n}(t)$ for $n = 1, 2, 3$, with knots at the integers.

b. Show that the B-spline basis functions satisfy the following recurrence:

$$N_{k,0}(t) = 1, \quad t_k \leq t < t_{k+1},$$

$$N_{k,n}(t) = \frac{t - t_k}{t_{k+n} - t_k} N_{k,n-1}(t) + \frac{t_{k+n+1} - t}{t_{k+n+1} - t_{k+1}} N_{k+1,n-1}(t).$$

27.7. Let $\{N_{k,n}(t)\}$ denote the degree $n$ B-spline basis functions defined in Exercise 27.6. Show that:

a. $\displaystyle\sum_{k=0}^{m} N_{k,n}(t) \equiv 1$ for $t_n \leq t \leq t_m$.

b. $0 \leq N_{k,n}(t) \leq 1$.

27.8. Let $\{N_{k,n}(t)\}$ denote the degree $n$ B-spline basis functions defined in Exercise 27.6. Show that if $P_0, \ldots, P_m$ are the control points for a degree $n$ B-spline curve $S(t)$ with knots $t_1, \ldots, t_{m+n}$, then

$$S(t) = \sum_{k=0}^{m} N_{k,n}(t) P_k.$$

27.9. Let $\{N_{j,m}(s)\}$ denote the degree $m$ B-spline basis functions for the knot sequence $s_1, \ldots, s_{m+\mu}$ and let $\{N_{k,n}(t)\}$ denote the degree $n$ B-spline basis functions for the knot sequence $t_1, \ldots, t_{n+\nu}$ (see Exercise 27.6). Suppose that $\{P_{i,j}\}$, $i = 0, \ldots, \mu$, $j = 0, \ldots, \nu$ are the control points for a tensor product B-spline surface $B(s, t)$ of bidegree $(m, n)$ with knots $s_1, \ldots, s_{m+\mu}$ and $t_1, \ldots, t_{n+\nu}$.

a. Show that

$$B(s, t) = \sum_{i=0}^{\mu} N_{i,m}(s) P_i(t) = \sum_{i=0}^{\mu} \sum_{j=0}^{\nu} N_{i,m}(s) N_{j,n}(t) P_{i,j},$$

where $P_i(t)$, $i = 0, \ldots, \mu$, are the degree $n$ B-spline curves with control points $P_{i,0}, \ldots, P_{i,\nu}$ and knots $t_1, \ldots, t_{n+\nu}$.

b. Show that

$$B^*(s,t) = \sum_{j=0}^{\nu} N_{j,m}(t)P_j^*(s) = \sum_{j=0}^{\nu}\sum_{i=0}^{\mu} N_{i,m}(s)N_{j,n}(t)P_{i,j},$$

where $P_j^*(s), j = 0, \ldots, \nu$, are the degree $n$ B-spline curves with control points $P_{0,j}, \ldots, P_{\mu,j}$ and knots $s_1, \ldots, s_{m+\mu}$.

c. Conclude that

$$B^*(s,t) = B(s,t),$$

and therefore that the two constructions for the tensor product B-spline surface given in Section 27.8 generate identical surfaces.

27.10. Show that tensor product B-spline surfaces:

a. Are affine invariant.

b. Lie in the convex hull of their control points.

c. Consist of patches that join smoothly along common knot lines.

27.11. Verify that the rational quadratic Bezier control points and weights for the quarter circle represented by Equation 27.4 with $0 \le t \le 1$ are given by the mass-points:

$$(m_0 P_0, m_0) = (1,0,1), \quad (m_1 P_1, m_1) = (1,1,1), \quad (m_2 P_2, m_2) = (0,2,2).$$

27.12. Using blossoming and the dual functional property for the denominator as well as for each coordinate in the numerator, find control points and weights to represent the circle in Equation 27.4 as a rational quadratic B-spline curve with knots at the integers.

27.13. Find the rational biquadratic Bezier control points and weights for the part of the sphere represented by Equation 27.5 with $0 \le s, t \le 1$.

# Chapter 28

## Knot Insertion Algorithms for B-Spline Curves and Surfaces

*and the crooked shall be made straight, and the rough ways shall be made smooth*

*– Luke 3:5*

### 28.1 Motivation

B-spline methods have several advantages over Bezier techniques. B-spline curves and surfaces are piecewise polynomials that meet smoothly at their common boundaries independent of the location of the control points. This guaranteed smoothness allows designers to use low degree polynomial pieces to construct complicated freeform shapes. In addition, a control point of a B-spline curve or surface has no influence on parts of the curve or surface that are far removed from the control point. Thus B-splines provide designers with local control over the shape of a curve or surface.

So far, however, we have no tools for analyzing the geometry of B-spline curves and surfaces. Bezier subdivision facilitates algorithms for rendering and intersecting Bezier curves and expedites procedures for shading and ray tracing Bezier patches. Subdivision also provides a proof of the variation diminishing property for Bezier curves. The goal of this chapter is to develop an analogous tool, called *knot insertion*, for analyzing B-splines curves and surfaces.

Knot insertion is a powerful device with many applications. In this chapter, we shall show how to apply knot insertion to convert from B-spline to piecewise Bezier form. We can then use Bezier subdivision to render and intersect the corresponding B-spline curves and surfaces. Alternatively, we shall explain how to apply knot insertion directly to develop fast, robust algorithms for rendering and intersecting B-spline curves and surfaces without converting to piecewise Bezier form. We shall also use knot insertion to prove the variation diminishing property for B-spline curves.

### 28.2 Knot Insertion

The control points of a B-spline curve describe a piecewise polynomial curve with respect to a fixed set of knots. A *knot* is simply a parameter value where the polynomials on either side meet smoothly at the join. But any parameter value, even parameter values that represent points internal to a polynomial segment, can be construed as a knot, since the polynomials on either side surely meet smoothly at this join. A knot where the same polynomial appears on both sides is called a *pseudo-knot*. Since every spline is a B-spline (see Chapter 27, Section 27.6), we should be able to represent a given spline with respect to a new set of knots, knots that include the original knots as well as a set of pseudo-knots. A *knot insertion algorithm* is an algorithm that finds these new B-spline control points from the original B-spline control points, the original knots, and a collection of pseudo-knots. Under knot insertion, the control points change, but the curve itself is not affected.

There are many reasons that we shall wish to perform knot insertion. After knot insertion, the control polygon of a B-spline curve is closer to the curve than the original control polygon, and in the limit as the distance between the knots approaches zero, the control polygon converges to the B-spline curve. Therefore knot insertion for B-spline curves is similar to subdivision for Bezier curves, so we can use knot insertion as a divide-and-conquer strategy for analyzing B-spline curves. We can also use knot insertion to convert B-spline curves to piecewise Bezier form. We can then apply Bezier subdivision to render and intersect B-spline curves. In addition, since knot insertion is a corner cutting procedure, we can use knot insertion to prove the variation diminishing property for B-spline curves.

We shall study two different types of knot insertion algorithms: local algorithms and global algorithms. Local knot insertion procedures insert new knots locally, within a single polynomial segment; global knot insertion procedures insert new knots simultaneously in all the polynomial segments of a spline. Local knot insertion procedures include Boehm's algorithm and the Oslo algorithm; global knot insertion procedures include the Lane–Riesenfeld algorithm for uniform B-splines and a new algorithm due to Scott Schaefer for nonuniform B-splines. We will begin with local knot insertion algorithms, and then go on to investigate global knot insertion procedures.

The main tool we shall use to derive all of these knot insertion algorithms is the dual functional property of the blossom. Recall that if $P(t)$ is a degree $n$ B-spline curve with knots $\{t_i\}$ and control points $\{P_j\}$, then

$$P_k = p(t_{k+1}, \ldots, t_{k+n}).$$

If you do not fully understand this dual functional property, now would be a good time to review Chapter 27, especially Section 27.2.

---

## 28.3   Local Knot Insertion Algorithms

There are two standard local knot insertion procedures: Boehm's algorithm and the Oslo algorithm. Boehm's algorithm inserts one new knot at a time; the Oslo algorithm computes one new control point at a time. Both Boehm's algorithm and the Oslo algorithm are valid for arbitrary knot sequences and both algorithms can be derived easily from blossoming.

### 28.3.1   Boehm's Knot Insertion Algorithm

Boehm's knot insertion algorithm inserts one new knot at a time. Consider a cubic B-spline curve with knots $t_1, \ldots, t_m$. Suppose that we wish to insert a new knot $u$ between $t_k$ and $t_{k+1}$—that is, we wish to replace the original knot sequence $\ldots, t_{k-1}, t_k, t_{k+1}, t_{k+2}, \ldots$ with the new knot sequence $\ldots, t_{k-1}, t_k, u, t_{k+1}, t_{k+2}, \ldots$. By the dual functional property (Chapter 27, Section 27.2), the B-spline control points are given by evaluating the blossom of the spline at sequences of consecutive knots. Thus if $P(t)$ represents the polynomial segment of the spline in the interval $[t_k, t_{k+1}]$, then locally we can write both the old and the new control points of the spline in terms of the blossom of $P(t)$.

*Original Control Points*

$$\ldots, p(t_{k-2}, t_{k-1}, t_k), p(t_{k-1}, t_k, t_{k+1}), p(t_k, t_{k+1}, t_{k+2}), p(t_{k+1}, t_{k+2}, t_{k+3}), \ldots$$

*New Control Points*

$$\ldots, p(t_{k-2}, t_{k-1}, t_k), p(t_{k-1}, t_k, u), p(t_k, u, t_{k+1}), p(u, t_{k+1}, t_{k+2}), p(t_{k+1}, t_{k+2}, t_{k+3}), \ldots$$

Our goal is to compute the new control points from the original control points, the original knots, and the one new knot. Since the blossom values of the new control points agree with the blossom values of the original control points in all but one parameter, we can compute the new control points from the original control points using the multiaffine property of the blossom. We illustrate this computation for cubic B-spline curves in Figure 28.1.

Boehm's knot insertion algorithm is actually one level of the local de Boor algorithm for the original spline evaluated at the parameter $u$ (compare Figure 28.1 with Figure 27.6 in Chapter 27). If we want to insert the same knot multiple times, we could run Boehm's algorithm once for each time we wish to insert the knot, but there is a more efficient way. To insert the same knot $u$ multiple times, we simply run the de Boor algorithm for multiple levels and take the new control points off the left, top, and right edges of the diagram. We illustrate this algorithm for cubic B-spline curves in Figures 28.2 and 28.3. Analogous algorithms hold for splines of arbitrary degree.

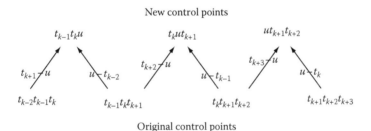

**FIGURE 28.1:**   Boehm's algorithm for inserting a new knot $u$ in the interval $[t_k, t_{k+1}]$ into a cubic B-spline curve. The original control points are placed at the base of the diagram; the new control points are computed by taking affine combinations of the original control points.

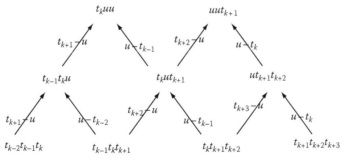

**FIGURE 28.2:**   Boehm's algorithm for inserting a double knot at $u$ in the interval $[t_k, t_{k+1}]$ into a cubic B-spline curve. The original control points are placed at the base of the diagram and the new control points emerge off the left, top, and right edges of the diagram.

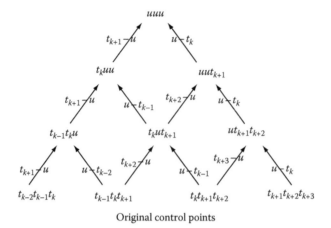

Original control points

**FIGURE 28.3:** Boehm's algorithm for inserting a triple knot at $u$ in the interval $[t_k, t_{k+1}]$ into a cubic B-spline curve. The original control points are placed at the base of the diagram and the new control points emerge off the left and right lateral edges of the diagram. Notice that this diagram is the same as the diagram for the de Boor algorithm in the interval $[t_k, t_{k+1}]$ evaluated at $t = u$. Compare to Figure 27.6 in Chapter 27.

Notice that for a spline of degree $n$, inserting a new knot $n$ times is much like the de Casteljau subdivision algorithm for Bezier curves: the new control points emerge off the left and right lateral edges of the diagram. In fact, the de Casteljau subdivision algorithm for Bezier curves is a special case of Boehm's $n$-fold knot insertion algorithm for B-spline curves because Bezier curves of degree $n$ over the parameter interval $[a,b]$ are B-spline curves of degree $n$ for the knot sequence $\underbrace{a, \ldots, a}_{n}, \underbrace{b, \ldots, b}_{n}$.

Boehm's knot insertion algorithm inserts one new knot at a time. To insert many new knots, simply apply Boehm's algorithm consecutively multiple times, once for each new knot.

### 28.3.2   The Oslo Algorithm

Boehm's knot insertion algorithm inserts one new knot at a time, but computes several new control points simultaneously. The Oslo algorithm inserts many new knots simultaneously in the same interval, but computes only one new control point at a time.

Consider again a cubic B-spline curve with knots $t_1, \ldots, t_m$. Suppose that we wish to insert the new knots $u_1, \ldots, u_p$ between the old knots $t_k$ and $t_{k+1}$—that is, we wish to replace the knot sequence $\ldots, t_{k-1}, t_k, t_{k+1}, t_{k+2}, \ldots$ with the knot sequence $\ldots, t_{k-1}, t_k, u_1, \ldots, u_p, t_{k+1}, t_{k+2}, \ldots$. Let $P(t)$ represent the polynomial segment of the spline in the interval $[t_k, t_{k+1}]$. Then once again, by the dual functional property, locally we can write both the old and the new control points of the spline in terms of the blossom of $P(t)$.

*Old Control Points*

$$\{p(t_{j+1}, t_{j+2}, t_{j+3})\}.$$

*New Control Points*

$$\{p(u_{i+1}, u_{i+2}, u_{i+3})\}.$$

To find the new control points from the original control points, we can simply apply the blossomed version of the de Boor algorithm (see Figure 28.4). Analogous algorithms apply in arbitrary degree.

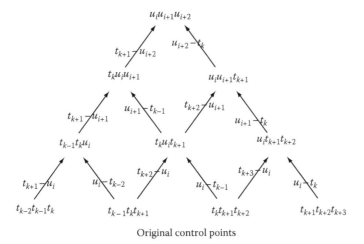

Original control points

**FIGURE 28.4:** The Oslo algorithm for computing a new control point $u_i u_{i+1} u_{i+2}$ in the interval $[t_k, t_{k+1}]$ for a cubic B-spline curve. The original control points are placed at the base of the triangle and a new control point emerges at the apex of the triangle. This diagram is just the blossomed version of the de Boor algorithm for the interval $[t_k, t_{k+1}]$. Compare to Figure 27.7 in Chapter 27.

The Oslo algorithm computes only one new control point at a time, so we must run the Oslo algorithm one time for each new control point. Notice, however, that the two initial new control points $p(t_{k-1}, t_k, u_1)$ and $p(t_k, u_1, u_2)$ emerge off the left lateral edge of the diagram during the computation of $p(u_1, u_2, u_3)$. Similarly, the two final new control points $p(u_{p-1}, u_p, t_{k+1})$ and $p(u_p, t_{k+1}, t_{k+2})$ emerge off the right lateral edge of the diagram during the computation of $p(u_{p-2}, u_{p-1}, u_p)$.

Many of the standard algorithms for Bezier and B-spline curves can be viewed as particular cases of local knot insertion procedures. Both the de Casteljau evaluation algorithm for Bezier curves and the local de Boor evaluation algorithm for B-splines are special cases of the Oslo algorithm because by the diagonal property of the blossom, evaluating a spline of degree $n$ is the same as inserting a knot of multiplicity $n$. In addition, we have already observed that the de Casteljau subdivision algorithm for Bezier curves is just a special case of Boehm's algorithm for inserting a multiple knot into a B-spline curve.

### 28.3.3 Conversion from B-Spline to Piecewise Bezier Form

Splines are piecewise polynomials. Therefore, one way to analyze B-spline curves and surfaces is to convert to piecewise Bezier form and then apply Bezier subdivision to analyze each of the polynomial pieces.

We can apply Boehm's knot insertion algorithm to convert B-spline curves to piecewise Bezier curves. Consider a B-spline curve of degree $n$ with knots $t_1, \ldots, t_m$. To convert to piecewise Bezier form, all we need to do is to insure that each knot has multiplicity $n$. Since each knot $t_k$ already has multiplicity at least one, we need only insert each knot at most $n - 1$ times. We illustrate this procedure in Figure 28.5 for cubic B-spline curves. Analogous procedures work for B-spline curves of arbitrary degree.

### 28.3.4 The Variation Diminishing Property for B-Spline Curves

We can use Boehm's knot insertion algorithm for converting from B-spline to piecewise Bezier form to prove that B-spline curves are variation diminishing. The key fact we shall need is that Boehm's knot insertion algorithm is a corner cutting procedure (see Figure 28.6).

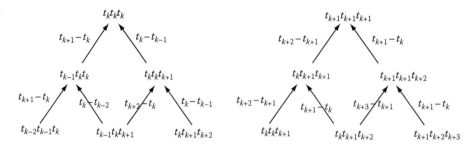

**FIGURE 28.5:** Conversion from B-spline to piecewise Bezier form. Here we illustrate the algorithm for cubic curves and the interval $[t_k, t_{k+1}]$. The knots for this B-spline segment are $t_{k-2}, \ldots, t_{k+3}$; the knots for the corresponding Bezier segment are $t_k, t_k, t_k, t_{k+1}, t_{k+1}, t_{k+1}$. On the left, the knot $t_k$ is inserted twice starting from the original B-spline control points. On the right, the output from the algorithm on the left (in particular the control point $t_k t_k t_{k+1}$) is used to insert the knot $t_{k+1}$ twice. The Bezier control points emerge along the right lateral edge of the diagram on the left—$t_k t_k t_k$ and $t_k t_k t_{k+1}$—and along the left lateral edge—$t_k t_{k+1} t_{k+1}$ and $t_{k+1} t_{k+1} t_{k+1}$—of the diagram on the right.

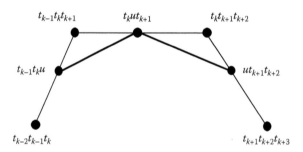

**FIGURE 28.6:** A geometric interpretation of Boehm's knot insertion algorithm for cubic curves. The control points $t_{k-1} t_k t_{k+1}$ and $t_k t_{k+1} t_{k+2}$ are replaced by the control points $t_{k-1} t_k u$, $t_k u t_{k+1}$, and $u t_{k+1} t_{k+2}$, which lie along the line segments of the original control polygon (see Figure 28.1). Thus Boehm's knot insertion algorithm is a corner cutting procedure.

**Theorem 28.1.** *B-spline curves satisfy the variation diminishing property.*

**Proof:** The proof is straightforward and follows directly from the following observations:

- Knot insertion is a corner cutting procedure.

- Therefore, since the Bezier control polygon for a B-spline curve can be generated by knot insertion, the piecewise Bezier control polygon is variation diminishing with respect to the original B-spline control polygon.

- Moreover, each Bezier segment is variation diminishing with respect to its Bezier control polygon (see Chapter 25, Section 25.4).

- Thus the piecewise Bezier curve must be variation diminishing with respect to the original B-spline control polygon.

- But since knot insertion does not alter the underlying curve, the B-spline curve and the piecewise Bezier curve are identical.

- Hence the original B-spline curve must be variation diminishing with respect to the original B-spline control polygon.

### 28.3.5 Algorithms for Rendering and Intersecting B-Spline Curves and Surfaces

Since we can easily convert from B-spline to piecewise Bezier form, we can use Bezier subdivision (see Chapter 25) to render and intersect B-spline curves and surfaces. We begin with algorithms for B-spline curves.

*Rendering Algorithm—B-Spline Curves*

Use Boehm's knot insertion algorithm to convert from B-spline to piecewise Bezier form.
Use de Casteljau's subdivision algorithm to render each of the Bezier segments.

*Intersection Algorithm—B-Spline Curves*

Use Boehm's knot insertion algorithm to convert each curve from B-spline to piecewise Bezier form.
Use de Casteljau's subdivision algorithm to intersect each pair of Bezier segments.

Similarly, we can use knot insertion to convert tensor product B-spline surfaces into a collection of tensor product Bezier patches. Here we proceed as follows. Recall that to build a tensor product B-spline surface $B(s,t)$ of bidegree $(m,n)$, we start with a rectangular array of control points $\{P_{ij}\}$, $i = 0, \ldots, \mu$, $j = 0, \ldots, \nu$ and two knot sequences $s_1, \ldots, s_{m+\mu}$ and $t_1, \ldots, t_{n+\nu}$. We let $P_i(t)$, $i = 0, \ldots, \mu$, be the degree $n$ B-spline curve with control points $P_{i,0}, \ldots, P_{i,\nu}$ and knots $t_1, \ldots, t_{n+\nu}$, and for each fixed value of $t$, we let $B(s,t)$ be the degree $m$ B-spline curve for the control points $P_0(t), \ldots, P_\mu(t)$ and the knots $s_1, \ldots, s_{m+\mu}$. Alternatively, we could begin with the B-spline curves $P_0^*(s), \ldots, P_\nu^*(s)$, where $P_j^*(s)$ is the degree $m$ B-spline curve with control points $P_{0,j}, \ldots, P_{\mu,j}$ and knots $s_1, \ldots, s_{m+\mu}$, and for each fixed value of $s$, let $B^*(s,t)$ be the B-spline curve for the control points $P_0^*(s), \ldots, P_\nu^*(s)$ and knots $t_1, \ldots, t_{n+\nu}$. To convert to piecewise Bezier patches, we first apply knot insertion to convert each B-spline curve $P_0(t), \ldots, P_\mu(t)$ into piecewise Bezier form. We then take the resulting curves $P_0^*(s), \ldots, P_\nu^*(s)$ generated by the new knots and the new control points and apply knot insertion to convert these curves into piecewise Bezier form. The resulting array of control points is the control polyhedron for the piecewise Bezier patches that represent the same surface as the original B-spline surface.

Now to render or ray trace a B-spline surface, we can proceed as follows.

*Rendering Algorithm—B-Spline Surfaces*

Use Boehm's knot insertion algorithm to convert from B-spline to piecewise Bezier form.
Use de Casteljau's subdivision algorithm to render each of the Bezier patches.

*Ray Tracing Algorithm—B-Spline Surfaces*

Use Boehm's knot insertion algorithm to convert from B-spline to piecewise Bezier form.
Use de Casteljau's subdivision algorithm to ray trace each of the Bezier patches.

Converting from B-spline to piecewise Bezier form solves many problems; we can always render B-spline curves and surfaces by first converting to piecewise Bezier form. From this point of view, B-spline methods are used for design because they have better geometric properties such as guaranteed smoothness, lower degree, and local control. Bezier techniques are used only for analysis, where we can take advantage of Bezier subdivision.

Nevertheless, converting B-splines to piecewise Bezier form for analysis seems a bit extravagant. Piecewise Bezier curves and surfaces have many more control points than the corresponding B-spline curves and surfaces. Surely it would be more efficient to work directly with these B-spline control points, rather than to convert to piecewise Bezier form. Also B-splines lie in the convex hull of their control points. If the convex hulls of two B-spline curves fail to intersect, then the two B-spline curves will not intersect. It seems wasteful to convert both B-spline curves to piecewise Bezier form and then check the convex hulls of every pair of Bezier control polygons. Surely it would be faster to work directly with the B-spline control polygons.

But there is a problem here. When we render Bezier curves and surfaces, we typically subdivide till the curve or surface can be approximated closely by its control polygon or control polyhedron. How should we proceed with B-spline curves and surfaces? If the spline is not approximated closely by its control polygon, we need to insert more knots to get a better approximation. But where should we insert these knots? Similarly, if the convex hulls of two B-spline curves intersect, then we may need to insert more knots to refine our approximation. But again, where should we insert these knots? Boehm's knot insertion algorithm and the Oslo algorithm are local knot insertion procedures. If we do not know where to insert the knots, we cannot effectively apply these procedures. To overcome this difficulty, we shall now turn our attention to global knot insertion algorithms.

## 28.4   Global Knot Insertion Algorithms

We are going to investigate two global knot insertion procedures: the Lane–Riesenfeld algorithm for uniform B-splines and Schaefer's new algorithm for B-splines with arbitrary knots. Both algorithms insert one new knot between each consecutive pair in the original knots. Iterating these knot insertion algorithms generates control polygons that converge to the original B-spline curve or surface.

### 28.4.1   The Lane–Riesenfeld Algorithm

To define a B-spline curve or surface, we need to specify both the control points and the knots. We place the control points in locations that roughly describe some desired shape; if we are unhappy with this shape, we can always adjust the location of the control points. But where should we place the knots? When there is no natural bias in our construction, we usually let the knots be evenly spaced, so typically we place the knots at the integers. The Lane–Riesenfeld algorithm is a knot insertion procedure that works whenever the knots are evenly spaced.

Given the control points for a spline with uniformly spaced knots, the Lane–Riesenfeld algorithm finds new control points for the same spline where one new knot is inserted midway between each consecutive pair of the original knots. The Lane–Riesenfeld algorithm consists of two basic steps: splitting and averaging. The splitting is done once; the averaging is repeated $n$ times, where $n$ is the degree of the spline. We illustrate the Lane–Riesenfeld algorithm for quadratic and cubic splines in Figures 28.7 and 28.8.

We can apply blossoming to establish the validity of the Lane–Riesenfeld algorithm for low-degree splines. Figures 28.9 and 28.10 illustrate the blossoming interpretation of the Lane–Riesenfeld algorithm for quadratic and cubic B-splines.

For quadratic B-splines, we can see directly from Figure 28.9 that the Lane–Riesenfeld algorithm follows easily from the dual functional property together with the symmetry and the multiaffine properties of the blossom. For cubic B-splines, however, the derivation is not so straightforward; blossom values on the top row of Figure 28.10 above which we have placed the

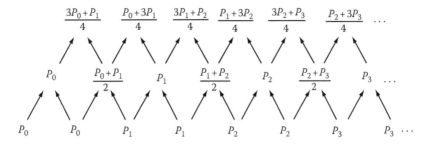

**FIGURE 28.7:** The Lane–Riesenfeld algorithm for quadratic B-spline curves. At the bottom of the diagram, each control point is split into a pair of points. Adjacent points are then averaged and this averaging step is repeated twice. The new control points for the spline with one new knot inserted midway between each consecutive pair of the original knots emerge at the top of the diagram.

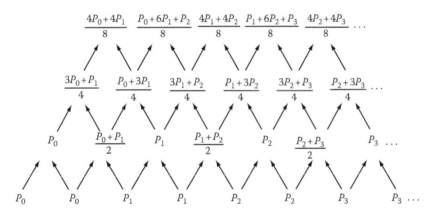

**FIGURE 28.8:** The Lane–Riesenfeld algorithm for cubic B-spline curves. At the bottom of the diagram, each control point is split into a pair of points. Adjacent points are then averaged and this averaging step is repeated three times. The new control points for the cubic spline with one new knot inserted midway between each consecutive pair of the original knots emerge at the top of the diagram.

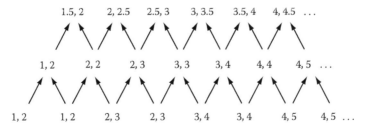

**FIGURE 28.9:** The blossoming interpretation of the Lane–Riesenfeld algorithm for quadratic B-splines. Here we adopt the notation $u, v$ for the blossom $s(u, v)$.

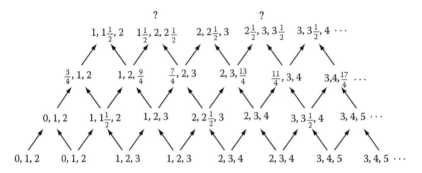

**FIGURE 28.10:** The blossoming interpretation of the Lane–Riesenfeld algorithm for cubic B-splines. Here we adopt the notation $u, v, w$ for the blossom $s(u, v, w)$.

symbol "?" are not an immediate consequence of the multiaffine property of the blossom. From the multiaffine property, for the spline $S(t)$,

$$1\frac{1}{2} = \frac{1}{2}\left(\frac{3}{4} + \frac{9}{4}\right) \Rightarrow s\left(1, 1\frac{1}{2}, 2\right) = \frac{1}{2}s\left(\frac{3}{4}, 1, 2\right) + \frac{1}{2}s\left(\frac{9}{4}, 1, 2\right),$$

but why is

$$s\left(1\frac{1}{2}, 2, 2\frac{1}{2}\right) = \frac{1}{2}s\left(\frac{9}{4}, 1, 2\right) + \frac{1}{2}s\left(\frac{7}{4}, 2, 3\right)?$$

To establish this result, observe that by the multiaffine property:

$$s\left(\frac{9}{4}, 1, 2\right) = \frac{s\left(1, 1\frac{1}{2}, 2\right) + s(1, 2, 3)}{2},$$

$$s\left(\frac{7}{4}, 2, 3\right) = \frac{s(1, 2, 3) + s\left(2, 2\frac{1}{2}, 3\right)}{2}.$$

Therefore, again by the multiaffine property:

$$\frac{1}{2}s\left(\frac{9}{4}, 1, 2\right) + \frac{1}{2}s\left(\frac{7}{4}, 2, 3\right) = \frac{1}{2}\left(\frac{s\left(1, 1\frac{1}{2}, 2\right) + s(1, 2, 3)}{2}\right) + \frac{1}{2}\left(\frac{s(1, 2, 3) + s\left(2, 2\frac{1}{2}, 3\right)}{2}\right)$$

$$= \frac{1}{4}s\left(1, 1\frac{1}{2}, 2\right) + \frac{3}{4}\left(\frac{2}{3}s(1, 2, 3) + \frac{1}{3}s\left(2, 2\frac{1}{2}, 3\right)\right)$$

$$= \frac{1}{4}s\left(1, 1\frac{1}{2}, 2\right) + \frac{3}{4}s\left(1\frac{1}{2}, 2, 3\right) = s\left(1\frac{1}{2}, 2, 2\frac{1}{2}\right).$$

Similar arguments hold for all the other blossom values above which we have placed the symbol "?".

The Lane–Riesenfeld algorithm works not only for knots at the integers, but also whenever the knots are evenly spaced. Indeed, if we divide each knot in Figure 28.9 or 28.10 by 2 or 4 or $2^n$, the algorithm remains valid. Thus we can iterate the Lane–Riesenfeld algorithm by taking the output of one stage as the input for the next stage. We shall see in Section 28.4.3 that the control polygons generated in this manner converge to the B-spline curve for the original control polygon. Thus we can use the Lane–Riesenfeld algorithm to render and intersect B-spline curves with uniformly spaced knots (see Section 28.4.4).

Although the Lane–Riesenfeld algorithm is valid in every degree, the blossoming proof gets more and more difficult as the degree increases. The standard proof of the Lane–Riesenfeld algorithm is not based on blossoming, but rather on continuous convolution of B-spline basis functions. This proof is well beyond the scope of this chapter, so we shall not pursue this proof here. Instead, we shall introduce a new global knot insertion algorithm that is not only compatible with blossoming but also works for arbitrary knots.

### 28.4.2 Schaefer's Knot Insertion Algorithm

Consider a B-spline curve with knots $t_0, \ldots, t_m$. The goal of Schaefer's algorithm is to insert one new knot $u_i$ between each consecutive pair $t_i, t_{i+1}$ of the original knots. As in the Lane–Riesenfeld algorithm, Schaefer's algorithm consists of two basic steps: splitting and averaging. The splitting is done once; the averaging is repeated $n$ times, where $n$ is the degree of the spline. The only difference here is that the averaging is not simply the arithmetic mean, adding and dividing by two, but rather is a weighted averaging, where the weights depend on the spacing between the knots. We illustrate Schaefer's algorithm using blossoming for quadratic and cubic B-splines in Figures 28.11 and 28.12.

To understand how Schaefer's algorithm works in the general case, consider first Schaefer's algorithm for quadratic B-splines illustrated in Figure 28.11. Notice that at each stage of this algorithm, every other point is promoted to the next stage with no additional computation (indicated by thicker arrows). For example, the control points represented by the blossom values $u_0 t_1$ and $u_1 t_2$ already appear at the first stage and are simply promoted to their upper left into the second stage; therefore the labels on the accompanying arrows are just zero and one. Every other small triangle in the algorithm introduces one new knot using the multiaffine property of the blossom.

To get from the quadratic algorithm to the cubic algorithm, just append the next original knot value to each of the blossoms on the first two stages. For example, on the first stage change $u_0 t_1$ to $u_0 t_1 t_2$. As in the quadratic algorithm, at each stage of the cubic algorithm, every other point is promoted to the next stage with no additional computation, and every other small triangle simply introduces one new knot using the multiaffine property of the blossom.

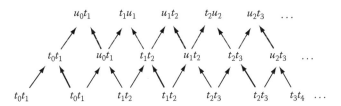

**FIGURE 28.11:** Schaefer's knot insertion algorithm for quadratic B-splines. As in Chapter 27, we adopt the multiplicative notation $uv$ for the blossom $s(u, v)$.

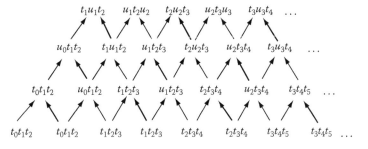

**FIGURE 28.12:** Schaefer's knot insertion algorithm for cubic B-splines. Again we adopt the multiplicative notation $uvw$ for the blossom $s(u, v, w)$.

**TABLE 28.1:** Explicit recurrences for Schaefer's algorithm and for the Lane-Riesenfeld algorithm. In the Lane-Riesenfeld algorithm the knots must be uniformly spaced; hence the knots do not appear explicitly in this algorithm.

| Schaefer Algorithm | Lane-Riesenfeld Algorithm |
|---|---|
| $p^0_{2i} = P_i$ | $p^0_{2i} = P_i$ |
| $p^0_{2i+1} = P_i$ | $p^0_{2i+1} = P_i$ |
| $p^k_{2i} = p^{k-1}_{2i+1}$ $\quad 1 \leq k \leq n$ | $p^k_{2i} = \dfrac{p^{k-1}_{i-1} + p^{k-1}_i}{2}$ $\quad 1 \leq k \leq n$ |
| $p^k_{2i+1} = \left( \dfrac{t_{i+n+1} - u_{i+k}}{t_{i+n+1} - t_{i+(k+1)/2}} \right) p^{k-1}_{2i+1} + \left( \dfrac{u_{i+k} - t_{i+(k+1)/2}}{t_{i+n+1} - t_{i+(k+1)/2}} \right) p^{k-1}_{2i+2}$ $\quad k \; odd \quad 1 \leq k \leq n$ | |
| $p^k_{2i+1} = \left( \dfrac{t_{i+n+1} - u_{i+k}}{t_{i+n+1} - t_{i+k/2}} \right) p^{k-1}_{2i+1} + \left( \dfrac{u_{i+k} - t_{i+k/2}}{t_{i+n+1} - t_{i+k/2}} \right) p^{k-1}_{2i+2}$ $\quad k \; even \quad 1 \leq k \leq n$ | |

You can get from the cubic to the quartic version of Schaefer's algorithm in much the same way: simply append the next knot value from the original knots to each of the blossoms in the first three stages. Every other point in the third stage is promoted to the fourth stage with no additional computation, and every other small triangle simply introduces one new knot using the multiaffine property of the blossom. That's it. It's easy. Better still, the proof that this approach works is immediate from the multiaffine and dual functional properties of the blossom. Unlike the Lane–Riesenfeld algorithm, Schaefer's algorithm is highly compatible with blossoming. In fact, Schaefer's algorithm is generated directly from blossoming.

Notice that both Schaefer and Lane–Riesenfeld build their algorithms for inserting knots into B-splines of degree $n + 1$ from their algorithms for inserting knots into B-splines of degree $n$. Lane–Riesenfeld simply append one more round of averaging on top of the $n$ rounds that appear in their algorithm for B-splines of degree $n$. Nothing changes in the first $n$ rounds. On the other hand, the first $n$ rounds of Schaefer's algorithm for inserting knots into B-splines of degree $n + 1$ are not identical to the $n$ rounds of Schaefer's algorithm for inserting knots into B-splines of degree $n$. The same knots are inserted, but unlike the Lane–Riesenfeld algorithm the (weighted) averages in each round change with the degree.

The diagrams for both the Lane-Riesenfeld algorithm and for Schaefer's algorithm translate into straightforward explicit recurrences. Let $\{P_i\}$ denote the original control points, let $\{t_j\}$ denote the original knots, and let $\{u_j\}$, with $t_j \leq u_j \leq t_{j+1}$, denote the knots to be inserted for a degree $n$ B-spline curve. Then the points $\{p^k_i\}$ at the *kth* level of each of these algorithms are given by the explicit recurrences in Table 28.1.

When the knots are evenly spaced, Schaefer's algorithm is not identical to the Lane–Riesenfeld algorithm. We can see this difference even in the quadratic case. In the Lane–Riesenfeld algorithm, the labels on the arrows in the first stage, indeed on every stage, are everywhere $1/2$; in Schaefer's algorithm, the labels in the first stage for evenly spaced knots are $1/4$ and $3/4$. Notice, however, how efficient Schaefer's algorithm is, even as compared to the Lane–Riesenfeld algorithm. In Schaefer's algorithm, half the computations simply promote a value to the next stage—that is, half the nodes require no computation whatsoever!

### 28.4.3   Convergence of Knot Insertion Algorithms

Knot insertion is a corner cutting procedure. Therefore, since B-spline curves lie in the convex hull of their control points, if more and more knots are inserted, the control polygon will get closer and closer to the original B-spline curve. We shall now show that as the knot spacing approaches zero, the control polygons generated by any knot insertion procedure necessarily converge to the B-spline curve for the original control points. We shall then apply this result to construct new algorithms for rendering and intersecting B-spline curves and surfaces directly from knot insertion without converting to piecewise Bezier form. We will also apply this result to provide a straight-forward proof of the variation diminishing property for B-spline curves.

**Theorem 28.2.** *The control polygons generated by knot insertion converge to the B-spline curve for the original control polygon as the knot spacing approaches zero.*

**Proof:** Let $S(t)$ be a B-spline curve of degree $n$ with knots $t_1, \ldots, t_m$, and let $P_k(t)$ denote the polynomial segment of the spline $S(t)$ over the interval $[t_k, t_{k+1}]$. By the dual functional property, the B-spline control points for the spline $S(t)$ over the interval $[t_k, t_{k+1}]$ are given by the blossom values $p_k(t_{k-n+1}, \ldots, t_k), \ldots, p_k(t_{k+1}, \ldots, t_{k+n})$. Suppose that knots are inserted so that the distance between any two consecutive knots gets arbitrarily small. Let $t_1^n, \ldots, t_d^n$ denote the new knot sequence after $n$ iterations of knot insertion, and let $[t_j^n, t_{j+1}^n]$ be a sequence of subintervals of $[t_k, t_{k+1}]$ converging to a point $\tau$ in the interval $[t_k, t_{k+1}]$. Again by the dual functional property, the new B-spline control points over the interval $[t_j^n, t_{j+1}^n]$ are given by the blossom values $p_k(t_{j-n+1}^n, \ldots, t_j^n), \ldots, p_k(t_{j+1}^n, \ldots, t_{j+n}^n)$. Now $p_k$ is a polynomial, so $p_k$ is certainly a continuous function. Hence as the distance between consecutive knots approaches zero, $t_{j-n+1}^n, \ldots, t_{j+n}^n \to \tau$, so $p_k(t_{j-n+1+i}^n, \ldots, t_{j+i}^n) \to p_k(\tau, \ldots, \tau) = P_k(\tau)$. Therefore the control polygons generated by knot insertion converge to the B-spline curve for the original control polygon as the knot spacing approaches zero.

There is an analogue of Theorem 28.2 for tensor product B-spline surfaces. The proof is much the same, so we shall not repeat this proof here. In Figure 28.13, we illustrate the convergence of the control polygon under the Lane–Riesenfeld knot insertion algorithm for cubic B-spline curves.

We can now provide a straightforward proof of the variation diminishing property for B-spline curves without converting to piecewise Bezier form.

**Theorem 28.3.** *B-spline curves satisfy the variation diminishing property.*

**Proof:** Since knot insertion is a corner cutting procedure, the limit of knot insertion must be variation diminishing with respect to the original control polygon. But by Theorem 28.2, the B-spline curve is the limit curve generated by iterating knot insertion, so B-spline curves are variation diminishing.

**FIGURE 28.13:** Control polygons converging to B-spline curves. Here we illustrate the original control polygons (left) and the first three iterations of the Lane–Riesenfeld knot insertion algorithm for cubic B-spline curves with two polynomials segments (right).

### 28.4.4 Algorithms for Rendering and Intersecting B-Spline Curves and Surfaces Revisited

We do not need to convert from B-spline to piecewise Bezier form to generate algorithms for rendering and intersecting B-spline curves and surfaces. Instead we can apply global knot insertion procedures and rely on the convergence of the control polygon or polyhedron to the corresponding B-spline curve or surface as the knot spacing approaches zero. We begin by illustrating this approach for B-spline curves and then extend these techniques to tensor product B-spline surfaces.

*Rendering Algorithm—B-Spline Curves*

> Apply a global knot insertion procedure to generate a control polygon that approximates the B-spline curve to within the given tolerance.
> Render the control polygon.

*Intersection Algorithm—B-Spline Curves*

> If the convex hulls of the control points of two B-spline curves fail to intersect, then the curves themselves do not intersect.
> Otherwise:
>> If each B-spline curve can be approximated by the straight line segments joining the first and last points of each polynomial segment, then intersect these line segments.
>> Otherwise apply a global knot insertion procedure to insert one new knot midway between each consecutive pair of the original knots and intersect the B-spline curves recursively.

Finding and intersecting two convex hulls of two sets of control points can be quite difficult and time consuming. In practice, the convex hulls in the intersection algorithm are typically replaced by bounding boxes that are much easier to compute and intersect than the actual convex hulls. Since the knot insertion algorithm converges rapidly, not much time is lost in the intersection algorithm by replacing convex hulls with bounding boxes.

To determine whether a B-spline curve can be approximated to within tolerance either by its control polygon or by the straight line segments joining the first and last points of each polynomial segment, it is sufficient, by the convex hull property, to test for each polynomial segment whether the corresponding control points all lie within tolerance of the line segment joining the first and last points of the segment. We can compute the end points of these line segments using the de Boor algorithm. We can then use the formula for the distance between a point $P$ and a line $L$ (see Chapter 11, Section 11.4.1.2) to test whether the control points all lie within some given tolerance of the line determined by the first and last points of the polynomial segment. (As with Bezier curves, we need to be careful. A control point may lie near to the infinite line determined by the two end points of a line segment, but still be far from the bounded line segment. Thus if the control point is close to the infinite line, we still need to check either that the orthogonal projection of the control point onto the line lies on the line segment—see Chapter 25, Section 25.3.1—or that the distance from the control point to one of the end points of the line segment is within tolerance.) In practice, after just a few iterations of a global knot insertion procedure, most, if not all of the polynomial segments will be well approximated by their control polygons (see Figure 28.13). If there are still a few anomalous segments, we can switch from a global to a local knot insertion procedure to get a better approximation for these remaining segments of the B-spline curve.

We can also render or ray trace tensor product B-spline surfaces using global knot insertion. We begin with a triangulation algorithm.

*Triangulation Algorithm—B-Spline Surfaces*

> If each polynomial patch of the tensor product B-spline surface can be approximated to within tolerance by two triangles each determined by three of its four corner points and if the four boundary curves of the patch can be approximated by straight line segments joining

the four corner points of the patch, then triangulate the B-spline surface by the triangles determined by these corner points of each of the polynomial patches.

Otherwise apply a global knot insertion procedure to insert one new knot midway between each consecutive pair of the original knots in both $s$ and $t$ and triangulate the surface recursively.

*Rendering Algorithm—B-Spline Surfaces*

Triangulate the B-spline surface.

Apply your favorite shading and hidden surface algorithms to render the triangulated surface.

*Ray Tracing Algorithm—B-Spline Surfaces*

If the ray does not intersect the convex hull of the control points of the B-spline surface, then the ray and the surface do not intersect.

Otherwise:

If the B-spline surface can be approximated to within tolerance by triangles each determined by three of the four corner points of each polynomial patch and if the four boundaries of each patch can be approximated by straight line segments joining the four corner points, then intersect the ray with these triangles.

Otherwise apply a global knot insertion procedure to insert one new knot midway between each consecutive pair of the original knots in both $s$ and $t$ and ray trace the surface recursively.

Keep the intersection closest to the eye.

Again finding the convex hull of the control points of a B-spline surface can be quite difficult and time consuming. In practice, for surfaces, as with curves, the convex hull is replaced by a bounding box. Since the knot insertion algorithm converges rapidly, not much time is lost in the ray tracing algorithm by replacing convex hulls with bounding boxes.

To determine whether a polynomial patch on a tensor product B-spline surface can be approximated to within tolerance by a pair of triangles determined by the four corner points, it is sufficient, by the convex hull property, to test whether each control point lies within some tolerance of at least one of the two triangles. We can compute the corner points of these polynomial patches using the de Boor algorithm. We can then use the formula for the distance between a point $P$ and a plane $S$ (see Chapter 11, Section 11.4.1.3) to test whether the control points all lie within some given tolerance of the plane determined by the corner points of the polynomial patch. (Again as with Bezier surfaces, we need to be careful. A control point may lie near to the infinite plane determined by the three vertices of a triangle, but still be far from the triangle. Thus if the control point is close to the infinite plane, we still need to check either that the orthogonal projection of the control point onto the plane lies inside the triangle—see Chapter 25, Section 25.3.2—or that the distance from the control point to one of the three vertices of the triangle is within tolerance.) In practice, as with B-spline curves, after just a few iterations of a global knot insertion procedure, most, if not all of the polynomial patches will be well approximated by two triangles determined by their four corner points. Again if there are still a few anomalous patches, we can switch from a global to a local knot insertion procedure to get a better approximation for these remaining patches of the B-spline surface.

## 28.5 Summary

Knot insertion is a powerful technique for analyzing B-spline curves and surfaces. Techniques for converting to piecewise Bezier form, algorithms for rendering and intersecting B-spline curves and surfaces, and proofs of the variation diminishing property for B-spline curves can all be derived from knot insertion.

New control points

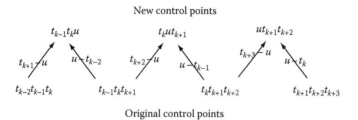

Original control points

**FIGURE 28.14:** Boehm's algorithm for inserting a new knot $u$ in the interval $[t_k, t_{k+1}]$ into a cubic B-spline curve. The original control points are placed at the base of the diagram and the new control points emerge at the top of the diagram.

There are two types of knot insertion algorithms: local and global. Boehm's algorithm and the Oslo algorithm are local procedures; the Lane–Riesenfeld algorithm and Schaefer's algorithm are global procedures. Below we summarize the main properties of each of these procedures.

*Boehm's Algorithm* (Figure 28.14)

- Local knot insertion

- Inserts one new knot at a time

- Calculation equivalent to one level of the de Boor algorithm

*Oslo Algorithm* (Figure 28.15)

- Local knot insertion

- Computes one new control point at a time

- Calculation equivalent to the blossomed version of the de Boor algorithm

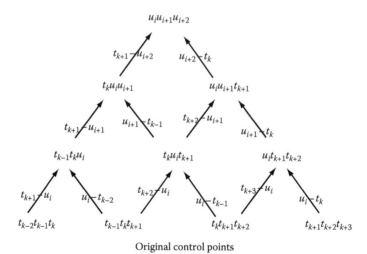

Original control points

**FIGURE 28.15:** The Oslo algorithm for computing a new control point $u_i u_{i+1} u_{i+2}$ in the interval $[t_k, t_{k+1}]$ for a cubic B-spline curve. The original control points are placed at the base of the diagram and the new control point emerges at the apex of the diagram.

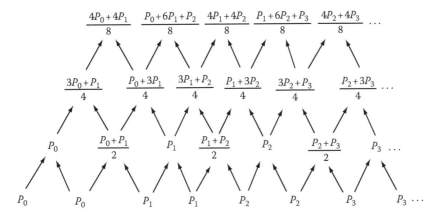

**FIGURE 28.16:** The Lane–Riesenfeld algorithm for cubic B-spline curves with uniform knots. The original control points are split and placed at the base of the diagram. Three rounds of midpoint averaging are performed and the new control points then emerge at the top of the diagram.

*Lane–Riesenfeld Algorithm* (Figure 28.16)

- Global knot insertion

- Works only for uniformly spaced knots

- Inserts one new knot midway between each consecutive pair of the original knots

- Two stage algorithm: split and average

*Schaefer's Algorithm* (Figure 28.17)

- Global knot insertion

- Works for arbitrary knots

- Inserts one new knot between each consecutive pair of the original knots

- Two stage algorithm: split and weighted average

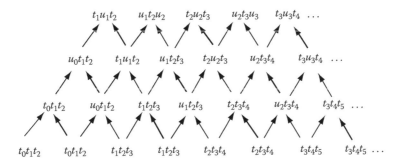

**FIGURE 28.17:** Schaefer's knot insertion algorithm for cubic B-spline curves with arbitrary knots. The original control points are split and placed at the base of the diagram. Three rounds of weighted averaging are performed and the new control points emerge at the top of the diagram.

## Exercises

28.1. Show that the procedure diagrammed in Figure 28.18 can be applied to convert from cubic B-spline to piecewise cubic Bezier form. Explain how this algorithm differs from the algorithm in Figure 28.5.

28.2. Show that the Lane–Riesenfeld algorithm for quadratic B-spline curves with uniform knots has the following three step interpretation:

    a. Split the control points.

    b. Convert to piecewise Bezier form.

    c. Apply Boehm's knot insertion algorithm to insert the new knots.

28.3. Based on Exercise 28.2:

    a. Develop an analogue of the Lane–Riesenfeld algorithm for quadratic B-spline curves with arbitrary knots.

    b. Extend this algorithm to cubic and then to arbitrary degree B-spline curves with arbitrary knots.

28.4. Draw a blossoming diagram of Schaefer's knot insertion algorithm for quartic B-spline curves.

28.5. Consider the algorithm diagrammed in Figure 28.19.

    a. Show that this algorithm correctly inserts one new knot $u_i$ between each consecutive pair $t_i, t_{i+1}$ of the original knots in a cubic B-spline curve.

    b. Show that this algorithm reduces to the Lane–Riesenfeld algorithm if the knots are evenly spaced.

28.6. Consider the algorithm diagrammed in Figure 28.20.

    a. Show that this algorithm correctly inserts one new knot $u_i$ between each consecutive pair $t_i, t_{i+1}$ of the original knots in a cubic B-spline curve.

    b. Show that this algorithm reduces to the Lane–Riesenfeld algorithm if the knots are evenly spaced.

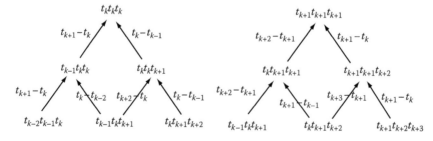

**FIGURE 28.18:**   An algorithm to convert from cubic B-spline to piecewise cubic Bezier form.

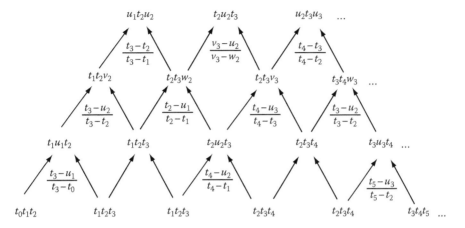

**FIGURE 28.19:** A global algorithm to insert knots into a cubic B-spline curve. The original knots are $t_0, \ldots, t_m$ and one new knot $u_i$ is inserted between each consecutive pair $t_i, t_{i+1}$ of the original knots. The expression inside each small triangle is the label on the left edge of the accompanying triangle; the label on the right edge is one minus the label one the left edge. Here $v_k = t_{k+1} - \frac{(t_{k+1} - u_{k-1})(t_{k+1} - u_k)}{(t_{k+1} - t_k)}$ and $w_k = u_k - \frac{(u_k - t_{k-1})(t_k - u_{k-1})}{(t_k - t_{k-1})}$.

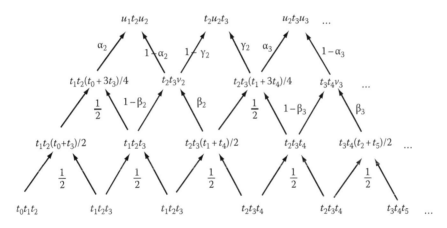

**FIGURE 28.20:** Another global algorithm to insert knots into a cubic B-spline curve. The original knots are $t_0, \ldots, t_m$ and one new knot $u_i$ is inserted between each consecutive pair $t_i, t_{i+1}$ of the original knots. Here $\alpha_k = \frac{4(t_{k+1} - u_{k-1})(t_{k+1} - u_k)}{(t_{k+1} - t_{k-2})(t_{k+1} - t_{k-1})}$, $\beta_k = \frac{2(u_k - t_{k-1})(u_{k-1} - t_{k-1})}{(t_{k+2} - t_{k-1})(t_{k+1} - t_{k-1})(1 - \alpha_k)}$, $\gamma_k = \frac{4(u_k - v_k)}{t_{k-1} + 3t_{k+2} - 4v_k}$, and $v_k = t_{k-1} + \frac{\beta_k(t_{k+2} - t_{k-1})}{2}$.

## Programming Project

### 28.1. *B-Spline Curves and Surfaces*

Implement a modeling system based on B-spline curves and surfaces in your favorite programming language using your favorite API.

a. Include algorithms for rendering B-spline curves and surfaces based on:

    i. Converting to piecewise Bezier form.

    ii. Global knot insertion algorithms.

b. Incorporate the ability to move control points interactively and have the B-spline curve or surface adjust in real time.

c. Create a new font using B-spline curves.

d. Build some interesting freeform shapes using B-spline surfaces.

# Chapter 29

## Subdivision Matrices and Iterated Function Systems

*The thing that hath been, it is that which shall be; and that which is done is that which shall be done: and there is no new thing under the sun*

*– Ecclesiastes 1:9*

### 29.1 Subdivision Algorithms and Fractal Procedures

Subdivision has the look and feel of a fractal procedure. In the standard fractal algorithm, we start with a compact set, iterate a collection of contractive transformations, and converge in the limit to a fractal shape. In Bezier subdivision, we start with a Bezier control polygon, iterate the de Casteljau subdivision algorithm, and converge in the limit to a Bezier curve (see Figure 29.1, bottom).

Are Bezier curves fractals? Bezier curves are polynomials, and polynomials are smooth almost everywhere, so Bezier curves do not look like the crinkly curves we have come to think of as fractals. Fractals are also self-similar. The Bezier curve in Figure 29.2a may be self-similar: any point on this polynomial curve is pretty much like any other point on this polynomial curve. But the Bezier curves in Figure 29.2b and 29.2c do not appear to be self-similar. These Bezier curves have an isolated cusp and an isolated self-intersection; clearly these curves cannot be decomposed into smaller segments that each has a cusp or a self intersection.

Moreover, one of the defining characteristics of fractals is that fractals are attractors. To generate a fractal, we can start the iteration with any compact set. As long as we iterate the same collection of contractive transformations, we will generate the same fractal regardless of the compact set we use to start the iteration. In contrast, with Bezier curves we are constrained to start with the Bezier control polygon in order to generate a particular Bezier curve using recursive subdivision.

Nevertheless, despite these apparent disparities, in this chapter we are going to show that Bezier curves are indeed fractals—in particular, we will show that Bezier curves are attractors. For each Bezier curve, we shall construct an iterated function system (IFS) so that starting with any compact set, not just the Bezier control polygon, and iterating the transformations, the resulting sets converge in the limit to the given Bezier curve. The transformation matrices in the iterated function system are constructed to mimic the subdivision procedure, so there is indeed a deep connection between subdivision algorithms and fractal procedures.

Are B-splines also fractals? Since each B-spline curve is a piecewise polynomial and since each polynomial piece is a Bezier curve, then if Bezier curves are fractals perhaps B-spline curves are also fractals? Actually, we do not need to think about the Bezier pieces of a B-spline curve. For B-splines, we have global knot insertion procedures and we know from Chapter 28, Theorem 28.2 that *the control polygons generated by knot insertion converge to the B-spline curve for the original control polygon as the knot spacing approaches zero.* Thus if we start with a B-spline control polygon and iterate a global knot insertion procedure, then in the limit the control polygons will

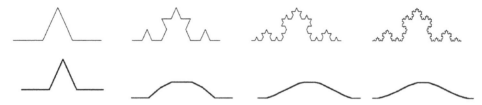

**FIGURE 29.1:**   The Koch curve (top row) starting with a triangular bump and performing three iterations of an iterated function system, and a quartic Bezier curve (bottom row) starting with a Bezier control polygon and performing three iterations of de Casteljau subdivision.

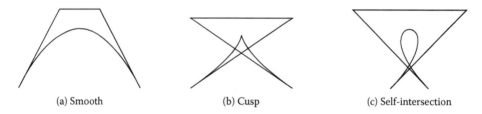

(a) Smooth        (b) Cusp        (c) Self-intersection

**FIGURE 29.2:**   Three cubic Bezier curves: (a) a smooth Bezier curve, (b) a Bezier curve with a cusp, and (c) a Bezier curve with a self-intersection.

converge to the B-spline curve for the original control points. For this reason, the Lane–Riesenfeld knot insertion procedure is often called *subdivision*. Hence knot insertion algorithms for B-splines also have the look and feel of fractal procedures. In this chapter we are going to show that B-spline curves are also attractors. Thus B-splines, like their Bezier kin, are also fractals.

## 29.2   Subdivision Matrices

The control points that subdivide a Bezier curve into two Bezier curves are affine combinations of the original Bezier control points. Thus we can represent Bezier subdivision by two square matrices whose entries are the coefficients in these affine combinations.

A similar approach works for uniform B-spline curves. Here we replace Bezier subdivision by the Lane–Riesenfeld knot insertion algorithm, which inserts one new knot between each pair of consecutive knots. The Lane–Riesenfeld algorithm converts a spline with $s$ polynomial segments into a spline with $2s$ polynomial segments or equivalently into two adjoining splines each with $s$ polynomial segments. Again the new control points are affine combinations of the original B-spline control points. Hence we can represent knot insertion for uniform B-splines by two square matrices whose entries are the coefficients in these affine combinations.

We restrict our attention here to uniform B-splines because we wish to iterate the knot insertion procedure using the same two matrices. For uniform B-splines, the expressions for knot insertion remain the same for each iteration of the Lane–Riesenfeld knot insertion algorithm; for nonuniform B-splines, however, the formulas for global knot insertion and hence too the two knot insertion matrices change with each iteration of the knot insertion procedure because each new iteration depends on the choice of the new knots.

We shall now derive explicit formulas for the entries of these subdivision matrices. We begin with Bezier curves and then go on to investigate uniform B-splines.

### 29.2.1 Subdivision Matrices for Bezier Curves

The de Casteljau evaluation algorithm for Bezier curves is also a subdivision procedure. Recall from Chapter 25, Section 25.2 that if we run the de Casteljau evaluation algorithm at the midpoint of the parameter interval $[a, b]$, then the control points that subdivide the Bezier curve at $t = (a + b)/2$ emerge along the left and right lateral edges of the de Casteljau triangle (see Figure 29.3). Moreover, the labels along each edge in this triangle are $1/2$, independent of the choice of the parameter interval.

Our initial goal is to develop explicit expressions for the control points $Q_0, \dots, Q_n$ and $R_0, \dots, R_n$ that subdivide a degree $n$ Bezier curve in terms of the original control points $P_0, \dots, P_n$ of the Bezier curve. Such formulas are easy to find. For example, from the de Casteljau algorithm, we see immediately that:

$$Q_0 = P_0, \qquad\qquad R_n = P_n,$$

$$Q_1 = \frac{P_0 + P_1}{2}, \qquad\qquad R_{n-1} = \frac{P_{n-1} + P_n}{2},$$

$$Q_2 = \frac{P_0 + 2P_1 + P_2}{4}, \qquad R_{n-2} = \frac{P_{n-2} + 2P_{n-1} + P_n}{4}.$$

In general, since all the paths from $P_j$ to $Q_k$ and from $P_{n-j}$ to $R_{n-k}$ are identical:

$$Q_k = \frac{1}{2^k} \sum_{j=0}^{k} p(k, j)P_j, \quad R_{n-k} = \frac{1}{2^k} \sum_{j=0}^{k} p(k, j)P_{n-j},$$

where $p(k, j)$ is the number of paths from $P_j$ to $Q_k$ or equivalently from $P_{n-j}$ to $R_{n-k}$. But the number of paths from the $j$th position at the base to the apex of a triangle is the same as the number of paths from the apex to the $j$th position at the base of a triangle. In Chapter 24, Section 24.3, we found that

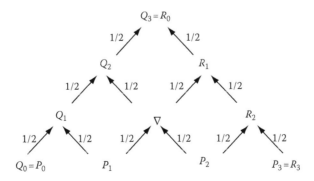

**FIGURE 29.3:** The de Casteljau subdivision algorithm for Bezier curves. The original control points $P_0, \dots, P_n$ of the Bezier curve are placed at the base and the control points $Q_0, \dots, Q_n$ and $R_0, \dots, R_n$ that subdivide the curve at the midpoint of the parameter interval emerge along the left and right lateral edges of the triangle. Here we illustrate the cubic case.

the number of paths from the apex to any node in a triangle is given by the values in the nodes of Pascal's triangle—that is, by binomial coefficients. Therefore, $p(k, j) = \binom{k}{j}$, so

$$Q_k = \frac{1}{2^k} \sum_{j=0}^{k} \binom{k}{j} P_j, \quad k = 0, \ldots, n,$$

$$R_{n-k} = \frac{1}{2^k} \sum_{j=0}^{k} \binom{k}{j} P_{n-j}, \quad k = 0, \ldots, n.$$

(29.1)

Using Equation 29.1, we can represent the points $Q_0, \ldots, Q_n$ and $R_0, \ldots, R_n$ that subdivide a Bezier curve using matrix multiplication. Let $S(n)$ be the $(2n + 1) \times (n + 1)$ matrix defined by

$$S(n) = \begin{pmatrix} 1 & 0 & \cdots & 0 \\ \frac{1}{2} & \frac{1}{2} & \cdots & 0 \\ \vdots & \vdots & \vdots & \vdots \\ \frac{1}{2^n} & \frac{n}{2^n} & \vdots & \frac{1}{2^n} \\ 0 & \frac{1}{2^{n-1}} & \cdots & \frac{1}{2^{n-1}} \\ \vdots & \vdots & \ddots & \vdots \\ 0 & 0 & \cdots & 1 \end{pmatrix}_{(2n+1)\times(n+1)}.$$

Then by Equation 29.1,

$$S(n) * \begin{pmatrix} P_0 \\ \vdots \\ P_n \end{pmatrix} = \begin{pmatrix} Q_0 \\ \vdots \\ Q_n = R_0 \\ \vdots \\ R_n \end{pmatrix}.$$

(29.2)

We can split $S(n)$ into two square $(n + 1) \times (n + 1)$ matrices $L(n)$ and $R(n)$, where $L(n)$ consists of the first $n + 1$ rows and $R(n)$ the last $n + 1$ rows of $S(n)$. Thus,

$$L(n) = \begin{pmatrix} 1 & 0 & \cdots & 0 \\ \frac{1}{2} & \frac{1}{2} & \cdots & 0 \\ \vdots & \vdots & \ddots & \vdots \\ \frac{1}{2^n} & \frac{n}{2^n} & \cdots & \frac{1}{2^n} \end{pmatrix} = \left( \frac{\binom{k}{j}}{2^k} \right), \quad j, k = 0, \ldots, n,$$

$$R(n) = \begin{pmatrix} \frac{1}{2^n} & \frac{n}{2^n} & \cdots & \frac{1}{2^n} \\ 0 & \frac{1}{2^{n-1}} & \cdots & \frac{1}{2^{n-1}} \\ \vdots & \vdots & \ddots & \vdots \\ 0 & 0 & \cdots & 1 \end{pmatrix} = \left( \frac{\binom{n-k}{n-j}}{2^{n-k}} \right), \quad j, k = 0, \ldots, n.$$

Now by Equation 29.1:

$$L(n) * \begin{pmatrix} P_0 \\ \vdots \\ P_n \end{pmatrix} = \begin{pmatrix} 1 & 0 & \cdots & 0 \\ \dfrac{1}{2} & \dfrac{1}{2} & \cdots & 0 \\ \vdots & \vdots & \ddots & \vdots \\ \dfrac{1}{2^n} & \dfrac{n}{2^n} & \cdots & \dfrac{1}{2^n} \end{pmatrix} * \begin{pmatrix} P_0 \\ \vdots \\ P_n \end{pmatrix} = \begin{pmatrix} Q_0 \\ \vdots \\ Q_n \end{pmatrix}, \tag{29.3}$$

$$R(n) * \begin{pmatrix} P_0 \\ \vdots \\ P_n \end{pmatrix} = \begin{pmatrix} \dfrac{1}{2^n} & \dfrac{n}{2^n} & \cdots & \dfrac{1}{2^n} \\ 0 & \dfrac{1}{2^{n-1}} & \cdots & \dfrac{1}{2^{n-1}} \\ \vdots & \vdots & \ddots & \vdots \\ 0 & 0 & \cdots & 1 \end{pmatrix} * \begin{pmatrix} P_0 \\ \vdots \\ P_n \end{pmatrix} = \begin{pmatrix} R_0 \\ \vdots \\ R_n \end{pmatrix}. \tag{29.4}$$

Thus $L(n)$ and $R(n)$ can be used to subdivide a degree $n$ Bezier curve. Notice that $L(n)$ and $R(n)$ share a common row: the last row of $L(n)$ is the same as the first row of $R(n)$ because $Q_n = R_0$.

We showed in Chapter 25, Section 25.2 that the Bezier control polygons generated by recursive subdivision converge to the original Bezier curve. Hence if we start with an initial control polygon $P = \{P_0, \ldots, P_n\}$ and generate new control polygons by iterating multiplication with the matrices $L(n)$ and $R(n)$ on the left as in Equations 29.3 and 29.4, then these new control polygons will converge to the original Bezier curve (see Figure 29.1, bottom). Thus the two square matrices $L(n)$ and $R(n)$ completely capture the subdivision procedure for degree $n$ Bezier curves.

**Example 29.1   Cubic Bezier Curves ($n = 3$)**

For cubic Bezier curves $n = 3$, so $S(3)$ is a $7 \times 4$ matrix and both $L(3)$ and $R(3)$ are $4 \times 4$ matrices. We display these matrices below. Since $L(3)$ consists of the first four rows and $R(3)$ the last four rows of $S(3)$, the last row of $L(3)$ is the same as the first row of $R(3)$.

$$S(3) = \begin{pmatrix} 1 & 0 & 0 & 0 \\ \dfrac{1}{2} & \dfrac{1}{2} & 0 & 0 \\ \dfrac{1}{4} & \dfrac{2}{4} & \dfrac{1}{4} & 0 \\ \dfrac{1}{8} & \dfrac{3}{8} & \dfrac{3}{8} & \dfrac{1}{8} \\ 0 & \dfrac{1}{4} & \dfrac{2}{4} & \dfrac{1}{4} \\ 0 & 0 & \dfrac{1}{2} & \dfrac{1}{2} \\ 0 & 0 & 0 & 1 \end{pmatrix}, \quad L(3) = \begin{pmatrix} 1 & 0 & 0 & 0 \\ \dfrac{1}{2} & \dfrac{1}{2} & 0 & 0 \\ \dfrac{1}{4} & \dfrac{2}{4} & \dfrac{1}{4} & 0 \\ \dfrac{1}{8} & \dfrac{3}{8} & \dfrac{3}{8} & \dfrac{1}{8} \end{pmatrix}, \quad R(3) = \begin{pmatrix} \dfrac{1}{8} & \dfrac{3}{8} & \dfrac{3}{8} & \dfrac{1}{8} \\ 0 & \dfrac{1}{4} & \dfrac{2}{4} & \dfrac{1}{4} \\ 0 & 0 & \dfrac{1}{2} & \dfrac{1}{2} \\ 0 & 0 & 0 & 1 \end{pmatrix}.$$

### 29.2.2   Subdivision Matrices for Uniform B-Spline Curves

The Lane–Riesenfeld knot insertion algorithm is the analogue of a subdivision procedure for uniform B-spline curves. Given the control points for a B-spline curve with uniformly spaced knots, the Lane–Riesenfeld algorithm finds new control points for the same spline where one new knot is

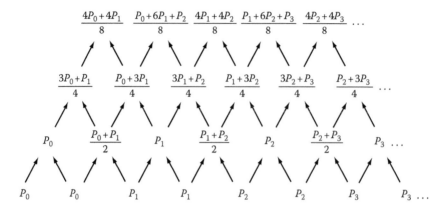

**FIGURE 29.4:** The Lane–Riesenfeld algorithm for cubic B-spline curves. At the bottom of the diagram each control point is split into a pair of points. Adjacent points are then averaged and this averaging step is repeated three times. The new control points for the cubic spline with one new knot inserted midway between each consecutive pair of the original knots emerge at the top of the diagram.

inserted midway between each consecutive pair of the original knots. We illustrate the Lane–Riesenfeld algorithm for cubic B-spline curves in Figure 29.4.

Again our initial goal is to find explicit expressions for the new control points $Q_0, \ldots, Q_k$ generated by the Lane–Riesenfeld knot insertion algorithm in terms of the original control points $P_0, \ldots, P_m$ of the given B-spline curve. For quadratic and cubic splines, we can simply read these formulas off the second and third levels of the diagram in Figure 29.4. In each case, we see that the coefficients of the original control points $P_0, \ldots, P_m$ in the new control points $Q_0, \ldots, Q_k$ are binomial coefficients divided by powers of two. For quadratic curves, we use the binomial coefficients for degree 3; for cubic curves, we use the binomial coefficients for degree 4. Notice that these binomial coefficients alternate as we scan from left to right. In the cubic case, the coefficients are: $4, 4$, then $1, 6, 1$, then $4, 4$, then $1, 6, 1 \ldots$ . The binomial coefficients of degree 4 are $1, 4, 6, 4, 1$; thus as we scan from left to right in the Lane–Riesenfeld algorithm for cubic B-splines, the coefficients alternate between the even and the odd entries on the fourth level of Pascal's triangle. This general pattern persists for splines of arbitrary degree. Indeed, it follows by induction on $n$ that for a spline of degree $n$ to find the coefficients of the original control points $P_0, \ldots, P_m$ in the new control points $Q_0, \ldots, Q_k$, we simply alternate between the even and the odd entries on the $(n + 1)$st level of Pascal's triangle and divide by $2^n$.

This straightforward pattern allows us to write down simple subdivision matrices $S_n$ for uniform B-splines of degree $n$, where the entries are just the binomial coefficients of degree $n + 1$ divided by $2^n$. For example, for quadratic splines:

$$S_2 = \frac{1}{4} \begin{pmatrix} 3 & 1 & 0 & 0 & 0 & \cdots \\ 1 & 3 & 0 & 0 & 0 & \cdots \\ 0 & 3 & 1 & 0 & 0 & \cdots \\ 0 & 1 & 3 & 0 & 0 & \cdots \\ 0 & 0 & 3 & 1 & 0 & \cdots \\ 0 & 0 & 1 & 3 & 0 & \cdots \\ \vdots & \vdots & \vdots & \vdots & \vdots & \ddots \end{pmatrix},$$

and

$$S_2 * \begin{pmatrix} P_0 \\ \vdots \\ P_m \end{pmatrix} = \frac{1}{4} \begin{pmatrix} 3 & 1 & 0 & 0 & 0 & \cdots \\ 1 & 3 & 0 & 0 & 0 & \cdots \\ 0 & 3 & 1 & 0 & 0 & \cdots \\ 0 & 1 & 3 & 0 & 0 & \cdots \\ 0 & 0 & 3 & 1 & 0 & \cdots \\ 0 & 0 & 1 & 3 & 0 & \cdots \\ \vdots & \vdots & \vdots & \vdots & \vdots & \ddots \end{pmatrix} * \begin{pmatrix} P_0 \\ \vdots \\ P_m \end{pmatrix} = \begin{pmatrix} Q_0 \\ \vdots \\ Q_k \end{pmatrix}. \tag{29.5}$$

Similarly, for cubic splines:

$$S_3 = \frac{1}{8} \begin{pmatrix} 4 & 4 & 0 & 0 & 0 & \cdots \\ 1 & 6 & 1 & 0 & 0 & \cdots \\ 0 & 4 & 4 & 0 & 0 & \cdots \\ 0 & 1 & 6 & 1 & 0 & \cdots \\ 0 & 0 & 4 & 4 & 0 & \cdots \\ 0 & 0 & 1 & 6 & 1 & \cdots \\ \vdots & \vdots & \vdots & \vdots & \vdots & \ddots \end{pmatrix},$$

and

$$S_3 * \begin{pmatrix} P_0 \\ \vdots \\ P_m \end{pmatrix} = \frac{1}{8} \begin{pmatrix} 4 & 4 & 0 & 0 & 0 & \cdots \\ 1 & 6 & 1 & 0 & 0 & \cdots \\ 0 & 4 & 4 & 0 & 0 & \cdots \\ 0 & 1 & 6 & 1 & 0 & \cdots \\ 0 & 0 & 4 & 4 & 0 & \cdots \\ 0 & 0 & 1 & 6 & 1 & \cdots \\ \vdots & \vdots & \vdots & \vdots & \vdots & \ddots \end{pmatrix} * \begin{pmatrix} P_0 \\ \vdots \\ P_m \end{pmatrix} = \begin{pmatrix} Q_0 \\ \vdots \\ Q_k \end{pmatrix}. \tag{29.6}$$

Similar matrices work for higher degree splines (see Exercise 29.1). Notice that as we descend in the rows, we shift the entries in the columns to pick up the appropriate control points in the product.

The number of rows in these subdivision matrices depends both on the degree and on the number of segments in the original B-spline curve. For a spline of degree $n$ with $s$ polynomial segments, we start with $n + s$ control points $P_0, \ldots, P_{n+s-1}$. Since the Lane–Riesenfeld algorithm inserts one new knot between each consecutive pair of the original knots, the Lane–Riesenfeld algorithm doubles the number of polynomial segments. Thus the Lane–Riesenfeld algorithm computes $n + 2s$ new control points $Q_0, \ldots, Q_{n+2s-1}$. Therefore the subdivision matrix must be of order $(n + 2s) \times (n + s)$. To indicate that the size of this matrix depends both on the degree $n$ and on the number of polynomial segments $s$, we shall denote this subdivision matrix by $S(n, s)$.

Now just as in the Bezier setting, we can split the subdivision matrix $S(n, s)$ into two square matrices $L(n, s)$ and $R(n, s)$, where $L(n, s)$ consists of the first $n + s$ rows and $R(n, s)$ the last $n + s$ rows of $S(n, s)$. Since

$$S(n, s) * \begin{pmatrix} P_0 \\ \vdots \\ P_{n+s-1} \end{pmatrix} = \begin{pmatrix} Q_0 \\ \vdots \\ Q_{n+2s-1} \end{pmatrix}, \tag{29.7}$$

it follows that

$$L(n, s) * \begin{pmatrix} P_0 \\ \vdots \\ P_{n+s-1} \end{pmatrix} = \begin{pmatrix} Q_0 \\ \vdots \\ Q_{n+s-1} \end{pmatrix} \quad \text{and} \quad R(n, s) * \begin{pmatrix} P_0 \\ \vdots \\ P_{n+s-1} \end{pmatrix} = \begin{pmatrix} Q_s \\ \vdots \\ Q_{n+2s-1} \end{pmatrix}. \tag{29.8}$$

Thus $L(n, s)$ generates the first $s$ segments and $R(n, s)$ last $s$ segments of the B-spline curve of degree $n$ with control points $Q_0, \ldots, Q_{n+2s-1}$. Therefore $L(n, s)$ and $R(n, s)$ can be used to insert knots into a uniform B-spline curve of degree $n$ with $s$ polynomial segments. Notice that $L(n, s)$ and $R(n, s)$ share $n$ common rows: the last $n$ rows of $L(n, s)$ are the same as the first $n$ rows of $R(n, s)$. This overlap occurs because two adjacent segments in a B-spline curve of degree $n$ share $n$ control points. Here the control points $Q_s, \ldots, Q_{n+s-1}$ are common to the $s$th and $(s + 1)$st segments of the B-spline curve of degree $n$ with control points $Q_0, \ldots, Q_{n+2s-1}$.

### Example 29.2   Cubic B-Spline Curves with Two Polynomial Segments ($n = 3$, $s = 2$)

For cubic B-spline curves with two polynomial segments $n = 3$ and $s = 2$, so $S(3, 2)$ is a $7 \times 5$ matrix and both $L(3, 2)$ and $R(3, 2)$ are $5 \times 5$ matrices. We display these matrices below. Since $L(3, 2)$ consists of the first five rows and $R(3, 2)$ the last five rows of $S(3, 2)$, the last three rows of $L(3, 2)$ are the same as the first three row of $R(3, 2)$.

$$S(3, 2) = \frac{1}{8} \begin{pmatrix} 4 & 4 & 0 & 0 & 0 \\ 1 & 6 & 1 & 0 & 0 \\ 0 & 4 & 4 & 0 & 0 \\ 0 & 1 & 6 & 1 & 0 \\ 0 & 0 & 4 & 4 & 0 \\ 0 & 0 & 1 & 6 & 1 \\ 0 & 0 & 0 & 4 & 4 \end{pmatrix}, \quad L(3, 2) = \frac{1}{8} \begin{pmatrix} 4 & 4 & 0 & 0 & 0 \\ 1 & 6 & 1 & 0 & 0 \\ 0 & 4 & 4 & 0 & 0 \\ 0 & 1 & 6 & 1 & 0 \\ 0 & 0 & 4 & 4 & 0 \end{pmatrix}, \quad R(3, 2) = \frac{1}{8} \begin{pmatrix} 0 & 4 & 4 & 0 & 0 \\ 0 & 1 & 6 & 1 & 0 \\ 0 & 0 & 4 & 4 & 0 \\ 0 & 0 & 1 & 6 & 1 \\ 0 & 0 & 0 & 4 & 4 \end{pmatrix}.$$

The matrices $L(n, s)$ and $R(n, s)$ subdivide a uniform B-spline curve of degree $n$ with $s$ polynomial segments into two uniform B-spline curves of degree $n$ with $s$ polynomial segments. Thus the knot insertion matrices $L(n, s)$ and $R(n, s)$ play the same role for uniform B-spline curves that the subdivision matrices $L(n)$ and $R(n)$ play for Bezier curves: both sets of matrices split the corresponding curves into two curves of the same type. The common rows in these matrices enforce either continuity or smoothness between adjacent polynomial segments.

To implement knot insertion, it is more convenient to use the two matrices $L(n, s)$ and $R(n, s)$ rather than the single matrix $S_n$ or $S(n, s)$. Each iteration of the Lane–Riesenfeld algorithm doubles the number of polynomial segments, so each time we iterate the Lane–Riesenfeld algorithm we would need to change the size of the matrix $S_n$ or $S(n, s)$. In contrast, the sizes of the matrices $L(n, s)$ and $R(n, s)$ remain fixed, since instead of dealing with a single spline with a varying number of segments, we deal here with collections of splines, each one with a fixed number $s$ of polynomial segments.

We showed in Chapter 28, Section 28.4.3 that the B-spline control polygons generated by knot insertion converge to the original B-spline curve as the knot spacing approaches zero. Hence if we start with an initial control polygon $P = \{P_0, \ldots, P_{n+s-1}\}$ and generate new control polygons by iterating multiplication with the matrices $L(n, s)$ and $R(n, s)$ on the left as in Equations 29.8, then these new control polygons will converge to the B-spline curve for the original control points. Thus the two square matrices $L(n, s)$ and $R(n, s)$ completely capture the Lane–Riesenfeld knot insertion procedure for B-spline curves of degree $n$ with $s$ polynomial segments.

## 29.3   Iterated Function Systems Built from Subdivision Matrices

Although we have built square matrices to represent subdivision both for Bezier curves and for uniform B-splines, these subdivision matrices do not represent iterated function systems for fractals in the plane. Iterated function systems are collections of transformations on affine space. If we use affine coordinates, then in two dimensions these transformations are represented by $3 \times 3$ matrices. But the subdivision matrices we developed in the previous section for Bezier curves are $(n + 1) \times (n + 1)$ matrices, where $n$ is the degree of the curve. Moreover, the subdivision matrices we built for uniform B-splines are $(n + s) \times (n + s)$ matrices, where $n$ is the degree of the spline and $s$ is the number of polynomial segments. Worse yet, these matrices multiply the control points from the left, whereas transformation matrices multiply points and vectors on the right. When we multiply the control points from the left, we are not computing a transformation of each control point; rather we are taking affine combinations of all the control points.

Nevertheless, we are going to construct iterated function systems to generate polynomial and piecewise polynomial curves built from these subdivision matrices. We shall investigate two approaches both based on projecting from higher dimensions. In the first approach, we lift the control points to higher dimensions and then apply orthogonal projections; in the second approach, we consider affine projections of normal curves lying in high-dimensional spaces.

### 29.3.1   Lifting the Control Points to Higher Dimensions

Consider first the case of quadratic Bezier curves. Here $n = 2$, so the subdivision matrices $L(2)$ and $R(2)$ are $3 \times 3$ matrices, which at least are the right size matrices for generating fractals in the plane. Unfortunately, it is not clear what happens to an arbitrary point when we multiply by these matrices on the right.

But suppose that instead of the matrices $L(2)$ and $R(2)$, we let

$$
P = \begin{pmatrix} P_0 & 1 \\ P_1 & 1 \\ P_2 & 1 \end{pmatrix} = \begin{pmatrix} x_0 & y_0 & 1 \\ x_1 & y_1 & 1 \\ x_2 & y_2 & 1 \end{pmatrix},
$$

be the matrix of control points and we consider the matrices

$$
\begin{aligned}
L_P &= P^{-1} * L(2) * P, \\
R_P &= P^{-1} * R(2) * P.
\end{aligned}
\tag{29.9}
$$

Applying these matrices to the control points $P$ on the right yields:

$$
\begin{aligned}
P * L_P &= P * \left( P^{-1} * L(2) * P \right) = L(2) * P, \\
P * R_P &= P * \left( P^{-1} * R(2) * P \right) = R(2) * P.
\end{aligned}
$$

Furthermore, if we iterate this procedure by continuing to apply the matrices $L_P$ and $R_P$ on the right, then we get exactly the same points that are generated by applying the matrices $L(2)$ and $R(2)$ to the control polygon $P$ on the left—that is, we generate the control polygons in the binary tree in Figure 29.5 with $L = L(2)$ and $R = R(2)$. But the polygons in this tree converge to the original Bezier curve, since these polygons are precisely the control polygons generated by recursive subdivision. Thus the points generated by applying the matrices $L_P$ and $R_P$ on the right to the control points $P$ converge to the Bezier curve for the control points $P$.

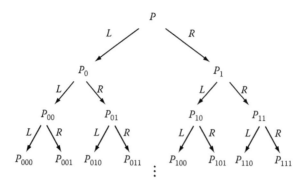

**FIGURE 29.5:** The binary tree of control polygons generated by recursive subdivision. The points $P_{b_1 \ldots b_n}$ are defined recursively. In the base case, $P_0 = L(P)$ and $P_1 = R(P)$; in the inductive step, $P_{b_1 \ldots b_n 0} = L(P_{b_1 \ldots b_n})$ and $P_{b_1 \ldots b_n 1} = R(P_{b_1 \ldots b_n})$.

The matrices $L_P$ and $R_P$ represent affine transformations, since the third column in both matrices is $(0, 0, 1)^{\mathrm{T}}$. We can prove this assertion in the following fashion. We know that

$$P^{-1} * P = \text{Identity}.$$

Since $(1, 1, 1)^{\mathrm{T}}$ is the last column of $P$, it follows that

$$P^{-1} * \begin{pmatrix} 1 \\ 1 \\ 1 \end{pmatrix} = \begin{pmatrix} 0 \\ 0 \\ 1 \end{pmatrix}.$$

But since the entries in each of the rows of $L(2)$ and $R(2)$ sum to one,

$$L(2) * \begin{pmatrix} 1 \\ 1 \\ 1 \end{pmatrix} = \begin{pmatrix} 1 \\ 1 \\ 1 \end{pmatrix} = R(2) * \begin{pmatrix} 1 \\ 1 \\ 1 \end{pmatrix}.$$

Therefore the last column of both $L_P$ and $R_P$ is $(0, 0, 1)^{\mathrm{T}}$ because

$$P^{-1} * L(2) * \begin{pmatrix} 1 \\ 1 \\ 1 \end{pmatrix} = P^{-1} * \begin{pmatrix} 1 \\ 1 \\ 1 \end{pmatrix} = \begin{pmatrix} 0 \\ 0 \\ 1 \end{pmatrix} \quad \text{and} \quad P^{-1} * R(2) * \begin{pmatrix} 1 \\ 1 \\ 1 \end{pmatrix} = P^{-1} * \begin{pmatrix} 1 \\ 1 \\ 1 \end{pmatrix} = \begin{pmatrix} 0 \\ 0 \\ 1 \end{pmatrix}.$$

The matrices $\{L_P, R_P\}$ form an iterated function system. The eigenvalues of $L(2)$ and $R(2)$ are $1.0, 0.5, 0.25$, so the matrices $L_P$ and $R_P$, which are similar to $L(2)$ and $R(2)$ and therefore have the same eigenvalues, are contractive maps. (The eigenvalue 1 corresponds to translation—that is, to the column $(0, 0, 1)^{\mathrm{T}}$, so it is only necessary to consider the other two eigenvalues—that is, the eigenvalues of the upper $2 \times 2$ submatrices of $L_P$ and $R_P$.) The limit curve—in this case a quadratic Bezier curve—is thus a fixed point of this iterated function system. Thus quadratic Bezier curves are fractals! Using an iterated function system, we can start with any compact set and still converge to the same fixed point. Thus we are no longer constrained to start the iteration with the Bezier control polygon; we can start with any compact set and still converge to the original Bezier curve. We illustrate this convergence for quadratic Bezier curves in Figure 29.6.

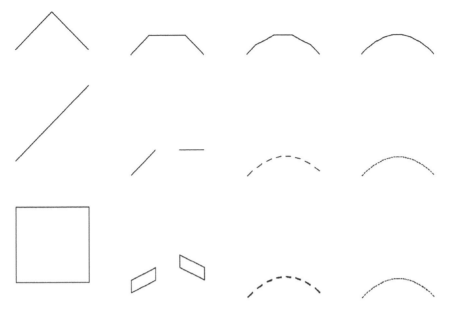

**FIGURE 29.6:** A quadratic Bezier curve as an attractor generated by an iterated function system derived from recursive subdivision. The top row contains the control polygon on the left and the first three levels of subdivision on the right. The middle row starts with a slanted line on the left and levels 1, 3, and 5 of the iteration are displayed on the right. The bottom row starts with a square on the left and levels 1, 3, and 5 of the iteration are displayed on the right.

What about Bezier curves with degree $n > 2$? Here the problem is that the matrices $L(n)$ and $R(n)$ are too big and the matrix of control points $P$ is not square so we cannot form $P^{-1}$. We are going to solve both problems by lifting to a higher dimension.

Subdivision works for Bezier curves in any dimension. If the control points $P$ are affinely independent points in $n$-dimensional space, then we could form $P^{-1}$ and construct the iterated function system $\{P^{-1} * L(n) * P, P^{-1} * R(n) * P\}$ just as we did in the case $n = 2$. If the control points $P$ lie in a lower dimensional space, we shall simply introduce additional coordinates to lift the points to $n$ dimensions. For example, for cubics we introduce one additional coordinate and set

$$P = \begin{pmatrix} P_0 & 1 & 0 \\ P_1 & 1 & 0 \\ P_2 & 1 & 0 \\ P_3 & 1 & 1 \end{pmatrix} = \begin{pmatrix} x_0 & y_0 & 1 & 0 \\ x_1 & y_1 & 1 & 0 \\ x_2 & y_2 & 1 & 0 \\ x_3 & y_3 & 1 & 1 \end{pmatrix};$$

for quartics we introduce two additional coordinates and set

$$P = \begin{pmatrix} P_0 & 1 & 0 & 0 \\ P_1 & 1 & 0 & 0 \\ P_2 & 1 & 0 & 0 \\ P_3 & 1 & 1 & 0 \\ P_4 & 1 & 1 & 1 \end{pmatrix} = \begin{pmatrix} x_0 & y_0 & 1 & 0 & 0 \\ x_1 & y_1 & 1 & 0 & 0 \\ x_2 & y_2 & 1 & 0 & 0 \\ x_3 & y_3 & 1 & 1 & 0 \\ x_4 & y_4 & 1 & 1 & 1 \end{pmatrix}.$$

Notice that in each case $P$ is invertible provided that $P_0, P_1, P_2$ are not collinear, so we have not introduced any additional degeneracies into the matrix $P$ by lifting the control points to higher dimensions. We can now form the matrices $L_P = P^{-1} * L(n) * P$ and $R_P = P^{-1} * R(n) * P$.

To construct a Bezier curve using the iterated function system $\{L_P, R_P\}$, we start with any compact subset of $n$-dimensional affine space and iterate these transformations. When we want to see the curve, we project the result orthogonally into two dimensions. Since the coordinates behave independently under recursive subdivision and since we do not change the $x,y$ coordinates when we lift the control points to higher dimensions, these orthogonal projections converge to the original Bezier curve. We illustrate this approach for cubic curves in Figure 29.7.

Iterated function systems for uniform B-spline curves can be generated in much the same manner. For uniform B-spline curves of degree $n$ with $s$ polynomial segments, simply replace $L(2)$ and $R(2)$ in Equation 29.9 with $L(n, s)$ and $R(n, s)$—that is, set

$$
\begin{aligned}
L_P &= P^{-1} * L(n, s) * P, \\
R_P &= P^{-1} * R(n, s) * P.
\end{aligned}
\tag{29.10}
$$

where $P$ is the matrix of control points lifted to affine dimension $n + s - 1$. The matrices $\{L_P, R_P\}$ form an iterated function system for the B-spline curve of degree $n$ with the control points $P$. We illustrate this approach for uniform quadratic B-spline curves in Figure 29.8.

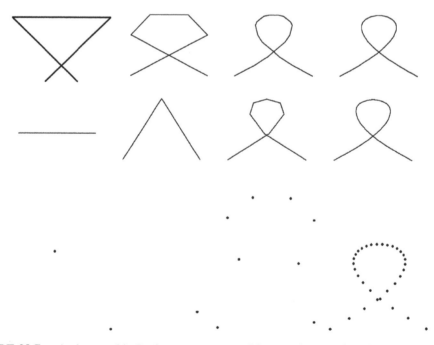

**FIGURE 29.7:**   A planar cubic Bezier curve generated from an iterated function system by lifting the control points to three dimensions, subdividing in three dimensions, and then projecting the resulting curve orthogonally back into the plane. The top row is generated in the usual manner starting from the control polygon; the middle row is generated starting from the chord joining the first and last control points; and the bottom row is generated from a single point. In each case, levels 0, 1, 3, and 5 of the iteration are illustrated.

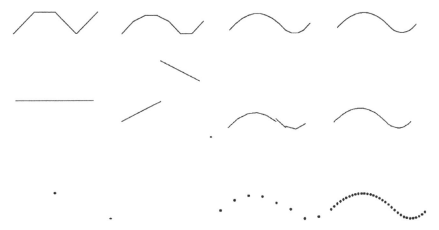

**FIGURE 29.8:** A uniform quadratic B-spline curve with three polynomial segments generated from an iterated function system by lifting the control points to four dimensions, inserting knots in four dimensions, and then projecting the resulting curve orthogonally back into the plane. The top row is generated in the usual manner starting from the control polygon; the middle row is generated starting from the chord joining the first and last control points; and the bottom row is generated from a single point. In the middle and bottom rows levels 0, 1, 3, and 5 of the iteration are illustrated.

### 29.3.2 Normal Curves

There is one set of control points $P$ for which the matrices $L_P$ and $R_P$ are especially simple: the points $P$ whose entries form the identity matrix. Curves generated by these control points are called *normal curves*. There is one normal Bezier curve for each fixed degree and one normal uniform B-spline curve for each fixed degree and each fixed number of polynomial segments.

Iterated function systems for these normal curves are particularly simple. For the normal Bezier curve of degree $n$, $L_P = L(n)$ and $R_P = R(n)$; for the normal uniform B-spline curve of degree $n$ with $s$ polynomial segments, $L_P = L(n, s)$ and $R_P = R(n, s)$. Thus, the iterated function system that generates the normal Bezier curve of degree $n$ is $\{L(n), R(n)\}$; similarly, the iterated function system that generates the normal uniform B-spline curve of degree $n$ with $s$ polynomial segments is $\{L(n, s), R(n, s)\}$. No projections are required here, since these normal curves live in high-dimensional affine spaces.

There is a simple explicit formula for the normal Bezier curve of degree $n$. Let $B_0^n(t), \ldots, B_n^n(t)$ be the Bernstein basis functions of degree $n$, and let $P(t)$ be the degree $n$ Bezier curve with control points $P_0, \ldots, P_n$. Then

$$P(t) = \sum_{k=0}^{n} B_0^n(t) P_k.$$

For the normal Bezier curve, the control points form the identity matrix. Therefore

$$P_k = (0, \ldots 0, 1, 0, \ldots, 0),$$

so the normal Bezier curve of degree $n$ is given by

$$P(t) = \left( B_0^n(t), \ldots, B_n^n(t) \right).$$

Thus the coordinate functions of the normal Bezier curve consist of the Bernstein basis functions. Similar results hold for normal uniform B-spline curves (see Exercise 29.4).

We are interested in normal curves because every Bezier curve and every uniform B-spline curve is an affine image of the corresponding normal curve. Recall that Bezier and B-spline curves are affine invariant: applying an affine transformation to the control points is equivalent to applying the affine transformation to the entire curve. Let $C_0, \ldots, C_k$ denote the control points of the normal curve, and consider a curve $P(t)$ of the same type with control points $P_0, \ldots, P_k$. In addition, let

$$C = \begin{pmatrix} C_0 \\ \vdots \\ C_k \end{pmatrix} \quad \text{and} \quad P = \begin{pmatrix} P_0 & 1 \\ \vdots & \vdots \\ P_k & 1 \end{pmatrix}.$$

Since the control points of the normal curve form the identity matrix,

$$C * P = P,$$

so

$$C_j * P = P_j, \quad j = 0, \ldots, k.$$

Thus the matrix $P$ maps the control points of the normal curve to the control points of $P(t)$. Therefore, by affine invariance, the matrix of control points $P$ maps the normal curve to the curve $P(t)$. Indeed for Bezier curves this result is easy to verify; see Exercise 29.3. From this perspective there is one canonical Bezier curve of degree $n$, and one canonical uniform B-spline curve of degree $n$ with $s$ polynomial segments—the normal curve. All the other Bezier and uniform B-spline curves can be viewed as affine images of these normal curves.

We can construct any degree $n$ Bezier curve $P(t)$ using the iterated function system $\{L(n), R(n)\}$ for the normal Bezier curve of degree $n$. Start with any compact subset of $n$-dimensional affine space and iterate the transformations $\{L(n), R(n)\}$. The shapes will converge to the normal curve. When we want to see any particular Bezier curve $P(t)$, we simply project the result into lower dimensions by applying the transformation $P$—the matrix consisting of the control points of $P(t)$—that maps the normal curve into the curve $P(t)$. A similar approach works for uniform B-spline curves. Here we replace the iterated function system $\{L(n), R(n)\}$ for the normal Bezier curve of degree $n$ by the iterated function system $\{L(n, s), R(n, s)\}$ for the normal uniform B-spline curve of degree $n$ with $s$ polynomial segments. These projections generate the same shapes as in Figures 29.7 and 29.8.

The advantage of using projections of normal curves is that we can use the same iterated function systems—the iterated function systems for the normal curve—to generate any degree $n$ Bezier curve or any degree $n$ uniform B-spline curve with $s$ polynomial segments. Thus we can reuse the same computations for all the different curves of the same type. The disadvantage of using projections of normal curves is that the projection here is a bit more complicated than the orthogonal projection we employed in the previous section because here we need to do more than just discard extraneous coordinates.

Normal curves have no cusps or self-intersections. Thus normal curves are self-similar, which explains why these curves are akin to fractals. The cusps and self-intersections we observe for Bezier curves in Figure 29.2 are artifacts of the projection of these normal curves into lower dimensions. Notice that not only are the normal curves self-similar, but the knot sequences generated by subdivision both for Bezier and for uniform B-spline curves are also self-similar.

One final caveat. The matrix consisting of the control points $P$ is an affine transformation—that is, a transformation between affine spaces. The standard affine space with which we are familiar is the space represented by affine coordinates, where the last coordinate for each point is one. This affine space is the space into which the matrix $P$ of control points maps points on the normal curve.

But the affine space of the normal curve is represented not by the space where the last coordinate for each point is equal to one, but rather by the affine space where the coordinates of each point sum to one (see Exercise 29.5). Indeed, by construction, the coordinates of each control point for the normal curve sum to one. Moreover, this constraint is maintained by the left and right subdivision matrices because the rows of these subdivision matrices sum to one (see Exercise 29.6). Therefore when we apply the iterated function system for the normal curve, we must represent points by coordinates that sum to one. The matrix $P$ of control points maps points whose coordinates sum to one to points whose last coordinate is equal to one because the last column of $P$ is $(1, \ldots, 1)^T$. Thus while the viewing is done in the standard affine space where the last coordinate of each point is equal to one, the fractal computations must be performed in an affine space where the coordinates of each point sum to one. Failure to satisfy this constraint will undermine the validity of the algorithm.

## 29.4   Fractals with Control Points

If splines behave like fractals, then perhaps fractals can also behave like splines. One advantage of splines is that splines have control points. Moving the control points allows us to adjust the shape of the spline. We shall now show how to introduce control points for fractals, allowing us to adjust the shape of a fractal in much the same way that we can adjust the shape of a spline.

While in principle we can introduce control points for any fractal, we shall restrict our attention here to bump fractals, where the notion of a control polygon is most natural. For a bump fractal, we shall take the control polygon to be the bump that generates the fractal. For example, the control polygon for the Koch curve is the triangular bump on the top left of Figure 29.1. The control points of a bump fractal are just the vertices of the corresponding bump.

Subdivision matrices for splines generate the new control polygon from the initial control polygon after one level of subdivision. To emulate this effect of subdivision matrices for bump fractals, we begin by representing the vertices of the bump after one iteration of the fractal algorithm as affine combinations of the vertices of the initial bump. The entries of the subdivision matrix are simply the coefficients of these affine combinations. Since at each level a bump fractal has more and more vertices, this subdivision matrix is not a square matrix. Therefore, just as with B-spline subdivision, we shall split this subdivision matrix into a collection of square matrices, each matrix having the same number of rows and columns as the number of points in the initial control polygon. Below we summarize this procedure for finding the subdivision matrices for a bump fractal.

*Procedure for Finding the Subdivision Matrices for a Bump Fractal*

1. Represent the vertices $Q_0, \ldots, Q_m$ of the bump after one iteration of the fractal algorithm as affine combinations of the vertices $P_0, \ldots, P_n$ of the initial bump—that is, compute

$$Q_j = \sum_{k=0}^{n} c_{jk} P_k, \quad j = 0, \ldots, m.$$

Note that the coefficients $c_{jk}$ are not necessarily unique. There may be many ways to express a point $Q_j$ in terms of the points $P_0, \ldots, P_n$. Pick one expression for each point $Q_j$.

2. Build the subdivision matrix $S$ by taking as entries the coefficients of these affine combinations—that is, set $S = (c_{jk})$. The matrix $S$ is of order $(m+1) \times (n+1)$.

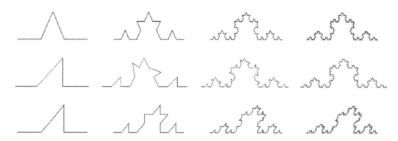

**FIGURE 29.9:** Three versions of the Koch curve after three iterations of an iterated function system. On the top row, the Koch curve starting from the original triangular bump, using the standard iterated function system. In the middle row, the initial bump is skewed to the right, but the standard iterated function system is still applied. The Koch curve emerges again on the right, since the Koch curve is an attractor. On the bottom row, the bump is once again skewed to the right, but here we use the subdivision matrices $M_1, \ldots, M_4$ conjugated with the control points $P$ lifted to four dimensions. When we project this curve to two dimensions, the curve is skewed to the right because the iterated function system depends on the initial control points.

3. Split the subdivision matrix $S$ into a collection of square $(n+1) \times (n+1)$ matrices $M_1, \ldots, M_d$. Note that as with B-splines, some of the rows of these square matrices may overlap.

   Once we have the subdivision matrices $M_1, \ldots, M_d$, then just as with B-splines there are two ways to proceed. In the first approach, we lift the control points $P = \{P_0, \ldots, P_n\}$ to higher dimensions so that $P$ is a square $(n+1) \times (n+1)$ matrix, and use the iterated function system $\{P^{-1} * M_1 * P, \ldots, P^{-1} * M_d * P\}$ to generate a fractal. We then project orthogonally to view the bump fractal in two dimensions. In the second approach, we use the iterated function system $\{M_1, \ldots, M_d\}$ to generate the corresponding normal curve for the bump fractal in $n$-dimensional affine space. Then, in order to view the bump fractal, we project to two dimensions by applying the affine transformation represented by the matrix of control points $P$. We demonstrate this approach to control points for bump fractals in Example 29.3 with the Koch curve (see also Figure 29.9).

**Example 29.3   Koch Curve**
   The initial triangular bump for the Koch curve has 5 vertices; after one iteration of the iterated function system the bump has 17 vertices (see Figure 29.1, top). Therefore the subdivision matrix for the Koch curve is of order $17 \times 5$. If we take as the initial five control points

$$
P = \begin{pmatrix} P_0 \\ P_1 \\ P_2 \\ P_3 \\ P_4 \end{pmatrix} = \begin{pmatrix} 0 & 0 & 1 \\ \dfrac{1}{3} & 0 & 1 \\ \dfrac{1}{2} & \dfrac{\sqrt{3}}{6} & 1 \\ \dfrac{2}{3} & 0 & 1 \\ 1 & 0 & 1 \end{pmatrix},
$$

then the 17 vertices $Q = (Q_0, \ldots, Q_{16})^{\mathrm{T}}$ after one iteration of the iterated function system are given by $Q = S * P$, where $S$ is the $17 \times 5$ matrix:

$$S = \begin{pmatrix} 1 & \frac{2}{3} & \frac{2}{3} & \frac{1}{3} & 0 & 0 & \frac{1}{3} & 0 & 0 & 0 & 0 & 0 & 0 & 0 & 0 & 0 & 0 \\[6pt] 0 & \frac{1}{3} & 0 & \frac{2}{3} & 1 & \frac{2}{3} & 0 & \frac{1}{3} & 0 & 0 & 0 & 0 & 0 & 0 & 0 & 0 & 0 \\[6pt] 0 & 0 & \frac{1}{3} & 0 & 0 & \frac{1}{3} & \frac{2}{3} & \frac{2}{3} & 1 & \frac{2}{3} & \frac{2}{3} & \frac{1}{3} & 0 & 0 & \frac{1}{3} & 0 & 0 \\[6pt] 0 & 0 & 0 & 0 & 0 & 0 & 0 & 0 & 0 & \frac{1}{3} & 0 & \frac{2}{3} & 1 & \frac{2}{3} & 0 & \frac{1}{3} & 0 \\[6pt] 0 & 0 & 0 & 0 & 0 & 0 & 0 & 0 & 0 & 0 & \frac{1}{3} & 0 & 0 & \frac{1}{3} & \frac{2}{3} & \frac{2}{3} & 1 \end{pmatrix}^{\mathrm{T}}.$$

Thus there are four $5 \times 5$ subdivision matrices:

$$M_1 = \begin{pmatrix} 1 & 0 & 0 & 0 & 0 \\[4pt] \frac{2}{3} & \frac{1}{3} & 0 & 0 & 0 \\[4pt] \frac{2}{3} & 0 & \frac{1}{3} & 0 & 0 \\[4pt] \frac{1}{3} & \frac{2}{3} & 0 & 0 & 0 \\[4pt] 0 & 1 & 0 & 0 & 0 \end{pmatrix}, \quad M_2 = \begin{pmatrix} 0 & 1 & 0 & 0 & 0 \\[4pt] 0 & \frac{2}{3} & \frac{1}{3} & 0 & 0 \\[4pt] \frac{1}{3} & 0 & \frac{2}{3} & 0 & 0 \\[4pt] 0 & \frac{1}{3} & \frac{2}{3} & 0 & 0 \\[4pt] 0 & 0 & 1 & 0 & 0 \end{pmatrix},$$

$$M_3 = \begin{pmatrix} 0 & 0 & 1 & 0 & 0 \\[4pt] 0 & 0 & \frac{2}{3} & \frac{1}{3} & 0 \\[4pt] 0 & 0 & \frac{2}{3} & 0 & \frac{1}{3} \\[4pt] 0 & 0 & \frac{1}{3} & \frac{2}{3} & 0 \\[4pt] 0 & 0 & 0 & 1 & 0 \end{pmatrix}, \quad M_4 = \begin{pmatrix} 0 & 0 & 0 & 1 & 0 \\[4pt] 0 & 0 & 0 & \frac{2}{3} & \frac{1}{3} \\[4pt] 0 & 0 & \frac{1}{3} & 0 & \frac{2}{3} \\[4pt] 0 & 0 & 0 & \frac{1}{3} & \frac{2}{3} \\[4pt] 0 & 0 & 0 & 0 & 1 \end{pmatrix}.$$

Notice that the first row of $M_{i+1}$ is the same as last row of $M_i$. We can now use the control points to adjust the shape of the Koch curve as in Figure 29.9, bottom.

## 29.5 Summary

The theme of this chapter is that at a deep, fundamental level, fractals and splines share the same underlying structure: B-splines are fractals, bump fractals are splines. B-splines are attractors; bump fractals have control points. B-splines can be generated from iterated function systems; bump fractals can be constructed via subdivision procedures. Both B-splines and bump fractals can be viewed as affine projections from higher dimensions of canonical normal curves.

These insights help to deepen our understanding of subdivision algorithms for Bezier curves and knot insertion procedures for uniform B-splines. We now have several different ways to perform these computations. We briefly summarize each of these techniques in the following subsections.

### 29.5.1   Bezier Curves

There are four distinct approaches to de Casteljau subdivision for Bezier curves: recursive procedures, explicit formulas, subdivision matrices, and iterated function systems.

*Subdivision Algorithm*

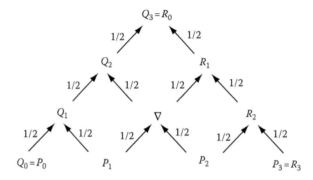

**FIGURE 29.10:**   The de Casteljau subdivision algorithm for Bezier curves. The original control points $P_0, \ldots, P_n$ of the Bezier curve are placed at the base and the control points $Q_0, \ldots, Q_n$ and $R_0, \ldots, R_n$ that subdivide the curve at the midpoint of the parameter interval emerge along the left and right lateral edges of the triangle. Here we illustrate the cubic case.

*Explicit Formulas*

$$Q_k = \frac{1}{2^k} \sum_{j=0}^{k} \binom{k}{j} P_j, \quad k = 0, \ldots, n,$$

$$R_{n-k} = \frac{1}{2^k} \sum_{j=0}^{k} \binom{k}{j} P_{n-j}, \quad k = 0, \ldots, n.$$

*Subdivision Matrices*

$$L(n) = \begin{pmatrix} 1 & 0 & \cdots & 0 \\ \dfrac{1}{2} & \dfrac{1}{2} & \cdots & 0 \\ \vdots & \vdots & \vdots & \vdots \\ \dfrac{1}{2^n} & \dfrac{n}{2^n} & \cdots & \dfrac{1}{2^n} \end{pmatrix} = \left( \dfrac{\binom{k}{j}}{2^k} \right), \quad R(n) = \begin{pmatrix} \dfrac{1}{2^n} & \dfrac{n}{2^n} & \cdots & \dfrac{1}{2^n} \\ 0 & \dfrac{1}{2^{n-1}} & \cdots & \dfrac{1}{2^{n-1}} \\ \vdots & \vdots & \ddots & \vdots \\ 0 & 0 & \cdots & 1 \end{pmatrix} = \left( \dfrac{\binom{n-k}{n-j}}{2^{n-k}} \right).$$

*Iterated Function System*

$$L_P = P^{-1} * L(n) * P,$$
$$R_P = P^{-1} * R(n) * P.$$

## 29.5.2 Uniform B-Splines

There are also four distinct approaches to the Lane–Riesenfeld knot insertion algorithm for uniform B-spline curves: recursive procedures, explicit formulas, subdivision matrices, and iterated function systems.

*Knot Insertion Algorithm*

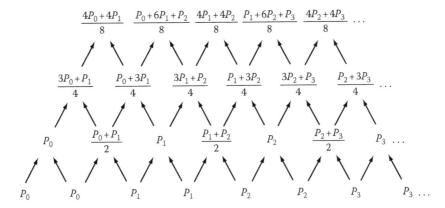

**FIGURE 29.11:** The Lane-Riesenfeld knot insertion algorithm for cubic B-spline curves. At the bottom of the diagram each control point is split into a pair of points. Adjacent points are then averaged and this averaging step is repeated three times. The new control points for the cubic spline with one new knot inserted midway between each consecutive pair of the original knots emerge at the top of the diagram.

*Explicit Formula*

$$Q_j = \sum_{k=\lfloor j/2 \rfloor}^{\left[\frac{j-n-1}{2}\right]} \frac{\binom{n+1}{j-2k}}{2^n} P_k.$$

*Subdivision Matrices*

$$S(n,s) = \frac{1}{2^n} \begin{pmatrix} n+1 & \binom{n+1}{3} & \binom{n+1}{5} & \cdots \\ 1 & \binom{n+1}{2} & \binom{n+1}{4} & \cdots \\ 0 & n+1 & \binom{n+1}{3} & \cdots \\ 0 & 1 & \binom{n+1}{2} & \cdots \\ \vdots & \vdots & \vdots & \ddots \end{pmatrix}_{(n+2s)\times(n+s)}$$

$L(n,s) = $ first $n+s$ rows of $S(n,s)$

$R(n,s) = $ last $n+s$ rows of $S(n,s)$

*Iterated Function System*

$$L_P = P^{-1} * L(n,s) * P,$$
$$R_P = P^{-1} * R(n,s) * P.$$

## Exercises

29.1. Compute the subdivision matrices $S(4, 3)$, $L(4, 3)$, and $R(4, 3)$ for uniform B-spline curves of degree 4 with three polynomial segments.

29.2. Consider the matrices $L(3, 2)$ and $R(3, 2)$ in Example 29.2.

    a. Compute the eigenvalues of $L(3, 2)$ and $R(3, 2)$.

    b. Conclude from part a that the matrices $L(3, 2)$ and $R(3, 2)$ represent contractive transformations.

29.3. Let $B(t) = (B_0^n(t), \ldots, B_n^n(t))$ be the normal Bezier curve of degree $n$, and let $P = (P_0, \ldots, P_n)^T$ be the control points for a degree $n$ Bezier curve $P(t)$.

    a. Show that $P(t) = B(t) * P$.

    b. Conclude from part a that $P(t)$ is the affine image of the normal curve $B(t)$ under the affine map represented by the matrix of control points $P$.

29.4. Let $N_{k,n}(t)$ denote the B-spline basis functions of degree $n$ defined in Chapter 27, Exercise 27.6. Show that the normal curve for the uniform B-splines of degree $n$ with $s$ polynomial segments is given by $N(t) = (N_{0,n}(t), \ldots, N_{n+s-1,n}(t))$.

29.5. Consider the collection of arrays $A = \{(a_0, \ldots, a_n) | a_0 + \cdots + a_n = 1\}$.

    a. Show that the set $A$ is an affine space—that is, show that the set $A$ is closed under affine combinations.

    b. Which arrays correspond to the vectors of this affine space?

29.6. a.  Show that each row of a matrix $M$ sums to one if and only if

$$M * \begin{pmatrix} 1 \\ \vdots \\ 1 \end{pmatrix} = \begin{pmatrix} 1 \\ \vdots \\ 1 \end{pmatrix}.$$

    b. Using part a, show that if each row of $M$ and each row of $N$ sum to one, then each row of $M * N$ also sums to one.

    c. Conclude from part b that if each row of a matrix $M$ sums to one and if the coordinates of a point $P$ sum to one, then the coordinates of the point $P * M$ also sum to one.

29.7. To what curve do the control polygons in Figure 29.5 converges if:

    a. $L = L(n)$ and $R = R(n)$.

    b. $L = L(n, s)$ and $R = R(n, s)$.

**FIGURE 29.12:**   Three bump curves that represent control polygons for three bump fractals.

29.8. a. Show that the two subdivision matrices for piecewise quadratic Bezier curves with two polynomial segments that meet continuously at their join are given by the $5 \times 5$ matrices:

$$
L = \begin{pmatrix} 1 & 0 & 0 & 0 & 0 \\ \frac{1}{2} & \frac{1}{2} & 0 & 0 & 0 \\ \frac{1}{4} & \frac{1}{2} & \frac{1}{4} & 0 & 0 \\ 0 & \frac{1}{2} & \frac{1}{2} & 0 & 0 \\ 0 & 0 & 1 & 0 & 0 \end{pmatrix}, \quad R = \begin{pmatrix} 0 & 0 & 1 & 0 & 0 \\ 0 & 0 & \frac{1}{2} & \frac{1}{2} & 0 \\ 0 & 0 & \frac{1}{4} & \frac{1}{2} & \frac{1}{4} \\ 0 & 0 & 0 & \frac{1}{2} & \frac{1}{2} \\ 0 & 0 & 0 & 0 & 1 \end{pmatrix}.
$$

   b. Generalize the result in part a to piecewise quadratic Bezier curves with $s$ polynomial segments.

29.9. Find the two $4 \times 4$ subdivision matrices for piecewise quadratic Bezier curves with two polynomial segments that meet with one continuous derivative at their join.

29.10. Find subdivision matrices for the three bump fractals corresponding to the three bump curves illustrated in Figure 29.12.

29.11. Using induction on the degree $n$, derive the explicit formula for the new control points $Q_j$ and the subdivision matrices $S(n, s)$ generated by the Lane-Riesenfeld algorithm for uniform B-spline curves of degree $n$ with $s$ polynomial segments given in Section 29.5.2.

## Programming Projects

*29.1. Fractal Algorithms for Bezier and Uniform B-Spline Curves*

   Implement de Casteljau subdivision for Bezier curves and Lane–Riesenfeld knot insertion for uniform B-spline curves using fractal algorithms (i.e., iterated function systems). Include both of the following techniques:

   a. Constructing Bezier and B-spline curves in higher dimensions by lifting the control points to higher dimensions, and then projecting orthogonally to view the curves in two dimensions.

   b. Constructing normal curves in high dimensions and then viewing their affine images in two dimensions.

29.2. *Control Polygons for Bump Fractals*

Implement algorithms for generating bump fractals with control points by constructing iterated function systems using the vertices of the initial bump curve as control points. Include both of the following techniques:

a. Constructing bump fractals in higher dimensions by lifting the control points to higher dimensions, and then projecting orthogonally to view the bump fractals in two dimensions.

b. Constructing normal bump fractals in high dimensions and then viewing their affine images in two dimensions.

# Chapter 30

## Subdivision Surfaces

*To give subtilty to the simple . . .*

*– Proverbs 1:4*

## 30.1 Motivation

Subdivision algorithms are similar to fractal procedures. In the standard fractal algorithm, we begin with a compact set $C_0$ and iterate over a collection of contractive transformations $W$ to generate a sequence of compact sets $C_{n+1} = W(C_n)$ that converge in the limit to a fractal shape $C_\infty = Lim_{n \to \infty} C_n$. In subdivision procedures, we start with a set $P_0$ (usually either a control polygon or a control polyhedron, or as we shall see shortly a quadrilateral or triangular mesh) and recursively apply a set of rules $S$ to generate a sequence of sets $P_{n+1} = S(P_n)$ of the same general type as $P_0$ that converge in the limit to a smooth curve or surface $P_\infty = Lim_{n \to \infty} P_n$.

Although one of the themes of Chapter 29 is that classical subdivision algorithms are essentially fractal procedures, nevertheless the goals of subdivision algorithms and fractal procedures are fundamentally different. The goal of fractal procedures is to construct extraordinary shapes, unconventional forms from conventional origins; the goal of subdivision algorithms is to construct smooth shapes, differentiable functions from discrete data. The de Casteljau subdivision algorithm for Bezier curves and surfaces and the Lane–Riesenfeld algorithm for uniform B-splines are examples of subdivision procedures that start with coarse control polygons or polyhedra and build refined control polygons and polyhedra that converge in the limit to smooth curves and surfaces.

We are interested in subdivision algorithms for four basic reasons:

1. Subdivision algorithms replace complicated formulas with simple procedures. Thus subdivision is more in the spirit of modern computer science than classical mathematics.

2. Subdivision algorithms are easy to understand and simple to implement.

3. Subdivision algorithms can generate a large class of smooth functions, not just polynomials and piecewise polynomials.

4. Subdivision algorithms on polyhedral meshes can produce shapes with arbitrary topology, unlike tensor product schemes which can generate only surfaces that are topologically equivalent to a rectangle, or by identifying edges to a cylinder or a torus.

Thus subdivision provides a simple approach to generating a wide variety of smooth shapes.

We are going to focus on subdivision algorithms for freeform surfaces. Most of this chapter is based on Chapters 2 and 7 of *Subdivision Methods for Geometric Design: A Constructive Approach*, by Warren and Weimer. We shall devote our attention here to the algorithmic details of different subdivision paradigms. Readers interested in proofs of convergence, differentiability, and other properties of these subdivision procedures should consult the book by Warren and Weimer.

---

## 30.2    Box Splines

Box spline surfaces are generalizations of uniform tensor product B-spline surfaces. The fundamental step in the Lane–Riesenfeld subdivision algorithm for uniform tensor product B-spline surfaces is averaging adjacent control points along canonical directions in a rectangular array. Box spline surfaces are generated by subdivision procedures that allow averaging along arbitrary directions in a rectangular array. To understand how this works in practice, we begin with a review of the Lane–Riesenfeld algorithm for uniform B-splines.

### 30.2.1    Split and Average

Recall that the Lane–Riesenfeld algorithm (Chapter 28, Section 28.4.1) for uniform B-spline curves consists of two basic steps: splitting and averaging (see Figure 30.1).

To extend the Lane–Riesenfeld knot insertion procedure to uniform tensor product B-spline surfaces $P(s, t)$, we need to insert knots in both the $s$ and $t$ parameter directions. Consider a rectangular array of control points $\{P_{ij}\}$. To insert knots midway between each pair of consecutive knots, we first split and average in the $s$-direction, and then split and average the result of this computation in the $t$-direction (see Figures 30.2 through 30.4). Alternatively, instead of initially splitting each control point into two control points, we can start by splitting each control point into four control points. We can then average consecutively in the $s$ and $t$ directions; now there is no need to split the output of the averages in the $s$-direction before averaging in the $t$-direction (see Figure 30.5). The result of both approaches is the same control polyhedron, and it makes no difference whether we average first in the $s$-direction and then in the $t$-direction or first in the $t$-direction and then in the $s$-direction; the result is always the same (see Exercise 30.1a).

### 30.2.2    A Subdivision Procedure for Box Spline Surfaces

Box spline surfaces are built by a generalization of the split and average approach to subdivision for uniform tensor product B-spline surfaces. The only difference is that for box splines, in addition to averaging along canonical directions in the rectangular array of control points, we also allow averaging along arbitrary directions in the rectangular array.

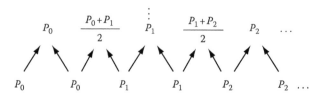

**FIGURE 30.1:**    The Lane–Riesenfeld algorithm for uniform B-spline curves. At the base of the diagram each control point is split into a pair of points. Adjacent points are then averaged and this averaging step is repeated $n$ times, where $n$ is the degree of the curve. The new control points for the spline with one new knot inserted midway between each consecutive pair of the original knots emerge at the top of the diagram. Iterating this procedure generates a sequence of control polygons that converge in the limit to the B-spline curve for the original control points.

$$\vdots \qquad\qquad\qquad\qquad\qquad \vdots$$

$$P_{02} \quad P_{02} \quad P_{12} \quad P_{12} \quad P_{22} \quad P_{22} \; \ldots \qquad\qquad P_{02} \quad \frac{P_{02}+P_{12}}{2} \quad P_{12} \quad \frac{P_{12}+P_{22}}{2} \quad P_{22} \; \ldots$$

$$P_{01} \quad P_{01} \quad P_{11} \quad P_{11} \quad P_{21} \quad P_{21} \; \ldots \qquad\qquad P_{01} \quad \frac{P_{01}+P_{11}}{2} \quad P_{11} \quad \frac{P_{11}+P_{21}}{2} \quad P_{21} \; \ldots$$

$$P_{00} \quad P_{00} \quad P_{10} \quad P_{10} \quad P_{20} \quad P_{20} \; \ldots \qquad\qquad P_{00} \quad \frac{P_{00}+P_{10}}{2} \quad P_{10} \quad \frac{P_{10}+P_{20}}{2} \quad P_{20} \; \ldots$$

(a) Split in the *s*-direction $\qquad\qquad\qquad$ (b) Average in the *s*-direction

**FIGURE 30.2:** Subdivision for uniform tensor product B-spline surfaces. First step: (a) split and (b) average in the (horizontal) *s*-direction. Here we illustrate the linear case. Higher degree surfaces can be generated by taking additional averages.

$$\vdots$$

$$P_{02} \quad \frac{P_{02}+P_{12}}{2} \quad P_{12} \quad \frac{P_{12}+P_{22}}{2} \quad P_{22} \quad \ldots$$

$$P_{02} \quad \frac{P_{02}+P_{12}}{2} \quad P_{12} \quad \frac{P_{12}+P_{22}}{2} \quad P_{22} \quad \ldots$$

$$P_{01} \quad \frac{P_{01}+P_{11}}{2} \quad P_{11} \quad \frac{P_{11}+P_{21}}{2} \quad P_{21} \quad \ldots$$

$$P_{01} \quad \frac{P_{01}+P_{11}}{2} \quad P_{11} \quad \frac{P_{11}+P_{21}}{2} \quad P_{21} \quad \ldots$$

$$P_{00} \quad \frac{P_{00}+P_{10}}{2} \quad P_{10} \quad \frac{P_{10}+P_{20}}{2} \quad P_{20} \quad \ldots$$

$$P_{00} \quad \frac{P_{00}+P_{10}}{2} \quad P_{10} \quad \frac{P_{10}+P_{20}}{2} \quad P_{20} \quad \ldots$$

Split in the *t*-direction

**FIGURE 30.3:** Subdivision for uniform tensor product B-spline surfaces continued. The output of the first step (Figure 30.2b) is split—that is, the points are doubled—in the (vertical) *t*-direction.

$$\vdots$$

$$P_{02} \qquad \frac{P_{02}+P_{12}}{2} \qquad P_{12} \qquad \frac{P_{12}+P_{22}}{2} \qquad P_{22} \quad \ldots$$

$$\frac{P_{01}+P_{02}}{2} \quad \frac{P_{01}+P_{02}+P_{11}+P_{12}}{4} \quad \frac{P_{11}+P_{12}}{2} \quad \frac{P_{11}+P_{12}+P_{21}+P_{22}}{4} \quad \frac{P_{21}+P_{22}}{2} \quad \ldots$$

$$P_{01} \qquad \frac{P_{01}+P_{11}}{2} \qquad P_{11} \qquad \frac{P_{11}+P_{21}}{2} \qquad P_{21} \quad \ldots$$

$$\frac{P_{00}+P_{01}}{2} \quad \frac{P_{00}+P_{01}+P_{10}+P_{11}}{4} \quad \frac{P_{10}+P_{11}}{2} \quad \frac{P_{10}+P_{11}+P_{20}+P_{21}}{4} \quad \frac{P_{20}+P_{21}}{2} \quad \ldots$$

$$P_{00} \qquad \frac{P_{00}+P_{10}}{2} \qquad P_{10} \qquad \frac{P_{10}+P_{20}}{2} \qquad P_{20} \quad \ldots$$

Average in the *t*-direction

**FIGURE 30.4:** Subdivision for uniform tensor product B-spline surfaces. Final step: average the points in Figure 30.3 in the (vertical) *t*-direction. Here we illustrate the bilinear case. Higher bidegree surfaces can be generated by taking additional averages.

$$\vdots \qquad\qquad\qquad\qquad \vdots$$

$$P_{02}\ P_{02}\ P_{12}\ P_{12}\ P_{22}\ P_{22}\ \ldots \qquad\qquad P_{02}\ \dfrac{P_{02}+P_{12}}{2}\ P_{12}\ \dfrac{P_{12}+P_{22}}{2}\ P_{22}\ \ldots$$

$$P_{02}\ P_{02}\ P_{12}\ P_{12}\ P_{22}\ P_{22}\ \ldots \qquad\qquad P_{02}\ \dfrac{P_{02}+P_{12}}{2}\ P_{12}\ \dfrac{P_{12}+P_{22}}{2}\ P_{22}\ \ldots$$

$$P_{01}\ P_{01}\ P_{11}\ P_{11}\ P_{21}\ P_{21}\ \ldots \qquad\qquad P_{01}\ \dfrac{P_{01}+P_{11}}{2}\ P_{11}\ \dfrac{P_{11}+P_{21}}{2}\ P_{21}\ \ldots$$

$$P_{01}\ P_{01}\ P_{11}\ P_{11}\ P_{21}\ P_{21}\ \ldots \qquad\qquad P_{01}\ \dfrac{P_{01}+P_{11}}{2}\ P_{11}\ \dfrac{P_{11}+P_{21}}{2}\ P_{21}\ \ldots$$

$$P_{00}\ P_{00}\ P_{10}\ P_{10}\ P_{20}\ P_{20}\ \ldots \qquad\qquad P_{00}\ \dfrac{P_{00}+P_{10}}{2}\ P_{10}\ \dfrac{P_{10}+P_{20}}{2}\ P_{20}\ \ldots$$

$$P_{00}\ P_{00}\ P_{10}\ P_{10}\ P_{20}\ P_{20}\ \ldots \qquad\qquad P_{00}\ \dfrac{P_{00}+P_{10}}{2}\ P_{10}\ \dfrac{P_{10}+P_{20}}{2}\ P_{20}\ \ldots$$

(a) Split　　　　　　　　　　　　　　(b) Average

**FIGURE 30.5:** Subdivision for uniform tensor product B-spline surfaces—alternative approach. First (a) split each point into four points and then (b) average in the (horizontal) s-direction. The resulting points are the same as the points in Figure 30.3. We can now average the output in the (vertical) t-direction to get the points in Figure 30.4. No additional splitting is required.

Thus to define a box spline surface, in addition to a rectangular array of control points $\{P_{ij}\}$ in three-space, we need to specify a collection of vectors $V = \{v = (v_1, v_2)\}$ in the plane. The only restriction on the collection $V$ is that the components $v_1, v_2$ of each of the vectors $v$ must be integers, and $V$ must contain at least one copy of the vectors $(1, 0)$ and $(0, 1)$. Note that vectors $v$ in the collection $V$ can appear multiple times.

The subdivision algorithm for a box spline surface begins by splitting each control point into four control points, just as in the subdivision algorithm for uniform tensor product B-spline surfaces (see Figure 30.5a). We then remove one copy of the vectors $(1, 0)$ and $(0, 1)$ from the collection $V$. Thus the initial copies of $(1, 0)$ and $(0, 1)$ correspond to quadrupling the control points. Now we make one averaging pass for each remaining vector $v = (v_1, v_2)$ in the collection $V$ by computing the averages of $v$-adjacent points—that is, by computing all averages of the form

$$Q_{i,j} = \frac{P_{i,j} + P_{i+v_1, j+v_2}}{2}. \tag{30.1}$$

The output of one averaging pass is used as input to the next averaging pass. The order in which we compute these averages does not matter, since averaging $v$-adjacent points and then averaging $w$-adjacent points is equivalent to averaging $w$-adjacent points and then averaging $v$-adjacent points (see Exercise 30.1b).

Iterating this procedure generates in the limit piecewise polynomial surfaces of degree $d - 2$, where $d$ is the number of vectors, counting multiplicities, in the collection $V$. Moreover, these surfaces have $\alpha - 2$ continuous derivatives, where $\alpha$ is the size of the smallest set $A \subset V$ such that the vectors in $V - A$ are all multiples of a single vector (see Warren and Weimer, Chapter 2).

### Example 30.1　Uniform Tensor Product B-Spline Surfaces

Let $V$ consist of the vector $(1, 0)$ repeated $m$ times and the vector $(0, 1)$ repeated $n$ times. The initial copies of $(1, 0)$ and $(0, 1)$ correspond to quadrupling the control points. Thus, repeating the vector $(1, 0)$ $m$ times corresponds to taking $m - 1$ averages in the s-direction, and repeating the vector $(0, 1)$ $n$ times corresponds to taking $n - 1$ averages in the t-direction. Therefore the corresponding box spline surface is a uniform tensor product B-spline surface of bidegree $(m - 1, n - 1)$. The total degree of this surface is $d = (m - 1) + (n - 1) = m + n - 2$ and this surface has $\alpha - 2$ continuous derivatives, where $\alpha = \min{(m, n)}$.

$$\vdots$$

| $P_{02}$ | $\dfrac{P_{02}+P_{12}}{2}$ | $P_{12}$ | $\dfrac{P_{12}+P_{22}}{2}$ | $P_{22}$ | $\cdots$ |

| $\dfrac{P_{01}+P_{02}}{2}$ | $\dfrac{P_{01}+P_{12}}{2}$ | $\dfrac{P_{11}+P_{12}}{2}$ | $\dfrac{P_{11}+P_{22}}{2}$ | $\dfrac{P_{21}+P_{22}}{2}$ | $\cdots$ |

| $P_{01}$ | $\dfrac{P_{01}+P_{11}}{2}$ | $P_{11}$ | $\dfrac{P_{11}+P_{21}}{2}$ | $P_{21}$ | $\cdots$ |

| $\dfrac{P_{00}+P_{01}}{2}$ | $\dfrac{P_{00}+P_{11}}{2}$ | $\dfrac{P_{10}+P_{11}}{2}$ | $\dfrac{P_{10}+P_{21}}{2}$ | $\dfrac{P_{20}+P_{21}}{2}$ | $\cdots$ |

| $P_{00}$ | $\dfrac{P_{00}+P_{10}}{2}$ | $P_{10}$ | $\dfrac{P_{10}+P_{20}}{2}$ | $P_{20}$ | $\cdots$ |

**FIGURE 30.6:** One level of subdivision for the three-direction, piecewise linear box spline in Example 30.2. These control points are generated by averaging the control points in Figure 30.5a along the direction $(1,1)$—that is, by using Equation 30.1 with $v_1 = v_2 = 1$. For an example, see Figure 30.8.

$$\vdots$$

| $\dfrac{3P_{01}+P_{02}+P_{11}+3P_{12}}{8}$ | $\dfrac{P_{01}+3P_{11}+3P_{12}+P_{22}}{8}$ | $\dfrac{3P_{11}+P_{12}+P_{21}+3P_{22}}{8}$ $\cdots$ |

| $\dfrac{P_{00}+3P_{01}+3P_{11}+P_{12}}{8}$ | $\dfrac{P_{00}+P_{01}+P_{10}+10P_{11}+P_{12}+P_{21}+P_{22}}{16}$ | $\dfrac{P_{10}+3P_{11}+3P_{21}+P_{22}}{8}$ $\cdots$ |

| $\dfrac{3P_{00}+P_{01}+P_{10}+3P_{11}}{8}$ | $\dfrac{P_{00}+3P_{10}+3P_{11}+P_{21}}{8}$ | $\dfrac{3P_{10}+P_{11}+P_{20}+3P_{21}}{8}$ $\cdots$ |

**FIGURE 30.7:** One level of subdivision for the three-direction, piecewise quartic box spline in Example 30.3. These control points are generated by averaging the control points in Figure 30.4 twice along the direction $(1,1)$. For an example, see Figure 30.9.

**Example 30.2 Three-Direction Linear Box Splines**

Let $V = \{(1,0),(0,1),(1,1)\}$. Then $d = 3$ and $\alpha = 2$. Thus the corresponding surface is a continuous, three-direction, piecewise linear box spline. The vectors $(1,0)$ and $(0,1)$ correspond to quadrupling the control points. Averaging these points in the direction $(1,1)$ yields the points in Figure 30.6.

**Example 30.3 Three-Direction Quartic Box Splines**

Let $V = \{(1,0),(1,0),(0,1),(0,1),(1,1),(1,1)\}$. Then $d = 6$ and $\alpha = 4$. Thus the corresponding surface is a three-direction, piecewise quartic box spline with two continuous derivatives. The initial vectors $(1,0)$ and $(0,1)$ correspond to quadrupling the control points. Averaging these points once in each of the directions $(1,0)$ and $(0,1)$ yields the points in Figure 30.4. Averaging twice more in the direction $(1,1)$ generates the points in Figure 30.7 (see Exercise 30.2).

## 30.3 Quadrilateral Meshes

A quadrilateral mesh is a collection of quadrilaterals, where each pair of quadrilaterals is either disjoint or share a common edge and each edge belongs to at most two quadrilaterals. One way to form a quadrilateral mesh is to start from a rectangular array of control points. Joining points with adjacent indices generates a regular quadrilateral mesh, a mesh where each interior vertex has four adjacent faces, edges, and vertices (see Figures 30.8 through 30.10). Tensor product B-splines are built from rectangular arrays of control points. Therefore uniform tensor product B-splines are an example of surfaces generated by subdivision starting from a quadrilateral mesh.

The goal of this section is to construct smooth surfaces starting from quadrilateral meshes of arbitrary topology. Unlike tensor product or box spline schemes, the vertices of an arbitrary

**FIGURE 30.8:** A three-direction, piecewise linear box spline surface. The initial control polyhedron is on the left; the first two levels of subdivision are illustrated in the middle and on the right.

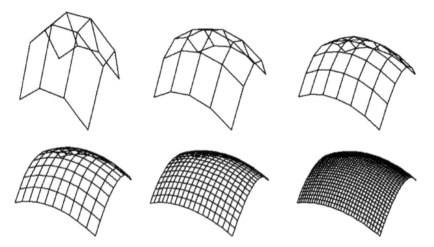

**FIGURE 30.9:** A three-direction, quartic box spline surface. The initial control polyhedron is in the upper left. The first five levels of subdivision are illustrated from left to right and from top to bottom.

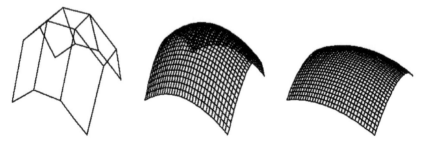

**FIGURE 30.10:** The $C^1$ biquadratic B-spline (center) and $C^2$ three-direction quartic box spline (right) for the same control polyhedron (left). Both surfaces employ four rounds of averaging for each level of subdivision.

quadrilateral mesh are not constrained to form a rectangular array; all that is required is that the vertices can be joined into nonoverlapping quadrilateral faces, where each pair of faces is either disjoint or share a common edge and each edge belongs to at most two faces. For example, the faces of a cube form a quadrilateral mesh, even though the vertices of the cube do not form a regular rectangular array, since in the cube each vertex has only three adjacent faces (see Figure 30.14). Two additional examples of quadrilateral meshes that are not generated by rectangular arrays of points are illustrated in Figures 30.13a and 30.15.

Next we shall provide two methods for constructing smooth surfaces from arbitrary quadrilateral meshes via subdivision: centroid averaging and stencils.

## 30.3.1   Centroid Averaging

Centroid averaging is an extension of the Lane–Riesenfeld subdivision algorithm for uniform tensor product bicubic B-spline surfaces from rectangular arrays of control points to quadrilateral meshes with arbitrary topology. Therefore we begin by revisiting the Lane–Riesenfeld algorithm for bicubic B-splines.

### 30.3.1.1   Uniform Bicubic B-Spline Surfaces

To simplify our discussion, let us start with uniform B-spline curves. For cubic B-splines, the Lane–Riesenfeld algorithm can be separated into two distinct stages: topology (connectivity) and geometry (shape). In the topological stage, corresponding to the first, piecewise linear step of the Lane–Riesenfeld algorithm, we introduce new control points at the midpoints of the original control polygon and we change the connectivity of the control polygon by adding edges joining these new points to adjacent control points (see Figures 30.1 and 30.11b). Notice, however, that in the topological phase we do not alter the shape of the control polygon. In the geometric stage, we reposition the control points by taking two successive averages. Notice that computing two successive averages is equivalent to taking the midpoints of the midpoints of these control points (see Figure 30.11c). Thus, in the geometric phase, we change the shape of the control polygon, but we do not alter the connectivity of the vertices and edges.

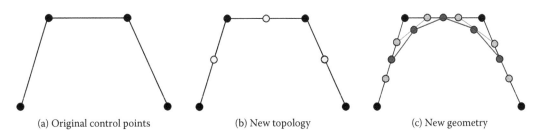

(a) Original control points          (b) New topology          (c) New geometry

**FIGURE 30.11: (See color insert following page 354.)**  One level of the Lane–Riesenfeld algorithm for uniform cubic B-spline curves. In (a), a segment of the original control polygon (black) is illustrated. In the topological stage (b), new control points (yellow) are introduced at the midpoints of the edges of the control polygon. Notice that there are now two types of control points: edge points (yellow) and vertex points (black). In the geometric stage (c), the control points (black and yellow) are repositioned to the midpoints (blue) of the midpoints (green) of these control points. Edge points (yellow) are relocated along the same edge (in fact, to the same position along the edge), but vertex points (black) are relocated to new positions off the original control polygon. The new control polygon is illustrated in blue.

A similar interpretation applies to the Lane–Riesenfeld subdivision algorithm for uniform tensor product bicubic B-spline surfaces. Once again the Lane–Riesenfeld algorithm can be separated into two distinct stages: topology (connectivity) and geometry (shape).

In the topological stage, we keep the original control points and we insert new control points at the midpoints of the edges and at the centroids of the faces of the control polyhedron (see Figures 30.4 and 30.12b). Thus we alter the connectivity of the control polyhedron, adding new edges and faces by connecting the centroid of each face to the midpoints of the edges surrounding the face. Notice again that in the topological phase, we do not alter the shape of the control polyhedron.

In the geometric stage, we reposition these control points by taking two successive averages in the $s$-direction and the $t$-direction. But successive averaging in the $s$ and $t$ directions is equivalent to a single (bilinear) average of four adjacent control points, since

$$\frac{1}{2}\left(\frac{Q_{i,j} + Q_{i+1,j}}{2}\right) + \frac{1}{2}\left(\frac{Q_{i,j+1} + Q_{i+1,j+1}}{2}\right) = \frac{Q_{i,j} + Q_{i+1,j} + Q_{i,j+1} + Q_{i+1,j+1}}{4}.$$

Therefore two successive averages in the $s$ and $t$ directions are equivalent to two successive centroid averages of four adjacent control points. Computing the centroid of adjacent centroids sends

$$Q_{ij} \to \frac{1}{4}\left(\frac{Q_{i-1,j-1} + Q_{i,j-1} + Q_{i-1,j} + Q_{i,j}}{4}\right) + \frac{1}{4}\left(\frac{Q_{i,j-1} + Q_{i+1,j-1} + Q_{i+1,j} + Q_{i,j}}{4}\right)$$
$$+ \frac{1}{4}\left(\frac{Q_{i-1,j} + Q_{i-1,j+1} + Q_{i,j+1} + Q_{i,j}}{4}\right) + \frac{1}{4}\left(\frac{Q_{i,j+1} + Q_{i+1,j+1} + Q_{i+1,j} + Q_{i,j}}{4}\right).$$

Therefore each control point is repositioned to the centroid of the centroids of the faces adjacent to the control point (see Figure 30.12c). Thus, in the geometric stage, we change the shape of the control polyhedron, but we do not alter the connectivity of the vertices, edges, and faces.

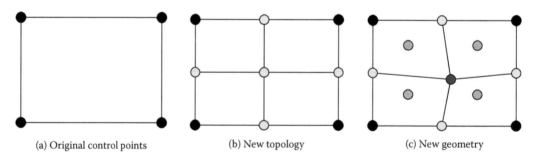

(a) Original control points        (b) New topology        (c) New geometry

**FIGURE 30.12: (See color insert following page 354.)**  One level of the Lane–Riesenfeld algorithm for bicubic B-spline surfaces. In (a), we illustrate one face of the control polyhedron. In the topological stage (b), new control points (yellow) are inserted at the midpoints of the edges and at the centroids of the faces of the control polyhedron; each edge is then split into two edges and each face is split into four faces. In the geometric stage (c), each control point—yellow or black—is repositioned to the centroid (blue) of the centroids (green) of the faces adjacent to the control point. Thus, after the geometric stage, each black and each yellow control point is repositioned based on the location of the green centroids of adjacent faces, but the topology of the polyhedron is not changed. The green centroids are used only to reposition vertices and are not themselves vertices of the refined polyhedron.

### 30.3.1.2 Arbitrary Quadrilateral Meshes

As with bicubic B-splines, there are two main phases to centroid averaging for arbitrary quadrilateral meshes: topology (connectivity) and geometry (shape). In the topological phase, we refine the mesh by introducing new vertices, edges, and faces, but we do not alter the underlying shape of the mesh. In the geometric phase, we reposition the vertices to change the geometry of the mesh, but we do not alter the underlying topology of the mesh; rather we maintain the connectivity of the mesh—the edges and faces—introduced during the topological phase.

To refine the topology of the mesh, we proceed exactly as we did for bicubic B-splines: we keep the original vertices and we insert new *edge vertices* at the midpoints of the edges and new *face vertices* at the centroids of the faces of the quadrilateral mesh. Each edge is then split into two edges and each face is split into four faces by connecting the centroid of the face to the midpoints of the edges surrounding the face (see Figure 30.13b).

To alter the geometry of the mesh, we reposition each vertex of the mesh to the centroid of the centroids of the adjacent faces. The only difference between subdivision for bicubic B-splines and subdivision for arbitrary quadrilateral meshes is that for arbitrary quadrilateral meshes there need not be four faces adjacent to each interior vertex. For example, in the cube there are only three faces adjacent to each of the vertices. An interior vertex of a quadrilateral mesh where the number of adjacent faces, edges, or vertices is not equal to four is called an *extraordinary vertex*, and the number of faces, edges, or vertices adjacent to a vertex is called the *valence* of the vertex. Thus, in the geometric phase, each vertex $Q$ of the mesh is repositioned by the formula

$$Q \longrightarrow \frac{1}{n} \sum_{k=1}^{n} C_k,$$

where

$C_1, \ldots, C_n$ are the centroids of the faces adjacent to $Q$
$n$ is the valence of $Q$.

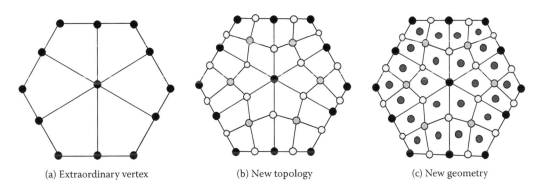

| (a) Extraordinary vertex | (b) New topology | (c) New geometry |

**FIGURE 30.13: (See color insert following page 354.)** One level of centroid averaging. In (a), six faces of a quadrilateral mesh surround an extraordinary vertex. In the topological stage (b), new vertices are inserted at the midpoints (yellow) of the edges and at the centroids (green) of the faces, and the centroid of each face is joined to the midpoints of the surrounding edges, splitting each face into four new faces. In the geometric stage (c), each vertex—yellow, green, or black—is repositioned to the centroid of the centroids (blue) of the faces adjacent to the vertex, but the topology of the mesh is not changed. Notice that the extraordinary vertex at the center remains an extraordinary vertex and is repositioned as the centroid of six adjacent centroids. All the other vertices—black, yellow, and green—are ordinary vertices and their new location is computed as the centroid of four blue points, just as in the Lane–Riesenfeld algorithm for bicubic B-splines.

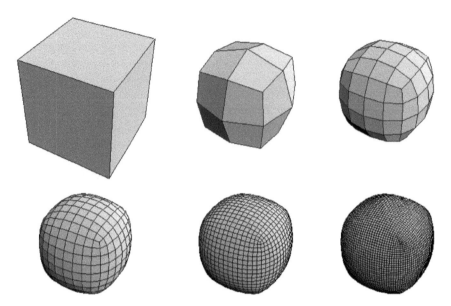

**FIGURE 30.14: (See color insert following page 354.)**   Five levels of centroid averaging applied to a cube. The corners are rounded, but the limit shape is not a perfect sphere. Notice the extraordinary vertices of valence three at the corners of the cube.

**FIGURE 30.15: (See color insert following page 354.)**   Centroid averaging applied to two cubes connected by a thin rectangular rod. The original quadrilateral mesh is on the left; the first and third levels of subdivision are illustrated in the middle and on the right.

After several levels of subdivision, most of the interior vertices of a quadrilateral mesh are ordinary vertices and the extraordinary vertices become more and more isolated topologically from one another. These features emerge because during the topological phase, each face is subdivided into four faces by edges joining the centroid of the face to the midpoints of the surrounding edges. Thus the new edge vertices and the new face vertices all have valence four (see Figure 30.13b), so all the new vertices introduced by subdivision are ordinary vertices. Also the valence of each of the original vertices is unchanged, since the number of faces surrounding a vertex does not change during subdivision. Hence, during subdivision, ordinary vertices remain ordinary vertices and extraordinary vertices remain extraordinary vertices with the same valence (see Figure 30.13b). Since all the new vertices have valence four, locally the new quadrilateral mesh looks exactly like a rectangular array of control points; only vertices from the original mesh that did not have valence four do not have four adjacent faces. Therefore, just like uniform bicubic B-splines, the surfaces generated in the limit by iterating centroid averaging have two continuous derivatives everywhere; the only exceptions are at the limits of the extraordinary vertices, where these surfaces are guaranteed to have only one continuous derivative (see Warren and Weimer, Chapter 8).

### 30.3.2 Stencils

Stencils are an alternative way to generate smooth surfaces via subdivision starting from quadrilateral meshes of arbitrary topology. Centroid averaging separates the computation of new vertices during subdivision into two phases: topology (connectivity) and geometry (shape). After the insertion of new vertices in the topological phase, all the vertices, old and new, are repositioned by a single formula—centroid averaging. This formula computes the new positions of the vertices in terms of the known positions of the vertices *after* the topological phase. In contrast, stencils compute the positions of all the new vertices in terms of the positions of the original vertices. Thus stencils combine the topological and geometric computations into a single phase.

#### 30.3.2.1 Stencils for Uniform B-Splines

Since centroid averaging is an extension of the Lane–Riesenfeld subdivision algorithm for tensor product bicubic B-spline surfaces from rectangular arrays of control points to quadrilateral meshes with arbitrary topology, we shall adapt stencils for bicubic B-spline surfaces to quadrilateral meshes of arbitrary topology. As usual, to simplify matters, we will begin our study of stencils with stencils for cubic B-spline curves.

The Lane–Riesenfeld algorithm for cubic B-spline curves is illustrated in Figure 30.16 (see also Chapter 28, Section 28.4.1). Notice that there are essentially two distinct explicit formulas for the new control points that emerge at the top of Figure 30.16:

$$Q_{i+1/2} = \frac{P_i + P_{i+1}}{2}, \tag{30.2}$$

$$Q_{i+1} = \frac{P_i + 6P_{i+1} + P_{i+2}}{8}. \tag{30.3}$$

The stencils for this algorithm are simply the coefficients $(1/2, 1/2)$ and $(1/8, 6/8, 1/8)$ that appear in these two formulas.

To understand the geometry behind these two formulas, recall from Figure 30.11 that after one level of subdivision topologically the new control polygon for a cubic B-spline curve consists of two types of control points: edge points and vertex points. Equation 30.2 repositions edge points; Equation 30.3 repositions vertex points. These stencils are illustrated schematically by the diagrams in Figure 30.17.

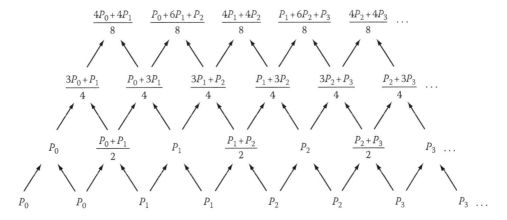

**FIGURE 30.16:** The Lane–Riesenfeld algorithm for cubic B-spline curves. At the bottom of the diagram each control point is split into a pair of points. Adjacent points are then averaged and this averaging step is repeated three times. The new control points for the cubic spline with one new knot inserted midway between each consecutive pair of the original knots emerge at the top of the diagram.

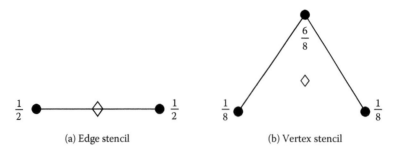

(a) Edge stencil                    (b) Vertex stencil

**FIGURE 30.17:**   Stencils for subdivision of cubic B-spline curves: (a) the edge stencil and (b) the vertex stencil. The position of the new control point (clear diamond) is computed by multiplying the original control points (black disks) by the associated fractions and summing the results.

There is a similar interpretation using stencils of the Lane–Riesenfeld subdivision algorithm for bicubic B-spline surfaces. Topologically there are now three kinds of points: face points, edge points, and vertex points. To find the stencils for these points, we need to find explicit formulas for the control points of the new control polyhedron after one level of subdivision. We can calculate these points by taking two rounds of centroid averaging of the control points in Figure 30.4. The results are presented in Figure 30.18.

As expected, Figure 30.18 illustrates three kinds of stencils:

$$Q_F = \frac{P_{00} + P_{01} + P_{10} + P_{11}}{4}, \tag{30.4}$$

$$Q_E = \frac{P_{00} + P_{10} + 6P_{01} + 6P_{11} + P_{02} + P_{12}}{16},$$

$$Q_E = \frac{P_{00} + P_{01} + 6P_{10} + 6P_{11} + P_{20} + P_{21}}{16}, \tag{30.5}$$

$$Q_V = \frac{P_{00} + P_{20} + 6P_{01} + 6P_{10} + 36P_{11} + 6P_{21} + 6P_{12} + P_{02} + P_{22}}{64}. \tag{30.6}$$

$$\vdots$$

| $\dfrac{9P_{01} + 3P_{02} + 3P_{11} + P_{12}}{16}$ | $\dfrac{3P_{01} + P_{02} + 9P_{11} + 3P_{12}}{16}$ | $\dfrac{9P_{11} + 3P_{12} + 3P_{21} + P_{22}}{16}$ | $\cdots$ |
|---|---|---|---|
| $\dfrac{3P_{00} + 9P_{01} + P_{10} + 3P_{11}}{16}$ | $\dfrac{P_{00} + 3P_{01} + 3P_{10} + 9P_{11}}{16}$ | $\dfrac{3P_{10} + 9P_{11} + P_{20} + 3P_{21}}{16}$ | $\cdots$ |
| $\dfrac{9P_{00} + 3P_{01} + 3P_{10} + P_{11}}{16}$ | $\dfrac{3P_{00} + P_{01} + 9P_{10} + 3P_{11}}{16}$ | $\dfrac{9P_{10} + 3P_{11} + 3P_{20} + P_{21}}{16}$ | $\cdots$ |

(a) First round of centroid averaging

$$\vdots$$

| $\dfrac{P_{00} + P_{10} + 6P_{01} + 6P_{11} + P_{02} + P_{12}}{16}$ | $\dfrac{P_{00} + P_{20} + 6P_{01} + 6P_{10} + 36P_{11} + 6P_{21} + 6P_{12} + P_{02} + P_{22}}{64}$ | $\cdots$ |
|---|---|---|
| $\dfrac{P_{00} + P_{01} + P_{10} + P_{11}}{4}$ | $\dfrac{P_{00} + P_{01} + 6P_{10} + 6P_{11} + P_{20} + P_{21}}{16}$ | $\cdots$ |

(b) Second round of centroid averaging

**FIGURE 30.18:**   The control points for one level of subdivision of a uniform bicubic B-spline surface constructed by applying two rounds of centroid averaging to the control points in Figure 30.4: (a) first round of centroid averaging and (b) second round of centroid averaging.

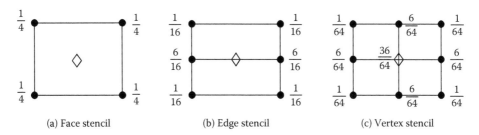

(a) Face stencil          (b) Edge stencil          (c) Vertex stencil

**FIGURE 30.19:** Stencils for subdivision of bicubic B-spline surfaces: (a) face stencil, (b) edge stencil, and (c) vertex stencil. The position of the new control point (clear diamond) is computed by multiplying the original control points (black discs) by the associated fractions and summing the results. Note that there are two edge stencils: one for horizontal edges and one for vertical edges. Only the horizontal edge stencil is illustrated here, since the vertical edge stencil is much the same, except that the fractions 6/16 lie along horizontal edges instead of along vertical edges.

Equation 30.4 represents the face stencil; Equation 30.5 represents the two edge stencils, one for vertical edges and one for horizontal edges; and Equation 30.6 represents the vertex stencil. We illustrate these stencils schematically in Figure 30.19.

### 30.3.2.2 Stencils for Extraordinary Vertices

To extend stencils from uniform bicubic B-spline surfaces to subdivision surfaces built from arbitrary quadrilateral meshes, we need not alter the face stencil or the edge stencil; all we need to do is to generalize the vertex stencil to extraordinary vertices. This vertex stencil should be affine invariant (the fractions should sum to one), symmetric with respect to the surrounding vertices, and reduce to the vertex stencil for bicubic B-spline surfaces for ordinary vertices with valence four. Also, we want the stencil to generate smooth surfaces—that is, surfaces with at least one continuous derivative—at the limits of the extraordinary vertices. Two such stencils are provided in Figure 30.20; both satisfy all of our constraints, but the Catmull–Clark stencil tends to generate more rounded surfaces.

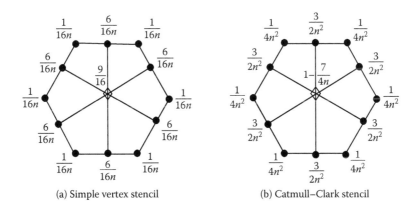

(a) Simple vertex stencil          (b) Catmull–Clark stencil

**FIGURE 30.20:** Two vertex stencils for extraordinary vertices with valence $n$: (a) a simple vertex stencil and (b) the Catmull–Clark stencil. Notice that both stencils reduce to the vertex stencil for bicubic B-splines at ordinary vertices—that is, vertices where the valence $n = 4$ (see Figure 30.19c).

## 30.4    Triangular Meshes

A triangular mesh is a collection of triangles, where each pair of triangles are either disjoint or share a common edge and each edge belongs to at most two triangles. Thus a triangular mesh is much like a quadrilateral mesh, except that the faces are triangles instead of quadrilaterals. A rectangular array of control points generates a regular quadrilateral mesh. Splitting each face in this quadrilateral mesh along a fixed diagonal direction generates a regular triangular mesh, a mesh where each interior vertex is adjacent to six vertices, edges, and faces. The control points for the three-direction box splines form a triangular mesh because the vector $(1, 1)$ splits each quadrilateral in the mesh generated by the control points into a pair of triangles. Therefore three-direction box splines are examples of surfaces generated by subdivision starting from a triangular mesh.

Triangular meshes are quite common in Geometric Modeling, perhaps even more common than quadrilateral meshes. Unlike the control points for three-direction box splines, the vertices of an arbitrary triangular mesh are not constrained to be generated from a rectangular array; all that is required is that the vertices can be joined into nonoverlapping triangular faces, where each pair of faces is either disjoint or share a common edge and each edge belongs to at most two faces. For example, the faces of an octahedron form a triangular mesh, even though the vertices of the octahedron do not form a regular mesh, since in the octahedron each vertex has only four adjacent faces (see Figure 30.24). Two additional examples of triangular meshes that are not regular are illustrated in Figures 30.23a and 30.25.

The goal of this section is to construct smooth surfaces via subdivision starting from triangular meshes of arbitrary topology. We shall explore the same two paradigms for subdivision of triangular meshes that we investigated for subdivision of quadrilateral meshes: centroid averaging and stencils.

### 30.4.1    Centroid Averaging for Triangular Meshes

Three-direction quartic box spline surfaces play the same basic role for subdivision algorithms of triangular meshes that uniform bicubic tensor product B-spline surfaces play for subdivision algorithms of quadrilateral meshes. Therefore we shall begin our investigation of subdivision algorithms for triangular meshes by taking another look at the subdivision algorithm for three-direction quartic box splines.

#### 30.4.1.1    Three-Direction Quartic Box Splines

Much like the subdivision algorithm for uniform bicubic B-spline surfaces, the subdivision algorithm for three-direction quartic box spline surfaces can be separated into two distinct stages: topology and geometry. In Section 30.2.2, we built the subdivision algorithm for one level of the three-direction quartic box spline by averaging the control points generated by one level of the subdivision algorithm for the three-direction piecewise linear box spline along the vectors $(1, 0)$, $(0, 1)$, $(1, 1)$. Computing the control points for the three-direction piecewise linear box spline is equivalent to changing the topology of the mesh; averaging these control points along the vectors $(1, 0)$, $(0, 1)$, $(1, 1)$ is equivalent to altering the geometry of the mesh.

In the topological stage, we keep the original control points and we insert new control points at the midpoints of the edges of the triangular mesh (see Figures 30.6 and 30.22b). Thus we alter the connectivity of the mesh, adding new edges and faces by connecting the midpoints of adjacent edges. Each edge is split into two edges and each face is split into four faces, but as usual in the topological phase we do not alter the shape of the mesh.

In the geometric stage, we reposition these control points by averaging the control points along the vectors $(1,0)$, $(0,1)$, $(1,1)$. Thus, in the geometric stage, we change the shape of the mesh, but we do not alter the connectivity of the vertices, edges, and faces (see Figure 30.22c).

There is, however, another somewhat easier way to reposition the vertices. We can generate the same points by applying the stencil in Figure 30.21 to the control points for the three-direction piecewise linear box spline. This result is easy to verify: simply overlay the stencil in Figure 30.21 on top of the points in Figure 30.6 and multiply. As the stencil in Figure 30.21 moves through the array of control points in Figure 30.6, these products generate the points in Figure 30.7 (see Exercise 30.4).

Serendipitously, the stencil in Figure 30.21 can be decomposed into the average of six simpler stencils. Indeed if we represent each stencil as a $3 \times 3$ matrix, then we have the following identity:

$$\begin{pmatrix} 0 & \frac{1}{8} & \frac{1}{8} \\ \frac{1}{8} & \frac{2}{8} & \frac{1}{8} \\ \frac{1}{8} & \frac{1}{8} & 0 \end{pmatrix} = \frac{1}{6}\begin{pmatrix} 0 & \frac{3}{8} & \frac{3}{8} \\ 0 & \frac{2}{8} & 0 \\ 0 & 0 & 0 \end{pmatrix} + \frac{1}{6}\begin{pmatrix} 0 & \frac{3}{8} & 0 \\ 0 & \frac{2}{8} & \frac{3}{8} \\ 0 & 0 & 0 \end{pmatrix} + \frac{1}{6}\begin{pmatrix} 0 & 0 & \frac{3}{8} \\ 0 & \frac{2}{8} & \frac{3}{8} \\ 0 & 0 & 0 \end{pmatrix}$$

$$+ \frac{1}{6}\begin{pmatrix} 0 & 0 & 0 \\ \frac{3}{8} & \frac{2}{8} & 0 \\ \frac{3}{8} & 0 & 0 \end{pmatrix} + \frac{1}{6}\begin{pmatrix} 0 & 0 & 0 \\ \frac{3}{8} & \frac{2}{8} & 0 \\ 0 & \frac{3}{8} & 0 \end{pmatrix} + \frac{1}{6}\begin{pmatrix} 0 & 0 & 0 \\ 0 & \frac{2}{8} & 0 \\ \frac{3}{8} & \frac{3}{8} & 0 \end{pmatrix}. \quad (30.7)$$

Equation 30.7 has a geometric interpretation. Consider any interior vertex $v$ in the triangular mesh generated by the subdivision algorithm for the three-direction piecewise linear box spline. Each of the six matrices on the right-hand side of Equation 30.7 represents a stencil for one of the six triangles in the mesh with $v$ as a vertex. To reposition $v$ for the three-direction quartic box spline, take the weighted centroid of each of these triangles with $v$ assigned the weight $2/8$ and the other two vertices assigned the weights $3/8$. Then take the centroid of these weighted centroids (see Figure 30.22c). Note that unlike centroid averaging for bicubic B-splines, the triangle centroids are weighted centroids, so we cannot reuse the same triangle centroids to reposition different vertices; rather we must compute

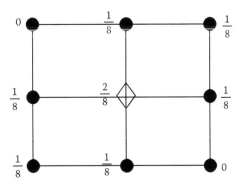

**FIGURE 30.21:** A stencil for generating the control points for one level of subdivision for the three-direction quartic box spline from the control points for one level of subdivision of the three-direction piecewise linear box spline. The point (clear diamond) at the center is repositioned by multiplying the points (black disks and clear diamond) by adjacent fractions and adding the results.

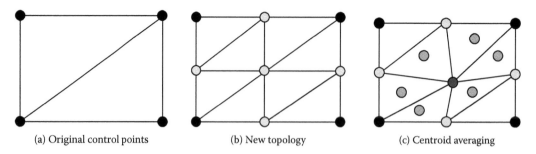

| (a) Original control points | (b) New topology | (c) Centroid averaging |

**FIGURE 30.22: (See color insert following page 354.)**   One level of subdivision for the three-direction quartic box spline. In (a), we illustrate two faces of the original mesh. In the topological stage (b), new control points (yellow) are inserted at the midpoints of the edges, and new edges and faces are added to the mesh by connecting vertices along adjacent edges. Thus each edge is split into two edges and each face is split into four faces. In the geometric stage (c), each control point—yellow or black—is repositioned to the centroid (blue) of the weighted centroids (green) of the faces adjacent to the control point, but the topology of the mesh is not changed. Note that since the green centroids are weighted centroids, we cannot reuse the same centroids to reposition different control points; rather we must recompute these centroids for each vertex. Compare to Figure 30.12 for bicubic B-splines.

different weighted centroids for each triangle in order to reposition different vertices. Despite this anomaly, the advantage of this somewhat convoluted approach is that we can easily extend this technique to arbitrary triangular meshes, even to triangular meshes with arbitrary topology.

### 30.4.1.2   Arbitrary Triangular Meshes

As with three-direction quartic box splines, there are two main phases to centroid averaging for arbitrary triangular meshes: topology and geometry.

To refine the topology of the mesh, we proceed exactly as we did for three-direction quartic box splines: we keep the original vertices and we insert new *edge vertices* at the midpoints of the edges. Thus each edge is split into two edges. We also introduce new edges to connect the midpoints of adjacent edges and new faces surrounded by these new edges, splitting each of the original faces into four faces (see Figure 30.23b).

To alter the geometry of the mesh, we reposition each vertex of the mesh to the centroid of the weighted centroids of the adjacent faces. The only difference between subdivision for three-direction quartic box splines and subdivision for arbitrary triangular meshes is that for arbitrary triangular meshes there need not be six faces adjacent to each interior vertex. For example, in the octahedron there are only four faces adjacent to each of the vertices. An interior vertex of a triangular mesh where the number of adjacent faces, edges, or vertices is not equal to six is called an *extraordinary vertex*. Again, the number of faces, edges, or vertices adjacent to a vertex is called the *valence* of the vertex. We can reposition each vertex $Q$ in an arbitrary triangular mesh by the following rule: *Take the weighted centroid of each of the triangles containing $Q$ with $Q$ assigned the weight $2/8$ and the other two vertices assigned the weights $3/8$; then take the centroid of these weighted centroids.* Thus each vertex $Q$ of the mesh is repositioned by the formula

$$Q \rightarrow \frac{1}{n} \sum_{k=1}^{n} C_k,$$

where
   $C_1, \ldots, C_n$ are the weighted centroids of the faces adjacent to $Q$
   $n$ is the valence of $Q$.

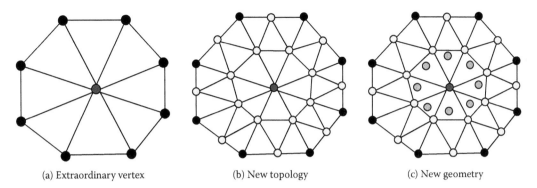

(a) Extraordinary vertex   (b) New topology   (c) New geometry

**FIGURE 30.23: (See color insert following page 354.)**   One level of centroid averaging. In (a), we illustrate eight faces of a triangular mesh surrounding an extraordinary vertex (blue). In the topological stage (b), new vertices (yellow) are inserted at the midpoints of the edges of the triangular mesh, and new edges and faces are added to the mesh by joining the midpoints of adjacent edges. In the geometric stage (c), each vertex—blue, yellow, or black—is repositioned to the centroid of the weighted centroids (green) of the faces adjacent to the vertex, but the topology of the mesh is not changed. Notice that all the new vertices (yellow) are ordinary vertices with valence six. Also, the extraordinary vertex (blue) at the center with valence eight remains an extraordinary vertex with valence eight and is repositioned to the centroid of eight adjacent weighted centroids (green). Since the green centroids are weighted centroids, we cannot reuse the same centroids to reposition different vertices; rather we must recompute these green centroids for each vertex. Compare to Figure 30.13 for quadrilateral meshes.

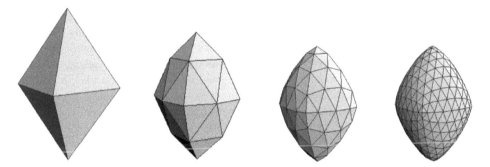

**FIGURE 30.24: (See color insert following page 354.)**   Three levels of centroid averaging applied to an octahedron (left). Notice the extraordinary vertices of valence four at the corners of the octahedron.

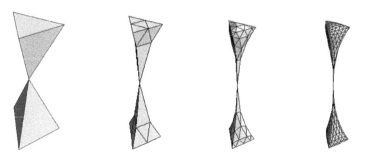

**FIGURE 30.25: (See color insert following page 354.)**   Centroid averaging applied to two tetrahedra meeting at a vertex. The original triangular mesh is on the left; the first three levels of subdivision are illustrated successively on the right.

After several levels of subdivision, most of the interior vertices of a triangular mesh are ordinary vertices and the extraordinary vertices become more and more isolated because during the topological phase, each face is subdivided into four faces by the edges joining the midpoints of the surrounding edges. Thus the new edge vertices and the new face vertices all have valence six (see Figure 30.23b), so all the new vertices introduced by subdivision are ordinary vertices. Also, the valence of each of the original vertices is unchanged, since the number of faces surrounding a vertex does not change during subdivision. These results are analogous to similar results for quadrilateral meshes (see Section 30.3.1.2). Hence, during subdivision, ordinary vertices remain ordinary vertices and extraordinary vertices remain extraordinary vertices with the same valence (see Figure 30.23b). Since all the new vertices have valence six, locally the new triangular mesh looks exactly like a triangular mesh for a three-direction quartic box spline surface; only vertices from the original mesh that did not have valence six do not have six adjacent faces. Therefore, just like three-direction quartic box spline surfaces, the surfaces generated in the limit by iterating centroid averaging for triangular meshes have two continuous derivatives everywhere; the only exceptions are at the limits of extraordinary vertices. At the limits of extraordinary vertices, these surfaces have only one continuous derivative, except at the limits of extraordinary vertices of valence three where these limit surfaces are continuous but not necessarily differentiable (see Warren and Weimer, Chapter 8).

## 30.4.2   Stencils for Triangular Meshes

For bicubic B-spline surfaces and more generally for quadrilateral meshes of arbitrary topology, we have seen that stencils provide an alternative to centroid averaging for generating smooth surfaces via subdivision. Centroid averaging separates the computation of new vertices during subdivision into two phases: topology and geometry. After the insertion of new vertices in the topological phase, all the vertices of the mesh are repositioned in the geometric phase by centroid averaging. But centroid averaging is a bit cumbersome for three-direction quartic box splines and more generally for triangular meshes of arbitrary topology because the centroids that must be averaged are weighted centroids. Thus we need to compute a different weighted centroid for each triangle in order to reposition each vertex. In contrast, stencils allow us to compute the positions of all the new vertices in terms of the positions of the original vertices without resorting to a different stencil for each vertex. For regular triangular meshes, all we need are a single edge stencil and a single vertex stencil. For triangular meshes of arbitrary topology, we shall also require additional stencils for extraordinary vertices. Thus, for three-direction quartic box splines and more generally for triangular meshes of arbitrary topology, stencils may be easier to implement than centroid averaging.

### 30.4.2.1   Stencils for Three-Direction Quartic Box Splines

For three-direction quartic box splines, Figure 30.7 provides essentially two distinct explicit formulas for computing the new control points from the original control points after one level of subdivision:

$$Q_{\mathrm{E}} = \frac{P_{00} + 3P_{01} + 3P_{11} + P_{12}}{8},$$

$$Q_{\mathrm{E}} = \frac{P_{00} + 3P_{10} + 3P_{11} + P_{21}}{8}, \tag{30.8}$$

$$Q_{\mathrm{E}} = \frac{3P_{10} + P_{11} + P_{20} + 3P_{21}}{8},$$

$$Q_{\mathrm{V}} = \frac{P_{00} + P_{01} + P_{10} + 10P_{11} + P_{12} + P_{21} + P_{22}}{16}. \tag{30.9}$$

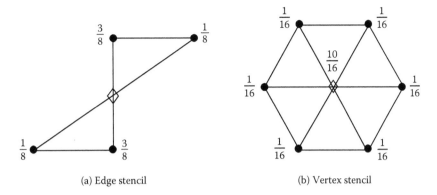

(a) Edge stencil         (b) Vertex stencil

**FIGURE 30.26:** Stencils for subdivision of three-direction quartic box splines: (a) edge stencil and (b) vertex stencil. The position of the new control point (clear diamond) is computed by multiplying the original control points (black discs) by the associated fractions and summing the results. There are three edge stencils, but they are all essentially the same, so only the vertical edge stencil is shown here.

Equation 30.8 represents the three edge stencils: one for horizontal, one for vertical, and one for diagonal edges. Notice that all three formulas are essentially the same. Equation 30.9 represents the vertex stencil. We illustrate these two stencils schematically in Figure 30.26.

### 30.4.2.2 Stencils for Extraordinary Vertices

To extend stencils from three-direction quartic box splines to subdivision surfaces built from arbitrary triangular meshes, we need not alter the edge stencil; all we need to do is to generalize the vertex stencil to extraordinary vertices. Similar to vertex stencils at extraordinary vertices of quadrilateral meshes, this vertex stencil should be affine invariant (the fractions should sum to one), symmetric with respect to surrounding vertices, and reduce to the stencil for three-direction quartic box splines for vertices with valence six. Also, we want the stencil to generate smooth surfaces—that is, surfaces with at least one continuous derivative—at the limits of the extraordinary vertices. Two such stencils are provided in Figure 30.27. Both stencils satisfy all of our criteria, but only the Loop stencil is guaranteed to generate surfaces with bounded curvatures.

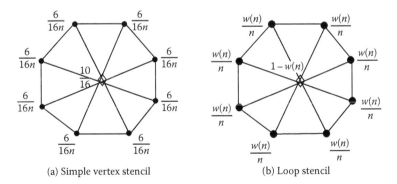

(a) Simple vertex stencil        (b) Loop stencil

**FIGURE 30.27:** Two vertex stencils for extraordinary vertices with valence $n$: (a) a simple vertex stencil and (b) the Loop stencil. For the Loop stencil, $w(n) = \frac{5}{8} - \left(\frac{3}{8} + \frac{1}{4}\cos\left(\frac{2\pi}{n}\right)\right)^2$. Notice that both stencils reduce to the vertex stencil for the three-direction quartic box spline at ordinary vertices—that is, vertices where the valence $n = 6$ (see Figure 30.26b).

## 30.5   Summary

Subdivision is an important tool in Geometric Modeling for several reasons:

1. Subdivision algorithms allow us to replace complicated mathematical formulas with simple computational procedures.

2. Subdivision algorithms are easy to understand and simple to implement.

3. Subdivision algorithms can generate a large class of smooth functions.

4. Subdivision algorithms on polyhedral meshes can produce smooth shapes with arbitrary topology.

Thus subdivision provides a simple approach to generating a wide variety of smooth curves and surfaces.

This chapter presents three distinct paradigms for generating subdivision surfaces: split and average (box splines), centroid averaging (meshes of arbitrary topology), and stencils (vertex, edge, and face stencils as well as special stencils for extraordinary vertices). Box spline surfaces are generalizations of uniform tensor product B-spline surfaces, constructed by a classical split and average approach to subdivision starting from a rectangular array of control points. These box spline surfaces are important because they provide a paradigm for constructing subdivision algorithms for triangular meshes of arbitrary topology. Indeed subdivision surfaces built by centroid averaging or by stencils starting from quadrilateral or triangular meshes of arbitrary topology are generalizations of standard subdivision procedures for bicubic tensor product B-splines and three-direction quartic box splines. One of the big advantages of subdivision algorithms for polyhedral meshes is that unlike tensor product B-spline or box spline schemes, subdivision procedures for polyhedral meshes can generate smooth surfaces with arbitrary topology.

An important leitmotif of this chapter is that subdivision procedures can be partitioned into two separate phases: refining the topology and altering the geometry—that is, changing connectivity and modifying shape. For box splines, we refine the topology by splitting each control point into four control points, leaving the position of each control point unchanged. For polyhedral meshes, we refine the topology by inserting new vertices and edges, and splitting each face into four faces, leaving the positions of the original vertices unchanged. In the geometric phase for box splines, we alter the positions of the control points by averaging these points along prespecified directions in the plane of their integer indices. In the geometric phase for polyhedral meshes, we change the positions of the vertices by taking the centroid of the centroids of adjacent faces. Stencils allow us to combine refining the topology and modifying the geometry of a polyhedral mesh into a single phase.

Next we summarize for easy access and comparison our three approaches to subdivision algorithms—split and average, centroid averaging, and stencils—for bicubic tensor product B-splines and three-direction quartic box splines, surfaces defined over regular meshes whose subdivision algorithms serve as models for subdivision procedures on quadrilateral and triangular meshes of arbitrary topology. We shall also review centroid averaging for quadrilateral and triangular meshes. Finally, we recall the stencils for extraordinary vertices in quadrilateral and triangular meshes.

### 30.5.1   Bicubic Tensor Product B-Splines and Three-Direction Quartic Box Splines

For bicubic tensor product B-splines and three-direction quartic box splines, there are three approaches to subdivision algorithms: split and average, centroid averaging, and stencils.

$$\vdots$$

$$P_{01} \quad P_{01} \quad P_{11} \quad P_{11} \quad \ldots$$

$$P_{01} \quad P_{01} \quad P_{11} \quad P_{11} \quad \ldots$$

$$P_{00} \quad P_{00} \quad P_{10} \quad P_{10} \quad \ldots$$

$$P_{00} \quad P_{00} \quad P_{10} \quad P_{10} \quad \ldots$$

**FIGURE 30.28:** Splitting each control point into four points to initiate subdivision for uniform B-spline and box spline surfaces.

### 30.5.1.1 Split and Average

*Split*

$$P_{2i,2j}^{0,0} = P_{2i+1,2j}^{0,0} = P_{2i,2j+1}^{0,0} = P_{2i+1,2j+1}^{0,0} = P_{i,j} \quad \text{(see Figure 30.28)}.$$

*Average—Bicubic B-Splines*

$$P_{i,j}^{k,0} = \frac{P_{i,j}^{k-1,0} + P_{i+1,j}^{k-1,0}}{2}, \quad k = 1, 2, 3 \quad \text{(horizontal direction)},$$

$$P_{i,j}^{3,k} = \frac{P_{i,j}^{3,k-1} + P_{i,j+1}^{3,k-1}}{2}, \quad k = 1, 2, 3 \quad \text{(vertical direction)}.$$

*Average—Three-Direction Quartic Box Splines*

$$P_{i,j}^{1,0} = \frac{P_{i,j}^{0,0} + P_{i+1,j}^{0,0}}{2} \quad \text{(horizontal direction)},$$

$$P_{i,j}^{1,1} = \frac{P_{i,j}^{1,0} + P_{i,j+1}^{1,0}}{2} \quad \text{(vertical direction)},$$

$$P_{i,j}^{k+1,k+1} = \frac{P_{i,j}^{k,k} + P_{i+1,j+1}^{k,k}}{2}, \quad k = 1, 2 \quad \text{(diagonal direction)}.$$

### 30.5.1.2 Centroid Averaging

*Bicubic B-Splines* (see Figure 30.29)

- *Refine the Topology*

$$P_{2i,2j}^{1} = P_{i,j} \quad \text{(original vertices)},$$

$$P_{2i+1,2j}^{1} = \frac{P_{i,j} + P_{i+1,j}}{2}, \quad P_{2i,2j+1}^{1} = \frac{P_{i,j} + P_{i,j+1}}{2} \quad \text{(new edge vertices)},$$

$$P_{2i+1,2j+1}^{1} = \frac{P_{i,j} + P_{i+1,j} + P_{i,j+1} + P_{i+1,j+1}}{4} \quad \text{(new face vertices)}.$$

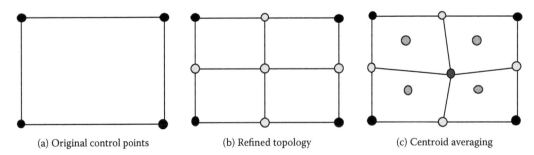

(a) Original control points          (b) Refined topology          (c) Centroid averaging

**FIGURE 30.29: (See color insert following page 354.)** Centroid averaging for bicubic B-splines. First (b) refine the topology by placing new edge vertices (yellow) at the center of each edge and new face vertices (yellow) at the centroid of each face; then connect these vertices with edges to split each face into four faces. The new control point (blue) in (c) is repositioned to the centroid of the centroids (green) of the adjacent faces.

- *Modify the Geometry*

$$P_{i,j}^k = \frac{P_{i,j}^{k-1} + P_{i+1,j}^{k-1} + P_{i,j+1}^{k-1} + P_{i+1,j+1}^{k-1}}{4}, \quad k = 2,3 \quad \text{(centroid averaging)}.$$

*Three-Direction Quartic Box Splines* (see Figure 30.30)

- *Refine the Topology (Insert New Edge Vertices)*

$$P_{2i,2j}^1 = P_{i,j},$$

$$P_{2i+1,2j}^1 = \frac{P_{i,j} + P_{i+1,j}}{2}, \quad P_{2i,2j+1}^1 = \frac{P_{i,j} + P_{i,j+1}}{2}, \quad P_{2i+1,2j+1}^1 = \frac{P_{i,j} + P_{i+1,j+1}}{2}.$$

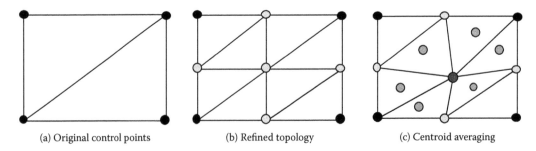

(a) Original control points          (b) Refined topology          (c) Centroid averaging

**FIGURE 30.30: (See color insert following page 354.)** Centroid averaging for three-direction quartic box splines. First (b) refine the topology by inserting a new edge vertex (yellow) at the center of each edge; then connect vertices on adjacent edges to split each face into four faces. The new control point (blue) in (c) is repositioned to the centroid of the weighted centroids (green) of the adjacent faces.

- *Modify the Geometry (Weighted Centroid Averaging)*

$$P^2_{i,j,k} = \frac{2P^1_{i,j} + 3Q_k + 3Q_{k+1}}{8}, \text{ where } Q_1, \ldots, Q_6 \text{ are the six vertices adjacent to } P_{ij}$$

$$P^3_{i,j} = \frac{P^2_{i,j,1} + P^2_{i,j,2} + P^2_{i,j,3} + P^2_{i,j,4} + P^2_{i,j,5} + P^2_{i,j,6}}{6}.$$

### 30.5.1.3 Stencils

Stencils represent explicit expressions in terms of the original control points for the new control points after one level of subdivision. For bicubic B-splines, these formulas come in three basic forms: expressions for face points, expressions for edge points, and expressions for vertex points. For three-direction quartic box splines, these formulas come in only two basic forms: expressions for edge points and expressions for vertex points.

*Bicubic B-Splines* (see Figure 30.31)

$$Q_{\text{Face}} = \frac{P_{00} + P_{01} + P_{10} + P_{11}}{4},$$

$$Q_{\text{Edge}} = \frac{P_{00} + P_{10} + 6P_{01} + 6P_{11} + P_{02} + P_{12}}{16} \quad \text{or} \quad \frac{P_{00} + P_{01} + 6P_{10} + 6P_{11} + P_{20} + P_{21}}{16},$$

$$Q_{\text{Vertex}} = \frac{P_{00} + P_{20} + 6P_{01} + 6P_{10} + 36P_{11} + 6P_{21} + 6P_{12} + P_{02} + P_{22}}{64}.$$

*Three-Direction Quartic Box Splines* (see Figure 30.32)

$$Q_{\text{Edge}} = \frac{P_{00} + 3P_{01} + 3P_{11} + P_{12}}{8} \quad \text{or} \quad \frac{P_{00} + 3P_{10} + 3P_{11} + P_{21}}{8} \quad \text{or} \quad \frac{3P_{10} + P_{11} + P_{20} + 3P_{21}}{8},$$

$$Q_{\text{Vertex}} = \frac{P_{00} + P_{01} + P_{10} + 10P_{11} + P_{12} + P_{21} + P_{22}}{16}.$$

## 30.5.2 Centroid Averaging for Meshes of Arbitrary Topology

Quadrilateral and triangular meshes are shown in Figures 30.33 and 30.34.

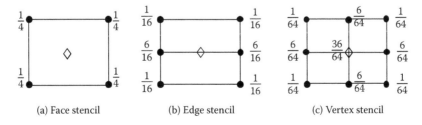

(a) Face stencil          (b) Edge stencil          (c) Vertex stencil

**FIGURE 30.31:** Stencils for bicubic B-splines: (a) face stencil, (b) edge stencil, and (c) vertex stencil.

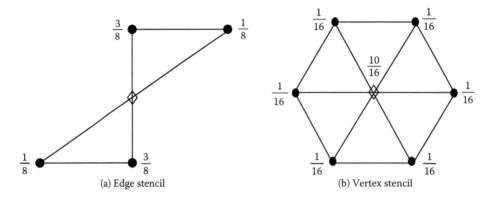

(a) Edge stencil             (b) Vertex stencil

**FIGURE 30.32:** Stencils for three-direction quartic box splines: (a) edge stencil and (b) vertex stencil.

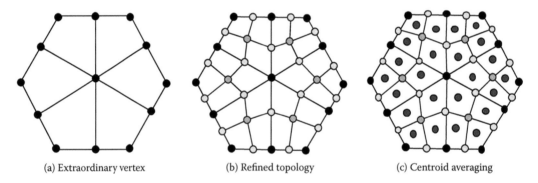

(a) Extraordinary vertex        (b) Refined topology        (c) Centroid averaging

**FIGURE 30.33: (See color insert following page 354.)** Quadrilateral meshes. Each vertex is repositioned to (c) the centroid of the centroids (blue) of the adjacent faces after (b) refining the topology by splitting each quadrilateral face into four quadrilateral faces.

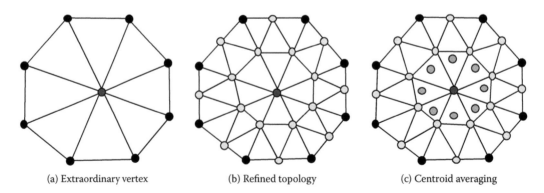

(a) Extraordinary vertex        (b) Refined topology        (c) Centroid averaging

**FIGURE 30.34: (See color insert following page 354.)** Triangular meshes. Each vertex is repositioned to (c) the centroid of the weighted centroids (green) of the adjacent faces after (b) refining the topology by splitting each triangular face into four triangular faces.

### 30.5.3  Stencils for Extraordinary Vertices

For quadrilateral meshes of arbitrary topology, the edge and face stencils are the same as the edge and face stencils for tensor product bicubic B-splines (see Figure 30.31a,b). Similarly, for triangular meshes, the edge stencils are the same as the edge stencils for three-direction quartic box splines (see Figure 30.32a). Also, for quadrilateral meshes, the vertex stencil at regular vertices (vertices with valence four) is the same as the vertex stencil for tensor product bicubic B-splines (see Figure 30.31c), and for triangular meshes, the vertex stencil at regular vertices (vertices with valence six) is the same as the vertex stencil for three-direction quartic box splines (see Figure 30.32b). Therefore, for meshes of arbitrary topology, we need only specify the vertex stencils at extraordinary vertices. These stencils are illustrated in Figures 30.35 and 30.36.

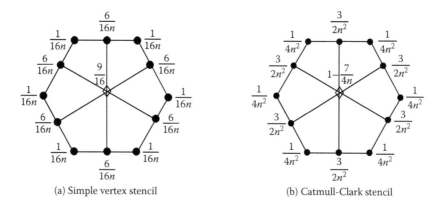

(a) Simple vertex stencil     (b) Catmull-Clark stencil

**FIGURE 30.35:** Quadrilateral meshes. Two vertex stencils for extraordinary vertices with valence $n$: (a) a simple vertex stencil and (b) the Catmull–Clark stencil.

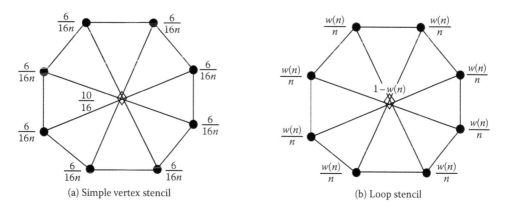

(a) Simple vertex stencil     (b) Loop stencil

**FIGURE 30.36:** Triangular meshes. Two vertex stencils for extraordinary vertices with valence $n$: (a) a simple vertex stencil and (b) the Loop stencil. For the Loop stencil, $w(n) = \frac{5}{8} - \left(\frac{3}{8} + \frac{1}{4} \cos\left(\frac{2\pi}{n}\right)\right)^2$.

## Exercises

30.1. Let $\{P_{ij}\}$ be a rectangular array of control points.

   a. Show that averaging adjacent control points in the $s$-direction and then averaging the resulting adjacent control points in the $t$-direction is equivalent to first averaging adjacent control point in the $t$-direction and then averaging the resulting adjacent control points in the $s$-direction by showing that in both cases the new control points are given by

$$Q_{ij} = \frac{P_{ij} + P_{i+1,j} + P_{i,j+1} + P_{i+1,j+1}}{4}.$$

   b. Generalize the result in part a by showing that averaging $v$-adjacent points and then averaging $w$-adjacent points is equivalent to first averaging $w$-adjacent points and then averaging $v$-adjacent points by showing that in both cases the new control points are given by

$$Q_{ij} = \frac{P_{ij} + P_{i+v_1,j+v_2} + P_{i+w_1,j+w_2} + P_{i+v_1+w_1,j+v_2+w_2}}{4}.$$

30.2. Consider the box spline surface generated by the control points $\{P_{ij}\}$ and the vectors $V = \{(1,0), (1,0), (0,1), (0,1), (1,1)\}$.

   a. What is the degree of this surface?

   b. How smooth is this surface?

   c. Show that the control points for this surface after one level of subdivision are given by the array of points in Figure 30.37.

   d. Using the result in part c, derive the control points for three-direction quartic box splines.

30.3. Consider the box spline surface generated by the control points $\{P_{ij}\}$ and the vectors $V = \{(1,0), (0,1), (1,1), (-1,1)\}$.

   a. What is the degree of this surface?

   b. How smooth is this surface?

   c. Show that the control points for this surface after one level of subdivision are given by the array of points in Figure 30.38.

$$\vdots$$

| $\dfrac{5P_{01} + P_{11} + P_{02} + P_{12}}{8}$ | $\dfrac{P_{01} + 2P_{11} + P_{12}}{4}$ | $\dfrac{5P_{11} + P_{21} + P_{12} + P_{22}}{8}$ | $\cdots$ |
| $\dfrac{P_{00} + 2P_{01} + P_{11}}{4}$ | $\dfrac{P_{00} + P_{10} + P_{01} + 5P_{11}}{8}$ | $\dfrac{P_{10} + 2P_{11} + P_{21}}{4}$ | $\cdots$ |
| $\dfrac{5P_{00} + P_{10} + P_{01} + P_{11}}{8}$ | $\dfrac{P_{00} + 2P_{10} + P_{20}}{4}$ | $\dfrac{5P_{10} + P_{20} + P_{11} + P_{21}}{8}$ | $\cdots$ |

**FIGURE 30.37:** The control points after one level of subdivision for the three-direction box spline in Exercise 30.2.

$$\vdots$$

$$\frac{P_{02} + 2P_{01} + P_{11}}{4} \qquad \frac{P_{01} + 2P_{11} + P_{12}}{4} \qquad \frac{P_{12} + 2P_{11} + P_{21}}{4} \qquad \cdots$$

$$\frac{P_{00} + 2P_{01} + P_{11}}{4} \qquad \frac{P_{01} + 2P_{11} + P_{10}}{4} \qquad \frac{P_{10} + 2P_{11} + P_{21}}{4} \qquad \cdots$$

$$\frac{P_{01} + 2P_{00} + P_{10}}{4} \qquad \frac{P_{00} + 2P_{10} + P_{11}}{4} \qquad \frac{P_{11} + 2P_{10} + P_{20}}{4} \qquad \cdots$$

**FIGURE 30.38:** The control points after one level of subdivision for the four-direction box spline in Exercise 30.3.

**Definition**: A stencil $S$ is called a *universal stencil* for a subdivision algorithm if all the control points—face points, edge points, and vertex points—after one level of subdivision can be generated by sliding the stencil $S$ over an array of points $Q$.

30.4. Show that the stencil in Figure 30.21 is a universal stencil for the three-direction quartic box spline by verifying that sliding this stencil on the array of control points in Figure 30.6 for one level of subdivision for the three-direction linear box spline generates the control points in Figure 30.7 for one level of subdivision for the three-direction quartic box spline.

30.5. Consider the stencil in Figure 30.39.

a. Show that this stencil is a universal stencil for the uniform tensor product bicubic B-splines by verifying that sliding this stencil on the array of control points in Figure 30.4 for one level of subdivision for bilinear B-splines generates the control points in Figure 30.18b for one level of subdivision for bicubic B-splines.

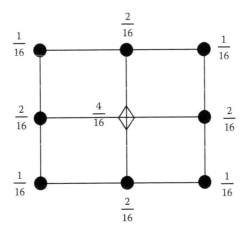

**FIGURE 30.39:** A universal stencil for generating the control points for one level of subdivision for bicubic B-splines from the control points for one level of subdivision for bilinear B-splines. The point (clear diamond) at the center is repositioned by multiplying the points (black disks and clear diamond) by the adjacent fractions and adding the results.

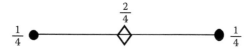

**FIGURE 30.40:** A universal stencil for generating the control points for one level of subdivision for cubic B-spline curves from the control points for one level of subdivision for linear B-spline curves. The point (clear diamond) at the center is repositioned by multiplying the points (black disks and clear diamond) by the adjacent fractions and adding the results.

    b. Verify that if we represent this stencil as a $3 \times 3$ matrix, then we have the following identity:

$$
\begin{pmatrix} \frac{1}{16} & \frac{2}{16} & \frac{1}{16} \\ \frac{2}{16} & \frac{4}{16} & \frac{2}{16} \\ \frac{1}{16} & \frac{2}{16} & \frac{1}{16} \end{pmatrix} = \frac{1}{4}\begin{pmatrix} \frac{1}{4} & \frac{1}{4} & 0 \\ \frac{1}{4} & \frac{1}{4} & 0 \\ 0 & 0 & 0 \end{pmatrix} + \frac{1}{4}\begin{pmatrix} 0 & \frac{1}{4} & \frac{1}{4} \\ 0 & \frac{1}{4} & \frac{1}{4} \\ 0 & 0 & 0 \end{pmatrix} + \frac{1}{4}\begin{pmatrix} 0 & 0 & 0 \\ \frac{1}{4} & \frac{1}{4} & 0 \\ \frac{1}{4} & \frac{1}{4} & 0 \end{pmatrix} + \frac{1}{4}\begin{pmatrix} 0 & 0 & 0 \\ 0 & \frac{1}{4} & \frac{1}{4} \\ 0 & \frac{1}{4} & \frac{1}{4} \end{pmatrix}
$$

    c. Conclude from part b that after bilinear averaging, each control point in the Lane–Riesenfeld algorithm for a bicubic B-spline is repositioned to the centroid of the centroids of the faces adjacent to the control point.

30.6. Consider the stencil in Figure 30.40.

    a. Show that this stencil is a universal stencil for uniform cubic B-spline curves by verifying that sliding this stencil along the array of control points in Figure 30.1 for one level of subdivision for the linear B-splines generates the control points in Figure 30.16 for one level of subdivision for the cubic B-splines.

    b. Show that this stencil can be decomposed into the average of two simpler stencils.

    c. Conclude from part b that after linear averaging, each control point in the Lane–Riesenfeld algorithm for a uniform cubic B-spline curve is repositioned to the midpoint of the midpoints of the edges adjacent to the control point.

30.7. Show that the stencil represented by the row vector $(\frac{1}{8}, \frac{3}{8}, \frac{3}{8}, \frac{1}{8})$ is a universal stencil for uniform cubic B-spline curves by verifying that sliding this stencil along the array where each of the original control points are doubled generates the control points for one level of subdivision for uniform cubic B-spline curves. Generalize this result to uniform B-spline curves of arbitrary degree.

30.8. *Four Point Scheme.* Let $\{P_i\}$ be a collection of control points, and let $\{P_i^3\}$ be the output of one level of the Lane–Riesenfeld algorithm for cubic B-spline curves. Define one additional level of subdivision for the four point scheme by setting

$$
P_i^4 = \frac{-P_{i-1}^3 + 4P_i^3 - P_{i+1}^3}{2}.
$$

    a. Show that

$$
P_{2i}^4 = P_i \quad \text{(interpolation)},
$$

$$
P_{2i+1}^4 = \frac{-P_{i-1}^3 + 9P_i^3 + 9P_{i+1}^3 - P_{i+2}^3}{16} \quad \text{(four point rule)}.
$$

    b. Describe the edge stencil and vertex stencil for the four point scheme.

c. Find two distinct universal stencils for the four point scheme.

d. Using the result of part a, conclude that the control polygons generated by iterating the subdivision procedure for the four point scheme converge in the limit to curves that interpolates the original control points.

e. Implement the four point scheme. Observe that the control polygons generated by this subdivision algorithm converge in the limit to smooth curves.

30.9. Verify that the Loop stencil in Figure 30.27b reduces to the vertex stencil for the three-direction quartic box spline in Figure 30.26b when the valence $n = 6$.

---

## Programming Projects

30.1. *Box Splines*

Implement subdivision for box spline surfaces in your favorite programming language using your favorite API.

a. Render uniform bicubic B-splines and four direction quartic box splines, using your favorite shading algorithm and hidden surface procedure.

b. Build some interesting freeform shapes using box spline surfaces.

30.2. *Quadrilateral Meshes*

Implement subdivision algorithms for quadrilateral meshes in your favorite programming language using your favorite API.

a. Include both centroid averaging and Catmull–Clark stencils at extraordinary vertices.

b. Render these subdivision surfaces, using your favorite shading algorithm and hidden surface procedure.

c. Build some interesting freeform shapes of arbitrary topology, using subdivision algorithms for quadrilateral meshes of arbitrary topology.

30.3. *Triangular Meshes*

Implement subdivision algorithms for triangular meshes in your favorite programming language using your favorite API.

a. Include both centroid averaging and Loop stencils at extraordinary vertices.

b. Render these subdivision surfaces, using your favorite shading algorithm and hidden surface procedure.

c. Build some interesting freeform shapes of arbitrary topology, using subdivision algorithms for triangular meshes of arbitrary topology.

30.4. *Split and Average—Geometric Mean*

Let $P_k = (x_k, y_k)$ be a collection of points in the $xy$-plane, where $x_k, y_k > 0$, and consider the following variation of the Lane–Riesenfeld algorithm:

Replace midpoint averaging for the $y$-coordinate by the geometric average—that is, replace the arithmetic mean $\frac{y_k+y_{k+1}}{2}$ by the geometric mean $\sqrt{y_k y_{k+1}}$—but use the arithmetic mean $\frac{x_k+x_{k+1}}{2}$ for averaging the $x$-coordinates. Thus the new averaging rule is

$$av(P_k, P_{k+1}) = \left( \frac{x_k + x_{k+1}}{2}, \sqrt{y_k y_{k+1}} \right).$$

a. Implement this new subdivision algorithm.

   i. Are the curves you construct always smooth?

   ii. Compare these new curves to the curves generated by the standard Lane–Riesenfeld algorithm for linear, quadratic, and cubic B-splines with the same control points.

b. Now replace midpoint averaging in the $x$-coordinate by the geometric mean, but use the arithmetic mean in the $y$-coordinate.

   i. Are the curves still smooth?

   ii. How are these curves related to the curves generated by the standard Lane–Riesenfeld algorithm for linear, quadratic, and cubic B-splines with the same control points?

   iii. How are these curves related to the curves generated in part a?

# *Further Readings*

**Computer Graphics—General References**

Angel, E., *Interactive Computer Graphics: A Top-Down Approach Using OpenGL*, 4th edn., Addison Wesley, Boston, MA, 2006.

Foley, J., van Dam, A., Feiner, S., and Hughes, J., *Computer Graphic: Principles and Practice in C*, 2nd edn., Addison Wesley, Reading, MA, 1995.

Hearn, D. and Baker, M., *Computer Graphics with OpenGL*, 3rd edn., Prentice Hall, Englewood Cliffs, NJ, 2003.

Hill, F.S., and Kelley, S.M., *Computer Graphics Using OpenGL*, 3rd edn., Prentice Hall, Englewood Cliffs, NJ, 2006.

**Graphics Gems—Tricks and Shortcuts in Computer Graphics**

*Graphics Gems*, Glassner, A. (ed.), Academic Press, New York, 1990.

*Graphics Gems II*, Arvo, J. (ed.), Academic Press, New York, 1991.

*Graphics Gems III*, Kirk, D. (ed.), Academic Press, New York, 1992.

*Graphics Gems IV*, Heckbert, P. (ed.), Academic Press, New York, 1994.

*Graphics Gems V*, Paeth, A. (ed.), Academic Press, New York, 1995.

**Turtle Graphics**

Abelson, H. and diSessa, A.A., *Turtle Geometry: The Computer as a Medium for Exploring Mathematics*, MIT Press, Cambridge, MA, 1986.

Papert, S., *Mindstorms: Children, Computers and Powerful Ideas*, 2nd edn., Basic Books, New York, 1993.

**Fractals**

Barnsley, M.F., *Fractals Everywhere*, 2nd edn., Morgan Kaufmann, Orlando, FL, 2000.

Prusinkiewicz, P. and Lindenmayer, A., *The Algorithmic Beauty of Plants*, Springer-Verlag, New York, 1996.

**Mathematical Methods—Geometric Algebra and Physical Geometry**

Dorst, L., Fontijne, D., and Mann, S., *Geometric Algebra for Computer Science: An Object Oriented Approach to Geometry*, Morgan Kaufmann, San Francisco, CA, 2007.

Kogan, B.Y., *The Application of Mechanics to Geometry*, translated from Russian by D. Sookne and R. Hummel, University of Chicago Press, Chicago, IL, 1974.

Uspenskii, V.A., *Some Applications of Mechanics to Mathematics*, translated from Russian by H. Moss and I. Sneddon, Blaisdell Publishing Company, New York, 1961.

**Radiosity**

Sillion, F. and Puech, C., *Radiosity and Global Illumination*, Morgan Kaufmann, San Francisco, CA, 1994.

### Freeform Curves and Surfaces

Farin, G., *Curves and Surfaces for Computer Aided Geometric Design: A Practical Guide*, 4th edn., Academic Press, San Diego, CA, 1996.

Goldman, R., *Pyramid Algorithms: A Dynamic Programming Approach to Curves and Surfaces for Geometric Modeling*, Morgan Kaufmann, San Francisco, CA, 2002.

Goldman, R. and Lyche, T. (eds.), *Knot Insertion and Deletion Algorithms for B-Spline Curves and Surfaces*, SIAM, Philadelphia, PA, 1993.

Ramshaw, L., *Blossoming: A Connect-the-Dots Approach to Splines*, SRC Research Report 19, Palo Alto, CA, 1987.

Warren, J. and Weimer, H., *Subdivision Methods for Geometric Design: A Constructive Approach*, Morgan Kaufmann, San Francisco, CA, 2002.

## Additional Relevant Papers by the Author and His Collaborators

### Turtle Graphics

Ju, T. and Goldman, R., Hodograph turtles, in *Proceedings of the IASTED Conference on Computer Graphics*, Kauai, HI, August 2004a.

Ju, T., Schaefer., S., and Goldman, R., Recursive turtle programs and iterated affine transformations, *Computers and Graphics* (2004b), 28, 991–1004.

Ju, T., Schaefer, S., and Goldman, R., Turtle geometry in computer graphics and computer-aided design, *Computer-Aided Design* (2004c), 36, 1471–1482.

### Mathematical Methods for Computer Graphics

Goldman, R., Illicit expressions in vector algebra, *Transactions on Graphics* (1985), 4(3), 223–243.

Goldman, R., Triangles, in *Graphics Gems*, A. Glassner (ed.), Academic Press, New York, 1990a, pp. 20–23.

Goldman, R., Intersection of two lines in three space, in *Graphics Gems*, A. Glassner (ed.), Academic Press, New York, 1990b, p. 304.

Goldman, R., Intersection of three planes, in *Graphics Gems*, A. Glassner (ed.), Academic Press, New York, 1990c, p. 305.

Goldman, R., Matrices and transformations, in *Graphics Gems*, A. Glassner (ed.), Academic Press, New York, 1990d, pp. 472–475.

Goldman, R., Area of planar polygons and volume of polyhedra, in *Graphics Gems II*, J. Arvo (ed.), Academic Press, New York, 1991a, pp. 170–171.

Goldman, R., Recovering the data from the transformation matrix, in *Graphics Gems II*, J. Arvo (ed.), Academic Press, New York, 1991b, pp. 324–331.

Goldman, R., Transformations as exponentials, in *Graphics Gems II*, J. Arvo (ed.), Academic Press, New York, 1991c, pp. 332–337.

Goldman, R., More matrices and transformations: Shear and pseudoperspective, in *Graphics Gems II*, J. Arvo (ed.), Academic Press, New York, 1991d, pp. 338–341.

Goldman, R., Cross product in 4-dimensions and beyond, in *Graphics Gems III*, D. Kirk (ed.), Academic Press, New York, 1992a, pp. 84–88.

Goldman, R., Decomposing projective transformations, in *Graphics Gems III*, D. Kirk (ed.), Academic Press, New York, 1992b, pp. 98–107.

Goldman, R., Decomposing linear and affine transformations, in *Graphics Gems III*, D. Kirk (ed.), Academic Press, New York, 1992c, pp. 108–116.

Goldman, R., The ambient space of computer graphics and geometric modeling, *IEEE Computer Graphics and Applications* (2000), 20, 76–84.

Goldman, R., Baseball arithmetic and the laws of pseudoperspective, *IEEE Computer Graphics and Applications* (2001), 21, 70–78.

Goldman, R., On the algebraic and geometric foundations of computer graphics, *Transactions on Graphics* (2002), 21, 1–35.

Goldman, R., Computer graphics in its fifth decade: Ferment at the foundations (Invited paper), in *Proceedings of Pacific Graphics 2003*, Canmore, Canada, October 2003a, pp. 4–21.

Goldman, R., Deriving linear transformations in 3-dimensions, *IEEE Computer Graphics and Applications* (2003b), 23, 66–71.

## Surface and Solid Modeling

Goldman, R., Two approaches to a computer model for quadric surfaces, *IEEE Computer Graphics and Applications* (1983), 3(6), 21–24.

Goldman, R., Quadrics of revolution, *IEEE Computer Graphics and Applications* (1983), 3(3), 68–76.

Goldman, R., The role of surfaces in solid modeling, in *Geometric Modeling: Algorithms and Trends*, G. Farin (ed.), SIAM, Philadelphia, PA, 1987, pp. 69–90.

Goldman, R., Curvature formulas for implicit curves and surfaces, *Special Issue of CAGD on Geometric Modeling and Differential Geometry* (2005), 22, 632–658.

## Freeform Curves and Surfaces

Barry, P., Beatty, J., and Goldman, R., Unimodal properties of B-splines and Bernstein basis functions, *Computer-Aided Design* (1992), 24, 627–636.

Barry P., DeRose, T., and Goldman, R., Pruned Bezier curves, *Proceedings Graphics Interface 90*, June 1990, pp. 229–238.

Barry, P. and Goldman, R., A recursive evaluation algorithm for a class of Catmull-Rom splines, in *Proceedings Siggraph '88*, ACM, New York, August 1988, pp. 199–204.

Barry, P. and Goldman, R., A recursive proof of Boehm's knot insertion technique, *Computer-Aided Design* (1988), 20, 4, 181–182.

Barry, P. and Goldman, R., Factored knot insertion, *Knot Insertion and Deletion Algorithms for B-spline Curves and Surfaces*, R. Goldman and T. Lyche (eds.), SIAM, 1993, pp. 65–88.

Barry, P. and Goldman, R., Knot insertion algorithms, *Knot Insertion and Deletion Algorithms for B-Spline Curves and Surfaces*, R. Goldman and T. Lyche (eds.), SIAM, 1993, pp. 89–133.

Barry, P. and Goldman, R., Knot insertion using forward differencing, in *Graphics Gems IV*, P. Heckbert (ed.), Academic Press, New York, 1994, pp. 251–255.

DeRose, T., Lounsbery, J., and Goldman, R., A tutorial introduction to blossoming, in *Geometric Modeling*, H. Hagen and D. Roller (eds.), Springer-Verlag, Berlin, 1991, pp. 267–286.

Goldman, R., Using degenerate Bezier triangles and tetrahedra to subdivide Bezier curves, *Computer-Aided Design* (1982), 14, 6, 307–311.

Goldman, R., Subdivision algorithms for Bezier triangles, *Computer-Aided Design* (1983), 15, 3, 159–166.

Goldman, R., An urnful of blending functions, *IEEE Computer Graphics and Applications*, (1983), 3, 7, 49–54. (First delivered orally at the *NASA Symposium on Computer Aided Geometry and Modeling*, Langley, Virginia, April 1983).

Goldman, R., Urn models and B-splines, *Constructive Approximation* (1988), 4, 265–288.

Goldman, R., Some properties of Bezier curves, in *Graphics Gems*, A. Glassner (ed.), Academic Press, New York, 1990a, pp. 587–593.

Goldman, R., Integration of Bernstein basis functions, in *Graphics Gems*, A. Glassner (ed.), Academic Press, New York, 1990b, pp. 604–606.

Goldman, R., Blossoming and knot insertion algorithms for B-spline curves, *Computer Aided Geometric Design* (1990c), 7, 69–81.

Goldman, R., Identities for the univariate and bivariate Bernstein basis functions, in *Graphics Gems V*, A. Paeth (ed.), Academic Press, New York, 1995a, pp. 149–162.

Goldman, R., Identities for the B-spline basis functions, in *Graphics Gems V*, A. Paeth (ed.), Academic Press, New York, 1995b, pp. 163–167.

Goldman, R., Polar forms in geometric modeling and algebraic geometry, *Topics in Algebraic Geometry and Geometric Modeling*, R. Goldman and R. Krasauskas (eds.), Vol. 334, AMS Contemporary Mathematics 2003, pp. 3–24.

Goldman, R., The fractal nature of Bezier curves, in *Proceedings of Geometric Modeling and Processing: Theory and Applications, GMP2004*, Beijing, China, 2004, pp. 3–11.

Heath, D. and Goldman, R., Linear subdivision is strictly a polynomial phenomenon, *Computer Aided Geometric Design* (1984), 1, 3, 269–278.

Krasauskas, R. and Goldman, R., Toric Bezier patches with depth, *Topics in Algebraic Geometry and Geometric Modeling*, R. Goldman and R. Krasauskas (eds.), Vol. 334, AMS Contemporary Mathematics, 2003, pp. 65–91.

Mann, S. and Goldman, R., Counting pruned Bezier curves, *Mathematical Methods for Curves and Surfaces: Tromso 2004*, M. Daehlen, K. Morken, and L. Schumaker, Nashboro Press, 2005, pp. 169–178.

Schaefer, S., Levin, D., and Goldman, R., Subdivision schemes and attractors, in *Proceedings of Symposium on Geometric Processing (SGP) 2005*, M. Desbrun and H. Pottmann (eds.), Vienna, Austria, July 2005, pp. 171–180.

Schaefer, S. and Goldman, R., Non-uniform subdivision for B-splines of arbitrary degree, *Computer Aided Geometric Design* (2009), 26, 75–81.

Schaefer, S., Vouga, E., and R. Goldman, R., Nonlinear subdivision through nonlinear averaging, *Computer Aided Geometric Design* (2008), 25, 162–180.

Vouga, E. and Goldman, R., Two blossoming proofs of the Lane-Riesenfeld algorithm, *Computing* (2007), 79, 153–162.

Warren, J. and Goldman, R., An extension of Chaiken's algorithm to B-spline curves with knots in geometric progression, *CVGIP: Graphical Models and Image Processing* (1993), 55, 58–62.

# *Index*

## A

Active edge list, 343–345
Affine combination, 120, 123, 126, 163,
      179, 212, 417, 423, 459, 478,
      485, 491, 496
Affine coordinates, 45–47, 50–52, 58, 62–63, 66,
      180–181, 191, 205, 271–272, 485, 490
Affine geometry, 41–42, 61–69, 208
Affine graphics, 36, 62–66, 118, 163
Affine invariance, 385, 396, 448, 453, 490
Affine matrix, 45–46, 48, 51, 55, 385
Affine space, 119–120, 123–124, 151, 208–212,
      216, 218–219, 228, 238, 244, 286,
      423–424, 451, 453, 485, 488–492, 496
Affine transformations, 5, 36, 39–59, 62–63,
      65–66, 73, 90, 94–96, 101, 103, 113,
      117–118, 124, 163–164, 171, 174,
      176–177, 179–184, 188–190, 193,
      198–199, 201, 205, 209, 220, 233, 244,
      268, 270, 319, 385, 486, 490, 492
   image of one point and two linearly
      independent vectors, 47–48, 50–51
   image of three non-collinear points, 47,
      50–51, 54, 66
Algebra, 43–44, 120, 131–137, 203, 205, 209,
      211–212, 223–224, 297, 424, 440
Ambient light, 250, 252–254, 307, 355, 371
Ambient reflection coefficient, 250–251, 253
Animation, 96, 203, 234–235, 239, 246
Anti-commutative, 122, 127, 147, 228, 230
Approximation, 379–380, 409
Archimedes' Law of the Lever, 213–214
Area, 82, 120, 122–123, 140, 142, 144–147,
      157, 311, 315, 317–318, 326,
      356–364, 366, 371
   of a parallelogram, 145–146, 157
   of a polygon, 146–147, 157
   of a triangle, 145–146, 157
Attraction, 29, 34–36, 74
Attractor, 36, 74, 76, 477–478, 487, 492–493

## B

Base case, 17, 19, 34–38, 74–76, 81–82, 91,
      93–94, 101, 103–110, 113, 439, 486
Bernstein basis, 384, 388, 396–398, 435,
      489–490
Bernstein representation, 383–384, 388, 392,
      394, 396
Bezier curve, 379–399, 401–415, 417–418, 421,
      423, 426–427, 429–431
Bezier patch, 309, 391–397, 399, 401, 404–405,
      407–409, 413–414, 449–450, 457,
      463
Bezier subdivision, 401–415, 422, 457–458,
      461, 463, 477–478
Bezier surface, 392, 395–396, 404–405,
      407–409, 413, 471
Bezout's theorem, 208–209
Bicubic, 392–393, 450, 505–514, 516,
      518–523, 525–527
Bidegree, 391, 396, 449, 454, 463, 501–502
Binary space partitioning tree (bsp-tree),
      341–342, 349, 351–353
Blending functions, 384
Blossom, 417–435, 438–439, 442–449, 453,
      458–461, 464–469
   affine, 417
   homogeneous, 423–427
Blossoming, 417–435, 437–438, 440, 442,
      451, 455, 458, 464–468, 474
Blossoming axioms, 418–419, 424, 433
Boehm's knot insertion algorithm, 458–464
Boolean operations, 309–313, 315–319,
      322–323
   difference, 309, 322–323
   intersection, 309, 316, 322–323
   union, 309, 322–323
Boundary representation (B-Rep), 309, 313–317
Bounded, 24–25, 27, 90–91, 202, 289–291,
      298, 303–304, 306, 314–317, 320,
      322, 350, 451, 453, 470, 517

Box spline, 500–504, 512–514, 516–525, 527
  three direction linear, 503, 525
  three direction quartic, 504, 512–522
B-REP, *see* Boundary representation
Brianchon's Theorem, 208
B-spline, 309, 431, 437–455, 457–478,
    481–484, 488–491, 495–497,
    500–502, 506, 509–512, 516,
    518, 526
  basis function, 454, 467
  bicubic, 450, 505–514, 516, 518–522
  cubic, 439, 441–446, 450, 452, 458–462,
    464–467, 469, 472, 475, 482, 484,
    495, 505, 509–510
  curve, 437, 439–446, 449–451, 457–464,
    466–470, 472, 475, 477–478,
    481–484, 488–490, 500, 509, 526
  surface, 449–450, 463, 470–471, 510
bsp-tree, *see* Binary space partitioning tree
Bump fractal, 17, 19–20, 22–23, 33, 36–38,
    109–112, 491–493, 497;
    *see also* Fractals

## C

Camera obscura, 308
Canonical position, 294–295, 297, 301,
    305, 313
Cantor set, 77
Catmull–Clark subdivision, 511, 523
Cauchy sequence, 82–83, 91
Cayley numbers, 224
*C*-curve, 19, 29–30, 37–38
Centroid, 29–30, 155, 505–516, 518–522,
    526–527
Centroid averaging, 505–516, 519–522
Chain rule, 269
CIFS, *see* Conformal iterated function
    system
Circle, 7, 10–11, 14–17, 21, 25, 29–30, 48, 64,
    66, 90, 154, 159, 163, 200, 227,
    271, 284–285, 287–288, 290–291,
    295–296, 299–300, 305, 358,
    362–363, 451, 453, 455
  inscribed in a triangle, 159
Clipping, 153, 214
Closed, 58, 90–91, 120, 146, 148, 154, 161,
    267, 283, 289, 303, 313, 383, 496
CODO, *see* COnnect the DOts
Coherence, 258, 341, 346–347, 352

Colors, 246, 249–262, 264, 287, 341–342,
    344–346, 372
  complementary, 249
  primary, 249–250, 252, 254–255
Compact, 90–94, 309, 313, 318–319, 477, 486,
    488, 490, 499
Complete space, 82–84, 89–92
Complex conjugate, 225
Complex numbers, 223–227, 238, 240–241,
    245–246, 415
Computational geometry, 3, 39
Computer graphics, 39, 51–52, 118, 120,
    163–164, 168, 173, 181, 183, 193,
    249, 325, 366, 381, 385–386
Condensation set, 74–79, 93, 104–105, 201
Cone, 214, 268, 279, 281, 284, 287, 290–292,
    299, 303–308, 310, 320, 379
  elliptical, 268, 292
  generalized, 279, 304
  right circular, 281, 284, 290, 292,
    294, 299
Conformal iterated function system (CIFS),
    101–111, 113, 245
Conformal transformation, 39–43, 46–47, 66,
    77, 101–108, 118, 163, 227, 230,
    232–235, 238–239, 270
Congruent triangles, 39
Conjugate, 225, 229, 238, 492
COnnect the DOts (CODO), 61–69, 71, 79, 118,
    163, 201
Constructive solid geometry, 309–313,
    315–316
Control point, 379–381, 383–399, 401–415,
    417–418, 421–422, 430–431,
    433–435, 437–455, 457–476,
    478–479, 481–485, 487–497,
    500–528
  Bezier, 390–391, 396, 402–403, 411,
    417–418, 422, 430–431, 437,
    451, 478
  box spline, 514
  B-spline, 442, 449–450, 458–463, 472
  bump fractal, 491–493, 497
Control polygon, 381–382, 384, 387, 391, 402,
    404, 406, 409, 413, 415, 449, 458,
    462–464, 468–470, 477–478, 481,
    484, 486–489, 491, 496–497,
    499–501, 504–506, 509, 527
Control polyhedron, 391–392, 405, 413,
    463–464, 499–500, 506, 510

Convex hull, 384, 386, 395–396, 398–399, 406–409, 439, 448, 450–451, 453, 455, 464, 468, 470–471
Convex set, 386
Coordinate-free, 117–129, 132, 163–177, 224, 228
Coordinates, 4–6, 9–10, 26, 40–42, 44–47, 49–52, 54–55, 58–59, 61–63, 66, 69, 107, 117–129
    affine, 51–52, 66, 190
    barycentric, 59, 330
    global, 10, 26
    homogeneous, 190–191, 194, 205–206
    local, 6, 26, 61, 107
    rectangular, 9, 22, 26–27, 40, 45, 66, 69, 117, 131–133, 136, 179, 190–191, 223–224, 228, 232, 238, 267
Corner cutting, 409–410, 458, 461–462, 468–469
Cross product, 120, 122–129, 133–137, 139–140, 144–147, 152–154, 156, 167, 182, 203, 209, 212, 223, 228, 230, 233, 238, 241, 269–270, 301, 326, 401
Cross ratio, 198
CSG, *see* Constructive solid geometry
Cubic B-splines, 439, 441–446, 450, 452, 458–461, 464–467, 472–475, 482, 484, 495, 505, 509–514, 516, 519–521, 523, 525–528
Curvature, 267, 274–277, 280, 305, 517
    Gaussian, 267, 274–278
    mean, 274–277
Cyclide, 305
Cylinder, 27, 267, 279, 281, 284, 287–292, 294, 298–300, 303–306, 309–310, 312, 320, 322, 379, 499
    bounded, 289–290, 298, 303, 320
    elliptical, 292
    generalized, 279
    right circular, 287–288

**D**

de Boor algorithm, 437–452, 454, 459–461, 470–472
    global, 441–443, 445–446
    local, 438–443, 448–449, 459
de Casteljau algorithm, 381–392, 394–396, 402, 404–405, 419–423, 426–427, 429–433, 437–439, 443, 445, 449, 451, 479

evaluation, 380–382, 384, 387, 393–394, 402, 405, 410, 412, 417, 432, 461, 479
    subdivision, 387, 401–405, 409–412, 415, 417, 423, 431, 460–461, 477–478, 494, 499
Dehomogenization, 423–424
Depth buffer, 342–343
Depth sort, 341–342, 347–352
Determinant, 59, 120, 123–124, 126–129, 134–135, 144, 147, 156, 203, 209, 212, 218, 235
Diagonal, 8, 11, 61, 64, 76, 88–89, 95, 118, 139, 196, 208, 236, 249, 369, 373, 418–419, 421, 423–424, 428, 443, 449, 461, 511–512, 517, 519
Diagonally dominant, 89, 95, 369, 373
Differentiability, 32–33, 499, 516
Differentiable, 29, 32–34, 36–37, 499, 516
Diffuse reflection, 250–256, 307, 325, 332–335, 337–338, 355–356, 358
Diffuse reflection coefficient, 251–253, 255, 356, 360, 371–373
Dimension, 29–32, 36–38, 77, 153, 162, 206, 211–212, 218, 224, 281, 440, 487–488; *see also* Fractal dimension
Directrix, 299–301, 303
Distance, 11, 14, 19, 42–43, 48, 54, 63, 68–91, 142–145, 152, 156, 165, 169, 184, 187, 202, 206, 214–216, 226, 249, 281, 285, 289, 295–296, 299, 307, 312, 357, 362, 364, 373, 404, 406, 408, 458, 469–471
    between a point and a line, 143, 157
    between a point and a plane, 143–144, 157
    between two lines, 144–145
    between two points, 142, 152, 157, 203
Distortion, 233–234, 239
Divergence, 274, 315–316, 322
Divergence theorem, 315–316, 322
Division algebra, 223–224
Dot product, 117–118, 120–122, 124–129, 133–135, 139–140, 142–143, 156, 162, 182, 197, 203, 209, 212, 218, 223, 228, 233, 238, 241, 271, 278, 335, 337, 348
Dual functional property, 418, 421–422, 435, 441–442, 448–449, 451, 453, 455, 458, 460, 464, 469

Bezier curves, 422–423, 441, 449, 451, 453, 460, 464, 469

B-splines, 445–446, 449, 451, 453, 457–458, 461, 464

**E**

Edge, 29–31, 56–57, 146, 155, 162, 249, 283, 313–315, 322, 326–327, 329, 331–332, 337, 343–345, 350, 367, 382, 387–388, 402–404, 411, 420, 422, 439, 447, 461–462, 475, 479, 503, 505–523, 525–526

 edge point, 314, 505, 509–510, 521, 525

 edge stencil, 510–511, 516–517, 521–523

 edge table, 343–345

 edge vertex, 520

Ellipse, 11, 25–27, 32, 48, 64–65, 163, 271

Ellipsoid, 202, 268, 292, 303

Emitter, 355–358, 360, 372–373

Euclidean geometry, 9, 15, 39

Euler's Formula, 227, 240, 315, 317, 321

Extraordinary vertex, 507, 514–515, 522

**F**

Face, 313–316, 322, 352, 364–365, 506–509, 511–514, 516, 518, 520, 522

 face stencil, 511, 521, 523

 face vertex, 314

First fundamental form, 277

Fixed point, 42, 48–49, 55, 63, 124, 173, 183, 193, 232, 279, 304, 322, 368–369, 374–375, 486

Fixed point theorem, 81–99, 110

Flat shading, 326

Form factors, 360–366, 369–373

Formula of Rodrigues, 165–167, 174, 184, 230

FORWARD, 4–5, 7, 9–10, 14, 17, 19, 22–26, 28, 39, 61, 66, 101, 106–110, 119–203, 297

Four point scheme, 526–527

Fractal dimension, 31–32

Fractals, 3, 7, 13–39, 65, 71–79, 81, 84, 89–94, 96, 98, 101, 104, 106, 109–113, 202, 240, 245, 268, 477–478, 491–492, 497, 499

 bump fractals, 17, 19–20, 109–110, 491–492

 *C*-curve, 19

 fractal bush, 20, 72–73

fractal flag, 13

fractal flower, 23, 72–73

fractal gaskets, 17–18

fractal hangman, 20, 72–73

fractal leaf, 23

fractal staircase, 74–76

fractal star, 23, 78

fractal tree, 20, 74–75

Koch curve, 492–493

Koch snowflake, 19, 71

Sierpinski triangle, 17–18, 74

Frame buffer, 341–342

Freeform curve, 379, 452

Freeform surface, 309, 379, 499

Frustum, 214–215, 365

**G**

Gasket, 17–18, 20, 22, 24, 29–32, 34–39, 71–75, 79, 90, 98, 107–108, 111–112, 201, 240, 245

Gathering, 368–370, 372, 375

Generator, 295–297, 321

Geometry, 3, 9, 15, 37, 39, 41–42, 44–45

 affine, 41–42, 61–69, 208

 extrinsic, 66

 global, 61, 401

 intrinsic, 61, 66

 local, 61, 66

 projective, 170–173

 turtle, 3, 61–62, 66

Global knot insertion algorithms, 464–471, 476

 Lane–Riesenfeld algorithm, 458, 464–467, 478, 481–484, 495, 497, 500

 Schaefer's algorithm, 467–468, 472–473

Gouraud shading, 325, 327–333, 335, 337, 340, 343, 366–367

Gradient, 269, 274, 293, 295, 297, 312

**H**

Haussdorf metric, 82, 90–92

Heading, 4, 6, 10, 17, 25, 27–28, 199, 202–203

Heedless painter, 342, 347

Hemi-cube, 363–366, 369–372

Hessian, 274, 276

Hidden surface algorithms, 215, 341–353, 366, 408, 471

 bsp-tree, 341–342, 349, 351–353

 depth sort, 341–342, 347–349, 352

 ray casting, 341–342, 346–347, 352

scan line, 341–346, 352–353
*z*-buffer, 341–346, 352, 365
Highlights, 170, 250, 252–255
Hilbert curve, 23–24
Hodograph turtle, 24–25, 113
Homogenization, 423–427, 431
Hyperboloid, 294, 303

## I

IFS, *see* Iterated function system
Illumination models, 249–250, 255, 325
Image space, 341–342, 347
Implicit equation, 142, 267, 279–280, 292,
    295–297, 299, 305–306, 312, 346
Index of refraction, 258, 264
In-out property, 445, 452
Intensity, 249–262, 264, 308, 326–329, 331–
    333, 335, 337, 342, 344–346, 355,
    360, 367, 372–373
Interpolation, 151–153, 158, 177, 234–235,
    239, 327–340, 379–381, 386–390,
    395, 409, 449, 526
Intersection, 86–87, 141, 148–151, 153–156, 158,
    160–161, 171, 203, 207–209, 211–214,
    257–265, 267, 272–278, 281–282
    line and plane, 282
    three planes, 149–151, 158, 203
    two lines, 32, 144, 148–150, 158, 203, 350
    two planes, 150–151, 158, 203, 289
Intersection algorithms, 258, 272–274,
    277–278, 284, 290, 294, 302, 374,
    402–409, 435
    line-circle, 285, 288, 291
    line-cone, 291
    line-cylinder, 288–289
    ray-Bezier patch, 401, 408
    ray-B-spline surface, 449, 463, 470
    ray-surface, 272–273, 276–278
    ray-torus, 298
Intersection strategies, 281–282
Inversion formulas, 285–288, 290–291, 305
Iterated function system, 36, 71–79, 81–82, 84,
    91–94, 96, 98, 101–113, 163,
    201–202, 268, 477–497

## J

Join, 32, 62, 388–391, 396, 401, 410–411, 437,
    441, 443–444, 446, 450–451, 453,
    455, 457

## K

Key frame animation, 153, 234–235, 239, 246
Knot(s), 437, 439–442, 444–446, 448–455
Knot insertion algorithms, 457–476, 478,
    481–482, 495
    Boehm's algorithm, 458–461
    global knot insertion algorithms,
        464–471, 476
    Lane–Riesenfeld algorithm, 458, 464–468,
        472–474
    local knot insertion algorithms, 458–464
    Oslo algorithm, 458, 464, 472
    Schaefer's algorithm, 467–468, 472–473
Koch curve, 19–20, 24, 29, 31–36, 71–74, 77,
    96, 98, 109, 111, 240, 245, 478,
    491–493
Koch snowflake, 13, 19, 37, 71

## L

Lagrange identity, 127–128, 135
Lambert's law, 251–252, 357–358
Lane Riesenfeld algorithm, 458, 464–468,
    472–474, 478, 481–484, 499–500,
    505–507, 509, 526–528
Law of cosines, 7–8, 11, 68, 139–140, 156
Law of sines, 139–141, 156
Length, 4, 6, 8–9, 11, 14, 16–17, 21, 26, 31, 33,
    39, 42, 48, 62, 64, 68, 77, 90, 102,
    118–122, 124–125, 127, 135,
    142–143, 146, 152–153, 164, 200,
    209, 227, 234–235, 270, 290, 292,
    295–296, 311, 315, 319–320, 335,
    362–363, 411, 413
Line(s), 62, 141, 143, 157, 159–160,
    284–290
Linear algebra, 39, 44, 52, 118, 120, 123, 131,
    150, 203, 205, 424, 440
Linear combination, 120, 131–132,
    249–250, 440
Linear interpolation, 151–153, 158, 177,
    234–235, 239, 327–337, 339–340,
    381, 395
Local control, 437, 452–453, 457, 463
Local knot insertion algorithms, 454–464
    Boehm's algorithm, 458–461
    Oslo algorithm, 458, 460–461, 464, 472
Logarithmic spiral, 17, 67, 69
LOGO, 3–4, 7, 9–10, 23, 25–28, 61–62, 64–67,
    71, 118, 163, 200–203

Looping Lemmas, 13–17, 21, 26
  Circle Looping Lemma, 14
  Spiral Looping Lemma, 14, 21
Loops, 13–15, 21–22, 313, 315
Loop subdivision, 517, 523

# M

Mach bands, 327, 329, 353
Manifold(s), 27–28, 202–203, 314–315, 341
Mass, 210–214, 216, 218, 228–229, 233, 244,
    286, 290–291, 309, 311, 424, 451
Mass-point(s), 205–218, 223–238, 286, 451,
    496–497
Matrix, 179–195
Mesh(es), 319, 366, 372, 499, 503–505,
    507–508, 512–516, 518
  quadrilateral, 503–512, 515–517, 522–523,
    527
  triangular, 499, 512–518, 521–523, 527
Midpoint, 39, 119, 402–404, 406, 408, 412, 479,
    484, 494, 505–508, 512, 514–516
Midpoint averaging, 473
Mirror, 167–168, 185–187
  mirror image, 164, 167–168, 170, 174, 177,
    181, 183, 185–186, 189,
    193–194, 203, 230–232, 238, 244,
    256, 265
  mirror plane, 167, 170, 185–186
Monte Carlo methods, 311, 315–316
MOVE, 4–5, 7, 66, 101, 103, 105
Multiaffine, 418–421, 424–426, 433, 438–439,
    459, 464, 466–468
Multilinear, 123, 129, 423–425, 428

# N

Natural quadrics, 281, 284–292, 294–295,
    303, 307
  right circular cones, 281, 284, 290, 292,
    294, 299, 303–304
  right circular cylinder, 281, 284, 287,
    292, 294, 299, 303
  sphere, 153, 200–202, 267–268, 280–281,
    284–287, 292, 294, 298–300,
    302–303, 306, 309–310, 312, 322,
    358, 363, 379, 451, 453, 455, 508
Neville's algorithm, 398
Newell's formula, 146–147, 326, 328, 331, 336
NEWSTAR, 10; *see also* STAR

Newton's method, 84–87, 95, 98, 333–334,
    337, 415
Nonuniform rational B-splines (NURBs),
    451, 453
Normal curve, 485, 489–493, 496–497
Normal vector, 141–145, 149–150, 160–161,
    167–168, 171–172, 175, 189, 195,
    202, 213, 231, 244, 251–252,
    264–265, 269–272, 274, 276
NURBS, *see* Nonuniform rational B-splines

# O

Object space, 341–342, 347, 352
Octree, 309, 317–319, 322–323
Orthogonal, 41, 48–49, 121, 131–133, 164, 175,
    199–200, 215, 228, 233–234, 249,
    270–271, 287–288, 292, 295–296,
    299, 301, 306, 319, 364, 488–489,
    492, 497
Orthogonal projection(s), 163–164, 170–171,
    174, 181, 188–189, 193, 195, 203,
    215, 244, 282, 285, 287, 290, 406,
    408, 470–471, 485, 488, 490
Oslo algorithm, 458, 460–461, 464, 472

# P

Paraboloid(s), 294, 303
Parallel, 56, 90, 121–122, 125, 128, 141,
    143–145, 148, 150, 157, 160–161,
    164–170, 172–173, 175–177, 182,
    184, 188, 195–197, 203, 207–208,
    216, 219, 230, 244, 253, 261, 263,
    282, 287–291, 293, 295–296,
    298–299, 301, 306, 311, 315, 317,
    319–321, 350, 364–365
Parametric equation(s), 69, 141–142, 267–268,
    272, 275, 280–282, 285, 293,
    296–297, 299–301, 305, 313, 328,
    346, 451
Pascal's theorem, 208
Pascal's triangle, 383–384, 480, 482
Perpendicular, 6, 11, 48, 121–122, 124, 128,
    142–143, 145, 149–150, 154,
    166–173, 175, 195, 199–200, 208,
    211, 224, 230–232, 238, 243–245,
    253, 256, 261, 263, 265, 270–271,
    273, 277, 282, 295–296, 300–301,
    305–306, 326, 347, 350, 362

Perspective, 36, 85–86, 106–107, 118, 163–164, 171–173, 189–192, 213–217
    plane, 192
    projection, 118, 163–164, 170–174, 181, 188–195, 203, 205, 207, 209, 211–215, 217, 219, 282, 286, 290
Phong shading, 325, 331–340, 342, 346, 353, 381
Piecewise Bezier, 437, 457–458, 461–464, 469–471, 474, 476
Pixel, 257–258, 311, 327–329, 331–337, 339, 341–347, 352
Pixel coherence, 341, 346–347, 352
Plane, 4, 7, 9–10, 24–25, 27–29, 39, 41, 46–47, 53–54, 141–144, 149–151
    affine plane, 39–59
    projective plane(s), 27–28, 206–208
Point, 4–5, 9–10, 20, 22, 33, 41, 49, 59, 75, 86–87
    affine point(s), 206, 209, 211–212, 218, 228, 232, 286, 291
    edge point(s), 314, 505, 509–510, 521, 525
    face point(s), 314, 510, 521, 525
    point at infinity, 206–209, 218
    vertex point, 314, 379, 505, 509–510, 521, 525
POLY, 14–18, 21, 25, 28, 37
POLYGON, 7
Polygons, 3, 7–8, 10–11, 13–14, 16, 18, 22, 24, 32, 61, 90, 145–147, 154, 163, 201, 282–284, 325–332, 341–342, 345–349, 351–352, 366, 381, 402, 404, 413, 415, 464, 466, 469–470, 477, 481, 484–486, 496–497, 500, 527, 530
    convex, 146–147, 283–284, 303, 346, 348
    non-convex, 146, 349
Polygon splitting, 349–351
Polynomial, 32, 84, 208–209, 241, 267–268, 272–273, 303, 379–380, 383–384, 386, 388–389, 391, 396, 398, 401–402, 409, 414–415, 417–435, 437–443, 445–446, 448–453
    homogeneous, 423, 434
Primitive solid(s), 309–311, 322
Principle of duality, 208
Projection, 163–177, 189, 191–192, 207, 212, 214–215, 217–219
    orthogonal, 163–164, 170–171, 174, 181, 188–189, 193, 195, 203, 215, 244, 282, 285, 287, 290, 406, 408, 470–471, 485, 488, 490

    parallel, 128, 175–177, 182, 188, 195–196, 244, 296
    perpendicular, 121
    perspective, 118, 163–164, 170–174, 181, 188–195, 203, 207, 209, 211–215, 217–219, 282, 286, 290
Projective plane, 27–28, 206, 208
Projective space, 205–221
Projective transformation(s), 163–177, 179–203, 205, 209, 212, 216, 218–219, 281–282, 293
Pseudodepth, 214–218, 220
Pseudo knot(s), 457–458
Pseudoperspective, 213–219, 341–342, 347

**Q**

Quadric surface, 292–295, 297, 303, 305
    cone, 281, 284, 290–292, 294, 303
    cylinder, 281, 287–290, 292, 294, 303
    ellipsoid, 292, 303
    hyperboloid, 294, 303
    paraboloid, 294, 303
    sphere, 284–287, 292, 294, 303
Quadrilateral mesh(es), 503–512, 515–517, 522–523, 527
Quadtree, 317–318
Quaternion(s), 153, 223–246

**R**

Radiosity, 89, 355–375
Radiosity equations, 89, 355–361, 366, 368–373
    continuous, 356–359
    discrete, 359–360
Raster display(s), 257
Raster graphics, 257
Rational Bezier, 451
Rational B-spline, 451, 453
Rational function(s), 173, 268, 302–303, 451
Ray casting, 148, 153–154, 259–261, 284, 311–313, 315–316, 341–342, 346–347, 352
Ray tracing, 257–269, 272–273, 281, 284–285, 290, 292, 294, 298, 303, 306, 308–309, 311–312, 315, 317–318, 325, 346, 355–356, 372, 375, 408–409, 457, 463, 471
Reciprocity relationship, 365–366, 370

Rectangular coordinates, 9, 26, 40, 45, 66, 69, 117, 131–132, 136–137, 179, 190–191, 218, 223–224, 228, 232, 238, 267

Recursive turtle program(s), 7, 13–28, 31–32, 34, 36–38, 71, 74, 81, 94, 101–113, 201

Reflection, 25, 202, 245, 250–256, 258, 260–261, 264–265, 307–308, 325, 331–335, 337–339, 355–356, 358, 360, 371–373

Reflector(s), 355–356

Refraction, 153, 258, 260–266, 307

Relaxation methods, 82, 84, 87–89, 95, 99, 369

　Gauss–Seidel, 87–89, 95–96, 99, 369, 375

　Jacobi, 87–89, 95, 369, 374

Rendering algorithms, 325, 349, 351, 366–367, 406, 408, 463, 470–471

　Bezier curve, 405–407

　Bezier patches, 407–409

　B-spline curves, 437–451

　B-spline surfaces, 437–451

Rendering equation, 355–356, 358–359, 372

RESIZE, 4–7, 9–11, 14–15, 17–19, 23, 26, 32, 37, 66, 101–103, 105–110, 200–201

RGB color model, 249–250, 254–255

Rigid motion, 39, 58, 95, 164–168, 183–186, 268, 270, 282, 294, 297, 301

Ron's algorithm, 106–108, 110–112

Root, 82, 84–87, 93, 98, 272–273, 276–277, 293, 296–297, 299, 310–312, 333–337, 339, 351, 413–415, 435

Rosette, 7–8, 11, 13, 24, 61, 64–65, 139

ROT, 62–63, 65, 102, 106–107, 109

Rotation, 5, 41–42, 46, 53, 165–167, 174, 184–185, 194, 238, 335

RTP, *see* Recursive turtle program(s)

## S

Sandwiching, 230–234, 238, 240, 242, 244

Scalar, 7, 10, 43, 47, 57, 59, 63, 118–121, 123–124, 126, 131–132, 134, 203, 207, 209–211, 218–219, 221, 223–224, 228, 232–233, 238–239, 284, 287, 292, 295, 298, 319–320, 332–334, 337, 379, 385

SCALE, 7–9, 23, 62–65, 69, 101–103, 106–109

Scaling, 5, 14, 22, 30, 39–40, 42–43, 46, 48–49, 51, 53–55, 64, 66, 71, 73, 82, 101–102, 108, 110, 112, 116, 163–164, 168–170, 174, 181, 187–188, 193–194, 200, 203, 227, 230, 232, 235, 238, 245, 270–271, 292, 297, 319

　non-uniform, 48–49, 51, 53, 57, 64, 66, 101, 163–164, 168–170, 174, 177, 181, 187, 193–194, 203, 245, 271, 292, 297

　uniform, 39–40, 42–43, 46, 48, 51, 53, 66, 73, 77, 82, 90, 102, 152, 163–164, 168–169, 174, 181, 187, 193–194, 200, 203, 230, 232, 234–235, 238–239, 270–271

Scan line, 257, 325, 327–335, 337, 341–346, 352–353

Scan line algorithm, 329–330, 332, 334, 343–346, 353

Schaefer's algorithm, 467–468, 472–473

Secondary ray, 260–262

Second fundamental form, 275, 277

Self-similar, 20, 31–32, 36, 39, 73–74, 76–77, 477, 490

Shading algorithms,153, 325, 327, 329, 337, 339, 341, 527

　flat, 326–327

　Gouraud, 325, 327–335, 337, 340, 343, 366–367

　Phong, 325, 331–340, 342, 346, 353, 381

　uniform, 326–327

Shadows, 250, 259–260, 264, 352, 355, 372, 409

Shape, 3, 7–8, 13, 25–26, 29, 31–32, 39, 63, 65–66, 69, 74, 76, 102, 117–118, 163, 202, 253, 268, 281, 309–310, 313, 325, 341, 379–382, 385, 391, 395, 401, 414–415, 437, 457, 464, 476–477, 490–491, 493, 499, 505–508, 513, 518, 527

Shear, 39, 56–57, 117, 175, 177, 195, 203, 282

SHIFT, 7–8, 64, 103

Shooting, 368, 370–372, 375

Sierpinski gasket, 17–18, 20, 24, 29–32, 34–36, 39, 71–75, 96, 98, 107–108, 111–112, 240, 245

Sierpinski triangle, 17–18, 74

Similar triangles, 39, 118

Skew symmetric, 123, 129

SLERP, *see* Spherical linear interpolation

Slope, 32–33

Snell's Law, 262–266

Solid, 267, 303, 309–323, 341, 356, 372
Solid modeling, 267, 303, 309–323
Sorting, 341, 347, 352
Specular exponent, 253
Specular highlights, 250, 252–255
Specular reflection, 252–256, 261, 307, 325, 331, 334–335, 337–339, 355, 372
Specular reflection coefficient, 253–255
Spherical linear interpolation (SLERP), 151–153, 158, 162, 177, 234–235, 239–240, 242, 263, 265, 335–337, 339–340
SPIN, 7–9, 11, 23, 39, 64–65, 78, 103
SPIRAL, 14, 17, 25, 28, 67–69
Spirals, 7, 13–15, 17, 21, 24–25, 28, 67–69, 85, 90, 201
    logarithmic, 17, 67, 69
    turtle, 7, 14–15, 24, 67–69, 201
Spokes, 64–65
Spotlight, 307–308
STAR, 10; *see also* NEWSTAR; TRISTAR
State, 101–103
Stencil, 505, 509–513, 516–518, 521–523, 525–527
    Catmull–Clark, 511, 523, 527
    edge, 510–511, 516–517, 521–523, 526
    face, 511, 521–523
    Loop, 517, 523, 527
    universal, 525–527
    vertex, 510–511, 516–518, 521–523, 526–527
Subdivision, 317–318, 331–332, 387, 401–415, 417, 422–423, 431, 437, 449, 457, 477–497, 499–528,
Subdivision algorithm, 387, 401–405, 407, 409–412, 415, 417, 423, 431, 460–461, 463, 477–479, 494, 499–500, 502, 505–506, 509–510, 512–513, 518, 525, 527–528
    Bezier–de Casteljau algorithm, 380–392, 394–396, 402, 404–405, 419–424, 426–433, 437–439, 443, 445, 448–449, 451, 479
    box spline, 500–504, 512–514, 516–525, 527
    B-spline–Lane Riesenfeld algorithm, 458, 464–469, 472–474, 478, 481–484, 495, 497, 499–500, 505–507, 509–510, 526–528
    Catmull–Clark, 511, 523

centroid averaging, 505–516, 518–522, 527
    four point scheme, 526–527
    Loop, 517, 523, 527
Subdivision matrix, 483, 491–492
Subdivision surface, 499–528
Surface(s), 267–308, 341–353, 379–399
    Bezier, 392, 395, 404–405, 407–409, 413, 471
    box spline, 500–504, 512–513, 516, 518, 527
    B-spline, 449–450, 453–455, 463–464, 469–471, 476, 500–506, 510–513, 516, 518
    cone, 290–292
    cyclide, 305–306
    cylinder, 287–290
    deformed, 268–270, 272–273, 275, 278
    implicit, 267, 269, 272–274, 276–278, 293
    parametric, 267–270, 272–276, 278–279, 282, 300–301, 304–305, 391
    plane, 282–284
    procedural, 268–269
    quadric, 292–295, 297, 303, 305
    ruled, 279, 305
    sphere, 284–287
    subdivision, 499–528
    subdivision surface, 499–518
    surface of revolution, 299–303
    torus, 295–299
Surface modeling, 303, 309, 311, 315
Surface normals, 261–262, 265, 267, 269–272, 281, 291–292, 294, 303–304, 357, 373, 401
Symmetry, 14, 411, 418, 420, 424, 433, 464

**T**

Tangent vector, 270–272, 363, 423
Tao's algorithm, 106–113
Tensor product surface, 449–450, 453–455, 463, 469–471, 500–502, 504–506, 509, 512, 518, 523, 525
    Bezier, 391–395
    B-spline, 449–450
Topology, 3, 313–317, 322, 499, 504–507, 509, 512, 514–516, 518–521, 523, 527
Torus, 27, 281, 295–299, 302, 305–306, 310, 312, 321, 499
Total reflection, 264–265

Trace, 61, 163, 196, 273–274, 283, 294, 301, 304, 306, 311, 401, 408–409, 463, 470–471

TRANS, 63–64, 102, 106–107, 109

Transformations, 39–59, 62–63, 65–66, 71–73
  affine, 39–59
  composite, 63, 95, 203
  conformal, 39–43, 46–47, 51, 56, 66, 230–232
  contractive, 82–84, 90–96
  linear, 5, 44–45, 52, 181–182, 190
  projective, 163–173, 179–204, 212

Translation, 5, 9, 39–40, 43–46, 51–53, 57–58, 66, 73, 77, 102, 108, 117, 163–165, 174, 177, 181–184, 193–194, 268, 270, 486

Translation invariance, 385

Triangles, 9, 13, 16–18, 34, 36, 39, 44–45, 51, 58, 68, 71, 74, 83, 91, 107, 118, 145, 157, 159, 164, 218, 283, 295, 330–332, 383–384, 386, 393–394, 397, 403–404, 407–408, 422, 426, 429, 443–445, 447, 450, 461, 467–468, 470–471, 475, 479–480, 482, 494, 512–514, 516
  congruent, 39
  similar, 39, 118

Triangular inequality, 83, 91

Triangular mesh, 512–518, 521–523, 527

TRISTAR, 10, 69; *see also* STAR

Trivial fixed point theorem, 81–93, 95, 97, 110

TURN, 4–11, 14–15, 17, 21, 24–26, 36, 39, 66, 101–103, 105–110, 154, 199–201

Turtle(s), 3, 24, 27–28, 39, 102, 154–155, 163, 199–203, 208, 303
  classical turtle, 23–28
  hodograph turtle, 24–25, 113
  left-handed turtle, 26, 113

Turtle commands, 4–7, 9, 11, 13, 19, 24, 26, 101–103, 105–108, 199
  FORWARD, 4–5, 7
  MOVE, 4–5, 7
  RESIZE, 4–7
  TURN, 4–7

Turtle geometry, 3–7, 51, 61–62, 66

Turtle graphics, 3–11, 13, 36, 40–42, 51, 61, 65–66, 118, 163

Turtle path, 3, 14–15

Turtle program, 7–11, 13–20, 67–68, 71, 101, 103, 109, 111–112, 117, 202–203

Turtle state, 7, 14, 102–105

**U**

Uniform shading, 325–327

Unit vector, 49, 55, 63, 131–133, 143–144, 151, 162, 164, 167, 169, 171, 173–177, 179, 185, 187, 190–191, 195–197, 207, 215, 220, 224–226, 228, 231, 233–235, 238, 242–243, 245, 251–252, 255, 260–263, 265, 282, 285–286, 288–289, 291, 299, 301, 306–307, 319–322, 326, 335–336

Universal stencil, 525–527

**V**

Valence, 507–508, 511, 514–517, 523, 527

Variation diminishing property, 384, 386–387, 395–396, 401, 409, 411, 448–449, 453, 457–458, 461–462, 469, 471
  Bezier curves, 384–388, 409–410
  B-spline curves, 461–463

Vector, 117–129, 131–132, 137, 139–162
  contravariant, 272
  covariant, 272
  normal, 141–145, 149–150, 160–162, 269–272
  tangent, 270–272

Vector algebra, 120, 139, 156, 164, 173, 179, 181, 203, 209

Vector geometry, 117–129, 139–163, 179

Vector products, 120–123, 132–134, 212, 228
  cross product, 120, 122–123, 133
  dot product, 121–122, 133
  triple product (*see* Determinant)

Vector space, 118–120, 123, 209–212, 218, 223–224, 249, 423–424, 451

Vertex, 8, 11, 16, 56–57, 63–64, 69, 153–154, 162, 290
  edge vertex, 327
  extraordinary vertex, 507, 514–515
  face vertex, 507–508, 516
  vertex point, 505, 509–510
  vertex stencil, 510–511, 516–517

Vertices, 8, 10, 14–17, 21, 29–30, 32, 61, 63–65, 67–69, 71, 117, 146, 148

Viewing frustum, 214–215
Virtual ray, 259–260
Volume, 120, 123, 142, 144, 147–148, 156,
    158, 160, 162, 309, 311–313,
    315–316, 318, 322–323
  of a CSG-tree, 309–313, 316
  of a parallelepiped, 144, 147–148, 158
  of a polyhedron, 160
  of a tetrahedron, 148, 158
Voxel, 317–319

**W**

Weighted centroid, 513–516, 520–522
Wheel, 7–8, 11, 64–65, 139
Winding number, 154–156, 158, 162, 284
Winged-edge data structure, 313–315

**Z**

*z*-buffer, 341–346, 352, 365

Viewing frustum, 214–215
Virtual ray, 259–260
Volume, 120, 123, 142, 144, 147–148, 156,
       158, 160, 162, 309, 311–313,
       315–316, 318, 322–323
    of a CSG-tree, 309–313, 316
    of a parallelepiped, 144, 147–148, 158
    of a polyhedron, 160
    of a tetrahedron, 148, 158
Voxel, 317–319

**W**

Weighted centroid, 513–516, 520–522
Wheel, 7–8, 11, 64–65, 139
Winding number, 154–156, 158, 162, 284
Winged-edge data structure, 313–315

**Z**

$z$-buffer, 341–346, 352, 365

Milton Keynes UK
Ingram Content Group UK Ltd.
UKHW052030141024
449569UK00017B/759